Springer Aerospace Technology

More information about this series at http://www.springer.com/series/8613

Jiaxing Liu

Spacecraft TT&C and Information Transmission Theory and Technologies

国防工業出版社
National Defense Industry Press

Springer

Jiaxing Liu
The 10th Institute of China Electronics
Technology Group Corporation
Chengdu, China

ISSN 1869-1730 ISSN 1869-1749 (electronic)
ISBN 978-3-662-51306-4 ISBN 978-3-662-43865-7 (eBook)
DOI 10.1007/978-3-662-43865-7
Springer Heidelberg New York Dordrecht London

Jointly published with National Defense Industry Press, Beijing
ISBN: 978-7-118-09523-4 National Defense Industry Press, Beijing

Softcover reprint of the hardcover 1st edition 2015

Printed on acid-free paper

Springer is part of Springer Science+Business Media (www.springer.com)

Foreword

"TT&C" is the abbreviation of tracking, telemetry and command. "Spacecraft information transmission" refers to the mutual transmission of information acquired from or generated by spacecraft and ground station. With the development of technologies, TT&C and information transmission have been organically linked, in that they are complementary in functions, unified in signal design and integrated in equipment. In addition, telemetry and telecommand are essentially a kind of information transmission. From the view of development, information warfare and integrated electronic information system require that TT&C and information transmission be designed on a unified basis and integrated into a single device. Therefore, this book highlights the characteristics of the combination of TT&C and information transmission with the latter being particularly emphasized. Based on the practices of the author in developing various TT&C equipment over many years and trying to combine theory with the practices, the book focuses on the basic technologies commonly applied for C&T system of various vehicles, with the intention to help the readers to obtain deeper understanding about TT&C systems and provide them with a theoretical basis for resolving practical engineering problems.

The book falls into two major parts of TT&C and Information Transmission, and is composed of five chapters. Chapter 1 introduces the past, present and future of TT&C and information transmission technologies. Chapter 2 introduces tracking and orbit measuring theories and technologies, with focus on CW radar's velocity-measuring, ranging, angle measuring technologies and locating technology. Chapter 3 (Information Transmission Technology) introduces analog transmission technology and digital transmission technology. In addition to the basic theories in combination with features of C&T, the telemetry, telecommand and remote sensing technologies are also provided. Chapter 4 (Spread Spectrum TT&C) introduces the application of spread spectrum technologies in C&T field. Chapter 5 introduces the development of TT&C frequency band in combination with special problems occurring in radio transmission channel, such as impacts of rain attenuation, atmospheric attenuation and radio wave propagation upon accuracy of orbit determination during transmission, impact of multi-path transmission, and polarization diversity synthesis.

With its main purpose being the introduction on basic theoretical knowledge and technology, the main feature of this book is the idea of combining TT&C with information transmission, and combining theory with the practical experiences in equipment development.

With the wider application of C&T system, particularly with its increasing significance in military application field, C&T system becomes the only information line between the spacecraft and ground, so there are more and more people engaged in the development, production, application and teaching of such engineering system. However, there are only a few books involving the technology. For this reason, the author, by incorporating the experiences from the practices of engaging in different engineering equipment research for many years, studies and summarizes his published papers and training materials to complete the writing of this book, hoping to provide support to the science-technological workers and university teachers and students who are involved in the development, design, production and application of the C&T system.

This book is compiled under great support from China Southwest Institute of Electronics Technology – the employer of the author – and National Defense Industry Press. Many contents in this book reflect the achievements made by China Southwest Institute of Electronics Technology from developing C&T systems for many years and so it is the fruit of all the participants' efforts. In addition, in the course of compiling this book, academician Yang Shizhong reviewed the manuscript, researchers Lei Li and Zhang Hansan put forward many valuable comments, Yang Hongjun provided a lot of materials, and Liu Yu and Liu Yanmeng completed numerous text processing works. The author hereby expresses his sincere gratitude to all of them for their great support.

Chengdu, China Jiaxing Liu
August, 2013

Contents

1 Introduction 1
1.1 General 1
1.2 TT&C and Information Transmission 2
1.3 Tasks, Functions, and Classification of TT&C and Information Transmission 3
1.4 Engineering Applications of TT&C and Information Transmission Technologies 9
1.4.1 Unified Carrier C&T System 9
1.4.2 Space-Based C&T System 10
1.4.3 Deep-Space C&T System 11
1.4.4 Phased Array C&T System 12
1.4.5 High-Accuracy Missile Range-Measuring System 13
1.4.6 Near-Space Vehicle TT&C and Information Transmission System 13
References 14

2 Theories and Technologies of Tracking and Orbit-Measuring 15
2.1 General 15
2.2 Localization and Orbit-Measuring 16
2.2.1 Localization 16
2.2.2 Station Distribution Geometry and Localization Accuracy 21
2.2.3 Trajectory Measurement System 30
2.3 Velocity-Measuring Theory and Technology: Two-Way, Three-Way, and Single-Way Velocity-Measuring Technologies 33
2.3.1 CW Velocity-Measuring 33
2.3.2 Doppler Frequency Measurement Method 36
2.3.3 Theoretical Analysis of Two-Way Coherent Doppler Velocity-Measuring Error 39
2.3.4 Three-Way Noncoherent Doppler Velocity Measuring and $\dot{S}$ Measuring 69

2.3.5 Two-Way Noncoherent Doppler Velocity Measuring 72
2.3.6 One-Way Noncoherent Doppler Velocity Measuring 79
2.3.7 Theoretical Calculation of Velocity-Measuring Accuracy 81
2.4 Ranging Techniques Theory and Technology: Two-Way, Three-Way, and Single-Way Ranging Technologies 86
2.4.1 Continuous Wave Two-Way Ranging Methods 86
2.4.2 Vector Analysis Method of Ranging Error 104
2.4.3 Group Delay Characteristics Analysis Method of Ranging Error 113
2.4.4 Random Ranging Error 124
2.4.5 Theoretical Calculation of Ranging Accuracy 131
2.4.6 One-Way Ranging Technique 135
2.4.7 Three-Way Ranging Technique 139
2.4.8 Deep-Space Ranging System 140
2.5 Angle Measurement Theory and Technology 143
2.5.1 Angle Measurement Using Antenna Tracking – Three-Channel, Dual-Channel, and Single-Channel Monopulse Technologies 143
2.5.2 Angle Measurement with Interferometer 155
2.5.3 Theoretical Calculation of Angle Tracking Accuracy 163
2.5.4 Theoretical Analysis of Angle Tracking of Wideband Signal – Cross-Correlation Function Method 167
2.5.5 Angle Measurement by Phased Array Tracking and Angle Measurement Accuracy – Space Window Sliding and Projective Plane Methods 178
2.5.6 Independent Guidance, Self-Guidance, and Multi-beam Guidance 199
2.5.7 Polarization Diversity-Synthesized Technology of Angle Tracking 217
References 225

3 Information Transmission Technologies 227
3.1 Analog Transmission Technology in C&T 228
3.1.1 Analog Signal Modulation 228
3.1.2 Demodulation of Analog Signal 236
3.1.3 Two-Way Carrier Acquisition in C&T – “Frequency Sweep to Acquisition” and “Following Sweep Slope Determination” 242
3.1.4 Combined Interference in the Unified Carrier C&T System – “Modulation/Demodulation Integration Characteristic Analysis Method” 250

3.2 Digital Signal Transmission Technology in C&T 259
3.2.1 Overview 259
3.2.2 Optimum Transmission Response of Digital Signal Transmission 261
3.2.3 Digital Modulation/Demodulation Technology 267
3.2.4 Channel Coding/Decoding Technology 277
3.2.5 Impacts of Noise on Data Transmission BER – Amplitude Noise Equivalent Method 291
3.2.6 Impacts of Linear Distortion on Data Transmission BER 297
3.2.7 Impacts of Non-linear Distortion on Data Transmission BER 308
3.3 Information Transmission Techniques for Telemetry, Command and Remote Sensing 322
3.3.1 Information Transmission Techniques for Telemetry 322
3.3.2 Command Information Transmission Technology 334
3.3.3 Remote Sensing Information Transmission Technique 347
References 359

4 Spread Spectrum TT&C 361
4.1 General 361
4.2 Features of Spread Spectrum TT&C 362
4.3 Basic Methods for Spread Spectrum TT&C 365
4.3.1 Direct Sequence Spread Spectrum (DSSS) 366
4.3.2 Frequency Hopping Spread Spectrum (FHSS) 372
4.3.3 Hybrid Spread Spectrum of DSSS and FHSS 375
4.3.4 Time Hopping Spread Spectrum 375
4.3.5 Code Hopping 376
4.4 Acquiring and Tracking of Spread Spectrum TT&C Signals 380
4.4.1 Acquiring and Tracking of DSSS Signals 380
4.4.2 Acquiring and Tracking of Frequency Hopping Spread Spectrum Signals 386
4.4.3 Velocity Measuring of Frequency Hopping Spread Spectrum—“Two-Step Method” Frequency Hopping Velocity Measuring 388
4.5 Measuring Accuracy and Tracking Threshold for Direct Spread Spectrum TT&C 391
4.5.1 Phase Error in Carrier Loop of Spread Spectrum Receiver 392
4.5.2 Range Measurement Error in Spread Spectrum TT&C 393
4.5.3 Rate Measurement Error in Spread Spectrum TT&C 395

4.6 "Double Spread Spectrum" and Its Application in TDRSS 395
4.6.1 Problems Raised 395
4.6.2 "Code Division Multiplexing" and "Double Spread Spectrum" 396
4.6.3 Main Technical Problems of "Double Spread Spectrum" . . 398
4.7 Chaotic Sequence and Chaotic Spread Spectrum TT&C 401
4.7.1 Characteristics of Chaotic Sequence 401
4.7.2 Type, Selection, and Generation of Chaotic Sequence 405
4.7.3 Synchronization and Ranging of Chaotic Spread Spectrum Signals 414
References 424

5 Special Issues on Radio Transmission Channel in C&T 425
5.1 C&T Frequency Band Developing to Ka-Band and Optical Bands 425
5.1.1 Principle for Selecting C&T Frequency Band 425
5.1.2 Development Trend 428
5.1.3 Background of Developing Ka-Band C&T System 430
5.1.4 Characteristics of Ka-Band C&T System 432
5.1.5 Main Technical Issues of Ka-Band C&T System 436
5.2 Rain Attenuation and Atmospheric Attenuation in Signal Transmission Channel 438
5.2.1 Significance of Rain Attenuation Study 438
5.2.2 Characteristics of Rain Attenuation 439
5.2.3 Calculation of Rain Attenuation 439
5.2.4 Increment in System Noise Temperature Caused by Rainfall 446
5.2.5 Rain Attenuation Countermeasure Technology 446
5.2.6 Atmospheric Attenuation in Signal Transmission Channel 450
5.3 Influence of Multipath Transmission 453
5.3.1 Three Types of Fast Fading Caused by Multipath Effect 455
5.3.2 Nature of Reflection Coefficient 458
5.3.3 Path Loss and Scintillation Fading in Case of Multipath Propagation 463
5.3.4 Orbit Determination Error Caused by Multipath Propagation 468
5.3.5 Effect of Multipath Interference on Data Transmission Bit Error Rate 472
5.3.6 Anti-multipath Interference Measures 475
5.4 New Methods for Simulation and Calibration 498
5.4.1 Dynamic Simulation Method Based on Motion Equation 498
5.4.2 Phase Calibration Using Radio Star Noise 503

5.4.3 On-Orbit Phase Calibration by Measuring "Cross-Coupling" Value 508
5.4.4 Geometric Optics Application for Range Calibration 511
5.4.5 Effect of Radio Wave Propagation Characteristic on Orbit Determination Accuracy 516
5.4.6 Tropospheric Radio Wave Refraction Error Correction . . . 522
5.4.7 Ionospheric Refraction Correction Methods 525
5.4.8 Factors Affecting Correction Accuracy 530
References 531

Chapter 1
Introduction

1.1 General

This book attempts to present the tracking, telemetry, and command (TT&C) of flight vehicles such as air vehicles (less than 20 km in altitude), near-space vehicles (20–100 km in altitude), and spacecrafts (greater than 100 km in altitude). Those flying at an altitude of more than 2×10^6 km are also known as deep-space vehicles. Specifically, the flight vehicles include missiles, satellites, spaceships, space stations, deep-space probes, near-space vehicles, UAVs, airships, or balloons, all of which are subject to TT&C. For TT&C and communication (C&T) systems, there are radio, optical, infrared, and other types. Among them, the radio continuous-wave C&T system is the most widely used one, which is mainly described in this book.

"Information transmission" for flight vehicles means transmission of information, acquired from or produced by vehicles or ground stations, between them through either point-to-point connection or network. An information transmission system is used to transmit information from a sending terminal to a receiving terminal at another space-time point. It can handle information resulted from telemetry, telecommand, remote sensing, reconnaissance, detection, guidance, scientific experiments and space environment observation, and traditional voice and images as well. The aforesaid information concept has broader implications than the traditional communications concept, for the former contains the telecommand and telemetry information necessary for normal operation of a vehicle platform (generally known as engineering TT&C), the application information acquired or forwarded by the payload (P/L) on the platform, and the information for P/L's telecommand and telemetry (generally known as service TT&C).

With the rapid development of aerospace technology, today presents a greater challenge for C&T. The traditional concept "C&T" is giving way to "communications and navigation," where "navigation" contains traditional tracking, orbit determination, orbit control, and orbit calculation, while "communication" means

J. Liu, *Spacecraft TT&C and Information Transmission Theory and Technologies*,
Springer Aerospace Technology, DOI 10.1007/978-3-662-43865-7_1

transmission of various types of information, and it has some new extensions such as information resulted from command and telemetry, voice, low-speed data, and the newly extended high-speed data.

1.2 TT&C and Information Transmission

Claude Elwood Shannon, the founder of information theory, defined "information" as "something used to eliminate the uncertainty," that is, a description of uncertainty of the motion state or existence of a thing. It is the unknown elements contained in messages, while signals are the carrier of messages. What we transmit actually are signals. In this sense, information transmission means transmission of unknown elements in messages carried by signals (namely, information).

Information exists in various forms, such as symbol, voice, and image. This book merely focuses on electronic information. An electronic information system is generally an electronic system that allows for generation, acquisition, transmission, processing, application, and countermeasure of information for operation, control, or utilization purposes, taking advantage of such major techniques as ET and IT. A space electronic information system refers to a system that allows for acquisition, transmission, processing, application, and countermeasure of information, using spacecraft platforms. A transmission subsystem is a significant component of an information system. In general, the part of a transmission subsystem that is mounted in a spacecraft is called a "payload" and that on ground is known as a "ground application system."

A TT&C information system acquires vehicle's orbit information (through tracking devices) and telemetry information (through sensors), generates command information through telecommand devices, and then converts and modulates such information to a signal carrier for long-distance transmission, which is eventually received, demodulated, and processed by a receiving terminal before being put into use. As a means of transportation, the information transmission system can be used to transmit TT&C information and others.

Technological progress has led to a close connection between TT&C and information transmission, which is mainly manifested in the following six aspects: (1) Functions. Information transmission depends on TT&C systems to enable air platforms and payloads to work normally and maintain a long service life and also relies on ground TT&C stations to align ground antenna with air platforms by tracking vehicle's orbits, providing a two-way information transmission path between vehicles and ground stations. (2) Signal design. TT&C signals and the signals carrying information have been unified under a common signal system. Unified carrier signals with multi-subcarrier frequency division and spread-spectrum signals with unified data stream are just the case. (3) Devices. TT&C systems and information transmission systems have been integrated together, and multifunctional integrated channel terminals and multifunctional integrated baseband devices have become available. (4) Techniques. Telemetry and telecommand are no more than long-distance transmission of information to realize "remote" measurement and control. With the advancement of software and radio technologies, telemetry and telecommand are performed with a common platform and software. (5) Application. As space technology was put into use after its successful R&D and test, the application of information transmission has been an increasing concern. For example, remote sensing stations, reconnaissance data-receiving stations, tracking and data relay satellite systems, and UAV TT&C systems mainly involve data transmission other than TT&C for payloads. (6) Future development. Information warfare, precision strike, and the development of integrated electronic information systems require the integration of TT&C and information transmission functions into an integrated system. The phrase "information transmission" is employed to describe such system so that it makes a stronger sense when it comes to information warfare and integrated information systems. Also, this coincides with the development trend of informationization.

Look at the "integrating" process of C&T techniques by history [1]. Prior to the 1960s, tracking, telemetry, and command functions were performed by separate devices. In mid-1960s, the S-band system (USB) came into being, having integrated TT&C, voice, and television into one system, using unified frequency-division carrier signals. In the 1980s, the tracking and data relay satellite system appeared, having incorporated TT&C and high-/medium-/low-speed data transmission functions into a single system, using spread-spectrum unified carrier signals with a time-division unified data stream, and now such functions can be performed by one software-supported radio device. Since spacecrafts were put into use, they have been chiefly used for information transmission. In China, C&T techniques evolved in a similar way. Before manned space flight projects were introduced, the TT&C system comprised many separate devices. After that, since voice and image transmission functions were incorporated, it was renamed "C&T system."

1.3 Tasks, Functions, and Classification of TT&C and Information Transmission

Major tasks include [2]:

(1) To perform a flight vehicle's tracking, orbit measuring, attitude measuring, and telecommand of its orbit and attitude to ensure that the vehicle takes off, operates, and returns in normal attitude and predetermined orbit
(2) To perform telemetry and telecommand of a vehicle and its payloads so that it is able to complete operations and perform functions as required
(3) To fulfill various information transmission tasks to provide relevant data for ground application systems
(4) To be used for weapon test range TT&C to provide test data for identifying and evaluating the technical performances of missiles and flight vehicles and to provide a basis for design improvement and finalization as well

(5) To perform safety control tasks in the circumstance that any vehicle or missile launching goes wrong, that is, to transmit commands to blow it up so as to ensure the safety of the launching region or flight course

A C&T system mainly consists of four subsystems for orbit measuring (range, velocity, and angle measuring), telemetry, telecommand, and information transmission. These subsystems are integrated in a single system with information transmission, feedback, and control functions to fulfill C&T tasks. For example, for a satellite C&T system, information acquisition and feedback transmission are done by its tracking and orbit-measuring and telemetry subsystems, transmission of command information and control of the vehicles by its telecommand subsystem, and weapon test and finalization by its TT&C system which performs measurement, control, and evaluation. Without the C&T system, its remote sensing satellite ground application system is sure to fail with regard to reception of high-speed data, telemetry and command of payloads, and satellite orbit determination based on its angle-measurement data. As soon as a spacecraft enters its orbit, it must be subjected to service TT&C and control by the C&T system. Small military satellites usually incorporate TT&C and service functions. As weapons, some spacecrafts and aircrafts must, when used in actual military operations, be equipped with military-purpose TT&C systems to conduct precise TT&C and evaluate strike effects. When the TT&C system is used for weapon guidance, navigation, and control, the three axial parameters of attitude, velocity, and position measured by the guidance/navigation/control subsystem should also be transmitted through the telemetry system to the facilities on ground for analysis, research, and computation before being converted into commands for attitude adjustment, pointing positions, and velocity increments, which are then sent back through the telecommand subsystem and executed on the flight vehicle. It can be seen that the C&T system is of significance in both civil and military fields.

C&T systems can be classified by TT&C object, orbit altitude, TT&C scheme, TT&C station carrier, etc. Shown in Fig. 1.1 are C&T systems classified by TT&C object.

Shown in Fig. 1.2 are C&T systems classified by orbit altitude.

Any spacecraft that flies 300 km – 2×10^6 km above the Earth is called an Earth-orbit spacecraft; that with its orbit being more than 2×10^6 km above the Earth is known as a deep-space probe. Any lunar probe is about 36.3×10^4 km above the Earth.

A TT&C scheme involves what signal form to be employed and how to transmit signals. The composition and performance for a C&T system are closely related to its employed scheme. As TT&C and information transmission systems evolved according to different signal forms and multifunctional integrated transmission schemes, several TT&C schemes have come into being, as shown in Fig. 1.3.

Shown in Fig. 1.4 are C&T systems classified by the carrier for a C&T station:

A C&T system consists of a control center, an onboard C&T subsystem, and a ground C&T subsystem, which form an organic network through the communication systems and timing systems between ground stations. (1) The control center

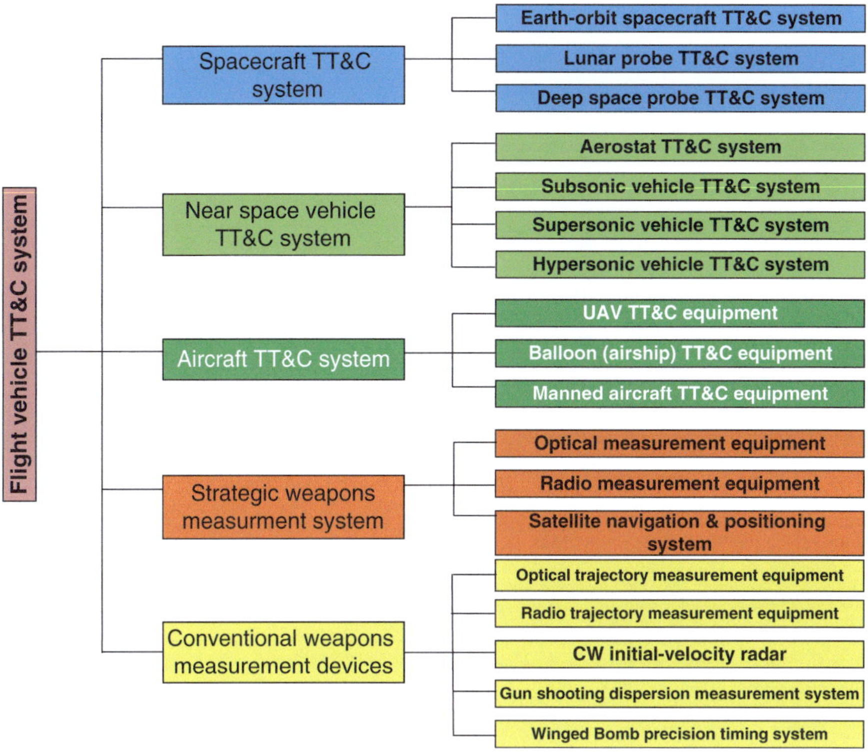

Fig. 1.1 C&T systems classified by TT&C object

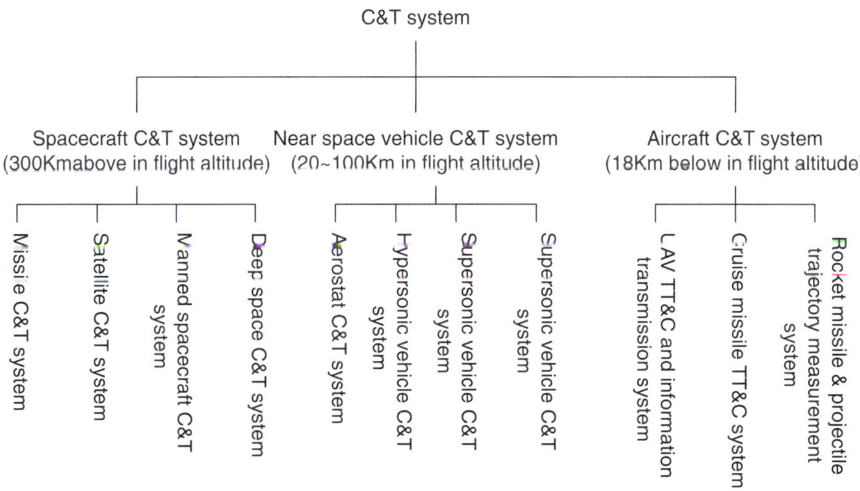

Fig. 1.2 C&T systems classified by flight altitude of a TT&C object

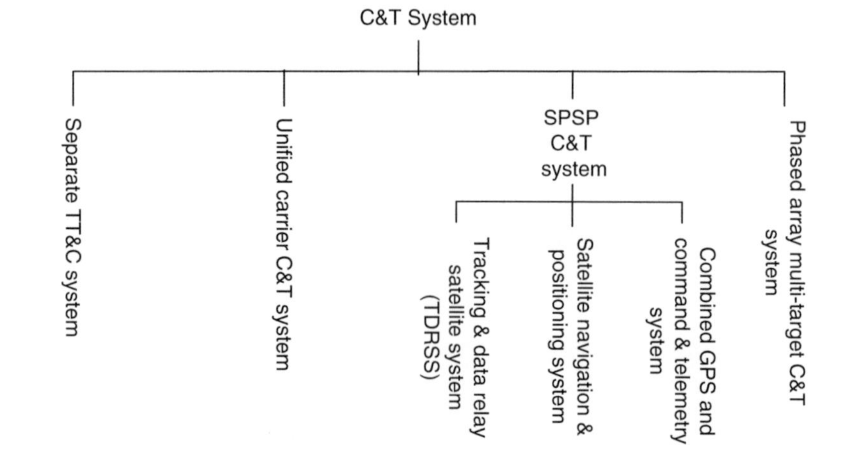

Fig. 1.3 C&T systems classified by TT&C scheme

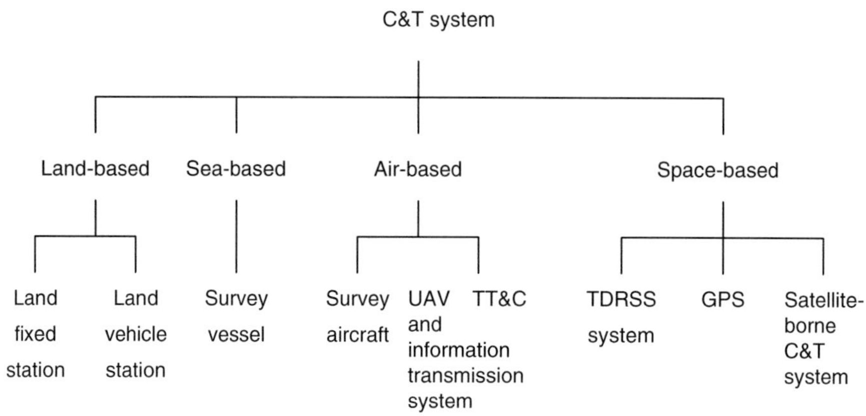

Fig. 1.4 C&T systems classified by platform

consists of a network operation and control branch and a flight control branch. The former is responsible for scheduling and assembling TT&C stations and facilities, setting operation modes and operating parameters, monitoring the current operating conditions of TT&C facilities, and coordinating multiple spacecrafts in sharing a TT&C system, while the latter is responsible for developing TT&C plans; scheduling and commanding flight missions and TT&C services; monitoring the orbit, attitude, equipment operation condition, and other telemetry data; processing its operation data; determining orbit parameters; and generating control commands and injection data. (2) The onboard C&T subsystem is constructed as shown in the block diagram of the transponder (Fig. 1.5). (3) The ground C&T subsystem is constructed as shown in the block diagram of the ground C&T station (Fig. 1.6).

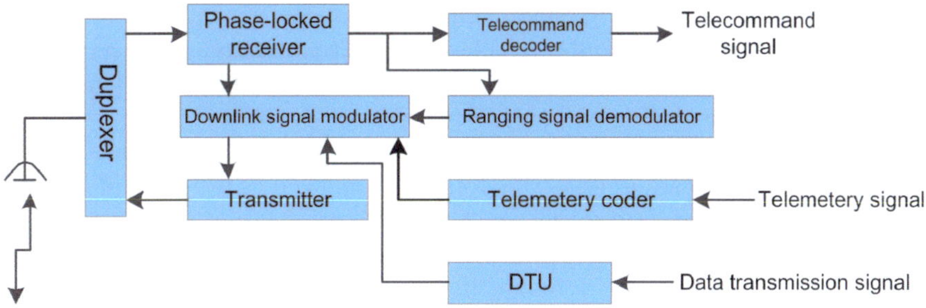

Fig. 1.5 Composition block diagram of transponder

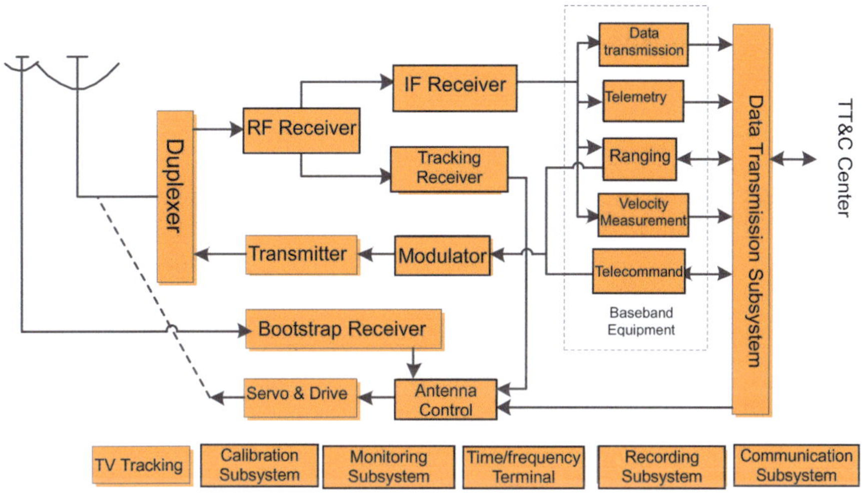

Fig. 1.6 Block diagram of ground C&T station composition

The ground C&T subsystem works together with the onboard C&T subsystem to perform functions such as tracking and orbit measuring, telemetry, telecommand, and space-to-ground information transmission.

(1) Tracking and orbit-measuring function: using the angle tracking loop, range tracking loop, and velocity tracking loop of a measuring station to measure target parameters such as angle, range, and velocity, so as to track and measure the flight trajectory of the vehicle.
(2) Telemetry function: the onboard TT&C system for the vehicle employs various technical means that allow it to approach the vehicle as close as possible to detect its internal operating condition, operating parameters, astronauts' bio-medical parameters, scientific research parameters, reconnaissance parameters, and environment parameters, which are then converted into radio signals and transmitted to the ground station for demodulating and processing before having original parameters retrieved, recorded, and displayed.

(3) Telecommand function: ground control commands are converted into radio signals and transmitted to the vehicle where they are demodulated, processed, and restored into commands to conduct the control of the vehicle for safety telecommand or vehicle telecommand purposes.
(4) Space-to-ground information transmission function: with the communication link established, various types of information obtained by the onboard measuring and observing instruments for the vehicle are transmitted to the ground C&T station via its digital communication system before being distributed to relevant users.

C&T is characterized by:

(1) High orbit-measuring accuracy. A CW radar scheme is generally used to achieve high accuracy in real-time measurement of velocity, range, and angle. This is also accompanied by certain problems like high short-term stability signals in velocity measurement, ambiguity resolution in ranging, and interferometer angle measurement.
(2) Information transmission is closely associated with TT&C. For "point-to-point" information transmission, since one transmission terminal is on the flight vehicle, it is required to guarantee the normal range of system values such as EIRP and G/T through TT&C. Moreover, as the flight vehicle and the ground station are in a relative motion state, ground station antenna and flight vehicle antenna have to get aligned with each other so as to establish an information transmission link, where flight vehicle orbit data acquired from TT&C system should be used.
(3) Highly safe and reliable telecommand. Special error control encryption and encoding/decoding techniques should be used to achieve high safety and reliability.
(4) Highly dynamic. For a satellite travels at the speed of about 7 km/s and a deep-space probe moves even faster, received signals have higher Doppler frequency and Doppler frequency change rate.
(5) Flight vehicles operate in a harsh environment, having strict restrictions on size, weight, and power consumption. The P/L (also known as a terminal station of the information transmission system) should be small in size, weight, and power consumption, with its transmitting power amplifier generally in a saturated state. This brings about a great challenge on how to reduce nonlinear impacts on data transmission for the system.
(6) Long distance and large time delay. The distance from the Earth to the geosynchronous satellite is about 4×10^4 km, to the moon is about 4×10^5 km, and to the deep-space probe is even farther. High path loss leads to a higher requirement for EIRP and G/T values.
(7) High data transmission rate. The more information the flight vehicle P/L obtains, the higher the data transmission rate is required.
(8) Power-limited system. Long distance, high data transmission rate, and relatively low EIRP value for the flight vehicle lead to the low power of signals received on ground, a bottleneck for information transmission.

(9) Generally unrestricted band. Generally, flight vehicles are designed for special purposes, with their P/L mostly operating in a single carrier with a constant envelope (currently except for SMA FDM in TDRSS and the new telemetry system) as well as in a high-frequency band with a wider available bandwidth, so its band is generally not limited nor the shaping filtering is needed. The carrier with a constant envelope allows the power amplifier to operate in a saturation state.

Due to the abovementioned special requirements, many TT&C and information transmission technologies turn out to be frontier technologies in the information transmission field, such as 2 Gb/s high-speed data transmission technology, cryogenic low-noise receiving technology that makes it possible to almost approach absolute zero (thermodynamic temperature scale), error correction coding/decoding technology that makes it possible to approach the Shannon limit, super-high-power transmitter technology, giant antenna technology, millimeter wave and laser communication technology, and high-accuracy ranging, velocity-measurement, angle-measurement, and positioning technology.

1.4 Engineering Applications of TT&C and Information Transmission Technologies [3]

As an emerging interdiscipline, TT&C and information transmission technologies involve continuous-wave radar, communication, automatic control, computer, etc. Accordingly, the C&T system for engineering facilities is also a highly integrated system based on information technology and benefiting from a variety of new technologies. It is a system engineering that allows for TT&C and information transmission and processing. The TT&C and information transmission system plays a vital role in various types of system engineering for flight vehicles. It is the only information link between a launched flight vehicle and the ground and is one of the three subsystems for a missile test project, one of the five subsystems for a satellite engineering system, and one of the eight subsystems for a manned space project. It is not only the indispensable support system for space missions but also an integral part of the integrated electronic information system. The C&T network, which is formed using the TT&C and information transmission system, is a subnet of the ground/space integrated information network. Taking into account different system engineering requirements, the flight vehicle C&T systems available at present can be chiefly grouped as follows:

1.4.1 Unified Carrier C&T System [4]

The unified carrier TT&C system is a system which enables multiple C&T functions with only one carrier, by incorporating three functions (tracking, telemetry, and command) into a single carrier.

Prior to the advent of the unified carrier TT&C system, a traditional TT&C system consisted of separate devices with different functions. This was attributable to low flight altitudes, low requirements for TT&C, and the original intention to make use of existing equipment as far as possible. With the rapid development of the aerospace industry, however, high-orbit spacecrafts and deep-space aircrafts appear constantly, which necessitate TT&C systems with multiple functions and long operating ranges, as well as onboard equipment in small size and with lightweight and good electromagnetic compatibility. As a result, the traditional TT&C system was phased out for its incompetence. Nevertheless, it is neither economical nor easy to satisfy these requirements by way of developing another single-function TT&C system comprising a variety of large equipment. In such context, the unified carrier system came into being. The unified carrier scheme contributed to the leap-frog advancement in TT&C technology. Consequently, this brought about a new generation of TT&C system which is particularly promising. With the development of the aerospace industry, it will as always play a big role in the TT&C field.

The first unified carrier TT&C system was based on the frequency-division multiplexing principle. The unified S-band (USB) system is typical of the first unified carrier TT&C system. It has been internationally recognized and has formed an international standard. For this reason, it is known as a standard unified TT&C system. Driven by demands and technical progress, the time-division spread-spectrum unified TT&C system came into being, which combines information representing various functions into a unified data flow before being modulated onto a carrier for transmission by spreading spectrum. Following deep-space exploration, the unified carrier TT&C technology has been also applied to this field, bringing about the so-called deep-space unified TT&C system (deep-space TT&C), which operates using residual carrier or suppressed carrier.

1.4.2 Space-Based C&T System [3]

The space-based C&T system consists of a tracking and data relay satellite system (TDRSS) and a "GPS + telemetry and telecommand" system.

TDRSS is a C&T system with data-relaying and TT&C functions, which enables medium- and low-orbit spacecraft tracking, orbit measuring, and data relay transmission from the ground terminal station, using relay transponders onboard geosynchronous satellites. As far as the parts of the system are concerned, the geosynchronous satellites are called tracking and data relay satellite, spacecraft in medium- or low-Earth-orbit called user satellite, and ground facilities called ground terminal station.

Characteristics of TDRSS include the following: when a geosynchronous satellite with its orbit being 36,000 km above the Earth is used for relaying, it can offer a high C&T coverage of medium- and low-orbit spacecrafts from the far to near. With two relay satellites apart 120° in longitude, it can cover 85 % orbits of medium- and low-orbit spacecrafts. With three relay satellites, it can cover almost 100 % of the orbits. Unlike a common communication satellite, a relay satellite is equipped with two 3–5 m high gain auto-tracking Ku (or Ka) band antenna, providing a wide frequency band and a high carrier-to-noise ratio which together allows for 300 Mb/s data transmission. The relay satellite is also furnished with an S-band multi-beam phased array antenna and employs the code division multiple access (CDMA) technology, making it possible for the satellite to simultaneously TT&C and communicate with dozens of targets. All these traits account for why TDRSS is developed as one of the key infrastructures in the aerospace industry. The original concept of TDRSS is induced from the application of space-based TT&C technology to improve TT&C coverage for manned space flight activities. With the development of space engineering, driven by the demand of information technology, data transmission has become one of the major functions of TDRSS. TDRSS is a typical TT&C and information transmission system.

The "GPS + telemetry and telecommand" system is a system that uses a spacecraft-borne high-dynamic GPS receiver for spacecraft positioning and then transmits the position data back to the ground through telemetry and communication.

1.4.3 Deep-Space C&T System [5, 6]

The deep-space C&T system is a system which enables C&T for the deep-space probes at a range of more than 2×10^6 km above the Earth. It is a unique information link between deep-space probes and the ground which is used to ensure that the probes are operating in correct orbits and their P/L is in a normal condition and which is also used to send the detected deep-space information back to the ground. Due to the long transmission distance, it incorporates many advanced C&T technologies.

A deep-space C&T system is composed of a ground deep-space TT&C station and a deep-space probe onboard transponder which is highly reliable and ultrasmall in size. The ground deep-space TT&C station consists of an S/X/Ka-band (or S/X-band) giant antenna servo feeder subsystem with antenna aperture of 35–70 m, a ultrahigh-power transmitter, a ultralow noise receiver, a digital baseband terminal with extremely low C/N_0 threshold and extremely narrow band phase-locked loop, a super-high stability atomic frequency standard, time-frequency equipment and interferometer angle-measuring subsystem, etc. It is mainly intended for orbit determination for the probes, using high-accuracy interferometer to measure angle, range, and Doppler frequency; telemetry of the probes, using downlink radio transmission channels to send the engineering data and mission data of the probes back to the ground for recording, displaying and processing, and real-time monitoring of the operating environment inside the probes and the operating status of equipment wherein; command of or data injection to the probes so that the deep-space probes act as predetermined and travel in orbits as designed; and telecommand of the on-time activation or deactivation of the probes, using uplink wireless transmission channel to transmit the ground control commands to the onboard command receiver for decryption. Taking advantage of digital communication technologies, the deep-space C&T system usually performs information transmission by sending detected deep-space information back to the ground for processing.

1.4.4 Phased Array C&T System [7]

The phased array TT&C system is a system which enables tracking, telemetry, and command (TT&C) and information transmission, using phased array antenna. After phased array antenna came into being, it was first applied to phased array radar and then in the communications field. This brought about a phased array communication system. In both fields, two different development directions can be seen. The former focuses on the technical issues concerning radar. The latter mainly involves smart antenna (e.g., adaptive array antenna), with its focus on the research and study of such technical issues as communication bit error rate and antijamming. C&T technology involves both CW radar and communications. Therefore, the phased array system incorporating this technology is known as a phased array C&T system. It is still mainly applied to: the CW radar filed, for example, the united carrier TT&C system requiring orbit determination, and the data transmission field, for example, SMA of TDRSS.

There are increasing demands for phased array C&T systems, mainly reflected in:

(1) Multi-target TT&C. Multi-beam of phased array antenna can be used for multi-target TT&C. In the aerospace field, the emergence of "one rocket launching multi-satellites," "constellation," and "satellite cluster" has made multi-target TT&C a research hotspot. In the aviation field, the vigorous development of UAVs has brought about research projects concerning TT&C of multiple UAVs.
(2) SMA service of TDRSS. TDRSS allows for multiuser data transmission and TT&C, taking advantage of the onboard satellite phased array antenna and the ground multi-beam forming technology.
(3) Antijamming demands. The antijamming function is critical to military TT&C systems. Phased array can be used to form adaptive spatial filter, adaptive beam, etc. to suppress the jamming from incoming waves in different directions.
(4) Demands of onboard phased array conformal antenna. In the TT&C and information transmission system for a flight vehicle, the return transmission bit rate is high, and the transmission choke point is EIRP of the onboard transponder. This is attributable to the antenna that cannot be made even bigger due to the constraints on the volume and weight of a flight vehicle. A good solution is to use an onboard phased array conformal antenna for the vehicle to increase the surface area of the antenna so as to improve antenna gain and obtain a wide scanning angle scope.

1.4.5 *High-Accuracy Missile Range-Measuring System [3]*

The missile TT&C system is a system which measures and controls various missiles. There are missiles such as strategic missile and tactical missile (including conventional rocket projectile, warhead, etc.). Missiles distinguish themselves from satellites due to their different missions. Accordingly, the TT&C system for missiles has distinct features, design method, and test method with respect to its technical requirements, composition, and scheme. The most stringent requirements are imposed on the TT&C system for flight test qualification of strategic and tactical missiles. This is mainly intended for: ① verifying the technical performances and accuracy of missiles and providing data for design improvement and finalization, ② providing real-time security control information for security control systems, ③ providing oversight and indication information to command systems at all levels, ④ providing relevant data to application systems.

Such system is mainly used for high-accuracy exterior ballistic measurement during a missile flight test. It can also be used for tracking and measurement of the powered phase of a rocket during a satellite launching process. It mainly provides high-accuracy measurement data for post data processing and real-time information for safety control systems as a basis to perform safety control or not. The measurement information includes range, velocity, and angle-measurement components, which can serve as an independent safety control information source or form multiple information sources together with other information sources contributed by other measurement devices.

The measurement data provided after the event have higher density and accuracy than those provided in real time. Through processing by personnel after the event, high-accuracy exterior ballistic measurement data can be provided and used for analyzing and separating the error coefficient in a missile or rocket guidance system, so as to improve the guidance system, the hit (injection) accuracy, and the lethality of weapons.

1.4.6 *Near-Space Vehicle TT&C and Information Transmission System [8]*

The near-space vehicle TT&C and information transmission system is a system that enables TT&C and information transmission for any near-space vehicle which travels at 20–100 km above the Earth. It is used for (1) transmitting information acquired by the vehicle and resulted from reconnaissance, remote sensing, communication, monitoring, etc.; (2) measuring the orbit of the vehicle; (3) performing telemetry, telecommand, and security control of the flight vehicle and its onboard devices.

According to the velocity of the flight vehicles subjected to TT&C, it can be divided into C&T systems for hypersonic vehicles, supersonic vehicles, subsonic vehicles, and aerostats.

(1) Hypersonic vehicle C&T system: Hypersonic vehicles fly five times faster than the velocity of sound. Their flights in the near space may cause the ionizing of the thin atmosphere there, leading to the generation of plasma sheath (blackout), which accounts for the attenuation or interruption of radio signals. To reduce the influence of blackout, therefore, equipment with a higher operating frequency is generally employed. For example, telemetry and tracking TT&C equipment operating in millimeter or higher operation band is used to perform tracking, FM telemetry, spread-spectrum telemetry, high alphabet safety control, spread-spectrum safety control, etc.
(2) Supersonic and subsonic flight vehicle C&T system: Supersonic and subsonic flight vehicles travel at about twice the speed of sound. Generally, the C&T systems for them must be reusable. Due to the strong multipath interference resulted from taking off, landing, and low-elevation long-distance covering, C&T systems with anti-multipath interference capability are generally employed. The UAVs flying in the near space are just the case. Their C&T system is generally composed of a launching/recovery subsystem, a mission control subsystem, a line-of-sight link, a beyond visual range link, a high-speed data transmission link, etc.
(3) Aerostat C&T system: Aerostats fly slow and are equipped with simple C&T systems. In certain circumstances, however, a high-speed data transmission capacity or multi-target C&T capacity is required for them.

References

1. Liu Jiaxing (1999) Past, presence and future of aerospace TT&C technology. Telecommun Eng 39(2):1–8
2. Shi Shuji, Liu Jiaxing et al (1999) Vehicle TT&C system. National Defense Industry Press, Beijing
3. Liu Jiaxing (2010) Spacecraft TT&C and communication engineering. National Defense Industry Press, Beijing
4. Liu Jiaxing (2005) Key technologies of the USB TT&C system for manned space flight. J Astronaut 26(6):743–747
5. Liu Jiaxing (2006) Features and main technical issues in deep space TT&C and telecommunication systems. J Spacecraft TT&C Technol 25(1):1–8
6. DSMS telecommunications link design handbook. JPL, http://deepspace.jpl.nasa.gov/dsndocs/810-005/
7. Liu Jiaxing (2007) Features and main technical issues in phased array TT&C system. J Spacecraft TT&C Technol 26(1):1–6
8. Liu Jiaxing (2008) Characteristics of TT&C system for near space spacecraft and major technical challenges. J Spacecraft TT&C Technol 27(1):1–7

Chapter 2
Theories and Technologies of Tracking and Orbit-Measuring

2.1 General

Tracking and orbit-measuring refers to tracking and measuring the flight trajectory of a flight vehicle by using measurement elements such as range, velocity, angle, and altitude measured by the measuring stations. "Trajectory" is a general term. The unpowered flight trajectory of spacecraft-like satellites refers to orbit, which follows the orbital dynamics. The powered flight trajectory of missiles and rockets refers to ballistic trajectory; the flight trajectory of aircrafts generally refers to track. Trajectory and track both follow the aerodynamics. Thus, orbit-measuring as discussed in this book refers to orbit measurement of spacecraft, trajectory measurement of missiles, and track measurement of aircrafts.

The flight trajectory of center of mass of a spacecraft revolving around the Earth refers to its motion orbit. Under ideal conditions, its motion around the Earth conforms to Kepler's laws, and its motion trajectory is an established Keplerian ellipse. So the path is called the orbit. In fact, the Earth is not a perfect sphere and its mass is unevenly distributed. Moreover, due to atmospheric drag, attraction of other celestial bodies to the spacecraft, and the acting forces of solar radiation pressure on the spacecraft, the spacecraft orbit becomes a complex curve approximating the Keplerian ellipse. The said acting forces are generally called "perturbing forces," and the deviation of spacecraft orbit from the Keplerian ellipse is called "perturbation." Since actual orbit deviates from the theoretical orbit, orbit measurement must be performed to control the spacecraft in order to maintain traveling in originally designed orbit or carry out necessary "orbital transfer." Methods of orbit determination mainly include:

(1) Geometric orbit determination: Also called kinematics orbit determination, it uses target angle, range, velocity, and other parameters measured by one or more tracking and measuring stations to calculate and obtain instantaneous position and velocity of the spacecraft according to relative geometric relationship and then obtain its flight trajectory through point fitting. This method is just a geometric approximation to the true orbit and does not reflect actual change law of spacecraft orbit. So it has poor precision.

J. Liu, *Spacecraft TT&C and Information Transmission Theory and Technologies*,
Springer Aerospace Technology, DOI 10.1007/978-3-662-43865-7_2

(2) Dynamics orbit determination: For orbiting spacecraft, first obtain the preliminary model of the spacecraft orbit according to the location, velocity, and direction of the injection point and then use the dynamics model of the spacecraft orbit to determine its orbit (e.g., six Kepler orbit elements) after taking into account various forces on the spacecraft in orbit during its motion. Since spacecraft orbit is determined by orbit dynamics model and its observation values relative to each measuring station can be determined by the model, there is no need of vast measured data to determine spacecraft orbit, and the data measured by the measuring stations are just used to improve the orbit model. That is, use the perturbation theory and statistical approach to solve the theoretical value of the orbit and figure out another orbit value based on the measurement element values obtained from the measuring stations, then take the difference between these two orbit values as the initial value for improved orbit and repeat the iteration to correct parameters and initial conditions of orbit's mathematical model until a high-precision orbit determination is achieved. Due to the application of constraint conditions specific to orbiting, the dynamic method has higher orbit determination precision than geometric orbit determination method. Moreover, it needs fewer measurement elements, so any one measurement element can be used for orbit determination, such as velocity, range, or angle.

For flight vehicles having driving force or more air drag such as missile and aircraft, it is not suitable to use dynamic orbit determination method due to its poor accuracy. Generally, geometric method is applied. Based on precise formula and sound numerical approaches, the data, such as angle, range, and velocity measured by the measuring stations, are used to determine the orbit of the spacecraft, in which the system errors are corrected and the random errors are smoothed to obtain accurate orbit. Since the flight vehicle is in motion, its orbit can be determined only when both its velocity and location are determined, that is, six orbit elements $x, y, z, \dot{x}, \dot{y}, \dot{z}$ should be determined. It can be seen that orbit measurement involves two aspects, namely, velocity measurement and location determination.

2.2 Localization and Orbit-Measuring

2.2.1 Localization

Spatial localization of the flight vehicle is a three-dimensional locating problem. At least three independent location parameters are needed to determine the spatial location of the target. Currently, location parameters which can be measured mainly include range R, azimuth angle A, elevation angle E, range sum S, range difference r, and direction cosine (including l, m, and n), just three of which are enough to determine the spatial location of the flight vehicle. The geometric principle of localization is to make three geometric surfaces by using these three location parameters. The intersection point of these surfaces is the spatial location of the vehicle. This geometric theory localizing the spacecraft is called location geometry. Three location parameters as measured can be used to name the localization system, for example, (A, E, R) localization system and (R, S, r) localization system. As any three of six location parameters can form a localization system, there are 56 combinations of three that can be drawn from six parameters, namely, 56 localization systems in total. Similarly, there are also 56 systems in velocity measurement, such as $(\dot{R}, \dot{A}, \dot{E})$. At present, there are many definition methods for the locating system. According to the localization geometry theory, the better definition method is using the name of measurement elements.

The geometric surfaces decided by seven major location parameters in engineering application are summarized as follows [1]:

(1) Range R: the distance between orbit-measuring station and the flight vehicle. The flight vehicle may be located at one point on a sphere which takes the measuring station as its center and R as its radius. In the (x, y, z) coordinate system, R meets the following expression:

$$x^2 + y^2 + z^2 = R^2 \tag{2.1}$$

The geometric figure of the sphere is shown in Fig. 2.1.

(2) Elevation angle ε: angle of elevation of the flight vehicle relative to the ground. In the (x, y, z) ground-transmitting coordinate system, elevation angle shall be determined by:

$$x^2 + y^2 = z^2 \cot^2 \varepsilon \tag{2.2}$$

Its geometric figure is a cone, as shown in Fig. 2.2.

As shown in Fig. 2.2, xOy is the horizontal plane. The vehicle may be located at any point on this cone.

(3) Azimuth angle $\varphi(\beta)$: angle of azimuth of the flight vehicle in the (x, y, z) coordinates. The azimuth angle should be determined according to the following expression:

$$y = x \tan \varphi \tag{2.3}$$

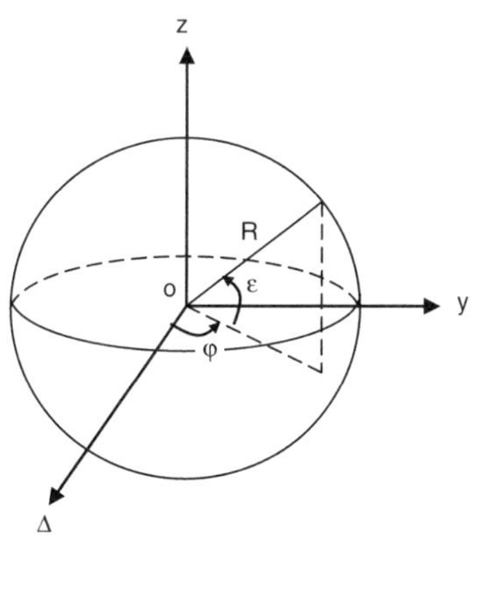

Fig. 2.1 Target location sphere determined by range R

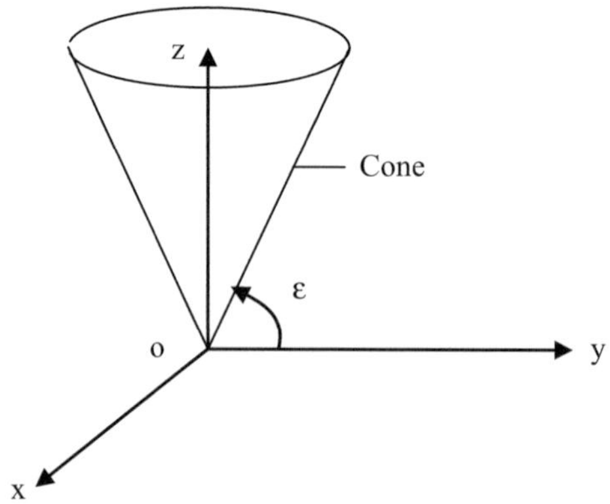

Fig. 2.2 Target location surface determined by elevation angle ε

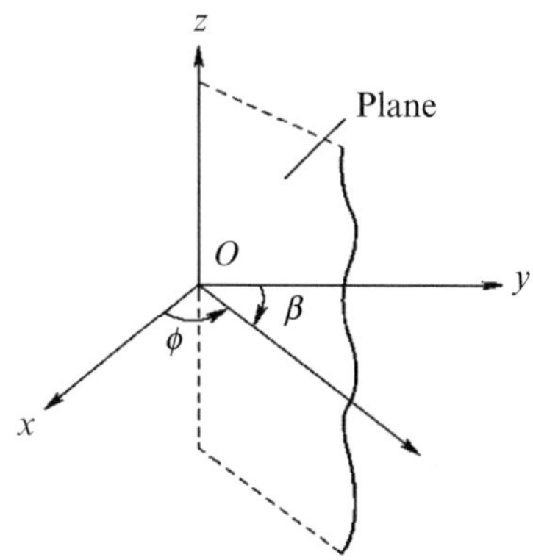

Fig. 2.3 Target location determined by azimuth angle φ

The geometric figure of azimuth angle is a plane perpendicular to xOy horizontal plane, as shown in Fig. 2.3.

The flight vehicle may be located at any point on this plane.

(4) "Range sum" $S = r_i + r_j$, where r_i refers to the distance between the master station and the target and r_j refers to the distance between the slave station and the target. At this point, the target is located on a rotational ellipsoid determined by the following expression:

$$\begin{aligned} S &= r_i + r_j \\ &= \left[(x - x_i)^2 + (y - y_i)^2 + (z - z_i)^2\right]^{\frac{1}{2}} \\ &\quad + \left[\left(x - x_j\right)^2 + \left(y - y_j\right)^2 + \left(z - z_j\right)^2\right]^{\frac{1}{2}} \end{aligned} \tag{2.4}$$

where $x_i = [x_i y_i z_i]^{\mathrm{T}}$, $x_j = [x_j y_j z_j]^{\mathrm{T}}$ are the ith and jth station locations, respectively, and $x = [x, y, z]^{\mathrm{T}}$ is spatial location of the target. The rotational ellipsoid determined by the said expression is as shown in Fig. 2.4.

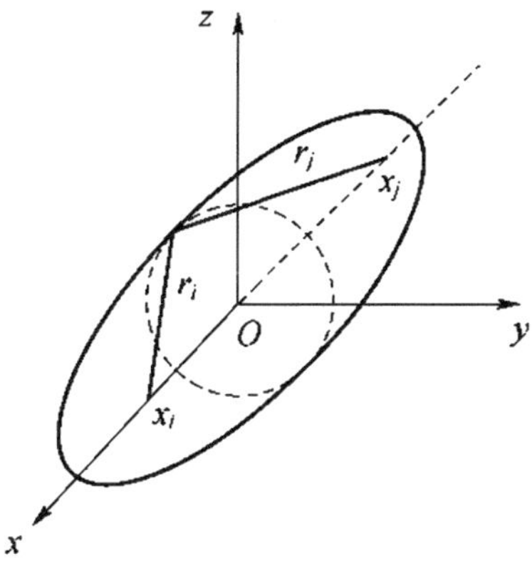

Fig. 2.4 Target location determined by range sum S

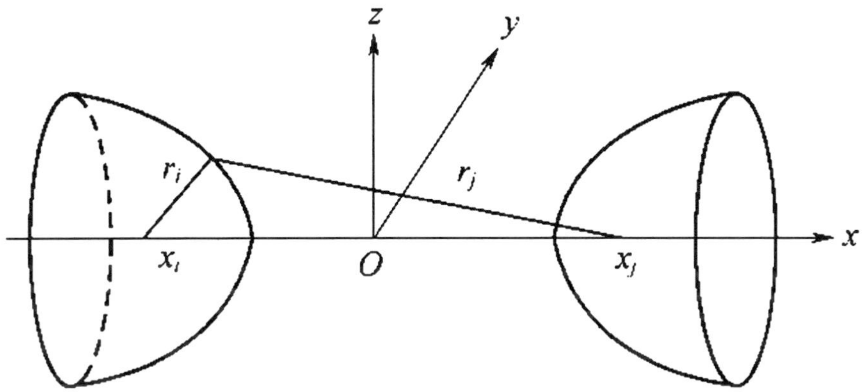

Fig. 2.5 Target location surface determined by range difference r

(5) Range difference $r = r_i - r_j$: at this point, the target is located on a rotational hyperboloid determined by the following expression:

$$\begin{aligned} r &= r_i - r_j \\ &= \left[(x - x_i)^2 + (y - y_i)^2 + (z - z_i)^2\right]^{\frac{1}{2}} \\ &\quad - \left[(x - x_j)^2 + \left(y - y_j\right)^2 + (z - z_j)^2\right]^{\frac{1}{2}} \end{aligned} \tag{2.5}$$

where all symbols have the same meaning as those in Expression (2.4). The rotational hyperboloid determined by the said expression is as shown in Fig. 2.5.

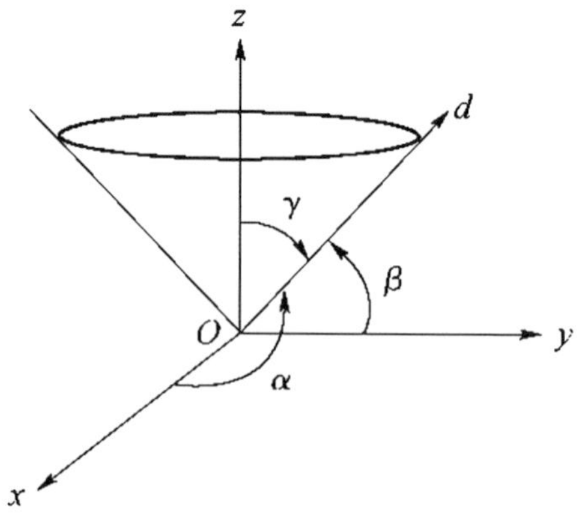

Fig. 2.6 Target location surface determined by direction cosine

(6) Direction cosine:

$$\begin{aligned} l &= \cos\alpha \\ m &= \cos\beta \\ n &= \cos\gamma = \sqrt{1 - l^2 - m^2} \end{aligned} \tag{2.6}$$

Target location surface determined by direction cosine is shown in Fig. 2.6.

In Fig. 2.6, three direction cosines ($l = \cos\alpha$, $m = \cos\beta$, $n = \cos\gamma$) correspond to three cones which take x-axis, y-axis, and z-axis as cone axis and α, β, and γ as semi-vertex angle, respectively (the figure shows an $n = \cos\gamma$ cone). Any two cones intersect in a line of direction d, i.e., the line of position (LOP).

It can be seen from Fig. 2.6 that target location surfaces determined by both direction cosine and elevation angle are cones (because they are measured angle parameters). The difference is that the geometric surface of target location determined by direction cosine is a cone taking the baseline between two stations as its axis and having an angle α (the angle between target angle direction and the baseline of two stations) while the target location surface determined by elevation angle is a cone taking the normal axis of the antenna as its axis and having an angle ε. The advantage of the method of measuring direction cosine is that higher accuracy of α measurement can be obtained.

(7) Target altitude h: the altitude is determined by the following expression:

$$z = h \tag{2.7}$$

Expression (2.7) represents a plane with altitude as h, as shown in Fig. 2.7.

According to the foregoing analysis, one location parameter can only determine that the target is located on a surface but can't determine exactly which point it is located on the surface. If there are two location parameters, the two surfaces determined by these two parameters are intersected on a curve. Then, we can only confirm that the target is located in this curve but could not confirm which point it is located on the curve. Only by getting the point of intersection of this curve

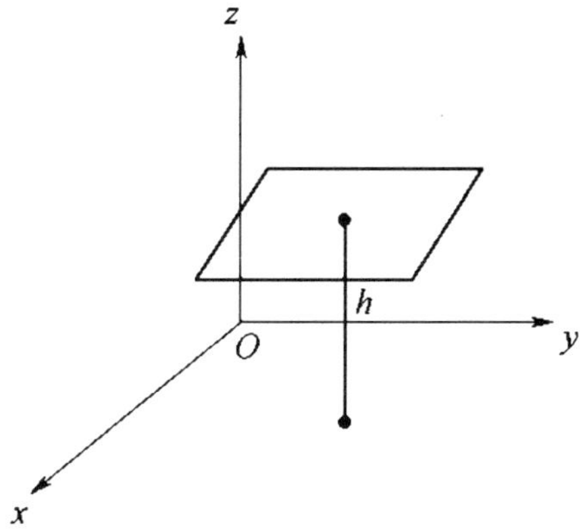

Fig. 2.7 Target location surface determined by altitude *h*

and the surface determined by the third location parameter can finally determine which point the target is located at. *R*, *r*, and *S* are ranging systems, among which one can be converted into another, and each of which is equivalent and can be converted into the pure ranging system.

Among the localization systems, some are suitable for LEO spacecraft, others are suitable for HEO or deep-space spacecraft; some have high accuracy, others have low accuracy; some are high in cost, others are low. So the selection of a localization system is very important in the TT&C system.

2.2.2 Station Distribution Geometry and Localization Accuracy

Localization methods of spacecraft can be classified into three types: range, azimuth, and elevation (RAE) localization method; angle intersection localization method; and range intersection location method. The point of intersection in space of three geometric surfaces determined by three location parameters is the location of the spacecraft. It is thus clear that the localization accuracy of a spacecraft is related to both measurement accuracy of TT&C equipment and geometry and accuracy of station distribution. Reference [2] gives an in-depth study and provides the relevant results.

Station distribution geometry refers to the relative geometric relationship between orbit-measuring stations and the target, which directly affect the localization accuracy of the target. With the same internal error and external error (or generally same), the different station distribution geometry would cause significant difference in orbit-measuring precision.

"Optimality" metric of station distribution geometry has many criteria. In general, the following minimum criterion is adopted for study of the optimal

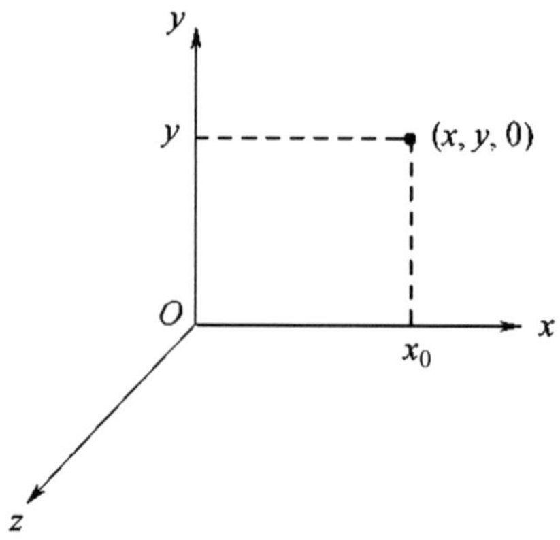

Fig. 2.8 Ground non-inertial launch coordinate system

localization geometry (error F expressed in this method is also called statistical module length):

$$F = \sigma_X^2 + \sigma_Y^2 + \sigma_Z^2$$

Total linear error is

$$\sigma_L = \sqrt{\sigma_x^2 + \sigma_y^2 + \sigma_z^2} = \sqrt{F}$$

where δ_x, δ_y, and δ_z are the location error in the directions of x-axis, y-axis, and z-axis in the ground non-inertial coordinate system, respectively.

It can also refer to velocity error. The study results are applicable to both. A selected coordinate system is shown in Fig. 2.8. The xOy surface and target launch surface coincide with each other; xOz is the ground surface; $(x, y, 0)$ represents the coordinate of the target; and $(x_0, 0, 0)$ refers to the subsatellite point of the target.

2.2.2.1 RAE Localization System

A ground non-inertial launch coordinate system is shown in Fig. 2.8.

When ranging RMS error is δ_R and angle measurement RMS error is $\delta_A = \delta_E$, the location error expressed by statistical module length F is

$$F = \sigma_R^2 + R^2\left(1 + \cos^2 E\right)\sigma_A^2 \tag{2.8}$$

where R is the distance between the measuring station and the target and E is the elevation.

A geometric relationship in single-station orbit determination is shown in Fig. 2.9.

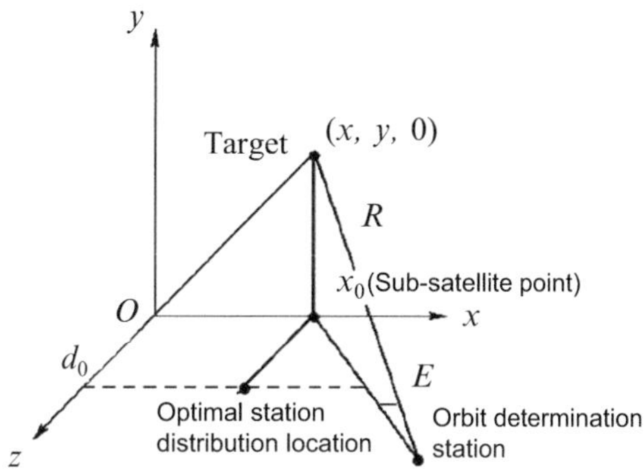

Fig. 2.9 Geometric relationship in single-station orbit determination

When the orbit determination stations are distributed in different locations, F has different values. The station distribution with F being the minimum value is the optimal distribution. According to the analysis of Reference [3], the target is located in the vertical line of Ox axis where $z = 0$, i.e., at subsatellite point (x_0, 0, 0). Then, the location error is

$$F = \sigma_R^2 + \left(y^2 + 2z_1^2\right)\sigma_A^2$$

When $z_1 = 0$, then

$$F = \sigma_R^2 + y^2\sigma_A^2 \tag{2.9}$$

where y refers to target altitude.

It is clear that the higher the target, the larger the location error will be.

Considering there is an unsafe area $|z_1| < d_0$ (d_0 is the distance to safety line) in the test range, optimal address of orbit determination station shall be at $(x_0, 0, \pm d_0)$, as shown in Fig. 2.9.

When two sets of RAE orbit-measuring equipment are used, optimal station distribution is located at $(x, 0, d_0)$ and $(x, 0, -d_0)$, respectively. Then, the location error is 1/2 less than that of single-station localization. The optimal station distribution properties of RAE localization are summarized in Table 2.1.

2.2.2.2 "Angle Intersection" Localization

This method adopts two or more high-precision angle measuring systems (such as optical theodolite or interferometer) to find the direction of the target and uses the point of intersection of two or more direction lines to determine the target location.

(1) Dual-station angle intersection localization

Given angle measurement accuracies of two stations are σ_1 and σ_2, respectively, $\sigma_1 = \sigma_2$. Optimal station address is located at the straight line which is perpendicular

Table 2.1 Optimal station distribution properties of RAE localization

S/N	Item	Single-station localization	Dual-station localization
1	Optimal station distribution location	On the vertical line of x-axis in xOz plane passing through the subsatellite point	On the vertical line of x-axis in xz plane passing through the subsatellite point
2	Vertical distance between orbit-measuring station and subsatellite point	$z_1 = d_0$ (or $z_1 = -d_0$)	$z_1 = d_0$ (or $z_2 = -d_0$)
3	Spatial included angle of the target		$\beta^* = 2^* \arctan(d_0/y)$
4	Elevation between orbit-measuring station and the target	$E = 90° - \alpha^*$	$E = 90° - \alpha^*$
5	Slant range between orbit-measuring station and the target	$R^* = \left[y^2 + d_0^2\right]^{\frac{1}{2}}$	$R^* = \left[y^2 + d_0^2\right]^{\frac{1}{2}}$
6	Minimum risk value	$F_{\min} = \sigma_R^2 + (2d_0^2 + y^2)\sigma_A^2$	$F_{\min} = 1/2F_{\min}$
7	Optimal estimation accuracy of target location	$\sigma_x = d_0\sigma_A$ $\sigma_y = \left[(y/R^*)^2\sigma_R^2 + d_0\sigma_A\right]^{\frac{1}{2}}$ $\sigma_z = \left[(d_0/y)^2\sigma_R^2 + y^2\sigma_A^2\right]^{\frac{1}{2}}$	$(\sigma_x, \sigma_y, \sigma_z) = \frac{1}{\sqrt{2}}(\sigma_x, \sigma_y, \sigma_z)$

Note: d_0 is the distance to safety line

to Ox axis and passes through the subsatellite point. Two stations are located at both sides of Ox axis, respectively, and the distance between station and Ox axis is:

$$d = 0.67183y \tag{2.10}$$

where y is the target altitude.

Then, optimal localization accuracy is

$$\begin{cases} \sigma_x = 0.4751y\sigma_1, & \sigma_y = 1.5276y\sigma_1 \\ \sigma_z = 1.0263y\sigma_1, & F_{\min} = 3.61235y^2\sigma_1^2 \end{cases} \tag{2.11}$$

Geometry of the station distribution is shown in Fig. 2.10.

(2) Tri-station angle intersection localization

Optimal station distribution is of a regular triangle, and the subsatellite point is located at the center of such regular triangle. The vertex of the regular triangle is located at a station distribution circle. The location error is

$$F = \sigma_a^2 R^4 \frac{\left(R^4 + y^2d^2 + 4d^4\right)}{\left[3d^2\left(R^4 + y^2d^2\right)\right]} \tag{2.12}$$

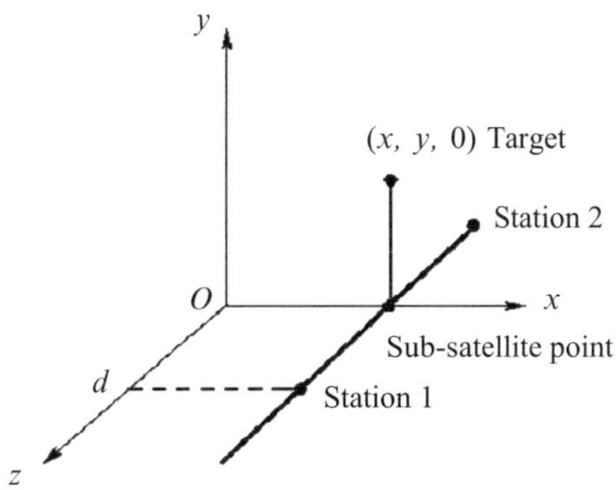

Fig. 2.10 Geometry of optimal station distribution in dual-station angle intersection localization

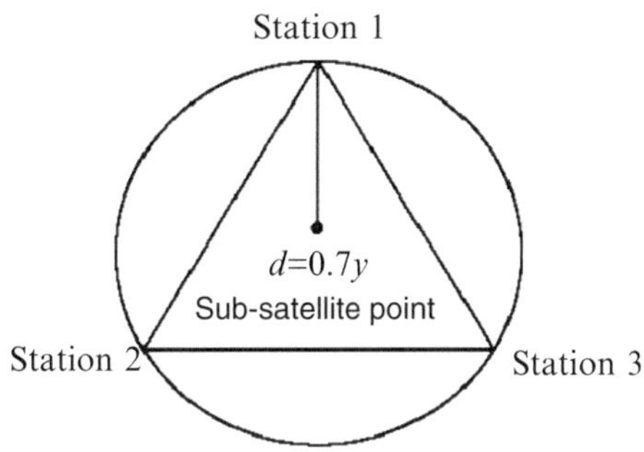

Fig. 2.11 Geometry of optimal station distribution in tri-station angle intersection localization

The optimal radius is

$$d = 0.70387y$$

Geometry of the station distribution is shown in Fig. 2.11.

Then, optimal localization accuracy is

$$\sigma_x = \sigma_z = 0.52y\sigma_1, \quad \sigma_y = 1.23y\sigma_1, \quad F_{\min} = 2.0454y^2\sigma_1^2 \tag{2.13}$$

where y is the target altitude and σ_a is the angle measurement error.

It is thus clear that orbit determination error in angle intersection method also increases with the increase of the target altitude. Station distribution properties of "angle intersection" orbit determination are listed in Table 2.2.

2.2.2.3 "Range Intersection" Localization

This method is to draw multiple spheres taking the stations as the center and the range measured by multiple high-precision ranging systems as the radius. The point

Table 2.2 Optimal station distribution properties of angle intersection localization

S/N	Item	Tri-station intersection localization	Dual-station intersection localization
1	Geometric figure of optimal station distribution (takes the subsatellite point as the center)	Regular triangle	The baseline between two stations passes through the subsatellite point and is perpendicular to *x*-axis, and the distance between stations and the subsatellite point is equal
2	Radius of station distribution circle	$d = 0.70387y$	$d = 0.67183y$
3	Space intersection angle between two stations and the target	$\beta_{12} = \beta_{13} = \beta_{23} = 59°48'$	$\beta_{12} = 67°48'$
4	Elevation between each station and the target	$E_1 = E_2 = E_3 = 20°18'$	$E_1 = E_2 = 56°6'$
5	Slant range between each station and the target	$R_1 = R_2 = R_3 = 1.2229y$	$R_1 = R_2 = 1.2047y$
6	Baseline between stations	$D_{12} = D_{13} = D_{23} = 1.21914y$	$D_{12} = 1.34366y$
7	Minimum risk value	$F_{\min} = 2.0454y^2\sigma_a^2$	$F_{\min} = 3.61235y^2\sigma_a^2$
8	Optimal estimation accuracy of target location	$\sigma_x = \sigma_z = 0.52y\sigma_a$	$\sigma_x = 0.4751y\sigma_a$
		$\sigma_y = 0.123y\sigma_a$	$\sigma_y = 1.5276y\sigma_a$
			$\sigma_z = 1.0263y\sigma_a$
9	Included angle between the radius of station distribution circle and *x*-axis	$\theta_1 = \theta_1$ (arbitrary, take 30° in practice)	$\theta_1 = 90°$
		$\theta_2 = \theta_1 + 120°$,	$\theta_2 = 270°$
		$\theta_3 = \theta_1 + 240°$	

of intersection of these spheres is the target location (three or more spheres must be provided).

(1) Tri-station range intersection localization

When three stations are located at the same station distribution circle,

$$F = \left\{ 1 + \left[\frac{2(1 - \cos\beta_{12}\beta_{13}\beta_{23})}{\Delta_1} \right] \right\} \sigma_R^2$$

where

$$\Delta_1 = 1 + 2\cos\beta_{12}\beta_{13}\beta_{23} - \cos^2\beta_{12} - \cos^2\beta_{13} - \cos^2\beta_{23}$$

where σ_R is ranging error of the measuring station, β_{12} is the space intersection angle of Station 1 and Station 2 at the target, β_{13} is the space intersection angle of Station 1 and Station 3 at the target, and β_{23} is the space intersection angle of Station 2 and Station 3 at the target.

Under the condition of $\beta_{12}=\beta_{13}$, the β_{12} which makes F minimum can be solved as

$$\cos\beta_{12} = \cos\beta_{13} = \cos\beta_{23}$$

Then,

$$\cos\beta_{12} = \cos\beta_{13} = \cos\beta_{23}$$

That is, the geometric figure of tri-station distribution is a regular triangle, and the subsatellite point is located at the center of the regular triangle, and

$$F = \sigma_R^2 + \sigma_R^2\left\{\frac{\left[8\left(y^2+d^2\right)-\left(2y^2-d^2\right)\right]^3}{27d^4y^2}\right\} \tag{2.14}$$

It is clear that F also relates to d, i.e., F value changes with d.

In order to compare three localization methods more intuitively, the localization accuracies of RAE method and angle intersection method for geostationary satellite localization are estimated as follows.

For RAE single-station localization method, the expression of total linear error in case of optimal station distribution is

$$\sigma_L = \sqrt{F} = \sqrt{\sigma_R^2 + \left(2d_0^2 + y^2\right)\sigma_a^2} \tag{2.15}$$

When ranging error $\sigma_R=4$ m, angle measurement error $\sigma_a=0.01\,^\circ=$ 0.0001745 rad, geostationary satellite altitude $y\approx3.69\times10^4$ km, and $d_0=0$, through calculation we can get $\delta_L=6{,}440$ m.

For tri-station angle intersection localization method, the expression of total linear error in case of optimal station distribution is

$$\sigma_L = \sqrt{F} = \sqrt{2.0454y^2\sigma_a^2} \tag{2.16}$$

When angle measurement error $\sigma_a=0.01^\circ=0.0001745$ rad and target altitude $y\approx3.69\times10^4$ km, then we get $\sigma_L=9{,}210$ m through calculation.

It can be seen from the above calculations that the tri-station range intersection localization method has the highest accuracy in geostationary satellite localization.

Let $\partial F/\partial d=0$, the radius of the optimal station distribution circle can be obtained as

$$d = \sqrt{2}y \tag{2.17}$$

and so

$$F_{\min} = 3\sigma_R^2 \tag{2.18}$$

It is thus clear that optimal station distribution geometry in tri-station range intersection localization method can be summarized: three stations are distributed in a circle taking the subsatellite point as its center; the radius of the circle is $d = \sqrt{2}y$; three stations are arranged like a regular triangle.

(2) Four-station range intersection localization

When four stations are distributed on the same circumference, if the four stations constitute a rectangle, compared with other inscribed quadrilateral distribution schemes, it will have a minimum value of F, namely,

$$F = \frac{\sigma_R^2 \left(y^2 + d^2\right)\left(4y^2 + d^2\right)}{4y^2 d^2} \tag{2.19}$$

from which it can be seen that F varies with d. By letting $\partial F / \partial d = 0$, the d that makes F minimum can be obtained, i.e.,

$$d = \sqrt{2}y \tag{2.20}$$

and the minimum F is

$$F_{\min} = \left(\frac{9}{4}\right)\sigma_R^2 \tag{2.21}$$

It is clear that $F_{\min}$ is 1.33 times less than that of tri-station localization.

According to the foregoing, an optimal station distribution geometry in a four-station range localization method can be summarized: four stations are distributed in a circle taking the subsatellite point as its center; the optimum radius of the circle is $d = \sqrt{2}y$; four stations are arranged like a square.

Geometric graph of optimal station distribution in "range intersection" localization is shown in Fig. 2.12.

Optimal station distribution properties of range intersection localization are listed in Table 2.3.

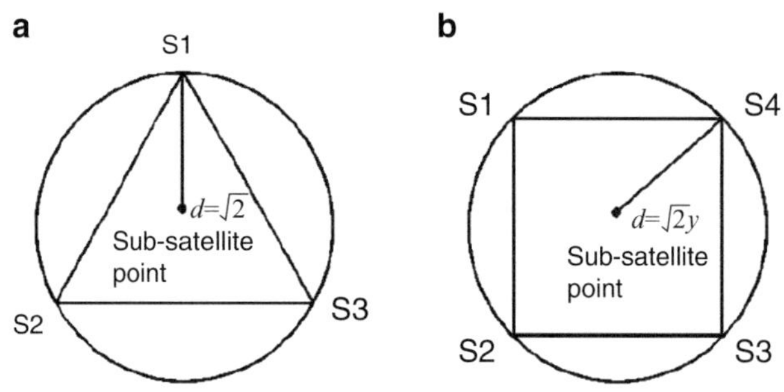

Fig. 2.12 Geometric graph of optimal station distribution in range intersection localization. (**a**) Tri-station intersection and (**b**) four-station intersection

Table 2.3 Optimal station distribution properties of range intersection localization

S/N	Item	Four-station intersection localization	Tri-station intersection localization
1	Geometric figure of optimal station distribution (takes the subsatellite point as the center)	Square	Regular triangle
2	Radius of station distribution circle	$d=\sqrt{2}y$	$d=\sqrt{2}y$
3	Space intersection angle between two stations and the target	$\beta_{12}=\beta_{14}=\beta_{23}=\beta_{34}=70°30'$	$\beta_{12}=\beta_{13}=\beta_{23}=90°$
		$\beta_{13}=\beta_{24}=109°30'$	
4	Elevation between each station and the target	$E_i=35°16'(i=1,2,3,4)$	$E_i=35°16'(i=1,2,3)$
5	Slant range between each station and the target	$R_i=\sqrt{3}y(i=1,2,3,4)$	$R_i=\sqrt{3}y(i=1,2,3)$
6	Baseline between stations	$D_{12}=D_{14}=D_{23}=D_{34}=2y$	$D_{12}=D_{13}=D_{23}=\sqrt{6}y$
		$D_{13}=D_{24}=2\sqrt{2}y$	
7	Minimum risk value	$F_{\min}=(9/4)\sigma_a^2$	$F_{\min}=3\sigma_a^2$
8	Optimal estimation accuracy of target location	$\sigma_x=\sigma_z=\sigma_y=\left(\sqrt{3}/2\right)\sigma_R$	$\sigma_x=\sigma_y=\sigma_z=\sigma_R$
9	Included angle θ_1 between the radius of station distribution circle and x-axis	$\theta_1=\theta_1$ (arbitrary, take 45° in practice)	$\theta_1=\theta_1$ (arbitrary, take 30° in practice)
		$\theta_2=\theta_1+\pi/2,$	$\theta_2=\theta_1+120°$
		$\theta_3=\theta_1+\pi$	$\theta_3=\theta_1+240°$
		$\theta_4=\theta_1+\frac{3}{2\pi}$	

According to the foregoing, it can be summarized as follows:

(1) Localization accuracy of RAE method and angle intersection method decreases with the increase of spacecraft altitude y. So both methods are more applicable to LEO spacecraft, instead of HEO spacecraft.
(2) Localization accuracy of range intersection method is independent of the spacecraft altitude and is only related to ranging accuracy σ_R in optimal station distribution. So it is most favorable to the localization of HEO spacecraft.
(3) The optimal station distribution in range intersection localization is of a circle with the diameter equal to $\sqrt{2}y$. When $y=4\times10^4$ km, the diameter is 5.6×10^4 km. It is clear that this could not be realized on the surface of the Earth. So it is impossible to realize optimal station distribution on the Earth. What can only be done is to expand the distance between orbit-measuring stations as much as possible. Then, specific station distribution geometry requires specific localization accuracy calculation.

2.2.3 Trajectory Measurement System

In order to measure the trajectory of the vehicle, at least six measurement elements must be measured. Several common systems composed of these measurement elements and their relevant characteristics are summarized as follows [4].

2.2.3.1 RAE System

The RAE system is also called single-station system or ranging and angle measurement system. Measurement elements R, A, and E are used to determine the location of the flight vehicle. Through differential smoothing, $\dot{R}$, $\dot{A}$, and $\dot{E}$ can also be used to determine the velocity of the flight vehicle. If higher velocity-measuring accuracy is required, the carrier Doppler velocity measurement can be used. Continuous wave radar, pulse radar, and optical theodolite with a laser ranging device can constitute this system. Due to limited angle measurement accuracy, this system has poor measurement accuracy for long-range targets. So it is a mid- and low-accuracy measurement system. It is a main system in satellite TT&C systems.

2.2.3.2 Multi-$R\dot{R}$ System

The multi-$R\dot{R}$ system is also called the range intersection system, in which each station independently measures R and $\dot{R}$ (the number of stations $\geq$3) and baseline is generally several hundred kilometers to several thousand kilometers. The transponder operating in this multi-receiver and multi-transmitter system must make a multichannel response in frequency division or code division method. So the multichannel transponder is one of the key issues. This system is applicable to intersection measurement using multiple pulse radars and continuous wave radars. Without need of baseline transmission, each station is relatively independent. So the station is distributed in a flexible way. According to the accuracy of the measuring station, a high- and mid-accuracy measurement system can be constituted. Currently, it has been widely used in missile high-accuracy measurement system, satellite TT&C system, and satellite navigation system and is a promising high-accuracy measurement system.

2.2.3.3 Rlm System

The Rlm system is also called short-baseline interferometer system. It consists of one master station and two slave stations. In general, the baseline is very short, only tens of meters to several kilometers. The range between the master station and the target is R; direction cosines of range difference between master station and slave station are l and m. Measurement elements of this system are R, l, m, $\dot{R}$, $\dot{l}$, and $\dot{m}$. It uses carrier frequency signals to measure range difference and has very high accuracy. However, due to a short carrier frequency period, the range ambiguity is very serious. So it is required to place multiple antennas on the baseline to solve the ambiguity. Under short-baseline conditions, ambiguity resolving is easy. But if the baseline is too long, the number of antennas will increase, the system will be complex, and the cost will be very expensive. Under short baseline conditions, by dividing the range difference by the baseline length, one can obtain direction cosines l and m. Signals between master station and slave station can be transmitted through phase-compensated cables or optical fibers. Due to lower measurement accuracy of direction cosines, this system is applicable to mid-accuracy measurement.

2.2.3.4 Rr$_i$ System

The Rr$_i$ system is also called medium- and long-baseline interferometer system. Measurement elements of medium- and long-baseline interferometer system are R, r_i, $\dot{R}$, and $\dot{r}_i$ ($i = 1, 2, \ldots n$, $n \geq 2$, r is the range difference between master station and slave station). Baseline length is generally dozens of kilometers, i.e., LOS. Signals can be transmitted through microwave communication or optical fibers. As the baseline length increases, range ambiguity when measuring range difference by the carrier will be more serious. So we have to use modulation signals to measure the range difference. The accuracy of measuring range difference in this system is lower than short-baseline system. But because the baseline is longer, equivalent angle measurement accuracy is higher than short-baseline system. When signals are transmitted via a microwave link between master and slave stations, it is required to build a baseline transmission tower. Standard frequency of master station should be sent to slave stations as their frequency reference; slave stations should convert and transmit received signals to the master station. The master station will extract, record, and transmit in real time information about velocity-measuring and ranging of master and slave stations to the data processing center, while the slave station will only receive and retransmit the measurement information, without need of timing and data recording devices.

2.2.3.5 Multi-AE System

The multi-AE system is also called angle intersection system. It can be constituted by multiple interferometers or by a combination of a cinetheodolite, electrooptic theodolite, and monopulse radar. Its measurement elements are multiple groups of A and E. Radar and optical observations are independent. There is no relation between devices. Associated instrumentation refers to combined utilization in data. In principle, arbitrary combination is available. At present, with technical development, the baseline uses optical fiber transmission and connected element interferometer (CEI), and very long baseline interferometer with intercontinental baseline appeared, thus significantly improving measurement and localization accuracy. Since only a beacon is installed on board to perform broadcast transmission and one-way measurement is performed on ground to implement passive tracking, equipment is simplified. Moreover, there is no rang system error introduced by the transponder in ranging, and radio star can be used for angle calibration, so another idea is provided for improving localization accuracy.

2.2.3.6 Multi-$S\dot{S}$ System

The multi-$S\dot{S}$ is also called $kS\dot{S}$ system. This system requires measuring the range sum and its rate ($k \geq 3$) only. Generally, it consists of one transmitting station and multiple receiving stations. It can be classified into medium-baseline system and long-baseline system. There are two methods in measuring range sum, namely, modulation signal ranging method and integral ranging method.

(1) Medium-baseline $kS\dot{S}$ system

Its system structure is very similar to that of a medium-baseline interferometer. The difference is that information transmission on baseline is in opposite direction. The transmitting station transmits time and frequency standards to the receiving station, and the receiving station performs $S\dot{S}$ measurement. The $kS\dot{S}$ system and interferometer have no essential distinction. They both use $S\dot{S}$ measured by the stations to calculate $r\dot{r}$.

(2) Long-baseline $kS\dot{S}$ system

This system features a single-carrier frequency signal and simple equipment. However, due to long baseline, if integral ranging method is used, other measures provided by external equipment or the internal system are required to provide the start point of integral, and no contingency interrupt is allowed for signals. All stations should provide timing, frequency reference, communication, and data recording devices to independently measure $\dot{S}$. No signal is transmitted between stations. Due to long baseline, this system is applicable to high-accuracy measurement.

The above-mentioned systems are quite distinct from each other. In selecting a system, one should take an overall consideration in terms of system engineering according to measurement accuracy requirements, operating requirements, and cost.

2.3 Velocity-Measuring Theory and Technology: Two-Way, Three-Way, and Single-Way Velocity-Measuring Technologies

2.3.1 CW Velocity-Measuring

The principle of measuring the target velocity is that, based on the Doppler effect in physics, when CW radar wave illuminates the target, the target will move and produce Doppler frequency which is in direct proportion to the velocity of the target relative to radar movement and in inverse proportion to wavelength λ. When the target flies to the radar station, Doppler frequency will be positive and the frequency of receiving signals is higher than that of transmitting signals. When the target departs from the radar, Doppler frequency will be negative and the frequency of receiving signal is lower than that of transmitting signals.

The velocity can be measured by measuring the Doppler frequency shift. The most basic method is the two-way carrier Doppler velocity-measuring. The two-way carrier Doppler velocity-measuring system is mainly composed of a ground transmitter, flight vehicle transponder, and ground receiver. Generally, a phase-lock coherent transponder is used. A noncoherent transponder with independent local oscillator is also available. "Coherent" refers to a certain phase relation between output and input signals; otherwise, it is "noncoherent." Velocity measurement corresponding to the former is called coherent Doppler velocity-measuring. Velocity measurement corresponding to the latter is called noncoherent Doppler velocity-measuring.

(1) Two-way Doppler velocity-measuring with coherent transponder

For coherent transponders, because after an uplink one-way Doppler frequency is retransmitted by a coherent transponder with a coherent turnaround ratio $1/\rho = f_{\text{uplink}}/f_{\text{downlink}}$, Doppler frequency is transformed $1/\rho$ times. The combined Doppler frequency of uplink and downlink is almost twice of downlink one-way Doppler frequency. It is well known that the radial velocity of the target is:

$$v = -\frac{f_d}{2f_r + f_d}c, \tag{2.22}$$

where f_r is the downlink receiving frequency on ground.

The coherent transponder is generally a phase-lock transponder. Figure 2.13 shows a common solution.

In Fig. 2.13, the receiving frequency of the transponder is

$$f_{r\text{Trans}} = \left(N_1 + N_2 + \frac{1}{2}\right)f_v$$

where f_v is the VCO frequency.

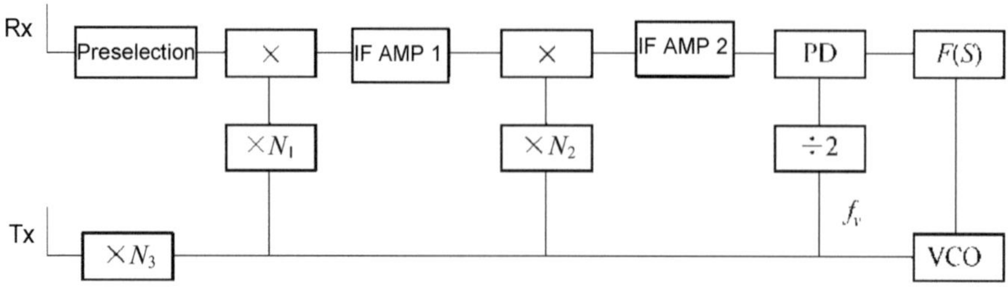

Fig. 2.13 Principle block diagram of phase-lock transponder

Transmitting frequency is

$$f_{b\text{Trans}} = N_3 f_v$$

Phase-lock transponder turnaround ratio is

$$\rho = \frac{f_{b\text{Trans}}}{f_{\text{Trans}}} = \frac{N_3}{N_1 + N_2 + 1/2} = \frac{f_{\text{downlink}}}{f_{\text{uplink}}} \tag{2.23}$$

(2) Two-way Doppler velocity-measuring with noncoherent transponder

A noncoherent transponder can also be used for a two-way Doppler velocity measurement system. But in order to improve velocity measurement accuracy, reasonable measures should be taken to eliminate noncoherent components from transponder noncoherent LO. According to different elimination schemes, the method can be classified as "ground station canceling method," "transponder canceling method," and "transponder measuring and ground processing method." These three methods will be introduced in detail below. Noncoherent transponder scheme is often used.

1) Mixed Repeating Noncoherent Transponder. As shown in Fig. 2.14, this transponder has an independent crystal oscillator with frequency of f_L. The input signal will be down-converted, amplified, filtered, and up-converted and then retransmitted via a power amplifier. Down-conversion LO is $(m-1)f_L$, up-conversion LO is mf_L, and the receiving frequency of the transponder is the sum of uplink transmitting frequency f_t and uplink Doppler frequency f_{td}, namely, $f_{r\text{Ttrans}} = f_t + f_{td}$.

Being converted twice, it becomes

$$f_{b\text{Trans}} = f_t + f_{td} + f_L$$

Transmitting frequency contains independent LO component f_L. In order to eliminate the incoherence, mf_L, the m-time multiplier of f_L, should be transmitted to the ground station to eliminate f_L incoherence in the process of extracting Doppler frequency. Figure 2.15 is the block diagram showing the ground station extracts of Doppler frequency.

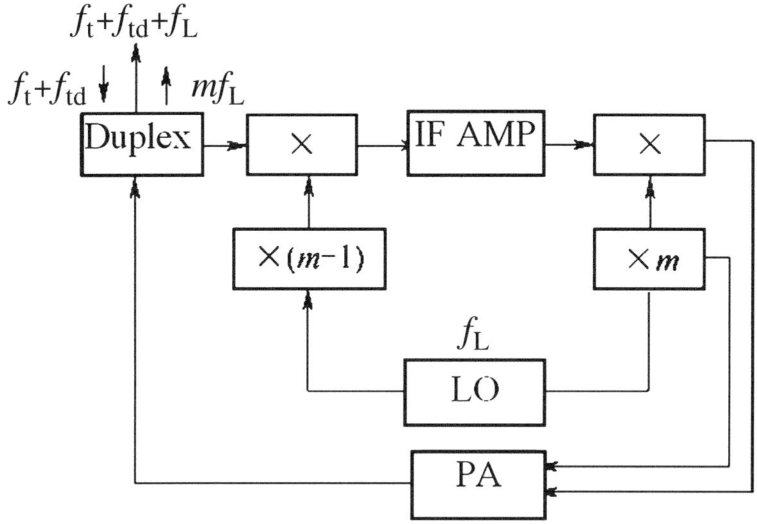

Fig. 2.14 Mixed repeating noncoherent transponder

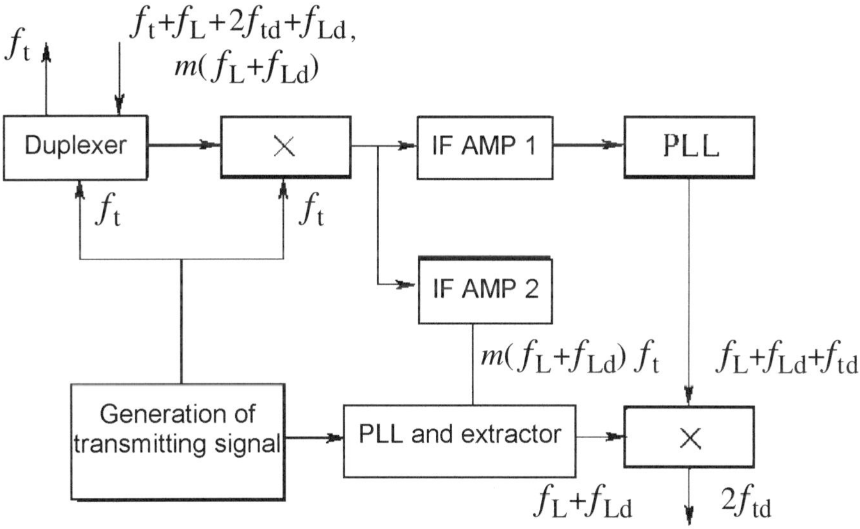

Fig. 2.15 Block diagram of Doppler frequency extraction in ground stations

Downlink signal returned from the transponder enters into the receiver, which contains two signal frequencies: $f_t + f_L + 2f_{td} + f_{Ld}$ and $m(f_L + f_{Ld})$, where f_{Ld} and f_{td} refer to one-way Doppler frequencies of the independent master oscillator of the transponder and of ground-transmitting frequency, respectively. Two signal frequencies are mixed with transmitting frequency f_t and filtered by two IF amplifiers at different frequencies and then extracted with the phase-locked loop. IF amplifier

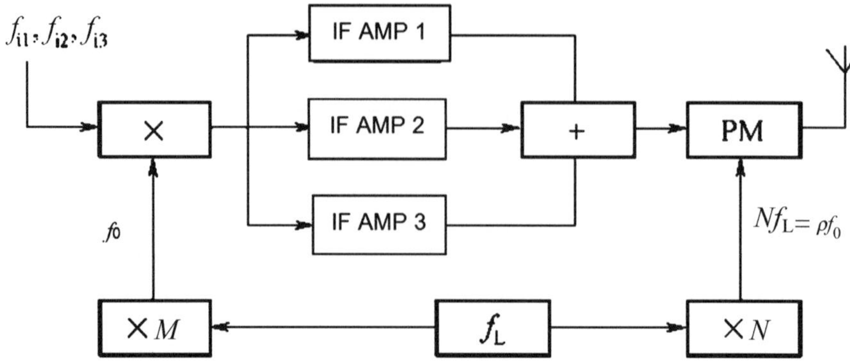

Fig. 2.16 IF modulation repeating noncoherent transponder

1 extracts and outputs $f_L + f_{Ld} + 2f_{td}$; the branch of IF amplifier 2, PLL and extractor, outputs $f_L + f_{Ld}$. After mixing and subtracting the two outputs, pure Doppler frequency $2f_{td}$ will be outputted, thus eliminating the incoherence caused by the independent crystal oscillator of the transponder.

2) Modulation Repeating Noncoherent Transponder. As shown in Fig. 2.16, the transmitting signal of the transponder is a constant envelope. So the power amplifier can operate in a saturated mode, thus avoiding the effect of intermodulation interference and amplitude/phase conversion on measurement accuracy. A transponder is used to repeat signals from three ground stations. The carrier frequency difference among three ground stations is small, several megahertz in general. The transponder receives transmitting frequencies from three ground stations, converting them into IF f_{i1}, f_{i2}, and f_{i3} via same mixer. Noncoherent LO frequency f_0 is M-time frequency multiplication of f_L. Three IF frequencies are filtered and amplified by three IF amplifiers in parallel and summed to modulate the phase of transmitting frequency of the transponder Nf_L and then transmitted to the ground stations. Since three phase modulation subcarriers (i.e., three intermediate frequencies) and the transmitting frequency of the transponder contain noncoherent components of master oscillator f_L, after phase-locking and extracting the main carrier and subcarrier in ground stations and proper frequency adding and subtracting, noncoherent components of master oscillator of the transponder f_L can be eliminated to obtain pure Doppler frequency f_D (Fig. 2.17).

2.3.2 *Doppler Frequency Measurement Method*

Currently, the digital phase-locked loop (PLL) velocity measurement method has been widely used. Previous methods such as "determination of integer ambiguity by fixed time," "time determination by fixed integer ambiguity," and "determination of integer ambiguity by basic fixed time" are used in certain special circumstances

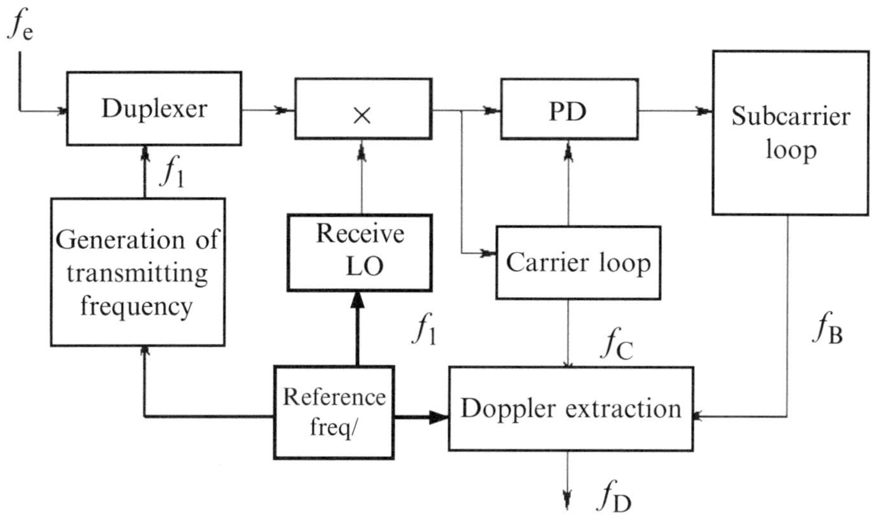

Fig. 2.17 Doppler extraction in ground station for IF modulation transponder

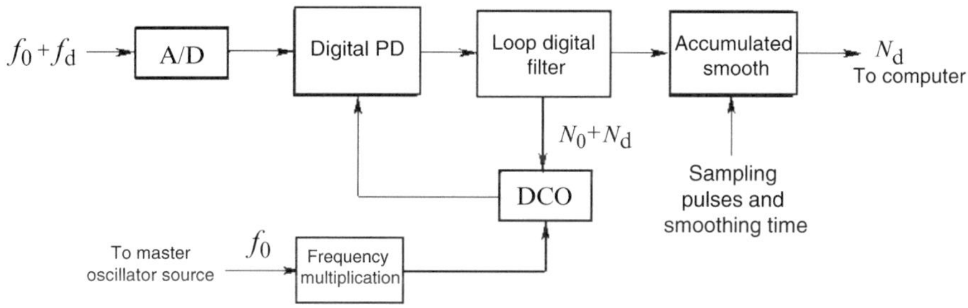

Fig. 2.18 The extraction and measuring of Doppler frequency in IF digital phase-locked loop

only. Now, the digital PLL of the phase-locked receiver captures and tracks the downlink signals and extracts and measures carrier Doppler frequency. This not only realizes software technology and multiple functions but also greatly simplifies velocity-measuring terminal equipment. Moreover, equipment simplification reduces additional phase noise and increases velocity measurement accuracy.

As shown in Fig. 2.18, IF digital PLL is mainly composed of digital circuits, which eliminate the slow drift in simulating parameters of phase discriminator, active filter, and VCO in the PLL with the change of time, temperature, and level. The phase discriminator of the digital PLL consists of a digital multiplier and an accumulator. The loop filter is a digital filter, generally a digital signal processor (DSP). The digital-controlled oscillator (DCO) is composed of a digital synthesizer driven by externally high-frequency clock. Output frequency and phase are under control by output of the loop filter. After loop-locking, the DCO frequency is equal to the frequency of input signals. The variation of DCO frequency control code fully corresponds to the frequency change of input signals, with a correct mathematical corresponding relation.

Multiplying the variation of frequency control code by determined conversion coefficient equals Doppler frequency shift of input signals. Then, output frequency control code finishes extraction and measurement of Doppler frequency, and by continuously outputting frequency control code (at an interval of 5-loop filter operation repetition period), the velocity value can be continuously outputted.

Certainly, frequency code will be affected and changed by input thermal noise. In order to eliminate the effect of thermal noise, frequency code should be smoothed and filtered. In general, accumulative smooth is used. Accumulative start time is under the control of sampling pulse, and smooth time is also controlled by external commands. After smooth, frequency code will be sent to the computer for conversion into velocity data.

Given the frequency control coefficient of the DCO is k_i, input frequency is $f_{io}+f_d$, center frequency f_{io}, and loop filter outputs control code N_0+N_d, then $N_0=f_{io}/k$, $N_d=f_d/k_i$. The average value of Doppler frequency code of the accumulated smoothing outputs is N_d, and average value of Doppler frequency $f_D=k_iN_d$.

In order to eliminate noncoherent changes of DCO-driven clock frequency and avoid noncoherent velocity-measuring error, high-frequency clock of DCO shall be generated by multiplication of master oscillator frequency f_0 of the transmitter to make transmitter frequency source and receiver Doppler extraction circuit form a complete coherent system. Moreover, master oscillator f_0 is of good short-term stability, thus decreasing velocity-measuring error caused by introduction of Doppler frequency measurement reference.

The digital method often has quantization error. In design, efforts should be made to reduce the quantization error to an extent much smaller than the error caused by inputting thermal noise. When the DSP is used as the loop filter arithmetic unit, a 16-bit operation or double-precision 32-bit operation is often used. 16-bit operation can be used in loop filter operation. Loop self-adjustment can be used to eliminate the effect of operation quantization error on DCO mean frequency. The resolution of frequency control code outputted after operation will be guaranteed jointly by loop filter output word length and DCO frequency adjustment accuracy. For design, first determine the shortest word length of f_d code according to velocity range and velocity resolution. Given maximum velocity of the flight vehicle 12 km/s, velocity resolution 0.1 cm/s, minimum capacity of f_d code 12 km/0.1 cm $=1.2\times10^7$, and $2^{24}=1.68\times10^7$, the shortest word length of f_d code is a 24-bit binary code. Then, derive DCO frequency control resolution from the velocity resolution.

Given DCO frequency adjustment step size ΔF and velocity resolution Δv, for a two-way Doppler, there is

$$\Delta F=\frac{2\Delta V}{c}f_r \tag{2.24}$$

When $f_r=2$ GHz, $\Delta V=0.1$ cm/s, we can obtain a maximum DCO frequency adjustment step size of 0.013 Hz.

DCO can be implemented by the DDS which consists of binary phase accumulator and digital sine (cosine) table. Frequency control resolution depends on bits of phase accumulator and clock frequency, namely,

$$\Delta F = \frac{f_{\text{op}}}{2^n} \tag{2.25}$$

The above discussion about the quantization is a sufficient condition, but not a necessary condition. When the DCO frequency adjustment step size is longer or loop filter operation word length is shorter, DCO mean frequency is always equal to input signal frequency due to the loop self-adjustment. During locking process, DCO control code fluctuates around the control value of the mean frequency, and time average of the control code will be a mixed decimal. Thus, when DCO or loop filter operation resolution is not high enough, output resolution can be improved through time accumulation of frequency control code. Certainly, instantaneous or short-time high resolution could not be obtained. Time accumulation of frequency control code can both eliminate the effect of thermal noise and improve velocity resolution in times equal to the number of smooth.

2.3.3 Theoretical Analysis of Two-Way Coherent Doppler Velocity-Measuring Error

In the radio velocity measurement technology field, continuous wave coherent Doppler velocity measurement is a velocity measurement system with highest accuracy. When a higher accuracy is required, short-term stability of velocity measurement signals is an important factor affecting the velocity measurement accuracy. This is a special issue in continuous wave velocity measurement technology. This section will first give an introduction to the issues on signal short-term stability.

2.3.3.1 Phase Noise Power Spectrum Density and Allan Variance [4, 6]

Phase noise is often characterized by several spectrum density functions:

(1) $S_{\Phi}(f)$: baseband spectrum of phase noise $\Phi(t)$
(2) $P_{\text{RF}}(f)$: RF signal amplitude spectrum observed on the spectrum analyzer, by which P_s/N_0 can be obtained
(3) $£(\Delta f)$: phase noise spectrum value displayed on the phase noise tester, also called normalized spectrum
(4) $S_y(f)$: baseband spectrum of frequency fluctuation noise

The said spectrums are common one-sided spectrums in engineering. $S_{\Phi}(f)$ is a low-pass one-sided spectrum of phase noise modulation signal $\Phi(t)$, used for

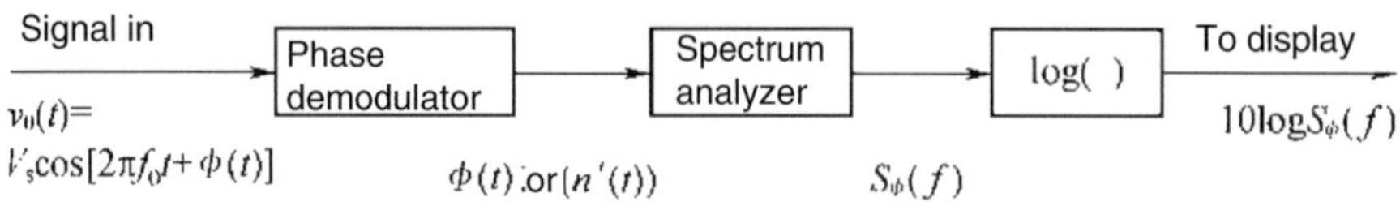

Fig. 2.19 Block diagram of $S_\Phi(f)$ phase noise analyzer

measuring $S_\Phi(f)$, as shown in Fig. 2.19. The measuring instrument consists of a $\Phi(t)$ phase demodulator, a low-spectrum analyzer receiving $S_\Phi(f)$ signals, and a logarithmic converter used for display. The logarithmic scale is usually used to display the relation between $10\log S_\Phi(f)$ and frequency. $S_\Phi(f)$ is expressed in rad^2/Hz, which should be explained to be dB relative to 1 rad^2/Hz, with respect to the dB scale.

In engineering practice, although $S_\Phi(f)$ is really measured, $10\log £(\Delta f)$ is always displayed in the ordinate. This value is considered to be a phase noise within 1 Hz bandwidth in a sideband at the frequency deviation. If phase noise amplitude is small enough, $£(\Delta f) \approx S_\Phi(f)/2$. When it is expressed in logarithm, by increasing 3 dB in low-pass spectrum of $10\log[£(\Delta f)]$, one can obtain a correct $10\log[S_\Phi(f)]$.

The above-mentioned spectrum density functions maintain a relationship, namely,

$$S_\Phi(f) = S_y(f)\left(\frac{f_0}{f}\right)^2 \tag{2.26}$$

$$£(\Delta f) \approx \frac{S_\Phi(f)}{2} \tag{2.27}$$

When white Gaussian noise is added, phase noise spectrum density is

$$N_\Phi(f) = W_n'(f) = \frac{2N_0}{V_s^2} = \frac{N_0}{P_S} \tag{2.28}$$

where P_s is signal power, N_0 is noise power spectrum density, $N_\phi(f)$ is the spectrum density of random variable $n'(t)$ at output end of phase demodulator, and V_s is signal amplitude.

Another characterization method of phase noise spectrum density is Allan variance. The Allan variance $\sigma_y^2(\tau)$ and phase noise power spectrum density are two features both used to describe the short-term frequency stability of one signal. $S_\phi(f)$ is used for frequency domain and $\sigma_y^2(\tau)$ is for time domain. However, with respect to their physical essence, they are indexes used to describe the same short-term stability. They are related to each other through Fourier transform.

The relationship between phase noise power density $S_\phi(\omega)$ and Allan variance $\sigma_y^2(\tau)$ will be represented in three methods.

(1) Integral representation method

A lot of documents have given the relational expression of $\sigma_y^2(\tau)$ and $S_y(\omega)$ [4]:

$$\sigma_y^2(\tau) = \int_0^{\infty} S_y(f)\left[\frac{2\sin^4(\pi f\tau)}{(\pi f\tau)^2}\right]df = \int_0^{\infty} S_\Phi(f)\frac{f^2}{f_0^2}\left[\frac{2\sin^4(\pi f\tau)}{(\pi f\tau)^2}\right]df \quad (2.29)$$

where τ is sampling interval (*s*) of measuring Allan variance; *f* is frequency interval (Hz) deviating from carrier frequency f_0; $S_y(f)$ is power spectrum density of relative frequency fluctuation $\Delta f/f_0$, short for relative frequency fluctuation spectrum density; and $S_\Phi(f) = S_y(f)(f_0/f)^2$ is phase fluctuation spectrum density.

Expression (2.29) shows that the solution of Allan variance in time domain is equivalent to $H_y(f) = (2\sin^4(\pi f\tau)/(\pi f\tau)^2)$ characteristic filtering in frequency domain $S_\phi(\omega)$, where $H_y(f)$ is the transfer function of $S_y(f)$ Allan variance and $H_\Phi(f) = \left[\frac{2\sin^4(\pi f\tau)}{(\pi f\tau)^2}\left(\frac{f}{f_0}\right)^2\right]$ is the transfer function of $S_\phi(f)$ Allan variance.

(2) Analytic representation method

In theoretical analysis, $S_\phi(f)$ is often of a power law spectrum model. It is an approximate assumption. The model is

$$S_\phi(f) = h_{-2}f_0^2f^{-4} + h_{-1}f_0^2f^{-3} + h_0f_0^2f^{-2} + h_{+1}f_0^2f^{-1} + h_{+2}f_0^2 \quad (2.30)$$

$$S_y(f) = h_{-2}f^{-2} + h_{-1}f^{-1} + h_0f^0 + h_{+1}f^{+1} + h_{+2}f^{+2} \quad (2.31)$$

where $S_\phi(f)$ is expressed in rad^2/Hz and $S_y(f)$ is expressed in (rad/s)^2/Hz.

Items in the above expression are in direct proportion to powers of *f*, so it is called power law spectrum. The relationship between these five noise types and spectrum density is shown in Table 2.4.

In Table 2.4, α is each power component of power law spectrum expression in the frequency domain, representing the type of frequency fluctuation noise. Similar to the frequency domain, in the time domain, power law expression of Allan variance is $\sigma_y^2(\tau) = \sum_{a=-2}^{2} D_a\tau^{\mu}$, where μ is the power, $\mu = -\alpha - 1$ (if $\alpha > 1$, $\mu = -2$), representing the type of frequency fluctuation noise in the time domain and D_α and h_α represent the intensity of corresponding noise. A different oscillator has a different noise type. Quartz crystal oscillator mainly has three types of noise, namely, PM white noise, FM white noise, and FM flicker noise. For atomic clocks, the phase noise spectrum is greatly narrowed.

Table 2.4 Spectrum densities of five types of noise

Noise type	α	$S_y(f)$	$S_\varphi(f)$	μ
PM white noise	2	h_2f^2	$h_2f_0^2$	−2
PM flicker noise	1	h_1f	$h_1f_0^2f^{-1}$	−2
FM white noise	0	h_0	$h_0f_0^2f^{-2}$	−1
FM flicker noise	−1	$h_{-1}f^{-1}$	$h_{-1}f_0^2f^{-3}$	0
Frequency random walk	−2	$h_{-2}f^{-2}$	$h_{-2}f_0^2f^{-4}$	1

$S_\phi(f)$ is expressed in a logarithmic way:

$$\left[S_\phi(f)\right]_{-2} = 10\lg h_{-2}f_0^2 - 40\lg f, \quad (f_1 \sim f_2 \text{ interval, frequency random walk noise}) \tag{2.32}$$

$$\left[S_\phi(f)\right]_{-1} = 10\lg h_{-1}f_0^2 - 30\lg f, \quad (f_2 \sim f_3 \text{ interval, FM flicker noise}) \tag{2.33}$$

$$\left[S_\phi(f)\right]_{0} = 10\lg h_0 f_0^2 - 20\lg f, \quad (f_3 \sim f_4 \text{ interval, FM white noise}) \tag{2.34}$$

$$\left[S_\phi(f)\right]_{+1} = 10\lg h_{+1}f_0^2 - 10\lg f, \quad (f_4 \sim f_5 \text{ interval, } PM \text{ flicker noise}) \tag{2.35}$$

$$\left[S_\phi(f)\right]_{+2} = 10\lg h_{+2}f_0^2, \quad (f_5 \sim f_6 \text{ interval, } PM \text{ white noise}) \tag{2.36}$$

where $S_\phi(f)$, expressed in dBc/Hz, is the relative spectrum density relative to the carrier.

The physical essence of the above expression is to use broken lines to approximate actual phase noise curves (represented by logarithmic coordinates). Each broken line segment has 40, 30, 20, and 0 dB/10 octave slope, representing one phase noise type respectively. h_{-2}, h_{-1}, h_0, h_{+1}, and h_{+2} are the heights of each broken line, respectively, representing phase noise amplitude, called intensity coefficient, of each type of phase noise.

The curve graph approximated by phase noise power law spectrum broken line is shown in Fig. 2.20 (approximation of three segments $[S_\phi(f)]_{-1}$, $[S_\phi(f)]_0$, and $[S_\phi(f)]_{+2}$).

The above logarithmic expressions are common approximate representations of phase noise index in engineering. If the above curves of signals are measured, when $\lg f = 0$, one can solve h_{-2}, h_{-1}, h_0, h_{+1}, and h_{+2} in Expressions (2.32), (2.33), (2.34), (2.35), and (2.36), namely,

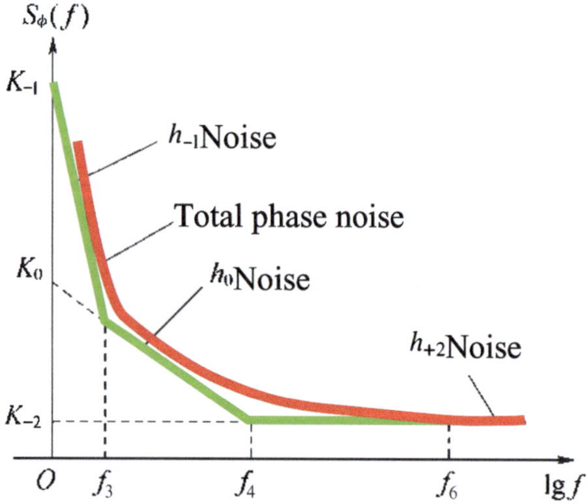

Fig. 2.20 Phase noise power law spectrum density curve

$$h_{-2} = f_0^{-2}\lg^{-1}(0.1K_{-2}) \tag{2.37}$$

$$h_{-1} = f_0^{-2}\lg^{-1}(0.1K_{-1}) \tag{2.38}$$

$$h_0 = f_0^{-2}\lg^{-1}(0.1K_0) \tag{2.39}$$

$$h_{+1} = f_0^{-2}\lg^{-1}(0.1K_{+1}) \tag{2.40}$$

$$h_{+2} = f_0^{-2}\lg^{-1}(0.1K_{+2}) \tag{2.41}$$

where $K_{-2}, K_{-1}, K_0, K_{+1}$, and K_{+2} refer to the values of the intersection points of each approximate broken line of $h_{-2}, h_{-1}, h_0, h_{+1}$, and h_{+2} with the vertical axis at $\lg f = 0$, respectively, which can be solved by the phase noise spectrum density curve. They can either be given by the analytic expression of Expressions (2.32), (2.33), (2.34), (2.35), and (2.36) or directly solved by Expressions (2.37), (2.38), (2.39), (2.40), and (2.41). Furthermore, it can be seen from the said expression that when f_0 increases N times, dB in the first item of the expression will increase 20 lg$N$. Therefore, when the frequency reference source is the same (e.g., 10 MHz frequency standard), due to its $h_{-2}, h_{-1}, h_0, h_{+1}$, and h_{+2} are fixed, its phase noise index will be different after N times of ideal frequency multiplication to generate f_0, i.e., will increase by 20 lg$N$.

By integral solving through Expression (2.29), one can obtain Allan variance corresponding to each item of $S_\phi(f)$. For example, for h_{+2} item (PM white noise), $S_\Phi = h_{+2}f_0^2$.

$$\left[\sigma_y(\tau)\right]_{+2}^2 = \frac{2h_{+2}}{\pi^2\tau^2}\int_{f_5}^{f_6} \sin^4(\pi\tau f)\mathrm{d}f = \frac{2h_{+2}}{\pi^2\tau^2}\left\langle\frac{3(f_6 - f_5)}{8} - \frac{1}{4\pi\tau}\cdot\left\{\sin^3\left[\pi\tau(f_6 - f_5)\right]\cos\left[\pi\tau(f_6 - f_5)\right] + \frac{3}{4}\sin\left[2\pi\tau(f_6 - f_5)\right]\right\}\right\rangle \tag{2.42}$$

where f_6 is cutoff frequency of integral operation in calculating short-term stability, corresponding to effective bandwidth the phase noise may give an effect on.

For example, in a phase-locked receiver, it refers to a single-side bandwidth of the phase-locked loop. This relationship shows that short-term stability relates to the phase-locked receiver bandwidth. Since short-term stability affects the velocity measurement accuracy, velocity measurement accuracy relates to the phase-locked loop bandwidth. When $f_6 - f_5 \gg 0.2/\tau$, PM white noise is

$$\left[\sigma_y(\tau)\right]_{+2}^2 \approx \frac{3h_{+2}(f_6 - f_5)}{4\pi^2\tau^2} \tag{2.43}$$

Similarly, the following items can be solved by integral [5], that is,
PM flicker noise:

$$\left[\sigma_y(\tau)\right]_{+1}^2 \approx \frac{1.038 + 3[\ln 2\pi\tau(f_5 - f_4)]}{4\pi^2\tau^2}h_{+1}, \quad \text{if } 2\pi\tau(f_5 - f_4) \gg 1 \tag{2.44}$$

FM white noise:

$$[\sigma_y(\tau)]_0^2 \approx \frac{h_0}{2\tau} \tag{2.45}$$

FM flicker noise:

$$[\sigma_y(\tau)]_{-1}^2 = 2h_{-1}\ln 2 \tag{2.46}$$

Frequency random walk:

$$[\sigma_y(\tau)]_{-2}^2 \approx \frac{2\pi^2\tau h_{-2}}{3} \tag{2.47}$$

After solving the five items of Allan variance, total Allan variance is

$$[\sigma_y(\tau)]_\Sigma^2 = [\sigma_y(\tau)]_{-2}^2 + [\sigma_y(\tau)]_{-1}^2 + [\sigma_y(\tau)]_0^2 + [\sigma_y(\tau)]_{+1}^2 + [\sigma_y(\tau)]_{+2}^2 \tag{2.48}$$

The above six expressions have the following meanings in the engineering application:

1) For each type of phase noise, when $\sigma_y(\tau_1)$ of a τ_1 value is given, one can solve its corresponding h_{-2}, h_{-1}, h_0, h_{+1}, and h_{+2} and then obtain $\sigma_y(\tau_x)$ of any other τ_x value.

2) By measuring $\sigma_y(\tau_1)$, $\sigma_y(\tau_2)\ldots$ at different $\tau_1, \tau_2 \ldots$, one can obtain the $\sigma_y^2(\tau)$ to τ curve and then approximately obtain the distribution frequency interval (corresponding to $f = 1/2\tau$) of each type of phase noise from the above expression and thus determine the type of phase noise. For example, when $\sigma_y^2(\tau)$ is in direct proportion to $1/\tau$, the noise is h_0 phase noise (if the phase noise could not be determined due to too large interval of $\tau_1, \tau_2\ldots$, reduce $\Delta\tau$ until the phase noise can be determined), and then solve h_0. Similarly, h_{+2}, h_{+1}, and h_{-2} can be solved, and thus phase noise characteristic $S_\phi(f)$ will be obtained.

3) If the power law spectrum characteristic curve is given, one can obtain Allan variance $\sigma_y(\tau)$.

The following is an example by using measured phase noise curve of a crystal oscillator to solve corresponding $\sigma_y(\tau)$.

As shown in Fig. 2.21, spectrum density of measured phase noise is £(f). As $S_\Phi(f) = 2$£(f) has been described in previous sections, use a power law spectrum broken line 3 dB more than £(f) in the figure to obtain an approximate $S_\Phi(f)$ (refer to Fig. 2.20, shown in a dashed line). Then, the values of intersection points of the extended line of each broken line segment with $\lg(f) = 0$ axis are

$$K_{-1} = -88 \text{ dBc/Hz}, \quad K_0 = -98 \text{ dBc/Hz}$$
$$K_{+1} = -118 \text{ dBc/Hz}, \quad K_{+2} = -158 \text{ dBc/Hz}$$

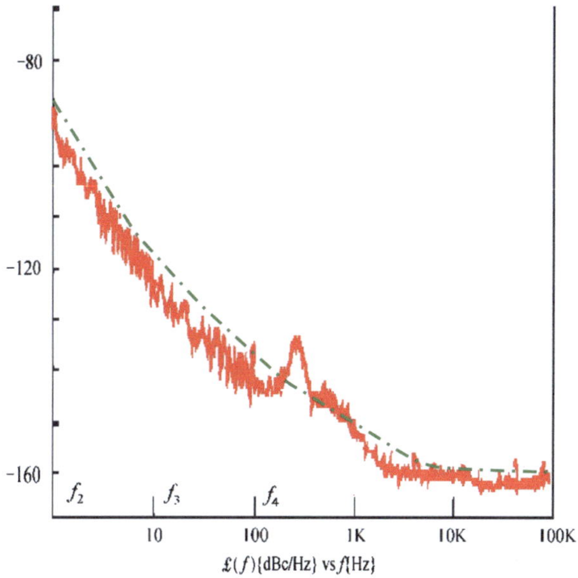

Fig. 2.21 Measured phase noise and short-term stability of a 10 MHz crystal oscillator

and the corner frequencies of each broken line are

$$f_2 = 1\ \text{Hz},\ f_3 = 10\ \text{Hz},\ f_4 = 100\ \text{Hz},\ f_5 = 4\ \text{kHz}$$

From Fig. 2.21, one can get $f_6 = 6.3 \times 10^3$ Hz since $\sigma_y(\tau)$ is measured at a cutoff frequency of 6.3×10^3 Hz.

According to the above known parameters, one can use Expression (2.39) to obtain

$$h_0 = f_0^{-2}\lg^{-1}(0.1K_0) = \left(10 \times 10^6\right)^{-2}\lg[0.1 \times (-98)] = 1.58 \times 10^{-24}(\text{rad})^2/\text{Hz}$$

Similarly,

$$h_{-1} = 1.58 \times 10^{-23}(\text{rad})^2$$
$$h_{+1} = 1.58 \times 10^{-26}(\text{rad})^2/\text{Hz}^2,$$
$$h_{+2} = 1.58 \times 10^{-30}(\text{rad})^2/\text{Hz}^3$$

Then, by Expression (2.45) one can obtain

$$\left[\sigma_y(20\ \text{ms})\right]_0^2 = \frac{h_0}{2\tau} = 39 \times 10^{-24}(\text{rad})^2$$
$$\left[\sigma_y(20\ \text{ms})\right]_0 = 6.24 \times 10^{-12}(\text{rad})$$

Similarly,

$$\left[\sigma_y(20\ \text{ms})\right]_{+1} = 4.2 \times 10^{-12}(\text{rad})$$
$$\left[\sigma_y(20\ \text{ms})\right]_{+2} = 0.85 \times 10^{-12}(\text{rad})$$
$$\left[\sigma_y(20\ \text{ms})\right]_{-1} = 4.7 \times 10^{-12}(\text{rad})$$

By Expression (2.48), one can obtain

$$\left[\sigma_y(20\ \text{ms})\right]_{\Sigma} = 8.8 \times 10^{-12}(\text{rad})$$

The measured value is 6.625×10^{-12}, rather consistent with the calculation result. The larger result is caused by bigger $S_\Phi(f)$ approximate broken line.

Similarly, the short-term stability can be calculated according to phase noise indexes given in the specifications. For example, for the following phase noise spectrum density $S_\Phi(f)$ indexes:

$$\begin{array}{ll} -27 - 30\lg f(\text{dBc/Hz}), & 10\ \text{Hz} \le f \le 100\ \text{Hz} \\ -32 - 20\lg f(\text{dBc/Hz}), & 100\ \text{Hz} \le f \le 1\ \text{kHz} \\ -52 - 10\lg f(\text{dBc/Hz}), & 1\ \text{kHz} \le f \le 100\ \text{kHz} \end{array}$$

Solving steps of short-term stability are:

1) If $f_0 = 2.1$ GHz, by Expressions (2.37), (2.38), (2.39), (2.40), and (2.41), one can obtain

$$K_{-1} = -27\ \text{dBc/Hz}$$
$$K_0 = -32\ \text{dBc/Hz}$$
$$K_{+1} = -52\ \text{dBc/Hz},$$

Then, solve the short-term stability of noise through Expressions (2.43), (2.44), (2.45), (2.46), (2.47), and (2.48).

2) When the unilateral loop of phase-locked receiver is 1 kHz, there are only h_{-1} and h_0 noise.

$$\left[\sigma_y(20\ \text{ms})\right]^2_{-1} = 2h_{-1}\ln 2 = 6.3 \times 10^{-22}, \quad \left[\sigma_y(20\ \text{ms})\right]_{-1} = 2.52 \times 10^{-11}$$
$$\left[\sigma_y(20\ \text{ms})\right]^2_0 = \frac{h_0}{2\tau} = 3.6 \times 10^{-22}, \quad \left[\sigma_y(20\ \text{ms})\right]_0 = 1.9 \times 10^{-11}$$
$$\left[\sigma_y(20\ \text{ms})\right]_{\Sigma} = 3.1 \times 10^{-11}$$

3) When cutoff frequency is 6.3 kHz, h_{+1} noise exists, $f_5 = 6.3$ kHz, $f_4 = 1$ kHz.

$$\left[\sigma_y(20\ \text{ms})\right]^2_{+1} = \frac{1.038 + 3\ln\left[(2\pi\tau(f_5 - f_4)\right]}{4\pi^2\tau^2} h_{+1} = 18.8 \times 10^{-22}$$

$$\left[\sigma_y(20\ \text{ms})\right]_{+1} = 4.3 \times 10^{-11}$$

$$\left[\sigma_y(20\ \text{ms})\right]_{\Sigma} = 5.37 \times 10^{-11}$$

(3) Feature point correspondence method

Analytic function representation method can be used for design calculation and index allocation but is still complex for adjustment, testing, and fault isolation, and its physical conception is less intuitive. Therefore, the "feature point" correspondence method is proposed. It refers to measured data correspondence method. "Feature point" refers to the point imposing the largest impact.

It is obvious from Expression (2.29) that $\sigma_y^2(\tau)$ is an output of $S_\phi(\omega)$ through a transfer function of $H_\Phi(f) = \sin^4(\pi\tau f)$. The filtering characteristic curve of $\sin^4(\pi\tau f)$ is shown in Fig. 2.22.

In Fig. 2.22, $S_\phi(f)$ is phase fluctuation spectrum density, $|H_\phi(f)|^2$ is the equivalent transfer function of Allan variance when inputting $S_\phi(f)$, $S_y(f)$ is phase-frequency fluctuation spectrum density, and $|H_y(f)|^2$ is the equivalent transfer function of Allan variance when inputting $S_y(f)$.

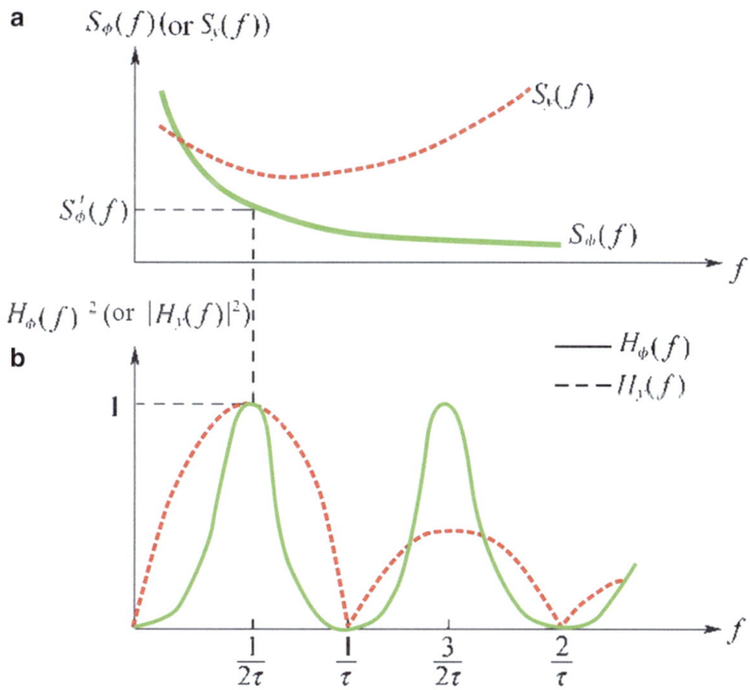

Fig. 2.22 Filtering characteristic of Allan variance $\sigma_y(\tau)$

It can be seen from Fig. 2.22 that filtering characteristic has a peak value at $f = 1/2\tau$. $S_\phi(\omega)$ sharply decreases with f increase, so phase noise power spectrum density at $f = 1/2\tau$ will impose greatest impact on the output (i.e., the frequency window at $1/2\tau$ corresponds to the time window at τ). According to this feature point, one can roughly solve the relationship between output Allan variance and phase noise at $f = 1/2\tau$ frequency point, that is to say, measured $S_\phi(1/2\tau)$ can obtain an approximate $\sigma_y(\tau)$ value. For example, if measured data shown in Fig. 2.21 are given, measured $\sigma_y(20\ \text{ms})$ is 6.63×10^{-12} and £(f) corresponding to $1/2\tau = 25$ Hz is -130 dBc/Hz (relevant $S_\phi(f)$ increases 3 dB, namely, -127 dBc/Hz), only the phase noise is to be measured in future test to obtain approximate Allan variance. That is, only phase noise at $f = 25$ Hz is lower than -130 dBc/Hz and $\sigma_y(20\ \text{ms})$ will be better than 6.63×10^{-12}. This result is approximate but is very convenient in application during development, and its physical concept is also clear and definite. If you want to represent it more accurately, additional feature points such as $3\tau/2$ and $5\tau/2$ can be used.

2.3.3.2 Theoretical Analysis of Two-Way Coherent Doppler Velocity-Measuring Error

Time Domain Analysis Method for Velocity Measurement Accuracy

(1) Fundamental principle of time domain analysis method

Before fundamental principle description, it is firstly assumed that the additional frequency fluctuation is not induced by either the uplink and downlink channels or the transponder for a two-way coherent Doppler velocity measurement and the receiving/transmitting frequency fluctuations are only induced by master oscillator source, with a spatial time delay of T between the two frequency fluctuations. The velocity measurement model is shown in Fig. 2.23.

It can be seen from Fig. 2.23 that Doppler frequency is equal to the difference between output frequencies of a transmitter in different times. The integration time is τ which is approximately equal to sampling time. The average smooth output value is

$$\overline{f_d(t)} = \overline{f_1(t+T) - f_1(t)}$$

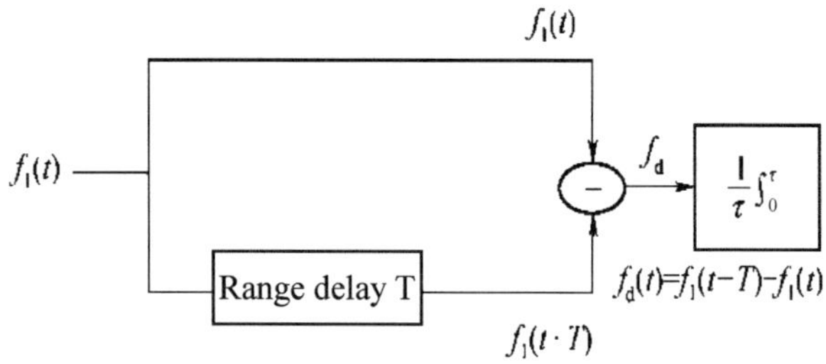

Fig. 2.23 Two-way coherent Doppler velocity measurement model

According to the theorem of mathematical expectation, the average of differences is equal to difference between average values,

$$\overline{f_d(t)} = \overline{f_1(t+T)} - \overline{f_1(t)} \tag{2.49}$$

where the first item and second item in the right side of the equation are two measured values of transmitting frequencies with T time difference. Corresponding time relationship is shown in Fig. 2.24.

In Fig. 2.24, if measuring at t_1, delay time of R/T signal is T. It is obvious that measurement of R/T signal as shown in Fig. 2.24 is equivalent to measurement of transmitting signal as shown in Fig. 2.25.

It can be seen in Fig. 2.25 measurement of frequency difference Δf between R/T signals with delay time T is equivalent to measurement of frequency difference $(f_2 - f_1)$ of the transmitting signal between frequency f_2 at $(t_1 - T)$ and frequency f_1 at t_1. For this measurement, two times in a group and multiple group method are typically employed. The frequency fluctuation variance is calculated and then averaged to obtain the Doppler frequency output variance of $\left\langle \left(\overline{f_2} - \overline{f_1}\right)^2{}_m \right\rangle$,

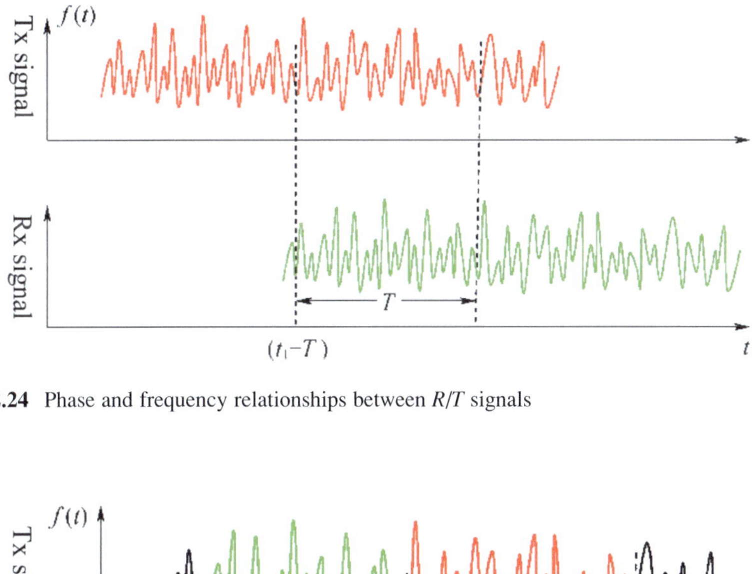

Fig. 2.24 Phase and frequency relationships between R/T signals

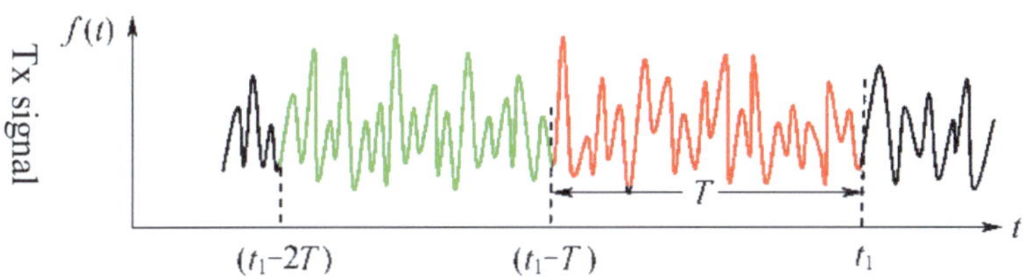

Fig. 2.25 Equivalent measurement of transmitting signal

where, <…> represents an average value of m groups of measurements. According to the definition of gapped Allan variance $\langle\sigma_y^2(2,T,\tau)\rangle$:

$$2\left\langle\sigma_y^2(2,T,\tau)\right\rangle = \langle(\bar{y}_2-\bar{y}_1)\rangle = \frac{\left\langle\left(\bar{f}_2-\bar{f}_1\right)_m^2\right\rangle}{f_0^2} \tag{2.50}$$

where $y=\Delta f/f_0$ is the relative frequency instability; y_2 and y_1 are the value of two measurements, respectively; T is measurement time interval; and τ is integration time. Expression (2.50) is the expression of gapped Allan variance of short-term stability. So the measurement of Doppler frequency variance is equivalent to the measurement of short-term stability of transmitter signals, i.e., if the $\langle\sigma_y^2(2,t_0,T)\rangle$ is measured, then the velocity can be measured.

Short-term instability of signal frequency is typically expressed by Allan variance $\langle\sigma_y^2(2,\tau,\tau)\rangle=\sigma_y^2(\tau)$. If the relation between $\langle\sigma_y^2(2,t_0,T)\rangle$ and Allan variance is solved, velocity-measuring error can be calculated by Allan variance. Reference [6] gives this relation, namely, Banes second bias function

$$B_2(\gamma.\mu) = \frac{\left\langle\sigma_y^2(2,T,\tau)\right\rangle}{\sigma_y^2(\tau)} \tag{2.51}$$

where $\gamma=T/\tau$ and μ is the noise type index. So the velocity-measuring error caused by short-term stability is:

$$\begin{aligned}\sigma_{\dot{R}}^2 &= \left(\frac{C}{2}\right)^2\frac{\left\langle\left(\bar{f}_2-\bar{f}_1\right)_m^2\right\rangle}{f_0^2} = \left(\frac{C}{2}\right)^2\times 2\left\langle\sigma_y^2(2,t_0,T)\right\rangle \\ &= \frac{1}{2}C^2B_2(\gamma,\mu)\sigma_y^2(\tau) \\ \sigma_{\dot{R}} &= \frac{\sqrt{B_2(\gamma,\mu)}}{\sqrt{2}}C\sigma_y(\tau)\end{aligned} \tag{2.52}$$

References [6] and [7] have given B_2 of various types of noise:
1) For $\mu=0$, i.e., FM flicker noise with $\alpha=-1$,

$$B_2 = \begin{cases} 0 & \text{when } \gamma=0 \\ \dfrac{-2\gamma^2\ln\gamma+(\gamma+1)^2\ln(\gamma+1)+(\gamma-1)^2\ln(\gamma-1)}{4\ln 2} & \end{cases} \tag{2.53}$$

2) For $\mu=-1$, i.e., FM white noise with $\alpha=0$,

$$B_2 = \begin{cases} r, & \text{when } 0\le\gamma\le 1 \\ 1, & \text{when } \gamma>1 \end{cases} \tag{2.54}$$

Table 2.5 $B_2(\gamma,\mu)$ function table

μ \ γ	0	0.1	0.8	1	1.1	2	4	16	64	256	1024
0	0	0.0274	0.7667	1.000	1.089	1.566	2.078	3.082	4.082	5.082	6.082
–1	0	0.1	0.8	1.000	1.000	1.000	1.000	1.000	1.000	1.000	1.000
–2	0	0.6667	0.6667	1.000	0.6667	0.6667	0.6667	0.6667	0.6667	0.6667	0.6667
+1	0			1.000	1.15	2.5	5.5	23.5			

3) For $\mu=-2$, i.e., PM white noise with $\alpha=+2$ and PM flicker noise with $\alpha=+1$,

$$B_2=\begin{cases}0, & \gamma=0\\ 1, & \gamma=1\\ 2/3, & \gamma \text{ others}\end{cases} \tag{2.55}$$

$B_2(\gamma,\mu)$ function table is listed in Table 2.5.

In order to use Table 2.5 for lookup $B_2(\gamma,\mu)$, one should firstly find out γ and μ, where γ can be solved through $\gamma=T/\tau$ for a different target range time delay T and sampling time τ. It can be seen in the above table that, for $\mu=0$ noise type, $B_2(\gamma_1,0)$ will gradually increase as T increases. This results in increasing velocity-measuring error $\sigma\dot{R}$. The physical reason is that the noise is an FM flicker noise component with $\mu=0$ which is a slowly changed random fluctuation and also a non-stable random fluctuation. Within an infinitely long time, there must be infinitely great fluctuations, that is, the frequency source exits slow drift. The gapped Allan variance could not neutralize this slow drift as T increases, so the $\sigma_y^2(2,T,\tau)$ increases. It is worth noticing that Allan variance will not increase because the $T=\tau$ is not high. Based on the velocity-measuring principles, when reference frequency is drifted, it will be introduced to the Doppler value from measurement. As mentioned above, in order to minimize velocity-measuring error, the first is to reduce T/τ value (e.g., increase τ for constant T) and the second is to improve the stability of the frequency source (e.g., the Allan variance with given average sampling time τ). For that reason, a high-stability frequency source with higher τ and lower Allan variance should be used when T is high (e.g., deep-space TT&C). For the noise of $\mu=-1$ and $\mu=-2$, both are stable random fluctuations and their $B_2(\gamma,\mu)$ does not relate to T.

The following is an example to explain velocity-measuring error caused by phase noise (or short-term stability) of reference frequency source. Taking the phase noise of a 10 MHz reference source in Fig. 2.21 as an example, at target range 2,500 km and sampling time 20 ms the velocity-measuring error can be calculated by the following steps:

1) Solve the power law spectrum coefficient. According to Fig. 2.21, there are four types of phase noise: h_{-1}, h_0, h_{+1}, and h_{+2}, which have been solved in Sect. 2.3.3.1,

$$h_{-1} = 1.58 \times 10^{-23} (\text{rad})^2$$
$$h_0 = 1.58 \times 10^{-24} (\text{rad})^2/\text{Hz}$$
$$h_{+1} = 1.58 \times 10^{-26} (\text{rad})^2/\text{Hz}^2$$
$$h_{+2} = 1.58 \times 10^{-30} (\text{rad})^2/\text{Hz}^3$$

2) Solve the short-term stability of each type of phase noise at sampling time 20 ms, which has been solved in Sect. 2.3.3.1, namely, $[\sigma_y(20\ \text{ms})]_\alpha$ (here, noise type is represented by α).

3) Solve Banes second bias function $B_2(\gamma, \mu)$ of each type of phase noise. Given $T = 16.66$ ms for the range of 2,500 km, $\tau = 20$ ms, $\gamma = T/\tau = 0.8$. It can be solved by Expressions (2.53), (2.54), and (2.55) or searched in Table 2.5 (here, noise type is represented by μ. The correspondence between μ and α is listed in Table 2.4).

$B_2(0.8, 0) = 0.7667$ corresponds to h_{-1} noise (see Table 2.4)
$B_2(0.8, -1) = 0.8$ corresponds to h_0 noise (see Table 2.4)
$B_2(0.8, -2) = 0.667$ corresponds to h_{+1} noise (see Table 2.4)
$B_2(0.8, -2) = 0.667$ corresponds to h_{+2} noise (see Table 2.4)

4) Solve velocity-measuring error caused by various types of phase noise. By Expression (2.52), one can obtain

$$\left[\sigma_{\dot{R}}\right]_{-1}^2 = 75.9 \times 10^{-8} (\text{m/s})^2$$
$$\left[\sigma_{\dot{R}}\right]_{0}^2 = 140.5 \times 10^{-8} (\text{m/s})^2$$
$$\left[\sigma_{\dot{R}}\right]_{+1}^2 = 56.16 \times 10^{-8} (\text{m/s})^2$$
$$\left[\sigma_{\dot{R}}\right]_{+2}^2 = 2.1 \times 10^{-8} (\text{m/s})^2$$

Total velocity-measuring error is

$$\sigma_{\dot{R}}^2 = \left[\sigma_{\dot{R}}\right]_{-1}^2 + \left[\sigma_{\dot{R}}\right]_{0}^2 + \left[\sigma_{\dot{R}}\right]_{+1}^2 + \left[\sigma_{\dot{R}}\right]_{+2}^2 = 274.66 \times 10^{-8} (\text{m/s})^2$$
$$\sigma_{\dot{R}} = 16.57 \times 10^{-4}\ \text{m/s} = 1.6\ \text{mm/s}$$

If the vehicle continues to fly to the Mars, the range is 401.3×10^6 km, corresponding to $T = 2{,}675.3$ s, $\gamma = 133{,}765$. In this case, $B_2(133{,}765, 0)$ would be very big, thus causing larger velocity-measuring error. This is because the correlation of R/T signal frequency fluctuation at receiving time is reduced to decrease the counteract effect of frequency fluctuation noise. In deep-space TT&C, in order to reduce the velocity-measuring error, integration time τ should be increased ($\tau = 60$ s for some stations) to reduce γ and h_{-1} (improve short-term stability).

If the total short-term stability of each noise type is provided instead of short-term stabilities of individual noise type, the following method may be used for approximate calculation. It can be seen in Fig. 2.22, $|H_y(f)|^2$ at $1/2\tau$ frequency is maximum value. Typically, frequency fluctuation $\sigma_y(\tau)$ output at $1/2\tau$ is the main component of total fluctuation, i.e., $S_y(f)$ noise type corresponding to $1/2\tau$ is main noise type of output $\sigma_y(\tau)$. Hence, approximate calculation may be performed based on this noise type and total short-term stability. With the phase noise characteristics of a crystal oscillator shown in Fig. 2.21, it is FM white noise (corresponding $\tau = 1/2f = 50 - 5$ ms) for $f = 10$–100 Hz, PM white noise ($\tau < 0.125$ ms) for $f > 4$ kHz, and FM flicker noise ($\tau > 50$ ms) for $f < 10$ Hz.

Accordingly, velocity-measuring error at different integration time τ can be calculated approximately:

① For $\tau = 50$–5 ms, FM white noise mainly occurs. In this case, the corresponding $\mu = -1$ and $B_2 = 1$ with $\gamma \geq 1$ as lookup for Table 2.5. So the following expression can be obtained according to Expression (2.52):

$$\sigma_{\dot{R}} = \frac{C}{\sqrt{2}} \sigma_y(\tau) \tag{2.56}$$

② For $\tau = 0.5$–10 s, FM flicker noise mainly occurs. In this case, corresponding $\mu = 0$ and B_2 will vary with T/τ as shown in Table 2.5 and the specific value can be obtained by lookup table.

③ For $\tau \leq 0.5$ ms, the errors are PM white noise and PM flicker noise components. In this case, corresponding $\mu = -2$ and as shown in Table 2.5:

If $\gamma = 1$, then $B_2 = 1$, and

$$\sigma_{\dot{R}} = \frac{C}{\sqrt{2}} \sigma_y(\tau) \tag{2.57}$$

When If $\gamma \neq 1$, then $B_2 = 2/3$, and

$$\sigma_{\dot{R}} = \frac{C}{\sqrt{3}} \sigma_y(\tau) \tag{2.58}$$

When $\gamma = 0$, $B_2 = 0$, and

$$\sigma_{\dot{R}} = 0 \tag{2.59}$$

The above calculation is performed for phase noise characteristics of the crystal oscillator shown in Fig. 2.21. This calculation also can be performed using the short-term stability specifications, such as short-term stability specifications of a hydrogen atomic clock as listed in Table 2.6.

In the range of $f = 0.5$–50 Hz, $\sigma_y(\tau)$ is in proportion to $1/\tau$.Thus according to Expressions (2.43), (2.44), (2.45), (2.46), and (2.47), the phase noise is h_0 noise (FM white noise). If $f > 50$ Hz, $\sigma_y(\tau)$ is almost independent of τ and the phase noise is h_{-1} noise (FM flicker noise). After the noise type is determined, corresponding

Table 2.6 Short-term stability specifications of hydrogen atomic clock

τ/s	1	10	100	1,000	1,000 above
$\sigma_y(\tau)$	1.5×10^{-13}	2×10^{-14}	2×10^{-15}	2×10^{-15}	2×10^{-15}
$f = \frac{1}{2\tau}/Hz$	0.5	5	50	500	500 above

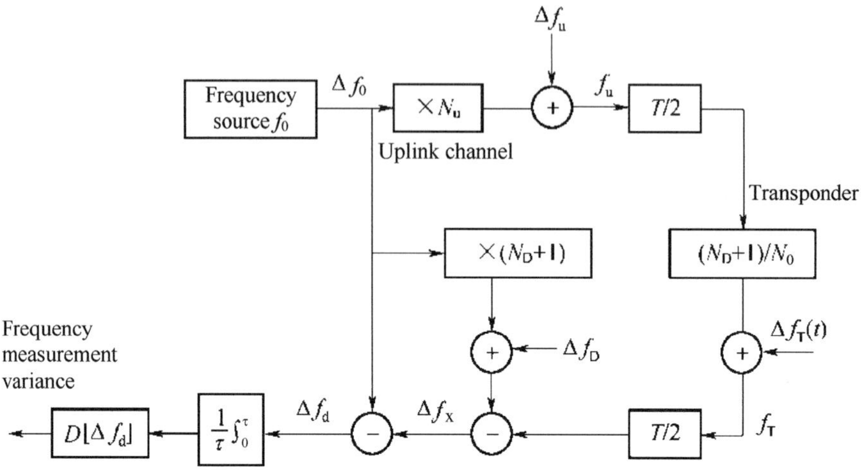

Fig. 2.26 Simplified frequency flow of a two-way coherent Doppler velocity measuring

B_2 can be looked up. Then, the velocity-measuring error can be calculated according to an appropriate expression.

According to the comparison of the above two types of oscillator phase noise, spectrum width of phase noise for atomic clock significantly reduces, and spectrum components of every phase noise move to low frequencies.

(2) Effect of additional phase noise [8]

The above analysis assumes that *R/T* frequency fluctuations are caused by the frequency reference source. However, the actual TT&C system includes multiple oscillators, such as frequency reference source (5 or 10 MHz source), local oscillator of transmitter (may be multiple), receiver local oscillator (may be multiple), and local oscillator of transponder. The above sections only give an analysis to effects of short-term stability of frequency reference source on Doppler velocity measurement accuracy but not to the effect of the thermal noise from other oscillators and the receiver. In practice, short-term stability of various coherent oscillators is different from that of the frequency reference source, which includes additional noncoherent phase noise. For example, many phase-locked frequency synthesizers have higher additional phase noise which result in its short-term stability lower than that of the frequency reference source. Moreover, these additional phase noises could not be offset in coherent velocity measurement. Thus, such additional phase noise would increase the velocity-measuring error. Specific analysis follows.

Simplified frequency flow of a two-way coherent Doppler velocity measuring is as shown in Fig. 2.26.

In Fig. 2.26, Δf_0 is frequency fluctuation noise caused by short-term instability of the frequency source (short-term stability is $\Delta f_0/f_0$); Δf_u, Δf_T, and Δf_D are frequency fluctuation noises added by the uplink channel, transponder, and downlink channel, respectively; τ is integration time; T is R/T delay time; f_u is uplink RF frequency; f_T is downlink RF frequency; and Δf_d is received two-way Doppler frequency shift. Due to frequency fluctuation noise, Δf_d is a random variable and $D[\Delta f_d]$ is its variance.

Velocity measurement is

$$\dot{R} = C\left(\frac{\Delta f_d}{f_u + f_T}\right) \tag{2.60}$$

Total frequency fluctuation noise at the output end of the receiver is

$$\begin{aligned}\Delta f_\Sigma(t) = (N_D + 1)\Delta f_0(t - T) + \left(\frac{N_D + 1}{N_u}\right)\Delta f_u(t - T) \\ + \Delta f_T\left(t - \frac{T}{2}\right) - (N_D)\Delta f_0(t) - \Delta f_D(t)\end{aligned} \tag{2.61}$$

Frequency fluctuation noise at the output end of a Doppler frequency extractor is

$$\begin{aligned}\Delta f_d(t) &= \Delta f_\Sigma(t) - \Delta f_0(t) \\ &= (N_D + 1)[\Delta f_0(t - T) - \Delta f_0(t)] + \left(\frac{N_D + 1}{N_u}\right)\Delta f_u(t - T) \\ &\quad + \Delta f_T\left(t - \frac{T}{2}\right) - \Delta f_D(t)\end{aligned} \tag{2.62}$$

Since random variables $\Delta f_0(t)$, $\Delta f_u(t)$, and $\Delta f_\tau(t)$ are independent from each another, according to digital characteristics theorem of random variables,

$$\begin{aligned}D[\Delta f_d(t)] = (N_D + 1)D\left[\Delta f_0\left(t - \frac{T}{2}\right) - \Delta f_0(t)\right] + \left(\frac{N_D + 1}{N_u}\right)D[\Delta f_U(t - T)] \\ + D\left[\Delta f_T\left(t - \frac{T}{2}\right)\right] - D[\Delta f_D(t)]\end{aligned} \tag{2.63}$$

In order to simplify the expression, the substitution is made:

$$D[\Delta f_d(t)] = D[\Delta f_d], \qquad D\left[\Delta f_u(t-T) = D[\Delta f_u],\right.$$
$$D\left[\Delta f_T\left(t-\frac{T}{2}\right)\right] = D[\Delta f_T], \quad D\left[\Delta f_D(t) = D[\Delta f_D],\right. \tag{2.64}$$

Since Δf_u, Δf_T, and Δf_D are independent of one another and are all stationary random processes, such substitution will not lose its general sense. Expression (2.63) can be expressed as

$$D[\Delta f_d] = (N_D+1)D[\Delta f_0(t)] + \left(\frac{N_D+1}{N_u}\right)D[\Delta f_u] + D[\Delta f_T] - D[\Delta f_D] \tag{2.65}$$

According to Expression (2.60), velocity-measuring error caused by frequency fluctuation noise is:

$$\sigma_{\dot{R}} = C\left(\frac{\sqrt{D[\Delta f_d]}}{f_u+f_T}\right)$$
$$= \frac{C}{f_u+f_T}\sqrt{(N_D+1)D[\Delta f_0(t-T)-\Delta f_0(t)] + \left(\frac{N_D+1}{N_u}\right)D[\Delta f_u] + D[\Delta f_T] - D[\Delta f_D]} \tag{2.66}$$

When $\Delta f_u = 0$, $\Delta f_T = 0$, and $\Delta f_D = 0$ (consider only the effect of frequency fluctuation noise of frequency source f_0),

$$D[\Delta f_d] = (N_D+1)D[\Delta f_0(t-T) - \Delta f_0(t)] \tag{2.67}$$

This is the short-term stability of the frequency source effect on the velocity measurement accuracy discussed in other references.

When $(N_D+1)D[\Delta f_0(t-T) - \Delta f_0(t)] \ll ((N_D+1)/N_u)D[\Delta f_u] + D[\Delta f_T] - D[\Delta f_D]$, i.e., when additional frequency fluctuation noise is higher than frequency fluctuation noise of the frequency source f_0,

$$D[\Delta f_d] = \frac{(N_D+1)}{N_u}D[\Delta f_u] + D[\Delta f_T] - D[\Delta f_D] \tag{2.68}$$

$\Delta f_0/f_0$ could not be improved by increasing short-term stability of the frequency source. These two conditions would occur in practical engineering and will be discussed as follows.

1) Effect of short-term stability of frequency source. When we only consider the effect of Δf_0 but don't consider the effect of Δf_u, Δf_T, and Δf_D, Expression (2.62) can be simplified as

$$\Delta f_{d0}(t) = (N_D+1)[\Delta f_0(t-T)] - (N_D+1)[\Delta f_0(t)] \tag{2.69}$$

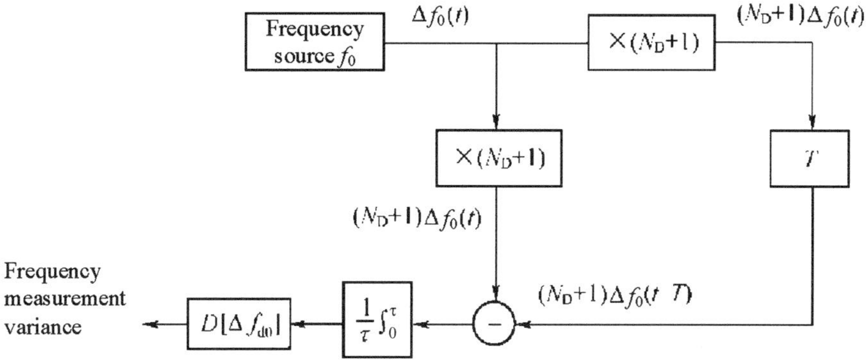

Fig. 2.27 Model of a two-way Doppler velocity-measuring error caused by short-term stability of frequency source

where $\Delta f_{d0}(t)$ is the output frequency fluctuation noise when only short-term stability of the frequency source f_0 is considered.

Expression (2.69) can be expressed as the form of Fig. 2.27.

In Fig. 2.27, $D[\Delta f_{d0}]$ refers to frequency measurement variance caused by short-term stability of frequency source f_0. Corresponding time domain process is shown in Fig. 2.28.

Figure 2.28a shows frequency fluctuation noise of transmitting RF signal. Figure 2.28b shows frequency fluctuation noise of receiving RF signal, and T is time delay of R/T signals. Figure 2.28c shows frequency difference $\Delta f_{d0}(t)$ obtained from subtraction between transmitting RF and receiving RF. Transmitting/receiving frequency noise has certain correlation effects, so $\Delta f_{d0}(t)$ after subtraction decreases. With T increasing, related effects become weak to make $\Delta f_{d0}(t)$ gradually increasing. When T increases to the level in which the T is not correlated with R/T Δf, $\Delta f_{d0}(t)$ will be the mean square sum of R/T $\Delta f(t)$ and is about twice of transmitting signal $\Delta f(t)$ variance. $\Delta f_{d0}(t)$ is the average of sampled values within τ time. The frequency measurement variance can be obtained by continuously sampling r N times and then calculating the variance. Figure 2.28d shows sampling time τ. It can be seen in the figure that averaging $\Delta f_{d0}(t)$ within τ time is equivalent to subtract mean value $(N_D+1)[\Delta f_0(t)]$ averaging $(N_D+1)[\Delta f_0(t)]$ within τ'' from mean value $(N_D+1)\left[\overline{\Delta f_0(t-T)}\right]$ averaging $(N_D+1)[\Delta f_0(t)]$ within τ' with sampling time $\tau'=\tau''=\tau$ and sampling interval of T. This is the gapped sampling variance of transmitting RF signal, which can be used to calculate velocity-measuring error.

Figures 2.27 and 2.28 show the physical processes of velocity measuring, which can be represented by mathematical expressions as follows.

Doppler frequency difference is:

$$\Delta f_{d0}(t) = \left[\Delta f_0(t-T) - \Delta f_0(t)\right](N_D+1)\Big] \tag{2.70}$$

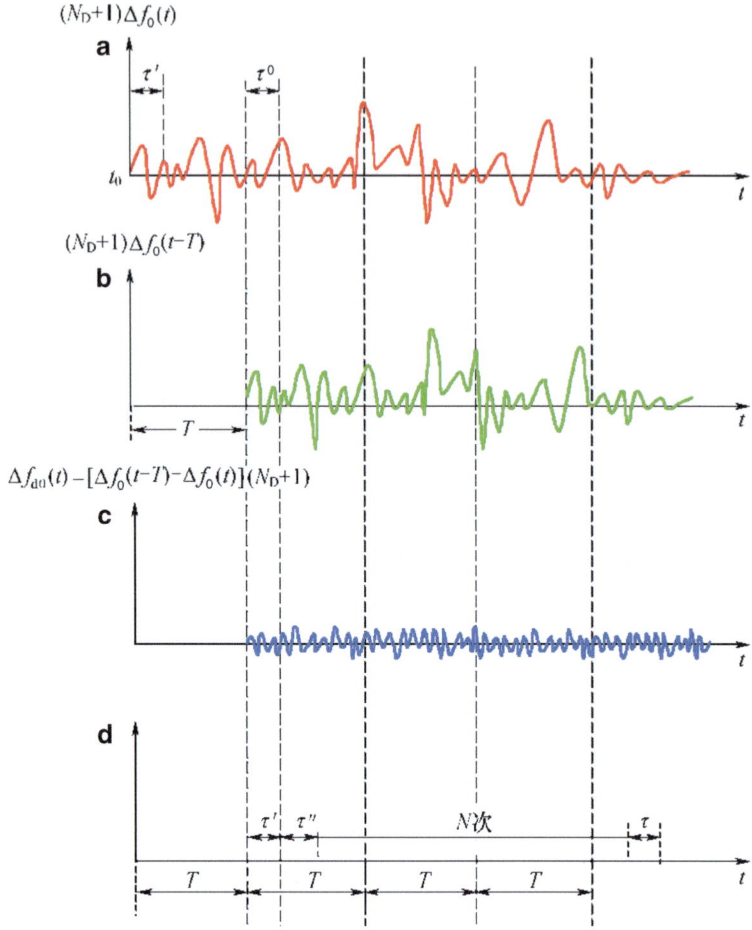

Fig. 2.28 Time domain relationship of Doppler velocity measuring

The integration over the integral time τ can be achieved through the averaging within time τ, i.e.,

$$\begin{aligned} \Delta\bar{f}_{d0}(t) &= \left[\Delta\bar{f}_0(t-T) - \Delta\bar{f}_0(t)\right](N_D+1)] \\ D[\Delta f_{d0}] &= \lim_{N\to\infty}\frac{1}{N}\sum_{n=1}^{N}\left[(\Delta f_{d0})_n - \frac{1}{N}\sum_{n=1}^{N}(\Delta f_{d0})_n\right]^2 \end{aligned} \tag{2.71}$$

According to the mathematical expectation theorem of random variable,

$$\begin{aligned} \Delta\bar{f}_{d0}(t) &= \left[\Delta\bar{f}_0(t-T) - \Delta\bar{f}_0(t)\right](N_D+1)] \\ D[\Delta f_{d0}] &= \left[\lim_{N\to\infty}\frac{1}{N}\sum_{n=1}^{N}(\Delta f_{d0})_n\right]^2 \end{aligned} \tag{2.72}$$

where $(\Delta f_{d0})_n$ is nth sampling value of Δf_{d0} (similarly, nth sampling value of any random variable X is expressed as $(X)n$).

$$\begin{aligned}\frac{1}{N}\sum_{n=1}^{N}(\Delta f_{d0})_n &= \frac{(N_D+1)^2}{N}\sum_{n=1}^{N}\left[\overline{\Delta f_0(t-T)-\Delta f_0(t)}\right]_n \\ &= (N_D+1)^2\left\{\frac{1}{N}\sum_{n=1}^{N}\left[\Delta\overline{f_0}(t-T)\right]_n - \frac{1}{N}\sum_{n=1}^{N}\left[\Delta\overline{f_0}\right]_n\right\}\end{aligned} \tag{2.73}$$

where both two items are mean values of frequency drift of frequency source f_0. Since the stability of frequency source f_0 is very high, both items are approximate to 0. For example, when $\tau = 20$ ms and $N = 1{,}000$ (mean value within 20 s), $\frac{1}{N}\sum_{n=1}^{N}(\Delta f_{d0})_n \approx 0$.

$$\begin{aligned}D[\Delta f_{d0}] &= \lim_{N\to\infty}\frac{(N_D+1)^2}{N}\sum_{n=1}^{N}\left[(\Delta f_{d0})_n\right]^2 \\ &= \lim_{N\to\infty}\frac{(N_D+1)^2}{N}\sum_{n=1}^{N}\left[\Delta\bar{f}_0(t-T)-\Delta\bar{f}_0(t)\right]_n^2\end{aligned} \tag{2.74}$$

According to the definition of gapped Allan variance,

$$\left[\Delta\bar{f}_0(t-T)-\Delta\bar{f}_0(t)\right]_n^2 = 2f_0^2\sigma_y(2,T,\tau) \tag{2.75}$$

where $\sigma_y(2,T,\tau)$ is gapped sampling Allan variance of relative value $\Delta f_0(t)/f_0$ of frequency fluctuation noise $\Delta f_0(t)$ of frequency source f_0, where sampling interval is T, number of sampling is 2, and average time is τ.

In the above expression, T is the target time delay in actual velocity measuring and τ is integration time of velocity-measuring data processing. So

$$\begin{aligned}D[\Delta f_{d0}] &= 2\lim_{N\to\infty}\frac{(N_D+1)^2}{N}f_0^2\sum_{n=1}^{N}\left[\sigma_y(2,T,\tau)\right]_n \\ &= 2(N_D+1)^2 f_0^2\langle\sigma_y(2,T,\tau)\rangle\end{aligned} \tag{2.76}$$

where <...> represents mean value.

Then, substitute B_2 into Expression (2.51) to obtain

$$D[\Delta f_{d0}] = 2(N_D+1)^2 f_0^2 B_2\sigma_y^2(\tau),\quad D[\Delta f_0] = 2f_0^2 B_2\sigma_y^2(\tau) \tag{2.77}$$

The velocity-measuring error can be obtained by Expression (2.66):

$$\sigma_{\dot{R}} = \frac{C}{2}\frac{\sqrt{D[\Delta f_{d0}]}}{(N_D+1)f_0} = \frac{C}{2}\frac{\sqrt{2B_2(N_D+1)^2 f_0^2}}{(N_D+1)f_0}\sqrt{D[\Delta f_{do}]} = \frac{\sqrt{B_2}}{\sqrt{2}}C\sigma_y(\tau) \quad (2.78)$$

According to Expression (2.78), velocity-measuring error $\sigma_{\dot{R}}$ can be solved by B_2 obtained from Allan variance and lookup table. The result is the same as the result of Expression (2.52).

2) Effect of additional noise. When additional frequency fluctuation noise is a primary contributing factor, velocity-measuring error is expressed as follows according to Expressions (2.66) and (2.68):

$$\sigma_{\dot{R}} = \frac{C}{f_u + f_T}\sqrt{\frac{(N_D+1)}{N_u}D[\Delta f_u] + D[\Delta f_T] - D[\Delta f_D]} \quad (2.79)$$

According to Expressions (2.63), (2.64), (2.65), (2.66), (2.67), (2.68), (2.69), (2.70), (2.71), (2.72), (2.73), (2.74), (2.75), (2.76), and (2.77) and Reference [8]:

$$D[\Delta f_u] = 2B_2 f_u^2 \sigma_{yu}^2(\tau), \quad D[\Delta f_T] = 2B_2 f_T^2 \sigma_{yT}^2(\tau), \quad D[\Delta f_D] = 2B_2 f_D^2 \sigma_{yD}^2(\tau) \quad (2.80)$$

where $\sigma_{yu}^2(\tau)$, $\sigma_{yT}^2(\tau)$, and $\sigma_{yD}^2(\tau)$ are short-term stabilities of each unit.

Substituting it into Expression (2.79) then obtains:

$$\sigma_{\dot{R}} = C\sqrt{2B_2\left[\frac{(N_D+1)}{N_u}\left(\frac{f_u}{f_u+f_T}\right)^2\sigma_{yu}^2(\tau) + \left(\frac{f_T}{f_u+f_T}\right)^2\sigma_{yT}^2(\tau) + \left(\frac{f_D}{f_u+f_T}\right)^2\sigma_{yD}^2(\tau)\right]} \quad (2.81)$$

According to Expression (2.81), the lower the f_u, f_T, f_D, the smaller the effects on $\sigma_{\dot{R}}$ and thus lower short-term stability requirements. In practical engineering, "weighted allotment" or "equal contribution" allotment can be used to determine the short-term stability requirements of each unit. If "equal contribution" allotment is used,

$$\frac{(N_D+1)}{N_u}\left(\frac{f_u}{f_u+f_T}\right)^2\sigma_{yu}^2 = \left(\frac{f_T}{f_u+f_T}\right)^2\sigma_{yT}^2(\tau) = \left(\frac{f_D}{f_u+f_T}\right)^2\sigma_{yD}^2(\tau) \quad (2.82)$$

According to Expression (2.82), we can obtain

$$\frac{\sigma_{yT}(\tau)}{\sigma_{yD}(\tau)} = \frac{f_D}{f_T} \quad (2.83)$$

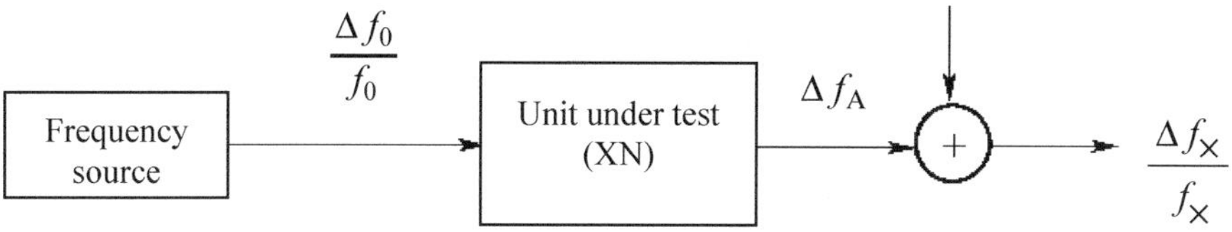

Fig. 2.29 Short-term stability contribution of additional phase noise

$$\frac{\sigma_{yT}(\tau)}{\sigma_{yD}(\tau)} = \sqrt{\frac{(N_D+1)}{N_u}\frac{f_u}{f_T}} \tag{2.84}$$

Expressions (2.83) and (2.84) can be used for performance allotment to each unit under "equal contribution criteria." The contribution of short-term stability of additional phase noise of each unit to the total short-term stability is shown in Fig. 2.29.

Given the short-term stability of frequency source f_0 is $\sigma_{y0}(\tau)$, the frequency after passing through a unit is converted into Nf_0, additional frequency fluctuation noise during conversion is Δf_A, its short-term stability is $\sigma_{yA}(\tau)$, and total short-term stability is $\sigma_{yN}(\tau)$.

Because

$$\Delta f_N = \sqrt{[N(\Delta f_0)]^2 + [\Delta f_0]^2}, \quad f_N = Nf_0$$

we have

$$\left(\frac{\Delta f_N}{f_N}\right)^2 = \left(\frac{\Delta f_0}{f_0}\right)^2 + \left(\frac{\Delta f_A}{f_N}\right)^2, \quad \left(\frac{\Delta f_A}{f_N}\right)^2 = \left(\frac{\Delta f_N}{f_N}\right)^2 - \left(\frac{\Delta f_0}{f_0}\right)^2 \tag{2.85}$$

Total short-term stability after adding frequency fluctuation noise Δf_A into this unit is

$$\sigma_{yN}(\tau) = \sqrt{\sigma_{yA}^2(\tau) - \sigma_{y0}^2(\tau)} \tag{2.86}$$

When $\sigma_{yN} \gg \sigma_{y0}$, $\sigma_{yN}(\tau) \approx \sigma_{yA}(\tau)$, i.e., when additional frequency fluctuation noise is higher, the short-term stability of output signal of this unit will depend on additional noise and can't be improved by increasing short-term stability $\sigma_{y0}(\tau)$ of frequency source. This is a phenomenon found in system ground joint test. In addition, in the ground joint test, $T \to 0$ results in $\gamma \to 0$ and thus $B_2 \to 0$ according to Table 2.5. According to Expression (2.78), velocity-measuring error caused by short-term stability shall be 0, but the actual measured result can't be zero. This is the result of additional frequency fluctuation noise.

Short-term stability of frequency source and effects of additional frequency fluctuation noise of each unit in the link have been discussed in the above paragraphs. Because they aren't correlated with each other, the comprehensive effects of them should be considered by using mean square sum to solve total frequency fluctuation noise caused by them and Allan variance. Since additional noise is PM white noise and hence $B_2 = 2/3$ is feasible, velocity-measuring error component caused by additional noise can be obtained using the Expression (2.78) as follows:

$$\sigma_{\dot{R}A} = \frac{C}{\sqrt{3}} \sigma_{yA}(\tau) \tag{2.87}$$

because total velocity-measuring error is expressed as

$$\sigma_{\dot{R}} = \sqrt{\sigma^2_{\dot{R}o} + \sigma^2_{\dot{R}A}} \tag{2.88}$$

Therefore, for $\mu = -2$ noise (PM white noise and PM flicker noise), it is expressed as

$$\sigma_{\dot{R}} = \frac{C}{\sqrt{3}} \sqrt{\sigma^2_y(\tau) + \sigma^2_{yA}(\tau)} \tag{2.89}$$

For $\mu = 0$, $\mu = -1$ noise (FM white noise and FM flicker noise), it is expressed as

$$\sigma_{\dot{R}} = C\sqrt{\frac{\sigma^2_{yA}(\tau)}{3} + \frac{B_2}{2}\sigma^2_y(\tau)} \tag{2.90}$$

where B_2 is obtained by lookup Table 2.5 with $\mu = -1$ and $\gamma = T/\tau$.

The above time domain analysis method can be used for performance allotment and accuracy calculation using short-term stability. However, the measurement of short-term stability during test and adjustment is very difficult and requires specific and expensive measuring instrument for short-term stability. When the "frequency domain analysis method" is used, the spectrum analyzer can be used to directly measure the phase noise. This "frequency domain analysis method" will be discussed in the following paragraph.

Frequency Domain Analysis Method of Velocity Measurement Accuracy [3]

(1) Relationship between velocity measurement error and spectral density of phase noise

In section above, the relationship between Allan variance and the velocity measurement error obtained by time domain analysis method is:

$$\sigma^2_{\dot{R}}(\tau) = \frac{c^2}{2} \times B_2(\gamma, \mu)\sigma^2_y(\tau) \tag{2.91}$$

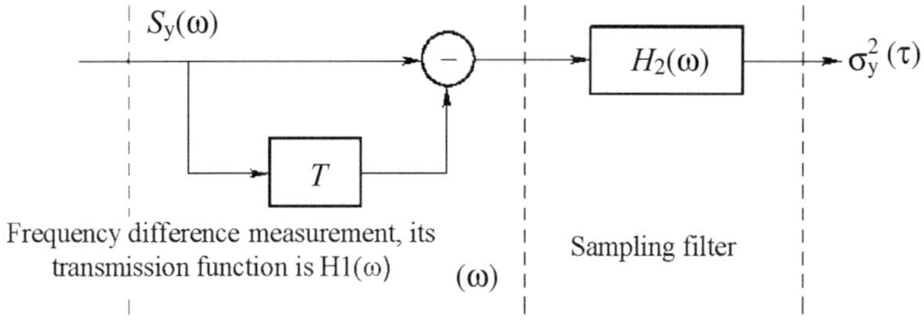

Fig. 2.30 The model of velocity measurement error caused by the phase noise of transmission signal

The power law spectrum model is hypothetical and is not fully consistent with actual phase noise curve; in conjunction with the $B_2(\gamma, \mu)$ obtained on the basis of some supposed conditions, Expression (2.91) is approximate. This is the disadvantage of time domain analysis method. Frequency domain analysis method can be used to overcome these disadvantages, in which the relationship between power spectral density of phase noise and velocity measurement error is derived as follows:

The model of velocity measurement error caused by the phase noise of transmission signal is shown in Fig. 2.30.

In Fig. 2.30, $H_1(\omega)$ is delay T cancellation, its power transmission function feature is expressed as:

$$|H_1(\omega)|^2 = 4\sin^2\frac{\omega T}{2} \tag{2.92}$$

Its characteristic curve is as shown as $|H_1(f)|$ in Fig. 2.31.

$H_2(\omega)$ is sampling smoothing filter within sampling time τ, and its power transmission function feature is [4]:

$$|H_2(\omega)|^2 = \sin^2\frac{(\omega\tau/2)}{(\omega\tau/2)^2} \tag{2.93}$$

Its feature curve is as shown as $|H_2(f)|$ in Fig. 2.31.

The sampling variance value of relative frequency fluctuating noise at the output end is

$$\sigma_y^2(\tau, T) = \frac{1}{2\pi}\int_0^{\infty} S_y(\omega)|H_1(\omega)|^2|H_2(\omega)|^2\mathrm{d}\omega \tag{2.94}$$

Create simultaneous expression out of combined Expression (2.92) with Expression (2.94) to yield:

$$\sigma_y^2(\tau, T) = \frac{1}{2\pi}\int_0^{\infty} S_y(\omega)\left[4\sin^2\left(\frac{\omega T}{2}\right)\right]\left[\sin^2\frac{(\omega T/2)}{(\omega T/2)^2}\right]\mathrm{d}\omega \tag{2.95}$$

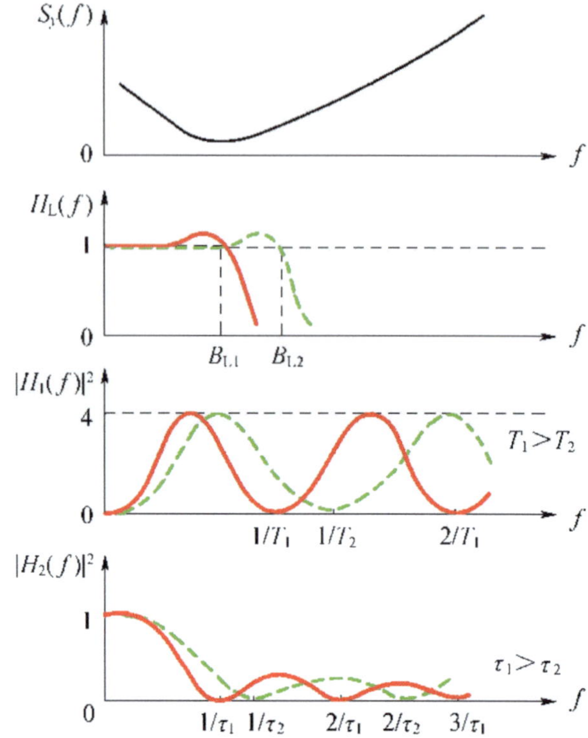

Fig. 2.31 Illustration of frequency domain analysis of velocity measurement error

where ω_0^2, T, τ are known values. As $S_y(\omega) = S_\Phi(\Phi)(\omega^2/\omega_0^2), \sigma_{\dot{R}}(\tau, T)$ can be obtained via computer using measured $S_\phi(\omega)$ (or using power law spectrum model). The multiplier ω^2 denotes that the spectral density of frequency fluctuation noise will increase as ω increases. In practical proposal, before the Doppler frequency difference is measured, the phase-locked loop typically is used to perform tracking filter to $S_\Phi(\omega)$. It supposes the single-side bandwidth of the phase-locked loop is $\omega_1(\omega_1 = 2\pi B_L)$; then, Expression (2.95) becomes:

$$\sigma_y^2(\tau, T, B_L) = \frac{1}{\pi}\int_0^{\omega_L} S_y(\omega)\left[\sin^2\left(\frac{\omega T}{2}\right)\right]\left[\sin^2\left(\frac{\omega\tau}{2}\right)/\left(\frac{\omega\tau}{2}\right)^2\right]\mathrm{d}\omega \qquad (2.96)$$

Expression (2.96) shows that $\sigma_y^2(\tau, T, B_L)$ is correlated with τ, T, and B_L. The smaller τ is (i.e., the wider the $H_2(\omega)$ is) and the bigger B_L is, and so the bigger $\sigma_{\dot{R}}^2(\tau, T, B_L)$ is.

The relationship between the gapped sampling variance and spectral density is as follows [4]:

$$\sigma_y^2(N, T, \tau) = \frac{N}{N-1}\int_0^{\infty} S_y(f)\left(\frac{\sin^2\pi\tau f}{\pi\tau f}\right)^2\left[1 - \left(\frac{\sin N\pi\tau r f}{N\sin\pi\tau r f}\right)^2\right]\mathrm{d}f$$

where $\gamma = T$ and $\omega = 2$. Substituting $N = 2$ into above expression, the gapped Allan variance within the bandwidth can be obtained:

$$\sigma_y^2(2, T, \tau, B_L) = \frac{1}{\pi}\int_0^{\omega_L} S_y(\omega)\left[\sin^2\left(\frac{\omega T}{2}\right)\right]\left[\sin^2\frac{(\omega\tau/2)}{(\omega\tau/2)^2}\right]\mathrm{d}\omega$$

Therefore,

$$\sigma_y^2(\tau, T, B_L) = 2\sigma_y^2(2, T, \tau, B_L)$$

and the velocity measurement error is:

$$\begin{aligned}\sigma_{\dot{R}}^2 &= \left(\frac{c}{2}\right)^2 \sigma_y^2(\tau, T, B_L) = \left(\frac{c}{2}\right)^2 2\sigma_y^2(2, T, \tau, B_L) = \left(\frac{c}{2}\right)^2 2B(\gamma)\sigma_y^2(\tau)\\ \sigma_{\dot{R}} &= \frac{c}{\sqrt{2}}\sqrt{B_2}\sigma_y(\tau)\end{aligned} \tag{2.97}$$

Expression (2.97) is in consistency with the velocity measurement error Expression (2.51) obtained by the time domain analysis method. But the frequency domain method is derived by filtering phase noise frequency spectrum, and its $\sigma_y^2(\tau)$ is the result after the signal is subject to phase-locked filtering (the bandwidth is B_L).The physical concept of such method is clear and applicable to solve actual problems in projects.

The physical implication is shown from Fig. 2.31: Suppose that input is frequency modulation white noise and its power spectrum density is $S_y(f)$.

It can be seen in Fig. 2.31 that for the given distance (corresponding to radio transmission delay T_1 in the figure), the smaller τ_2 is, the bigger $1/\tau_2$ is and the wider the corresponding $|H_2(f)|$ frequency window. Therefore, the bigger the frequency fluctuating noise is, the bigger the velocity measurement error becomes.

For a given sampling time (such as t_1 in Fig. 2.31), the smaller the T is, the value of $|H_1(f)|$ corresponding to frequency window of $1/\tau_1$ will be. Therefore, the smaller the frequency fluctuating noise is, the velocity measurement error will become.

For given τ and T, the wider B_L is, the greater the areas of $|H_1(f)|$ and $|H_2(f)|$ (equivalent noise bandwidths) will be. Therefore, the bigger the output frequency fluctuating noise is, the bigger the velocity measurement error will become.

n/τ, $n = 1, 2, 3\ldots$ positive integers are dumb points of the sampling filter. Therefore, τ value can be selected to eliminate interference. For example, for 50 Hz power supply jamming, $n/\tau = 50$ Hz can be selected, namely, $\tau = n \times 20$ ms (at this time, τ may be 20 ms, 40 ms, 80 ms, 120 ms...).

If noncoherent velocity measurement is used, there is no filtering of $H_1(f)$ and the velocity measurement error will increase. Especially, there is no null point of

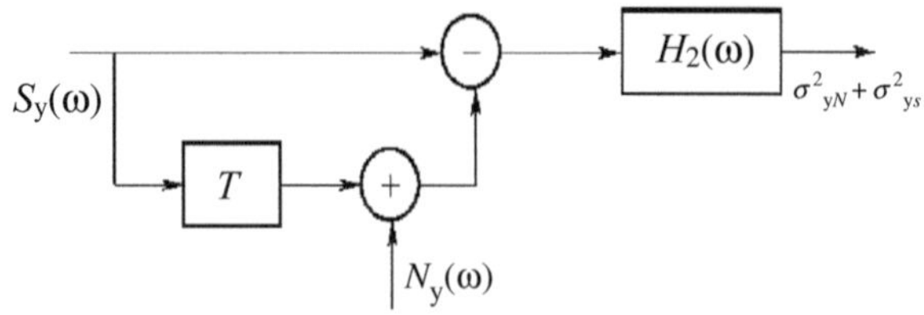

Fig. 2.32 Velocity-measurement error model caused by noise of the receiving system

$H_1(f)$ in the vicinity of $f = 0$, which strengthens the impact of frequency random walk and frequency modulation flash noise.

(2) Relationship between thermal noise of a receiver and velocity-measuring error

Velocity-measuring error is not only correlated with short-term frequency stability but also relevant to the thermal noise of the receiving system. The model of velocity measurement error caused by noise of receiving system is shown in Fig. 2.32.

In Fig. 2.32, $S_y(\omega)$ is the one-sided power spectrum density of relative frequency fluctuation noise (short-term frequency stability) of signals and $N_y(\omega)$ is generated by noise of the receiving system (thermal noise of the receiver is generally considered as additive normal white noise). Section 2.3.3.1 has provided the formula $N_y(\omega) = N_\phi(\omega)(\omega^2/\omega_0^2) = (N_0/P_s)(\omega^2/\omega_0^2)$ (P_s refers to the power of receiving signals and N_0 refers to power spectrum density of amplitude noise of the receiving system). Since $N_y(f)$ is non-correlated with $S_y(f)$, $N_y(f)$ and $S_y(f)$ power is added when the difference is solved. The output after $|H_2(f)|$ filtering includes two parts: σ_{yN}^2 (velocity variance generated by noise of receiving system) and σ_{ys}^2 (velocity variance generated by short-term frequency stability of transmitting signals). The total output is the mean square sum. The solution of σ_{ys}^2 has been provided above, and σ_{yN}^2 can be derived from Fig. 2.32, namely,

$$\sigma_{yN}^2(\tau, B_L) = \frac{1}{2\pi}\int_0^{\omega_L} N_y(\omega)|H_2(\omega)|^2 \mathrm{d}\omega \tag{2.98}$$

where $|H_2(\omega)|^2$ has been given in Expression (2.93), substituting it into Expression (2.98) to obtain:

$$\begin{aligned}\sigma_{yN}^2(\tau, B_L) &= \frac{1}{2\pi}\int_0^{\omega_L} \frac{N_0}{P_s}\frac{\omega^2}{\omega_0^2}\left[\frac{\sin^2(\omega\tau/2)}{(\omega\tau/2)^2}\right]\mathrm{d}\omega \\ &= \frac{1}{\pi\omega_0^2\tau^3}\frac{N_0}{P_s}(\omega_L\tau - \sin\omega_L\tau)\end{aligned} \tag{2.99}$$

When $\omega_L\tau \gg 1$, we can get

$$\sigma_{yN}^2(\tau, B_L) = \frac{1}{\sqrt{2}\pi f_0 \tau}\sqrt{\frac{N_0}{P_s}B_L} \tag{2.100}$$

The velocity-measuring error introduced by thermal noises is:

$$\sigma_{\dot{R}N}(\tau, B_L) = \frac{C}{2\sqrt{2}\pi f_0 \tau}\sqrt{\frac{N_0}{P_s}B_L} = \frac{C}{2\sqrt{2}\pi f_0 \tau}\sqrt{\frac{1}{\rho_L}} \tag{2.101}$$

where $\rho_1 = P_s/B_L N_0$ is SNR of the downlink carrier loop; $N_0 = KT$, K is Boltzmann constant and T refers to the noise temperature; and B_L is the loop sideband equivalent noise bandwidth.

According to Expression (2.101), by reducing B_L, the velocity measurement error introduced by thermal noise can be reduced accordingly. However, decrease in B_L will enlarge dynamic error. To meet dynamic error requirements, higher-type higher-order narrowband loop shall be employed ("type" refers to the number of integrators in a loop). According to the classification method of the phase-locked loop types in Reference [9], the transfer functions and the calculation formula of its bandwidth of several phase-locked loops are introduced in the section below:

Type II order II:

$$H(s) = \frac{2\zeta\omega_n s + \omega_n^2}{s^2 + 2\zeta\omega_n s + \omega_n^2} \tag{2.102}$$

Type II order III:

$$H(s) = \frac{K_{\tau_2}(s\tau_2 + 1)}{s^3\tau_2^3/b + s^2\tau_2^2 + Ks\tau_2^2 + K\tau_2} \tag{2.103}$$

Type III order III:

$$H(s) = \frac{K\left(s^2 + \frac{K_2}{K_1}s + \frac{K_3}{K_1}\right)}{s^3 + K\left(s^2 + \frac{K_2}{K_1}s + \frac{K_3}{K_1}\right)} \tag{2.104}$$

B_L of these loops can be calculated by the formula in Expression (2.7).

(3) Features of frequency domain method

According to the analysis above, the features of frequency domain analysis method are concluded that such method is more practical for equipment R&D because the power spectrum density of phase noise can be obtained through measuring with spectrum analyzer and the physical concept is more clear and intuitive and more favorable for the researcher to find and solve problems.

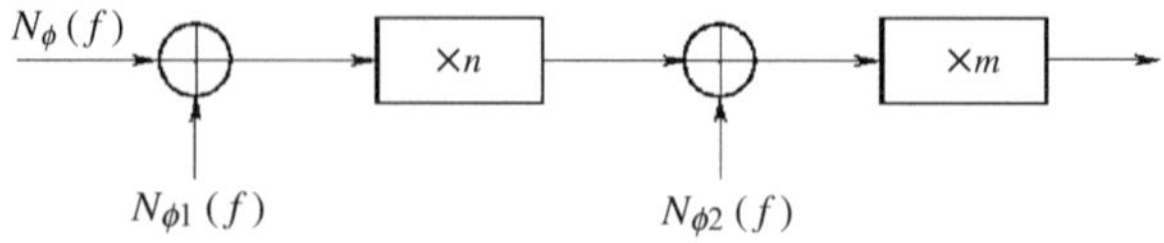

Fig. 2.33 Equivalent circuit of a frequency multiplier

Frequency domain method enjoys higher accuracy. The power spectrum density of phase noise measured by spectrum analyzer can work out the velocity measurement error, or the measured power spectrum density can be input into computers for accurate calculation or simulation.

"Phase domain transfer" and "amplitude domain transfer" are similar in many aspects, including:

1) Frequency doubling of "phase domain" is equivalent to "amplification" of "amplitude domain." Namely, n times of doubling of signal frequency equals to n times of amplification of phase noise. The signal frequency is equivalent to signal amplitude. The power spectrum density of output phase noise is n^2 times of input phase noise spectrum density, that is, dB is 20 lg$n$. When the frequency multiplier generates internal phase noise $N_\phi(f)$, the equivalent circuit is as shown in Fig. 2.33. Similar to low-noise amplifier, phase noise is mainly influenced at the first stage of a frequency multiplier; therefore, the frequency multiplier with low-phase noise shall be designed at the first stage. The subsequent order phase noise $N_{\Phi 2}(f)$ influence can be reduced by increasing n. Analysis of frequency divider is similar to that of frequency multiplier.

2) The "subtractor" in "phase domain" is equivalent to "down frequency mixer" in "amplitude domain." For phase, the down frequency mixer equals to the "subtractor." Similarly, for up-frequency mixing, it equals to "adder." When phase noise of ϕ_1 and ϕ_2 is not correlated mutually, both "subtractor" and "adder" implement mean square addition to phase noise.

3) Frequency multiplier and phase-locked frequency synthesizer: the power spectrum density of phase noise output from n-time frequency multiplier increases by n^2 times in theory, but in fact, partial internal noise will be added at the front stage of the frequency multiplier. If frequency multiplier with low-phase noise is designed, such additional internal noise is relatively small. When phase-locked loop frequency synthesizer is used to implement similar frequency doubling functions, the following phase noise will be generated in the phase-locked frequency synthesizer:

① Phase noise of frequency divider

② Phase discriminator of locked-phase loop and noise of its post amplifier, which will cause phase jitter of VCO

③ Phase noise of VCO, influencing phase noise outside of the phase-locked loop bandwidth

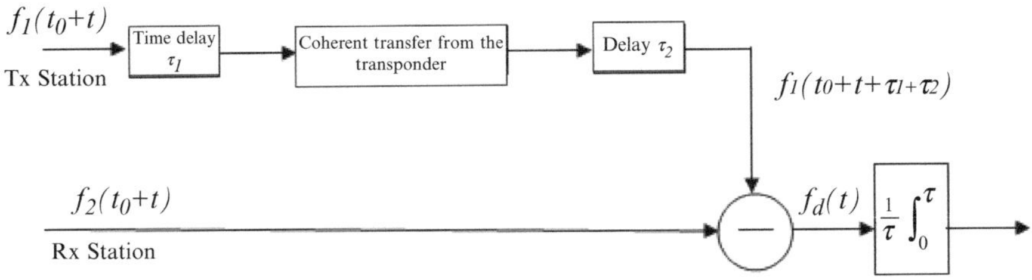

Fig. 2.34 The model of a three-way incoherent Doppler velocity measuring

2.3.4 Three-Way Noncoherent Doppler Velocity Measuring and $\dot{S}$ Measuring

"Three-way" is relative to "two-way." In "two-way" measuring, only one ground station is needed; the forward and backward signals of which are used to realize two-way measuring. Three-way measuring happens in the condition that another ground station is used to realize receiving in a third way; in this case, frequency sources of the transmitting and receiving stations are incoherent, that is the reason it is called noncoherent velocity measuring. The model of a three-way noncoherent Doppler velocity measuring is shown in Fig. 2.34.

In Fig. 2.34, $f_1(t)$ is the uplink frequency of a ground-transmitting station and $f_2(t)$ means the reference frequency of a ground receiving station. Using $f_2(t)$ to measure $f_d(t)$, we find that $f_1(t)$ and $f_2(t)$ are incoherent. t_0 is the starting time to measure frequency drifts of $f_1(t)$ and $f_2(t)$, $(\tau_1+\tau_2)$ is the time delay from the transmitting station to the receiving station, and t represents the continuous variable starting from 0. Compared with two-way coherent velocity measuring, the three-way incoherent measuring introduces the following velocity-measuring errors:

(1) The error caused by frequency drift $v(T)$

The $v(T)$ is defined as the change of frequency appears after time interval of T, which starts from t_0; and (t_0+T) here means the time of Doppler frequency measuring. Assume that frequency drifts of $f_1(t)$ and $f_2(t)$ are $v_1(T)$ and $v_2(T)$, respectively; then, at the time of (t_0+T), they will introduce an error of Doppler frequency measuring:

$$F_1(T)=\sqrt{v_1^2(T)+v_2^2(T)} \tag{2.105}$$

The reason why the above expression is in the form of mean square addition is because $f_1(t)$ and $f_2(t)$ are independent. Take f_2 as a reference, the velocity-measuring error resulted from F_1 is

$$\Delta_1=\frac{C}{2}\frac{\sqrt{v_1^2(T)+v_2^2(T)}}{f_2} \tag{2.106}$$

(2) The error caused by frequency accuracy $f_1(t_0)$ and $f_2(t_0)$ are respectively the actual values of $f_1(t)$ and $f_2(t)$ at the time of t_0, which have an error to the given nominal value. As velocity calculation is based on the nominal value, this error will directly bring in a Doppler frequency error. Assume that frequency accuracies of $f_1(t)$ and $f_2(t)$ are $A_1 = \Delta f_1/f_1$ and $A_2 = \Delta f_2/f_2$, respectively; then, the Doppler frequency error introduced by them is

$$F_2(t_0) = \sqrt{A_1^2 f_1^2 + A_2^2 f_2^2} \tag{2.107}$$

The velocity-measuring error caused therefrom is:

$$\Delta_2 = \frac{C}{2} \frac{\sqrt{A_1^2 f_1^2 + A_1^2 f_2^2}}{f_2} \tag{2.108}$$

(3) The velocity-measuring error caused by short-term instability

According to Expression (2.49), the mean value of $f_d(T)$ is:

$$\begin{aligned} \overline{f_d}(T) &= \overline{f_1(t_0 + T + \tau_1 + \tau_2)} - \overline{f_1(t_0 + T)} \\ &= \left[\overline{f_1(t_0 + T + \tau_1 + \tau_2)} - \overline{f_1(t_0)}\right] - \left[\overline{f_2(t_0 + T)} - \overline{f_2(t_0)}\right] \\ &\quad + \left[\overline{f_2(t_0)} - \overline{f_1(t_0)}\right] \end{aligned} \tag{2.109}$$

Expression (2.109) includes three components:

1) Item $\left[\overline{f_2(t_0)} - \overline{f_1(t_0)}\right]$

This is the system error caused by frequency accuracy, which can be calculated according to Expression (2.108).

2) Item $\left[\overline{f_2(t_0 + T)} - \overline{f_2(t_0)}\right]$

This is the random error resulted from $f_2(t)$ short-term instability. According to Expression (2.50), there is:

$$\left\langle \left[\overline{f_2(t_0 + t)} - \overline{f_2(t_0)}\right]_{m^2} \right\rangle = 2f_2^2 \left\langle \sigma_{y_2}^2 (2, T, \tau) \right\rangle$$

where $\sigma_{y_2}^2 (2, T, \tau)$ is the gapped Allan variance of $f_2(t)$ when the sampling interval is T and the integration time is τ. This error contains FM white noise, PM white noise, FM flicker noise, and frequency random walk noise. As T lasts too long (e.g., several hours or days) during a three-way velocity measuring, the last two items are contained in frequency drift. The other two errors can be calculated according to the following formula:

① FM white noise ($\mu = -1$ noise):

According to Expression (2.52), the variance of the velocity-measuring error caused by $f_2(t)$ short-term instability is:

$$\left[\sigma^2_{\dot{R}_2}\right]_{-1} = \frac{C^2}{2} B_2(\gamma, \mu)\sigma^2_{y_2}(\tau)$$

As the value of T is quite large, $\gamma = T/\tau$ is also huge; we look up B_2 $(\gamma, \mu) = 1$ from Table (2.5); then, the velocity-measuring error caused by PM white noise is

$$\left[\sigma_{\dot{R}_2}\right]_{-1} = \frac{C}{\sqrt{2}} \sigma_{y_1}(\tau) \tag{2.110}$$

where $\sigma^2_{y_2}(\tau)$ is the Allan variance of $f_2(t)$ at the sampling time τ.

② PM white noise (noise for $\mu = -2$)

In the same way the error of $\left[\sigma_{\dot{R}_2}\right]_{-2}$ is obtained, we get $B_2(\gamma, \mu) = 2/3$ from Fig. 2.5, thus the velocity-measuring error caused by PM white noise is

$$\left[\sigma_{\dot{R}_2}\right]_{-2} = \frac{C}{\sqrt{3}} \sigma_{y_1}(\tau) \tag{2.111}$$

3) Item $\left[\overline{f_1(t_0 + T + \tau_1 + \tau_2)} - \overline{f_1(t_0)}\right]$

Through analysis the same with that for $\left[\overline{f_2(t_0 + T)} - \overline{f_2(t_0)}\right]$, we get:

① FM white noise

$$\left[\sigma_{\dot{R}_1}\right]_{-1} = \frac{C}{\sqrt{2}} \sigma_{y_1}(\tau) \tag{2.112}$$

where $\sigma_{y_1}(\tau)$ is the Allan variance of $f_1(t)$ at the sampling time τ.

② PM white noise

$$\left[\sigma_{\dot{R}_2}\right]_{-2} = \frac{C}{\sqrt{3}} \sigma_{y_1}(\tau) \tag{2.113}$$

As the errors above are independent, the total velocity-measuring variance $\sigma_{\dot{R}}$ obtained in Fig. 2.34 is the mean square sum of the above errors.

$$\sigma_{\dot{R}}^{\ 2} = \Delta_1^2 + \Delta_2^2 + \left[\sigma_{\dot{R}_2}\right]^2_{-1} + \left[\sigma_{\dot{R}_2}\right]^2_{-2} + \left[\sigma_{\dot{R}_1}\right]^2_{-1} + \left[\sigma_{\dot{R}_1}\right]^2_{-2} \tag{2.114}$$

By substituting (2.106), (2.108), (2.110), (2.111), (2.112), and (2.113) into (2.114), we get:

$$\sigma_{\dot{R}} = \frac{C}{\sqrt{2}}\left[\frac{V_1^2(T) + V_2^2(T) + A_1^2 f_1^2 + A_2^2 f_2^2}{2f_2^2} + \frac{5}{3}\sigma_{y_2}^2(\tau) + \frac{5}{3}\sigma_{y_1}^2(\tau)\right]^{\frac{1}{2}} \tag{2.115}$$

Expression (2.114) indicates that the accuracy of a three-way incoherent velocity measuring is not as satisfactory as that of the two-way coherent velocity measuring, the reason of which is (1) errors caused by "frequency drift" and "accuracy" of frequency sources and (2) among error terms resulted from the short-term stability; the mean square of each error term is added in three-way measuring, while in two-way measuring, the mean squares can be partly counteracted. Though there are weak points as mentioned above, three-way incoherent velocity measuring is still applied in some special areas, such as:

(a) Deep-space TT&C: that deep-space probes are far away from the Earth results into long delay of TT&C signals from a probe to the ground station or vice versa. This means when the echo arrives at the Earth, the original transmitting ground station has lost its sight to the echo because it is moving with the Earth's rotation (namely, the station's location is below the horizon). In this case, another ground station to receive the echo is necessary. Although the probe is far away from the ground station, the visual angles of the transmitting and the receiving stations for the vehicle are basically the same, so do the forward and the backward velocity. Thus calculation can be done according to Expression (2.115).

(b) Multiple $\dot{S}$ measuring at a range: in this case, multiple slave ground stations for only receiving are needed to measure the velocity and ($\dot{S}$) from ground main station → transponder → slave ground station. It is also a kind of incoherent three-way Doppler velocity measuring. In this case, if the visual angle difference of the transmitting station and the receiving station for the vehicle is not very small, then the forward and the backward velocities are different; and the measured is the sum of the forward/backward Doppler frequencies, namely, the measured is $\dot{S}$, the expression of which is:

$$\sigma_{\dot{S}} = \sqrt{2}C\left[\frac{V_1^2(T) + V_2^2(T) + A_1^2 f_1^2 + A_2^2 f_2^2}{2f_2^2} + \frac{5}{3}\sigma_{y_2}^2(\tau) + \frac{5}{3}\sigma_{y_1}^2(\tau)\right]^{\frac{1}{2}} \tag{2.116}$$

2.3.5 *Two-Way Noncoherent Doppler Velocity Measuring*

The expression "two-way noncoherent" means a ground station is used to transmit and receive forward and backward signals, but a transponder is used for noncoherent responding. Compared with coherent responding, it will introduce a noncoherent system error, so the coherent Doppler velocity measurement provides highest accuracy. To reduce the velocity-measuring system error caused by noncoherent components, measures are taken to counteract the noncoherent components.

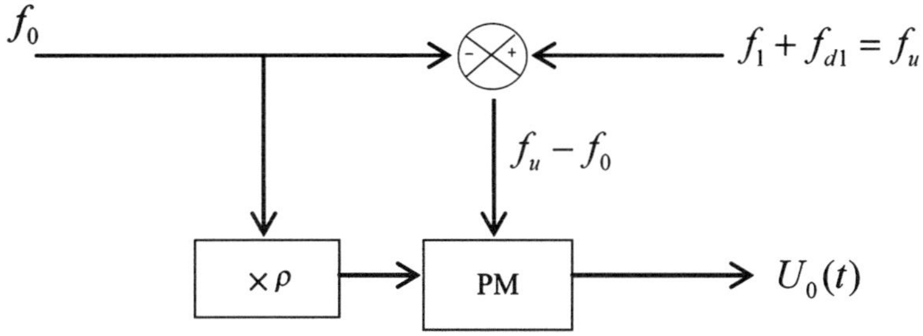

Fig. 2.35 Principle of transponder modulation retransmission

2.3.5.1 Ground Counteracting

These options include "mixing responding" and "modulation responding." The latter is comparatively better, the block diagrams of which are shown in Figs. 2.16 and 2.17. For the convenience of introduction of its counteracting principle of noncoherent components, Fig. 2.16 is simplified to the condition that the transponder responds to only one uplink signal, as shown in Fig. 2.35.

In Fig. 2.35, $\rho = N/M$ is the turnaround ratio, f_1 is the uplink frequency transmitted by the ground station, which is the reference in measuring of Doppler frequency drift. f_0 is the local oscillator (LO) frequency introduced by the transponder, which is noncoherent to f_1. When an onboard transponder moves at a speed of V, the received frequency is $f_1 + f_{d1}$, and f_{d1} is an uplink Doppler frequency. According to the Doppler effect, there is

$$\begin{aligned} f_1 + f_{d1} &= \frac{C-V}{C} f_1 = K_1 f_1 \\ K_1 &= \frac{C-V}{C} \end{aligned} \tag{2.117}$$

where K_1 is the coefficient of the uplink Doppler frequency drift.

$(f_1 + f_{d2})$ and f_0 is mixed to form subcarrier frequency $[(f_1 + f_{d1}) - f_0]$ which is used to conduct phase modulation (PM) for the noncoherent frequency (ρf_0), and the generated PM wave is $U_1(T)$:

$$\begin{aligned} U_1(t) = J_0(m)\sin\rho\omega_0 t + J_1(m)[\rho\omega_0 + (K_1\omega_1 - \omega_0)]t \\ - J_1(m)\sin[\rho\omega_0 - (K_1\omega_1 - \omega_0)]t + \cdots \end{aligned} \tag{2.118}$$

where m is a PM index and $J_0(m), J_1(m)$ are zero-order and first-order Bessel functions, respectively. The following only discuss $J_0(m), J_1(m)$. $U_1(t)$, when reaching the ground station and changing into $U_2(t)$, and under the influence of the Doppler effect, its frequency f_2 changes into:

$$f_2 + f_{d2} = f_2 \frac{C-V}{C} = K_2 f_2$$
$$K_2 = \frac{C}{C+V} \tag{2.119}$$

where K_2 is the coefficient of the downlink Doppler frequency drift. With this coefficient, we get the following from Expression (2.118):

$$\begin{aligned} U_2(t) = & J_0(m)\sin K_2\rho\omega_0 t + J_1(m)\sin K_2[\rho\omega_0 + (K_1\omega_1 - \omega_0)]t \\ & -J_1(m)\sin K_2[\rho\omega_0 - (K_1\omega_1 - \omega_0)]t \end{aligned}$$

Use a phase-locked loop to draw out residual carrier item $c(t)$:

$$c(t) = J_0(m)\sin K_2\rho\omega_0 t$$

where $\omega_c = K_2\rho\omega_0$ is the residual carrier frequency; use $c(t) \times U_2(t)$ to demodulate subcarrier item $B(t)$.

$$B(t) = J_1(m)\sin[K_2(K_1\omega_1 - \omega_0)]t = J_1(m)\sin\omega_B t$$

where $\omega_B = K_2(K\omega_1 - \omega_0)$ is a subcarrier frequency.

Measure out ω_c, ω_B and ω_1 at the ground station and then get the two-way carrier Doppler frequency ω_D by the following calculation:

$$\omega_D = \omega_B + \frac{\omega_C}{\rho} - \omega_1 = K_1K_2\omega_1 - \omega_1 \tag{2.120}$$

Substitute Expressions (2.117) and (2.119) into Expression (2.120) to obtain:

$$\omega_D = \left[\left(\frac{C-V}{C}\right)\left[\frac{C}{C+V}\right] - 1\right]\omega_1 = \frac{-2V}{C+V}\omega_1$$
$$V = \frac{\omega_D}{2\omega_1 + \omega_D}C = \frac{f_D}{2f_1 + f_D}C \tag{2.121}$$

By comparing Expression (2.121) with Expression (2.22) in Sect. 2.3.1, we find that they are completely the same. Thus we know that the application of this noncoherent measuring of "modulation responding" realizes coherent velocity measuring and their measuring accuracy analyses are also the same, but two additional errors will be introduced.

(1) The influence of additional non-coherent phase noise: In Expression (2.120), suppose ω_c/ρ of $c(t)$ and $K_2\omega_0$ of $B(t)$ are equal, so their difference is zero. This result is based on frequency accuracy, drift, short-term stability, etc. of f_0 generated coherent components in $c(t)$ and $B(t)$. However, if there are some noncoherent components added into $c(t)$ and $B(t)$, then their mean squares will be added rather than they are counteracted; these additional noncoherent phase noises include: ① phase noises introduced by different parts of transmission channels of $c(t)$ and $B(t)$; ② measurement error and phase noises introduced respectively while ω_c and ω_B are being measured; ③ after f_0 passes through $c(t)$ and $B(t)$, the cross-correlation weakens due to different time delays of the transmission channels of $c(t)$ and $B(t)$; and ④ phase noises added when frequency turnaround ratio ρ is generated, namely, in Fig. 2.16, the noncoherent phase noises added respectively when M-order and N-order frequency doubling are generated; this condition is equal to make random changes to ρ. In calculation of Expression (2.120), ρ is assumed to be a designed constant, so the calculation result will have an error. Above-mentioned factors in the development should be reduced as much as possible; it has been proved in practice that their influence can be reduced to a tiny extent, only occupying a very small part of velocity-measuring error.

(2) The influence of thermal noises: Phase noises caused by thermal noises in $c(t)$ and $B(t)$ are noncoherent. Assume that the carrier-to-noise ratio of $c(t)$ is J_0/N_0; then, the power spectral density of the relevant phase noise is N_0/J_0. The two are noncoherent, the power spectral density of the phase noise of their sum Φ_D is:

$$\Phi_D = \frac{N_0}{J_1} + \frac{N_0}{J_0} = \left(\frac{1}{J_1} + \frac{1}{J_0}\right)N_0 = \left(1 + \frac{J_1}{J_0}\right)\frac{N_0}{J_1} \tag{2.122}$$

It is clear from here that J_0/N_0 of the residual carrier will produce additional velocity-measuring error; in the expression, $(1 + J_1/J_0)$ is an additional factor, namely, the measuring error caused by thermal noises will deteriorate $\sqrt{1 + J_1/J_0}$ times.

One of the distinguishing good points of noncoherent "modulation responding" is that multiple subcarriers can be adopted to realize multi-station intersected measuring with frequency division multiple access (FDMA). So long as each subcarrier frequency is carefully designed to prevent their combination interferences from falling into useful bandwidth, it will achieve a high measuring accuracy. This technique has been applied in multiple $R\dot{R}$ high-accuracy measuring systems.

2.3.5.2 Measuring in the Air and Processing on the Ground

This method uses a transponder to measure the data of uplink Doppler frequency and transmits such data to the ground through a data channel, then jointly processes the above-mentioned data with the data of downlink Doppler frequency measured on the ground, and finally, a two-way Doppler is obtained. The principle is shown as follows:

For the same point on a flight locus, from simultaneous expressions of (2.117) and (2.119), we have:

$$v = \left[\frac{f_1 f_2 - (f_1 + f_{d1})(f_2 + f_{d2})}{f_1 f_2 + (f_1 + f_{d1})(f_2 + f_{d2})}\right] c \tag{2.123}$$

where f_1 and f_2 are uplink and downlink frequencies, respectively, and $(f_1 + f_{d1})$ and $(f_2 + f_{d2})$ are frequency values to be measured; when the TT&C system is in velocity measuring, f_1 will be regarded as the reference and f_2 is noncoherent with f_1. One of the conditions that make Expression (2.123) tenable is that the four frequency values in the expression are defined according to the same reference, generally, the ground clock is used as the reference; otherwise, Expressions (2.117) and (2.119) cannot be made simultaneous, and each piece of measured information cannot be integrated either.

In this case, the transponder and the ground station use different frequency measurement clocks. Commonly, $(f_1 + f_{d1})$ is measured out from f_1, f_2, and $(f_1 + f_{d2})$ with the ground F_g as the reference and $(f_1 + f_1)_s$ is measured with the transponder clock F_s as the reference; their relationship is:

$$(f_1 + f_{d1}) = \frac{F_s}{F_g}(f_1 + f_{d1})_s \tag{2.124}$$

Substitute Expression (2.124) into Expression (2.123) to obtain:

$$v = \frac{\left[f_1 f_2 - \frac{F_s}{F_g}(f_1 + f_{d1})_s (f_2 + f_{d2})\right]}{\left[f_1 f_2 + \frac{F_s}{F_g}(f_1 + f_{d1})_s (f_2 + f_{d2})\right]} c \tag{2.125}$$

As noncoherent F_s components are already contained when transponder F_s is taken as the reference to measure $(f_1 + f_{d2})_s$, and transponder downlink transmitting frequency f_2 is obtained with F_s as its reference, thus if they are jointly processed, a part of noncoherent components can be eliminated, but there are still some of them remaining. It is obvious from Expression (2.125) that there are six error factors that can affect v measuring accuracy, which are $(f_1 + f_{d1})_s$, $(f_2 + f_{d2})$, F_s, F_g, f_1 and f_2. Measuring error, accuracy, or changes of these factors will induce velocity-measuring error. Apply total differential to Expression (2.125) to acquire each velocity-measuring error.

Expression (2.125) can also be expressed as follows:

$$(f_1 + f_{d1}) = \frac{F_s(1 + \sigma_s)}{F_g(1 + \sigma_g)}(f_1 + f_{d1})_s$$

where σ_s and σ_g are the short-term stability of the transponder clock and the ground station frequency clock, respectively, and F_s and F_g are the sums of their frequency drifts and frequency accuracy, respectively.

From the above expression, it is clear that σ_s, σ_g and changes of F_s and F_g and will cause velocity-measuring error; accuracy of F_s and F_g will influence velocity-measuring accuracy; as $(f_1+f_{d1})_s$ is measured onboard (on the transponder), its error is comparatively large (e.g., it is difficult to use an extreme narrowband phase-locked loop). Besides, Expression (2.125) is obtained from simultaneous expressions of (2.117) and (2.119) under the same orbit point condition. Thus, above conditions will not be satisfied and a velocity-measuring error will be introduced if there is relative dislocation or different smoothing time lengths during measuring frequencies of $(f_1+f_{d1})_s$ and (f_2+f_{d2}).

From comparison between Expressions (2.125) and (2.22), it is clear that there are only two error factors f_2 and f_1 in coherent Doppler velocity measuring, while there are six error factors in this noncoherent measuring ($f_2, f_{d2}, f_1, f_{d1}, F_s$ and F_g). Thus, velocity-measuring error in this plan is more than that in coherent measuring; the good point of which is that no additional noncoherent f_0 beacon is needed, but uplink frequency needs to be measured on the transponder, which complicates aerial equipment. Another advantage is that technology of code division multiple access (CDMA) can be used in both up- and downlink to realize modes like multi-station measuring single satellite or single-station measuring multi-satellite and downlink can take broadcast transmission to save power, but CDMA interference to measuring accuracy should be taken into account.

2.3.5.3 Onboard Counteracting

The block diagram of this kind of transponder [10] is shown as follows:

In Fig. 2.36, f_L is an introduced noncoherent frequency. Generally, it makes f_2 noncoherent to f_1, but if an intermediate frequency $F_2 = \rho F_1^{'} = \frac{N}{M}F_1$ is generated in a digital signal processor, the output of such transponder will be

$$f_2 = Nf_L + \frac{N}{M}F_1 = Nf_L + \frac{N}{M}(f_1 - Mf_L) = \frac{N}{M}f_1 \qquad (2.126)$$

In this scheme, f_2 and f_1 are coherent, and (M/N) is the turnaround ratio. Its physical meaning is that signal frequency changes caused by F_L at terminals A and B of a transmitting mixer are completely coherent, which are entirely counteracted after frequency mixing. However, in an actual circuit, they may only be coherent in part, introducing velocity-measuring error. Reasons for noncoherence include:

(1) $\rho F_1^{'} \neq \frac{N}{M}F_1$. In the expression, M and N are designed deterministic values, and F_1 is a continuous variable. But this plan aims to use digital signal processing circuits to generate $\rho F_1^{'}$ which is disperse, having quantizing error. The error Δf produced by it is

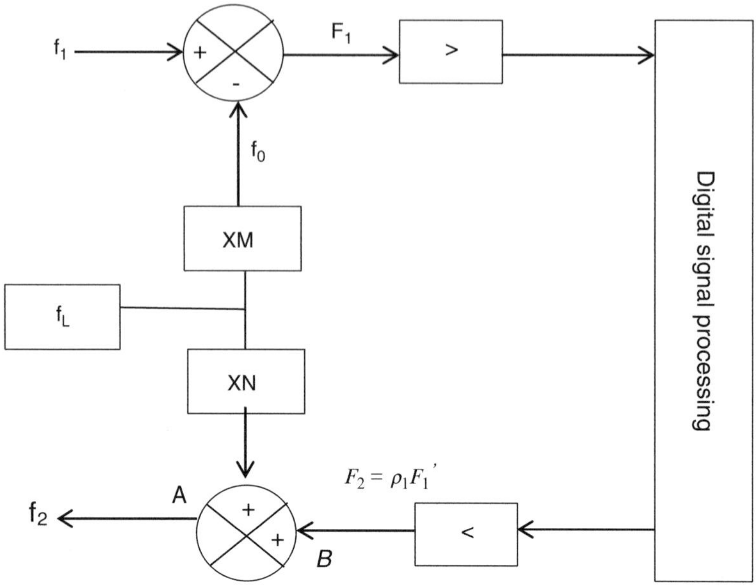

Fig. 2.36 Block diagram of the counteracting principle of noncoherent components in a transponder

$$\Delta f = \frac{N}{M} F_1 - \rho F^{'}$$

But Δf will produce velocity-measuring error Δv:

$$\Delta v = \frac{1}{2} \frac{\Delta f}{f_2} C \tag{2.127}$$

A common solution is to use DDS_1 of a digital carrier loop to measure F_1, getting a proximate frequency code $F_1{}'$. A binary number ρ is used to approximate M/N and is multiplied by the frequency code. Their product is filtered digitally and then used to control another DDS_2 to produce f_2. In this procedure, there exist three parts of error:

1) The error in measuring of F_1 with the digital carrier loop: its calculation is determined by phase accumulation digits of DDS_1. Please refer to Expressions (2.24) and (2.25) in this book.
2) The error between ρ and the digital filter: it is determined by its word length. Refer to Expression (2.121) for its calculation. As in computer the word length of binary numbers can be long, thus this is not a main component.
3) The quantizing error of DDS_2: it is mainly determined by multiples of the phase accumulator of DDS_2. Please refer to Expressions (2.24) and (2.25) in this book for its calculation.

(2) Additional phase noises: In circuit between Points A and B in Fig. 2.36, except f_L, additional phase noises produced by other parts (including those caused by onboard vibration) are noncoherent. They cannot be subtracted but their mean squares are added in the transmitting mixer, causing velocity-measuring error. The analysis about this can be found in Sect. 2.3.3.2 in this book; and refer to Expression (2.87) for its calculation.

(3) Influence from circuit delay time: Namely, make $\rho F' = \frac{N}{M} F_1$, but delay time from f_L to Point A and that from f_L to Point B is different, causing a bad cross-correlation of f_L with Point A and Point B. Thus, the influence from f_L cannot be totally counteracted. Its physical mechanism is the same with that of the influence of transmission signal short-term instability to velocity-measuring accuracy. Please refer to Sect. 2.3.3.2 in this book for its analysis and to Expression (2.52) for its calculation. In the expression $V = T/\tau$, if T is made very small, the error will be small too.

This scheme simplifies and digitalizes the transponder, bringing a series of good points of digitalization and software using. Its weak point is it can only "approximate coherence" rather than "achieve a complete one," hence introducing a noncoherent velocity-measuring error.

2.3.6 *One-Way Noncoherent Doppler Velocity Measuring*

The simplified model of a one-way noncoherent Doppler velocity measuring is shown in Fig. 2.37.

In Fig. 2.37, $f_1(t)$ is the frequency of velocity-measuring beacon onboard, $f_2(t)$ is the reference frequency used by the ground station to measure Doppler frequency f_d, t_0 is the starting time to measure frequency drifts of $f_1(t)$ and $f_2(t)$, delay time T is the time difference between t_0 and measuring time $(t_0 + T)$, and τ is the integration time of velocity measuring.

(1) The velocity-measuring error introduced by thermal noises

Thermal noises only add phase noises in $f_1(t)$, the analysis method of which is the same with that of the two-way coherent measuring system (see Sect. 2.3.3.2).

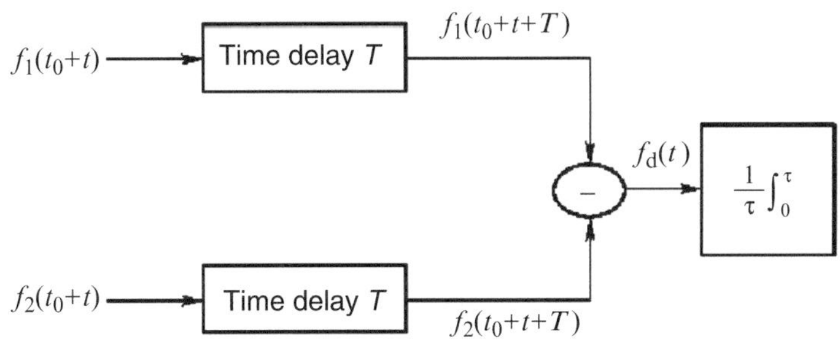

Fig. 2.37 The model of a one-way noncoherent Doppler velocity measuring

One-way Doppler frequency offset is 1 time smaller than that of the two-way one; thus, the error is 1 time larger, hence the expression:

When $\omega_L \tau \gg 1$, there is

$$\sigma_{\dot{R}1}(\tau, B_L) = 2\sigma_{\dot{R}N}(\tau, B_L) = \frac{c}{\sqrt{2}\pi f_0 \tau}\sqrt{\frac{N_0 B_L}{P_s}} = \frac{c}{\sqrt{2}\pi f_0 \tau \sqrt{\frac{1}{\rho_L}}} \tag{2.128}$$

where $\rho_1 = P_s/B_L N_0$ is the signal-to-noise ratio of a downlink carrier loop.

(2) The velocity-measuring error introduced under non-coherent conditions

From Fig. 2.37, it is obvious that:

$$\begin{aligned} f_d(t) &= [f_1(t_0 + t + T) - f_2(t_0 + t + T)] \\ &= [f_1(t_0 + t + T) - f_1(t_0)] - [f_2(t_0 + t + T) - f_2(t_0)] + [f_1(t_0) - f_2(t_0)] \end{aligned} \tag{2.129}$$

As $f_2(t)$ is on the ground, its short-term instability can be made better, and there are no phase noises added by its transmission channel; moreover, it has no Doppler changes introduced dynamically by targets, thus it can apply comparatively large time constant filtering. Therefore, generally speaking, the short-term instability of $f_2(t)$ is far better than that of $f_1(t)$, namely,

$$f_d(t) = [f_1(t_0 + t + T) - f_1(t_0 + t + T)] + \Delta F(t_0) \tag{2.130}$$

$$\Delta F(t_0) = [f_1(t_0) - f_2(t_0)] \tag{2.131}$$

From Expressions (2.129) and (2.130), one-way noncoherent Doppler velocity measuring is a special example of the three-way one. The first item of Expression (2.130) can be analyzed with the short-term stability of $f_1(t)$. It is a random error, the analysis method of which is the same with that of a three-way noncoherent Doppler velocity measuring. From Expression (2.115), there is:

$$\sigma_{\dot{R}} = \frac{2c}{\sqrt{2}}\left[\frac{V_1^2(T) + A_1^2 f_1^2}{2f_2^2} + \frac{5}{3}\sigma_{y1}^2(t)\right]^{\frac{1}{2}} \tag{2.132}$$

From Expression (2.132), it is clear that the error of one-way noncoherent Doppler velocity measuring is larger than that of two-way coherent Doppler measuring. The reasons are:

1) For a spacecraft not far from the ground, because T is comparatively large (in two-way coherent measuring, T is just the delay time of signal receiving/transmitting), $r = T/\tau$ increases. From Table 2.5, it is clear that the velocity-measuring error caused by $\mu = +1$ noise (frequency random walk noise) and $\mu = 0$ noise (frequency modulation flicker noise) becomes larger. But in deep-space exploration, as time delay of signal receiving/transmitting is large, the two measuring methods share comparatively small difference. Thus good points of one-way Doppler measuring appear more outstanding.

2) As the Doppler frequency offset of one-way measuring is one time smaller than that of the two-way one, there is one time larger error.

3) System error $\Delta F(t_0)$ brought by frequency drift and accuracy.

The short-term stability of velocity-measuring signals exerts comparatively large influence on error of one-way noncoherent velocity measuring, but if measuring error caused by short-term stability can be made smaller than that of thermal noises in the system design phase, then the influence on the system's total velocity-measuring accuracy is comparatively small.

From the analysis above, it is clear that two-way coherent Doppler velocity measuring can acquire Doppler data with the highest accuracy so far, because it takes the same ground-based frequency standard as the reference signal for Doppler frequency measuring of up- and downlink signals. However, as spacecraft oscillator stability improves, one-way Doppler velocity measuring can match the two-way one, especially in deep-space TT&C. One-way Doppler measuring enjoys simplified equipment and provides better signal-to-noise ratio (SNR) for receiving of spacecraft telemetry data. Reasons for SNR improvement: First, one-way transmission is above spacecraft receiver thermal noises; second, the ground station uses one-way receiving mode during one-way velocity measuring; however, the two-way measuring needs duplex mode which will raise the equivalent noise temperature of the ground receiving system; third, in deep-space TT&C, short-term stability of two-way transmission will be affected by the influence of solar plasma on uplink signals; and fourth, there is another far-reaching good point about one-way measuring, namely, one antenna is enough to receive at the same time one-way Doppler and telemetry information of multiple aircrafts and landers within the same beam, with no need for multiple uplink signals. Therefore, such configuration will economize ground-based resources more effectively and improve estimation accuracy of orbit solution and lander location by carrying out differential measuring.

2.3.7 Theoretical Calculation of Velocity-Measuring Accuracy

Use the following expression to calculate coherent Doppler velocity-measuring error:

$$V = -\frac{1}{2}\frac{f_d}{f_0 + f_d} \tag{2.133}$$

Use the following expression to calculate one-way noncoherent Doppler velocity-measuring error:

$$V = -\frac{c}{f_0} f_d \tag{2.134}$$

2.3.7.1 Random Error

(1) The velocity-measuring error caused by system short-term instability

During coherent Doppler velocity measuring, its expression has already been obtained from Expression (2.52), namely,

$$\sigma_{\dot{R}_1} = \frac{\sqrt{B_2}}{\sqrt{2}} C\sigma_y(\tau) \tag{2.135}$$

where $\sigma_y(\tau)$ is Allan variance and B_2 means Barnas second offset function; refer to Sect. 2.3.3.2 for value choosing of τ and B_2.

During single-way noncoherent Doppler velocity measuring, there is:

$$\sigma_{\dot{R}} = 1.8c\sigma_y(\tau) \tag{2.136}$$

(2) The velocity-measuring error introduced by thermal noises

1) Coherent Doppler velocity measuring. Expression (2.101) has already given its expression as follows:

$$\sigma_{\dot{R}_2} = \frac{C}{2\sqrt{2}\pi f_0 \tau}\sqrt{\frac{N_0 B_L}{P_s}} = \frac{C}{2\sqrt{2}\pi f_0 \tau}\sqrt{\frac{1}{\rho_L}} \tag{2.137}$$

where f_0 is the carrier frequency; τ means integration time; B_L represents carrier loop single-side bandwidth (refer to Table 2.7 for B_L calculation expression of different types of phase-locked loops); P_S is receiving power: N_0 stands for noise power spectral density; and ρ_L symbolizes the SNR of downlink carrier loop when there still exist residual carriers.

The value of ρ_L changes along with different carrier modulation modes. The expressions are as follows:

(a) Carriers are modulated with non-return-to-zero (NRZ) signals rather than subcarriers, but there still exist residual carriers. The carrier loop SNR will produce additional loss which arises from increased effective noise level of carrier

Table 2.7 Noise bandwidth of common PLL

Types of PLL	Noise bandwidth, B_L/Hz
Type I order I	$K/4$
Type I order II (lagging filter)	$K/4$
Type II order II	$\frac{\omega_n}{2}\left(\zeta + \frac{1}{4\zeta}\right)$
Type II order III	$\frac{K}{4}\left(\frac{1+1/K_{\tau_2}}{1-1/b}\right)$
Type III order III	$\frac{K}{4}\left(\frac{1+\frac{K_2}{KK_1}+\frac{K_1K_3}{KK_2^2}}{1-\frac{K_1K_3}{KK_2^2}}\right)$

synchronization due to overlapping of residual carriers in frequency domain by data sides. In this case, ρ_L shall be calculated according to the expression below:

$$\rho_L = \frac{P_C}{N_0}\bigg|_{D/L} \cdot \frac{1}{B_L} \cdot \frac{1}{1 + 2E_S/N_0} \tag{2.138}$$

where E_s/N_0 means the ratio between energy and noise spectral density of each symbol.

(b) In the case of suppressed carrier tracking, the carrier loop SNR is:

$$\rho_L = \frac{P_T}{N_0}\bigg|_{D/L} \cdot \frac{S_L}{B_L} \tag{2.139}$$

where $p_T|N_0|_{D/L}$ is the ratio between total signal power and noise spectral density of the downlink; S_L means the square loss of Costas loop.

$$S_L = \frac{2(E_S/N_0)}{1 + 2(E_S/N_0)} \tag{2.140}$$

(c) In the case of QPSK, carrier loop SNR is:

$$\rho_L = \frac{P_T}{N_0}\bigg|_{D/L} \cdot \frac{S_{LQ}}{B_L} \tag{2.141}$$

where S_{LQ} is the square loss of the QPSK Costas loop.

$$S_{LQ} = \frac{1}{1 + \frac{9}{2E_{SQ}/N_0} + \frac{6}{\left(E_{SQ}/N_0\right)^2} + \frac{3}{2\left(E_{SQ}/N_0\right)^3}} \tag{2.142}$$

where E_s/N_0 means ratio between energy and noise spectral density of each quaternary channcl symbol.

When telemetry data in non-return-to-zero (NRZ) form is used to directly modulate carriers (i.e., without subcarriers), and there exists imbalance in data (i.e., the number of logic 1 is not equal to that of 0), then the residual carrier loop will jitter. This jitter is a kind of error source for Doppler measuring. The influence of this error depends on the statistical distribution of telemetry data.

2) One-way noncoherent Doppler velocity measuring

$$\sigma_{\dot{R}_2} = \frac{C}{\sqrt{2}\pi f_0 \tau}\sqrt{\frac{N_0 B_L}{P_s}} = \frac{C}{\sqrt{2}\pi f_0 \tau}\sqrt{\frac{1}{\rho_L}} \tag{2.143}$$

3) Thermal noises introduced by the transponder

$$\sigma_{\dot{R}_3} = \frac{c}{2\sqrt{2\pi f_C T}}\sqrt{\frac{1}{\rho_L} + \frac{G^2 B_L}{P_C/N_0|_{U/L}}} \tag{2.144}$$

where T is the measured integration time (S), f_C is downlink carrier frequency (Hz), c is light speed in vacuum (mm/s), G is retransmission ration, B_L is single-side equivalent noise bandwidth of the downlink carrier loop, and $P_C/N_0|_{U/L}$ is the ratio between uplink carrier power and noise spectral density.

The fact that the carrier loop bandwidth of the transponder (uplink) is larger than that of the downlink is supposed in Expression (2.144).

(3) The velocity-measuring error introduced by clutters

Clutters of the transmitter power supply cause spurious phase modulation to the phase modulator and power amplifier; the introduced velocity-measuring error is

$$\sigma_{\dot{R}_4} = \frac{\lambda_i}{2}\frac{\sigma_\varphi}{2\pi T} \tag{2.145}$$

where λ_i is the uplink carrier wavelength, σ_φ is phase jitter caused by spurious phase modulation, and T is sampling time.

(4) The velocity-measuring error introduced by VCO short-term instability

In the receiver phase-locked loop, the measuring error introduced by loop output phase jitter caused by VCO short-term instability is:

$$\sigma_{\dot{R}_5} = \frac{\lambda_i}{2\pi T}\sigma_{\theta V} \tag{2.146}$$

where $\sigma_{\theta V}$ is the loop output phase jitter caused by VCO short-term instability and λ_i is downlink carrier wavelength.

(5) Quantization error in carrier loop velocity measuring

Directly extract Doppler frequency from the receiver digital carrier loop, smooth it, and then work out velocity-measuring error with the computer. In this case, there are no $\sigma_{\dot{R}_4}$ and $\sigma_{\dot{R}_5}$ but quantization error $\sigma_{\dot{R}_6}$:

$$\sigma_{\dot{R}_6} = \frac{1}{\sqrt{6}}\frac{\dot{R}}{N} \tag{2.147}$$

where $\dot{R}$ is flying target velocity and N is digital filter word length.

(6) The velocity-measuring error introduced by VCO short-term frequency instability

The velocity-measuring error caused by VCO short-term frequency instability in the coherent transponder is:

$$\sigma_{\dot{R}_7} = \frac{\lambda_i \sigma_v}{4\pi\tau} \tag{2.148}$$

where σ_v is phase jitter caused by VCO short-term instability.

Total velocity-measuring error is:

$$\sigma_R = \sqrt{\sigma_{\dot{R}_1}^{2} + \sigma_{\dot{R}_2}^2 + \sigma_{\dot{R}_3}^2 + \sigma_{\dot{R}_4}^2 + \sigma_{\dot{R}_5}^2 + \sigma_{\dot{R}_6}^2 + \sigma_{\dot{R}_7}^2 + \sigma_{\dot{R}_8}^2} \tag{2.149}$$

2.3.7.2 Velocity-Measuring System Error

(1) The error introduced by long-term frequency instability of the master oscillator

Find the differential of velocity-measuring formula with respect to f_i to obtain the caused velocity-measuring error:

$$\Delta V = \frac{Cf_d}{2f_i^2}\Delta f_i = \frac{Cf_d}{2f_i}\frac{\Delta f_i}{f_i} = V \times \frac{\Delta f_i}{f_i} \tag{2.150}$$

The error is equal to the product of the measured velocity value and the relative long-term instability of the master oscillator; long-term instability includes drift and accuracy error. When $V = 10$ km/s and long-term instability $= 1 \times 10^{-9}$, the caused velocity-measuring error is just 10^{-5} m/S. It is obvious that velocity-measuring system error complies to relevant requirement easily.

(2) Dynamic error

Velocity-measuring dynamic error (frequency error, unit rad/s) of different types of phase-locked loops can be calculated with Table 2.8 [9].

(3) The velocity-measuring error introduced by light speed ambiguity

The velocity-measuring error introduced by light speed ambiguity is

$$\sigma_{\dot{R}_8} = 3.33 \times 10^{-7}\dot{R} \tag{2.151}$$

Table 2.8 Dynamic error of carrier Doppler velocity measuring

	Constant range rate	Constant range rate derivative	Second derivative of constant range rate
Loop	Constant Doppler frequency drift	Constant Doppler rate	Constant Doppler acceleration
Type II loop with standard underdamping	0	0	$\left(\frac{9\pi\beta}{16B_L^2}\right)$
Type II loop with super-critical underdamping	0	0	$\left(\frac{25\pi\beta}{32B_L^2}\right)$
Type III loop with standard underdamping	0	0	0
Type III loop with super-critical underdamping	0	0	0

Note: B_l is equivalent loop bandwidth of carrier loop (single-side) noises; when continuous Doppler acceleration appears, Type II loop periodically carries out cycle skip; β is Doppler acceleration (Hz/s^2)

2.4 Ranging Techniques Theory and Technology: Two-Way, Three-Way, and Single-Way Ranging Technologies

2.4.1 Continuous Wave Two-Way Ranging Methods

Continuous wave ranging means certain forms of ranging signals are modulated on a continuous carrier to acquire the range information by comparing time delay between transmitted and received ranging signals. The above-mentioned certain forms of ranging signals refer to signals with particular time marks (such as the zero crossing point (ZCP) of sine waves) or phase marks. The characteristics of these continuous wave ranging signals are that there exists phase ambiguity, so the range de-ambiguity is the most special problem for continuous wave ranging. At present, there are mainly three types of continuous wave ranging signals: pure tone signal (single-frequency sine signal or square signal), pseudorandom code signal (for short, pseudo code signal, abbrev: PN code), and tone and code-mixed signal (the mixed signal of pure tone signal with pseudorandom code signal). Besides, based on the similar principles, there exist other ranging methods such as spread spectrum ranging, carrier ranging, and Doppler accumulative ranging. This section mainly introduces the first three ranging methods. Spread spectrum ranging will be described in the section of spread spectrum TT&C, Doppler accumulative ranging will be explained along with range acquisition, and carrier ranging has been applied in GPS, for which readers can refer to relevant materials.

2.4.1.1 Tone Ranging System

Tone means pure sine wave signal modulated on a carrier. It is also called side tones since its frequency spectrum is located at both sides of the carrier. The relationship between range R and tone time delay T and its phase shift Φ can be expressed as follows:

$$R = \frac{cT}{2} = \frac{c\varphi}{2\omega} = \frac{\Phi c}{720F} \tag{2.152}$$

where ω is the angular frequency of the ranging signal (rad/s), F is frequency of the ranging signal (Hz), Φ is phase shift in unit of degree (°), φ is phase shift in unit of radian (rad), and c is light speed (m/s).

Therefore, time delay T can be obtained by measuring the phase shift. This method is called phase measuring method. Another method is to directly measure ZCP time delay of the phase between the transmitted and received tones. This is called time measuring method. It is clear from Expression (2.152) that if a certain phase measuring error $\Delta\Phi$ exists, the higher ω is, the smaller the time-delay error will be. Thus the ranging accuracy can be improved by using high frequency ranging signal. As the phase of a sine periodic signal circulates by 2π, the real phase shift = the measured sine wave phase shift + $2n\pi$, where $n = 0, 1, 2, \ldots$ This leads to multi-valuedness of the real phase of sine waves. This phenomenon is called phase ambiguity, the corresponding range of which produces range ambiguity. One solution to this problem is to constitute a set of tone signals with multiple pure sine wave signals in different frequencies. The set of tone signals are used to measure the same range simultaneously. Use the signal $U_2(t)$ in a lower frequency F_2 to solve the phase ambiguity produced by signal $U_1(t)$ in a higher frequency F_1 and make F_1 be K_1 times as large as F_2, namely, $F_1 = K_1F_2$, where K_1 is called the frequency ratio of F_1 to F_2, also called matching ratio.

The most prominent feature of continuous wave ranging is the existence of range ambiguity; therefore, it is comparatively complicated to be realized. At present there are many ways for solving ambiguity, the main of which are listed as follows:

(1) Simultaneous multitone method

Its principle is shown as in Fig. 2.38 (where a uniform motion target is taken as an example).

In Fig. 2.38, the ordinate indicates the phase difference Φ between received/transmitted tones; the abscissa indicates the corresponding time delay t of target in range R; the solid line Φ_1 represents the phase change curve of F_1; the dotted line Φ_2 is the phase change curve of F_2; F_1 is the higher tone, and F_2 is the lower tone. In this example, $K_1 = (F_1/F_2) = 4$, it is obvious from the figure that when the target starts to move from zero range, the echo delay time t increases gradually, so does Φ_1 of F_1; when $t = 1/F_1$, Φ reaches to 360°. Because of sine wave periodicity, at the next moment, Φ_1 jumps to 0° from 360°. If this 0° is used for ranging, its corresponding range will be 0. Apparently, it is wrong. This phenomenon is called

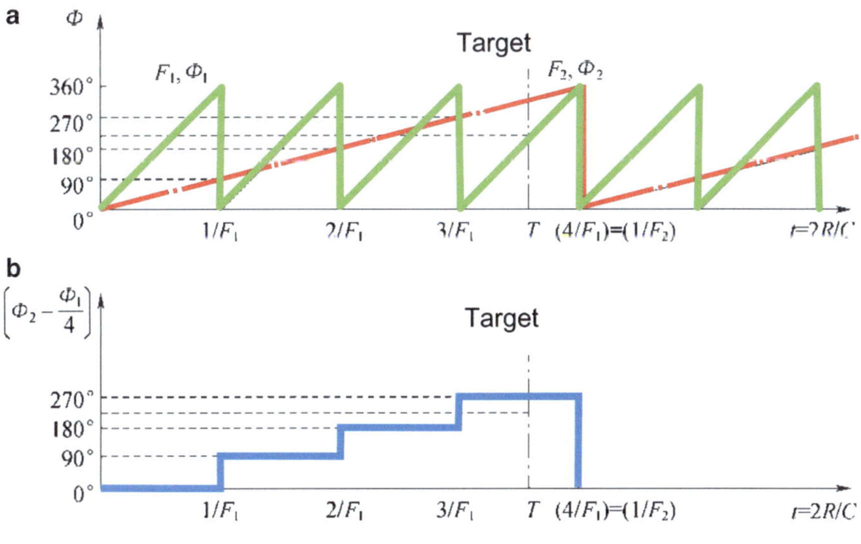

Fig. 2.38 Principle of multitone de-ambiguity

the appearance of range ambiguity. When the target continues to move, at time $t = 2/F_1$, the second step from 360° to 0° occurs. Its ambiguity number of times $n = 2$; when the target moves to time $t = 3/F_3$, a third ambiguity with $n = 3$ will happen.... When the target echo is at time T, the measured phase of F_1 tone is Φ_1. However, as it has already generated three times of ambiguity ($3 \times 360° + \Phi_1$), and in the foresaid example $n = 3$, thus when continuous wave ranging is underway, apart from "odd item" Φ_1, n is also needed to be worked out, which is the number of range space corresponding to $\Delta R = c/2F_1$ and "to solve n" is called de-ambiguity.

For de-ambiguity, a secondary tone F_2, which frequency is lower than that of high tone F_1, is needed. When F_1 has produced three times of ambiguity, the phase Φ_2 of F_2 remains in the m-th period at a range of 0°–360°. Thus, the following criteria can be used for de-ambiguity:

$$\begin{cases} \text{when } 0° < \Phi_2 < 90°, & n = 0 \\ \text{when } 90° < \Phi_2 < 180°, & n = 1 \\ \text{when } 180° < \Phi_2 < 270°, & n = 2 \\ \text{when } 270° < \Phi_2 < 306°, & n = 3 \end{cases} \tag{2.153}$$

In this way, the phase value of the secondary tone can be used to solve the phase ambiguity of the high tone, which is called "using F_2 to solve R ambiguity." At this time, the total phase shift $\Phi_\Sigma = n \times 360° + \Phi_1$. This is the basic principle of range de-ambiguity. When the target flies further, making F_2 produce new phase ambiguity, it is feasible to use tone F_3 whose frequency is lower than that of F_2 to solve F_2 ambiguity. In the same manner, multiple secondary tones can be used to solve further range ambiguity. A priori information of the orbit can be used to reduce numbers of secondary tones, so that de-ambiguity is to be conducted only on a section of orbit, simplifying the equipment.

The foresaid basic principle of de-ambiguity is theoretically tenable, but in engineering implementation, at the boundaries (e.g., 0°, 90°, 180°. and 270°) of above-mentioned criteria, random error (e.g., phase jitter) and system error (e.g., group delay time error) of Φ_2 may cause erroneous judgment. For example, assuming actual $\Phi_2 = 179°$ without above-mentioned error, then, according to the criteria, the corresponding $n = 1$. But if the error exists, making $\Phi_2 = 181°$, then the judgment will be $n = 2$, producing range jump of $\Delta R = C/2F_1$, which is called "mismatch error." The following criteria can be adopted to reduce such influence:

$$\begin{cases} \text{when } -45° < (\Phi_2 - \Phi_1/4) < +45°, & n = 0 \\ \text{when } 45° < (\Phi_2 - \Phi_1/4) < 135°, & n = 1 \\ \text{when } 135° < (\Phi_2 - \Phi_1/4) < 225°, & n = 2 \\ \text{when } 225° < (\Phi_2 - \Phi_1/4) < 360°, & n = 3 \end{cases} \tag{2.154}$$

The $(\Phi_2 - \Phi_1/4)$ curve of these criteria is shown as in Fig. 2.38b, the judgment of n mainly bases on this step trait. It is clear from this figure that corresponding to a period of Φ_1 without ambiguity, judgment value $(\Phi_2 - \Phi_1/4)$ is a constant value with an error tolerance of ±45°. Therefore, if each kind of phase error (mainly caused by inconsistent group delay of each tone and by time-delay variance of each tone after demodulation) does not exceed ±45°, the ranging value will not be affected by mismatch error. Thus the tolerance becomes much looser. To make plus-minus error share the same margin to maximize error tolerance, in equipment commissioning, mean value of $(\Phi_2 - \Phi_1/4)$ shall be adjusted to 0°, 90°, 180°, and 270°. This procedure is called "phase matching" adjustment of Φ_1 and Φ_2, also named as "range matching" adjustment among ranging tones (in broad terms, "phase matching" means the phase of each tone shares certain interrelation, e.g., align their zero phase points or align the last tone phase to the tolerance middle point of the next tone phase). In order to simplify engineering implementation, change the criteria of Expression (2.154) as follows: use 90° (judgment tolerance range $\Delta\Phi$) to divide the above criterion expression to obtain:

① When $-0.5 < ((\Phi_2 - \Phi_1/4)/\Delta\Phi) < +0.5$, $n = 0$. Round the calculated value of this criterion expression to obtain the value of n (here $n = 0$), and the tolerance zone middle point is 0.0.

② When $0.5 < ((\Phi_2 - \Phi_1/4)/\Delta\Phi) < 1.5$, $n = 1$. Round the calculated value of this criterion expression to obtain the value of n (here $n = 1$), and the tolerance zone middle point is 1.0.

③ When $1.5 < ((\Phi_2 - \Phi_1/4)/\Delta\Phi) < 2.5$, $n = 2$. Round the calculated value of this criterion expression to obtain the value of n (here $n = 2$), and the tolerance zone middle point is 2.0.

④ When $2.5 < ((\Phi_2 - \Phi_1/4)/\Delta\Phi) < 3.5$, $n = 3$. Round the calculated value of this criterion expression to obtain the value of n (here $n = 3$), and the tolerance zone middle point is 3.0.

According to the above expression, the range matching adjustment procedure is that under static state and in the case of strong S or large Φ, ① measure the values of Φ_2 and Φ_1; ② work out the value of $\left(\frac{\Phi_2 - \Phi_1/4}{\Delta\Phi}\right)$ and adjust the phase Φ_2 of the secondary tone F_2 (or Φ_1 of the main tone F_1) to make the digits of the mean value after the decimal point get close to "0," namely, "XX · 00..."; and ③ round the value gained in the last step to obtain the "large number" n (or add "0.5" to the value of $\left(\frac{\Phi_2 - \Phi_1/4}{\Delta\Phi}\right)$ and then "round" it).

The above example is for the case of $n = 3$, namely, there are three times of ambiguity; in the same manner, the general condition where $n = N$ can be obtained. Under this condition, there is:

$$F_2 = \frac{F_1}{(N+1)} = \frac{F_1}{K_1} \tag{2.155}$$

where K_1 is matching ratio and $K_1 = N + L$.

$$\Delta\Phi = \frac{360°}{(N+1)} = \frac{360°}{K_1} \tag{2.156}$$

The general expression of the judgment expression is:

$$\frac{(\Phi_2 - \Phi_1/K_1)}{360^\circ/K_1} = \frac{K_1\Phi_2 - \Phi_1}{360^\circ} \tag{2.157}$$

After the above-mentioned phase matching, $(K_1\Phi_2 - \Phi_1)/360^\circ$ shall be at the middle point of tolerance range ($360^\circ/K_1$), and its margin at plus and minus directions is symmetrical. In this case, the probability of matching error caused by a normal noise is:

$$P(\Phi_n > \pi/K_1) = 1 - \mathrm{erf}\left(\frac{\pi}{K_1}\sqrt{\left(\frac{S}{N}\right)_L}\right) \tag{2.158}$$

where Φ_n is tone phase error (rad) generally determined by Φ_2, $(S/N)_L$ is tone loop SNR, and erf is error function.

Expression (2.158) also determines the probability of "mismatch error."

From Expression (2.158), generally speaking, the calculation expression of the large number n is:

$$n = 1NT\left[\frac{K_1\Phi_2 - \Phi_1}{360^\circ} + 0.5\right] \tag{2.159}$$

where $1NT$ is "rounding" operation and K_1 is matching ratio.

The total phase shift is:

$$\Phi = n \times 360^\circ + \Phi_1 \tag{2.160}$$

The corresponding range value is:

$$R = \frac{\Phi C}{720F} \tag{2.161}$$

If there are multiple secondary tones used to solve range ambiguity, the method can be employed in like manner. Total phase of F_1 is:

$$\Phi = (n_m K_{m-1} K_{m-1} K_{m-2} \cdots K_2 + n_{m-1} K_{m-1} K_{m-2} \cdots K_2 + \cdots + n_2 K_2 + n_1)360^\circ + \Phi_1 \tag{2.162}$$

where $K_2 = F_3/F_2, K_m = F_{m+1}/F_m$ is matching ratio; $F_1, F_2, \ldots$ are secondary tones arranged in order from high frequency to low frequency; generally, F_1 is the main tone which determines range accuracy; $n_1, n_2, \ldots n_{\mathrm{m}}$ are the numbers of complete cycles contained by each secondary tones, among which, n_1 is the number of complete cycles of F_1 contained by F_2, and n_m is the number of complete cycles of F_m contained by F_{m+1}.

The selection of the highest fine measuring frequency F_1 depends on the requirement for ranging accuracy. The stricter the requirement is, the higher the frequency shall be:

$$F_1 \geq \frac{C}{\left(4\pi\sigma_R\sqrt{P/N}\right)} \tag{2.163}$$

where c is light speed, σ_R is required ranging accuracy, and S/N is signal SNR.

The selection of the lowest coarse measuring frequency depends on the requirement for the largest nonambiguous range; namely, the further the range is, the lower the coarse measuring frequency F_L shall be:

$$F_L \leq \frac{c}{R_{\max}} \tag{2.164}$$

where c is light speed and $R_{\max}$ is the largest nonambiguous range (the sum of a two-way range).

When the nonambiguous range is required to be comparatively far, F_L is pretty low. The tone spectrum after carrier which is directly modulated by this low frequency tone will get very close to carrier spectrum. This will not only affect the normal tracking for carrier but also make it difficult to demodulate the tone signals with such a low frequency. Therefore, a folded tone scheme is commonly adopted for solving this problem, that is, to conduct amplitude modulation to a certain secondary tone, then use its upper sideband-folded spectrum to modulate a carrier. Taking a unified S-band TT&C system as an example, besides an 100 kHz main tone and a 20 kHz secondary tone, 4 and 20 kHz are folded to generate 16 kHz folded tone, and then the 16 kHz tone is folded with an 800 Hz tone in turn to generate 16.8, 16.16, 16.032, and 16.08 kHz folded tones, which are used together with the main tone to collectively perform phase modulation to the carrier.

The foresaid basic principle for de-ambiguity is based on simultaneously transmitting a group of tones and measuring n and fine measured phase Φ_1 (corresponding to n and Φ_1 at the same range) at the same sampling time. So this method is called "simultaneous multitone method."

(2) Large Number Accumulation Method

From Fig. 2.38, it is also clear that each time when F_1 produces an ambiguity, F_1's phase Φ_1 generates a $360° \rightarrow 0°$ step, making relative n be added by "1." Therefore, phase change of Φ_1 can be detected by making the large number n added by "1" (in reverse motion direction, subtracted by "1") each time when phase step occurs. This is called the "large number accumulation method." It is mainly used under a condition that range has been captured and initial range R_0 has been obtained, for which the large number n in subsequent range changes can be acquired in this method. As detection of Φ_1 is conducted between two sampling points, assuming sampling time is τ and target radial motion velocity is $v(t)$, the moving distance in time τ is $\tau v(t)$ corresponding phase change is $2\tau v(t) \times 360°/c$, the resulting number of phase steps producing phase ambiguity is decreased to $[360° - 2\tau v F_1 \times 360°/c]$, and the phase difference $\Delta\Phi$ between two sampling points shall be within the following scope:

$$\frac{(2\tau v F_1 \times 360°)}{c} < \Delta\Phi < \left[360° - \frac{(2\tau v F_1 \times 360°)}{c}\right] \quad (2.165)$$

so as to make a judgment that a phase ambiguity has occurred. But a proper threshold value Φ_p should be chosen within this scope. Considering that $\Delta\Phi$ may be with plus and minus error, if the error influence makes $\Delta\Phi$ go beyond the foresaid boundary scope values, wrong n added by "1" (or n decreased by 1) will happen, leading to mismatch error of the ranging data. Therefore, it is optimal to place threshold Φ_p in the middle of abovesaid scope, making both sides have the same error margin. Thus, there is

$$\Phi_P = \frac{[360° - (2\tau \upsilon F_1 \times 360°)/C] + [(2\tau \upsilon F_1 \times 360°)/C]}{2} = 180° \quad (2.166)$$

Therefore, when $|\Delta\Phi| > 180°$, a time of phase ambiguity can be judged to have occurred and the large number n shall be added by "1" (or subtracted by 1).

However, if Expression (2.165) cannot be satisfied, the "large number accumulation method" shall not be applied. For instance, if the frequency of F_1 is too high (or target velocity is too fast, or sampling time τ is too long), it is possible that Expression (2.165) may become false. In this case, it may make $(2\tau \upsilon F_1 \times 360°) > 180°$ which will lead to a wrong judgment. Moreover, this method also has following disadvantages:

1) As large numbers n and phase Φ_1 are not measured every time when range values are read and the "large number is accumulated," the happening of ranging mismatch error may not get aware of, accumulation may be going on, which cannot be found and corrected in a timely manner. But the above-mentioned "simultaneous multitone method" is free from this disadvantage.

2) As "large number accumulation" is a sort of open-loop measuring technique, it does not have automatic adjustment and correction functions as the closed-loop mode does.

3) When this method is adopted, the range capture scheme for acquiring initial range is usually different from the subsequent scheme for range tracking. Connection between both schemes is required, which is apt to introduce ranging error (e.g., time is not aligned), resulting in different ranging values to appear at each capture.

4) Wrong $\Delta\Phi$ measurement will bring about ranging mismatch error, and the random mistake in $\Delta\Phi$ measurement may result from several influences, such as dropped frames of ranging data, time difference caused by inaccurate timing, cycle slip of F_1 phase-locked loop, interference, and strong power supply fluctuation.

However, this scheme also has its advantages, which include simple equipment and less power consumption due to the fact that only fine tone is transmitted in ranging process.

(3) Time-sharing multitone method

In the above concurrent multitone method, it requires measuring the receiving/transmission phase shift Φ at the same range (i.e., at the same time) to obtain corresponding n and calculate the true range. Such multitone at the same time may occupy some power, and each tone needs simultaneous phase measurement, which causes complexity to equipment. Time-sharing multitone method is employed in engineering, and only one tone is transmitted for each time. The following two schemes are usually adopted.

1) Signal replication method

Ranging by signal replication method is to replicate the received tone signal by frequency division chain under high-tone driving, and then conduct non-ambiguity ranging with the lowest tone. Its working principle is as follows.

Under frequency scale synchronization, a tone generator produces tone frequency and transmits it from high to low in a time-sharing mode. However, the high tone is constantly transmitted and initial phase of low tone can be controlled and regulated. IF phase-modulated signals and reference signals from a receiver shall be firstly subject to coherent demodulation so as to obtain basic low tones in a time-sharing mode. High tone is purified by phase-locked loop and low tone is replicated successively by a frequency demultiplier. The tone generator compares the phase difference of each received and replicated secondary tone, regulates state of frequency division chain at each level to make each pair of tones in the same phase. In this way, the lowest secondary tone output by the frequency division chain is the replication of the received lowest tone and its time delay is equal to transmission delay of radio wave. When a range counter is formed through filling of a high-stability frequency clock and is enabled from the lowest transmitted tone and disabled at the lowest tone of frequency division chain, its output data will be the target delay of frontier moment for transmitting lowest tone. Sinusoidal function generator is employed but not frequency division chain method for transmission of primary tone, which can greatly increase the bandwidth of primary tone filter, improve delay stability and introduce no ranging drift error. Due to effect of dispersion from channel transmission, different tones create different delays. Initial phase of the transmitted tones shall be regulated to make the received tones aligned at the zero point so as to avoid frequency division chain capturing high error. Such adjustment is accomplished in system joint test.

Because the frequency division chain replicates and records all tones and range change is also recorded through driving of received primary tone to the frequency division chain, ambiguity solution for each tone at the same range point can be realized even though tones are transmitted in a time-sharing mode when the aforesaid method is adopted.

The nearest frontier alignment method is usually employed for phase matching zero method of system tones, which regulates transmitted tone phase on the basis of received frequency division chain to make positive direction of received tones aligned crossing the zero point. The specific process is as follows.

Firstly set the initial phase of transmitted tones as zero; frequency division chain is under random state after primary tone is Loop-locked. The correlator measures the phase difference $\phi_\iota - \phi_{0i}$ between echo tone and frequency division chain, which will be taken as neighboring frequency ratio after (360°/a) modulus and remainder $\Delta\phi_\iota, a$, then

$$\Delta\phi_\iota = (\phi - \phi_{0i})\mathrm{mod}[360^\circ/a](^\circ) \tag{2.167}$$

After regulating the transmitted tone phase with remainder $\Delta\phi_\iota$, received tone zero point can be aligned. Such zero method is to align the lower tone to the nearest positive direction zero point, i.e., near matching. As the equipment does not eliminate the additional phase shift of high tone, the system may have high zero-range set constant (e.g., several kilometers). Maybe it is strange in theory, the second step of matching zero method is to eliminate high zero-range set constant. Its method is to calculate and determine which low tones will be given high phase shift from the system according to current zero-range set constant, change the initial phase of several transmitted tones through keyboard (change value is one or several phase interval (360°/a), and move forward the received low tone across the zero point to decrease zero-range set constant.

Feature of such signal replication method: use received primary tone to promote frequency divider and make phase matching by correlational method to replicate lowest tone signals. These lowest tone signals follow primary tone signals for successive tracking and possess the accuracy of primary tone phase as well as period of lowest tones, thereby realizing non-ambiguity and high-accuracy ranging at a long distance.

Its defect: Constant transmission of primary tone requires transmission of two tones at each time. Although capture time may be shorter, transmission energy is bigger. One possible improvement method is to correlate transmission tone with transmission carrier. If correlation ratio of primary tone and carrier is m_1, carrier correlation transmission ratio of transponder is m_2, then the correlation ratio of primary tone and received carrier is $m_1 \times m_2$, conduct $m_1 \times m_2$ time of frequency division of received carrier to make signal replication and ranging according to above method after received primary tone. But phase ambiguity from $m_1 \times m_2$ frequency division can be solved by real-time measurement on the phase difference of frequency division primary tone and received primary tone. Apply it into range calculation formula to obtain no-fuzzy range and the accuracy of high tone is determined by carrier C/N_0. The phase difference of high tone ambiguity solution can be decreased by increasing integral time to obtain high ranging accuracy. Such method can make time-sharing transmission of sequence tone, but memory and driving of tone is replaced by carrier frequency division.

2) Phase measurement calculation method

Such method makes time-sharing measurement of receiving/transmission phase shift of each tone and calculates ambiguity solution to obtain range value.

For example, transmitted tones are folded tones of 100, 20, 16, 16.8, 16.16, 16.032 Hz, and 16.008 kHz. No need to restore 16.008 kHz to 8 Hz at receiving end. We can directly measure 16.008 kHz phase shift, minus 16 kHz phase shift to obtain 8 Hz phase shift:

Sum frequency phase shift: $\Phi_{\Sigma n} = (\omega_B + \omega_n)\tau = \omega_B\tau + \omega_n\tau$

Base frequency phase shift: $\Phi_B = \omega_B\tau$

Secondary tone f_n phase shift: $\Phi_n = \Phi_{\Sigma n} - \Phi_B = \omega_n\tau$

It can directly measure phase shift at sum frequency, which is one advantage of tone phase calculation method.

Range finder transmits one tone for each time under time-sharing mode. After lowest tone of phase loop is locked, it transmits from low to high and finally accomplishes locking of primary tone. In case of digital signal processing, phase measurement can be done through digital phase-locked loop. Tone generator produces primary tone and various secondary tone waveforms, the order of which are sent to uplink modulator for carrier modulation.

Tone digital signal from receiver is sent to tone extracting loop for tone phase extraction and measurement of receiving/transmission phase difference. Tone frequency is produced and extracted from low to high and the receiving/transmission phase shift measurement of such tone is done after indication of tone extraction loop-lock is stable, phase ambiguity resolution is done in order. Receiving/transmission phase difference is sent to data processing computer for calculation of tone phase ambiguity resolution and calculation of target range. Time scale of ranging data is the time of tone receiving end.

Capture time of ranging depends on the locking time of tone loop. Wideband loop capture and narrowband loop tracking is employed to shorten capture time and ensure ranging accuracy. Tone loop employs Doppler compensation, which guides rapid locking of tone loop by use of Doppler information from R&C unit.

Range capture procedures are as follows:

① Process of tone transmission. In case of range capture, tone transmission is done in order and there is only one tone at any time. For transmission of one tone, wait for loop-locking of received tone and delay for some time; make *R*/*T* phase shift measurement after loop DCO phase is stable.

Firstly transmit the reference frequency of combined tone $F_1 = 16$ kHz, then transmit tone frequency from low to high one by one, so as to calculate the secondary tone phase shift from combined tone phase shift and provide phase ambiguity resolution in order.

② Phase calibration. Calculate and store the *R*/*T* phase zero of *R*/*T* equipment channel to tones, i.e.,

$$\theta_{0i} = \theta_{\text{Trans}0i} - \theta_{\text{Reciev}0i}, \quad i = 1 \sim 7 \tag{2.168}$$

where $\theta_{\text{Trans}0i}$ is the sample value of tone phase of transmitted tone generator in case of phase calibration; $\theta_{\text{Receiv}0i}$ is the sample value of tone phase of received tone loop DCO in case of phase calibration.

In case of range capture on target, tone *R/T* phase shift caused by target range delay is:

$$\varphi_i = \theta_i - \theta_{0i}, \quad \mathrm{i} = 1 \sim 7$$

where

$$\theta_i = \theta_{\mathrm{Trans}i} - \theta_{\mathrm{Recievi}}, \quad i = 1 \sim 7$$

③ Transmit 7 tones in order and record *R/T* phase shift at corresponding moments of $t_1 \sim t_7$. For ambiguity solution at the same range point, conduct range matching calculation after obtaining phase shift value of 7 tones at t_7. Phaseshift increment $\Delta\varphi_{di}$ from target movement in time periods of $t_1{-}t_7$ shall be calculated.

$$\Delta\phi_{di} = \int_{t_i}^{t_7} f_d(t)\mathrm{d}t \tag{2.169}$$

where $f_d(t)$ refers to Doppler frequency.

Expression (2.146) is called Doppler integration, which calculates $\Delta\phi_{di}$ and corresponding range variation value. Specific implementation methods include Doppler frequency single- point sampling method, multi-point average method, fragment accumulation method, in which fragment accumulation method has higher accuracy.

R/T phase shift of combined tone at the moment of 6 is:

$$\Phi_{\Sigma i} = \phi_i + \Delta\phi_{di}, \quad i = 1 \sim 7$$

④ Calculate the *R/T* phase shifts of basic tones.

R/T phase shift is $\Phi_1 = \Phi_{\sum 1} - \phi_1$ for $f_1 = 8$ Hz; *R/T* phase shift is $\Phi_2 = \Phi_{\sum 3} - \varphi_1$ for $f_2 = 32$ Hz; *R/T* phase shift is $\Phi_3 = \Phi_{\sum 3} - \varphi_1$ for $f_3 = 160$ Hz; *R/T* phase shift is $\Phi_4 = \Phi_{\sum 4} - \varphi_1$ for $f_4 = 800$ Hz; *R/T* phase shift is $\Phi_5 = \Phi_{\sum 5} - \varphi_1$ for $f_5 = 4$ kHz; *R/T* phase shift is $\Phi_6 = \varphi_6$ for $f_6 = 20$ kHz; *R/T* phase shift is $\Phi_7 = \varphi_7$ for $f_7 = 100$ kHz.

Feature of phase measurement calculation method: discrete capturing of phase and Doppler data of each tone in order. As ambiguity solution calculation shall be done at the same time, Doppler integration method is employed to calculate the phase of each tone to arrival time of final tone, and then conduct ambiguity solution calculation, which causes following problems.

① Doppler integration is impacted by measurement accuracy, sampling interval, integration time and integration algorithm, there is error for such phase calculation mode. When time is longer, speed and acceleration is higher, its error is bigger, which may cause matching mistake and mismatch error for range capture. There is serious defect for ranging of long range of deep space, etc. and high-speed target.

② Its phase measurement is accomplished by phase-locked loop. As there is such special problem as capture time, capture bandwidth, cycle slip for phase-locked loop, it may cause phase measurement error and matching error when design is improper (capture time is not sufficient).

③ Time difference, interference, etc. from frame-falling, inaccurate time scale may cause capture mistake. Its advantage is to transmit one tone at each time.

Doppler integration ranging in Expression (2.169) is also one range determination method, which has two problems: firstly, starting value of range must be known; secondly, Doppler integration error, the longer the integration time lasts, the bigger the accumulation error becomes. It can calculate range increment of high accuracy and high resolution to short time period by use of velocity integration with known original value of range and be applied to following cases.

① When ranging time and data sampling time is different and there is difference of AT time interval, make Doppler integration in AT to obtain the range value at sampling time.

② When the time interval of ranging is long (e.g., PN code period is long) and there is high requirement for data sampling rate, Doppler integration ranging can be used for high resolution compensation between two ranging moments.

③ It can modify the propagation delay error of ionosphere and plasma.

2.4.1.2 Pseudorandom Code (PN Code) Ranging System

Tone ranging method needs to use multiple tone signals and it has a disadvantage that the phase de-ambiguity will get very complicated. There is another approach called PN code ranging which uses only one signal rather than multiple signals. It makes use of long period of the PN code to solve ambiguity problem and takes code clock as accurate measurement signal similar to high tone. This is the original intention when the PN code ranging was conceived. With the development of technology, now it has also been endowed with other functions such as anti-interception, anti-interference, spread spectrum and code division multiple access (CDMA).For ranging, phase of transmitted carrier is under 0 or π phase modulated by binary PN code and target reflected echo is received by the receiver, in which there is also a same PN code generator served for reproducing the received PN code. The local pseudo code is multiplied with the modulated IF signal of the received pseudo code in a multiplier, which equivalently performs a modulation of 0 or π phase shift on the input received signal. If local pseudo code is completely consistent with input modulation pseudo code in time, the occurrence order of 0 and 1 will be exactly the same (reproduced) and the original 0 or π modulation will get removed by such counter modulation. So the receiving multiplier will eliminate the pseudo code modulation on input carrier to make the input signal become a carrier, which then can be extracted by a narrow band-pass filter. If the local pseudo code is not aligned to the modulated PN code of input signal in time, the multiplier cannot remove the modulation on PN code and will add a new modulation, so signal carrier cannot be restored and the band-pass filter will have no carrier signal output. The multiplier and band-pass filter constitute the IF correlator of PN code modulation. Output signal of the band-pass filter depends on the time relationship between local

PN code and input PN code. Analysis shows that when the time difference of 2 PN codes is more than 1 code element, output is close to zero ($1/p$ of the maximum output, where p is the number of code elements in a pseudo code period, called code length); when time difference is zero, i.e., when two sequences are completely aligned, output is maximum; when time difference is within plus/minus 1 code element, correlation value changes with time difference. In one sequence period, correlation value has obvious output only in one code element width, having a form like a triangle. Other time output is zero.

Figure 2.39 is the diagram of autocorrelation function of an m-sequence pseudo code.

The shape of the correlation peak is like a triangle. Its hypotenuse shows that in one code element width the correlation value is linearly related with time, which can be used to constitute the time discriminator for PN code tracking loop. Such discrimination feature is as shown in Fig. 2.40. In Fig. 2.40, where Δ is code element width, the voltage level change caused by time change is used to control local PN code phase to make it synchronous to input PN code.

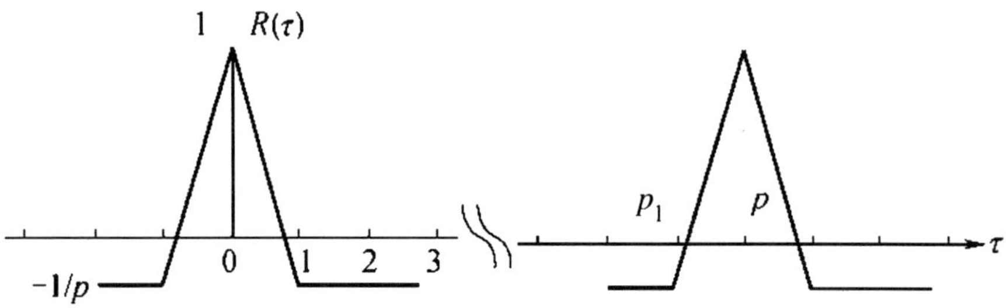

Fig. 2.39 Autocorrelation function of an *m*-sequence pseudo code

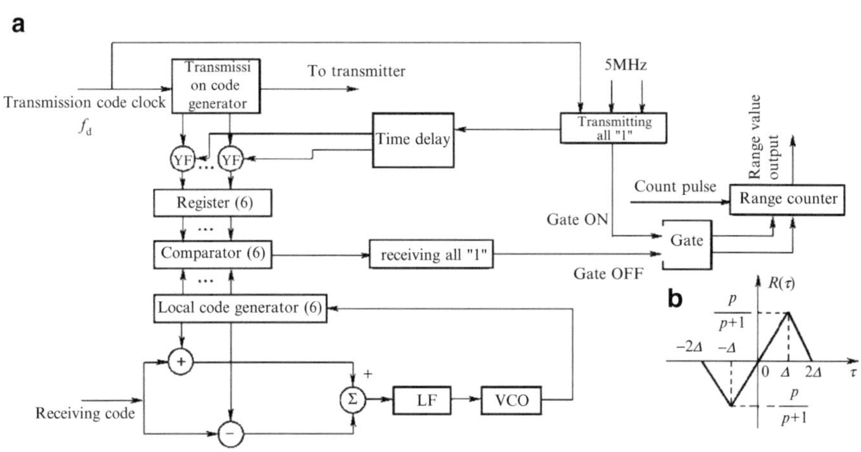

Fig. 2.40 Principle of pseudo code ranging

As the band-pass filter bandwidth of correlator is narrow and bandwidth of PN code loop is also narrow, it can filter much input noise. PN code tracking loop composed of correlator can work under low SNR environment and accomplish ranging task.

Its work process can be divided into following four steps.

(1) Use transmission end PN code to modulate carrier and transmit it to target. After sampling pulse for ranging arrives, its rising edge is used to put 0 and 1 states of all levels of shift register in transmission end PN code generator into the latch. Shift register states represent the phase of PN code sequence. At the same time, sampling pulse rising edge is used as transmitting "1" pulse to make range counter counts code clock.

(2) Produce a local code identical to transmitting code but with a certain delay to it at reception end, and make frequency and code sequence of the local code clock synchronous to receiving code, thus the local code can be served as the reproducing signal of receiving code. Track the delay change of input signal by PN code tracking loop. As both the initial states of local code and receiving code are random, a capture circuit is needed to roughly align local code and receiving code within one code element. Discriminator will have error output after the time difference is within the linear area of discriminator's characteristic curve. After filtering, the error output is used to control VCO, making local code align with receiving code and so realizing synchronous tracking, i.e., composing a delay locked loop.

(3) After synchronization, the states of each level of shift register in local PN code generator, which represent the phase of local PN code sequence, are sent to a multi-digit comparator for comparing with latched states of the shift register in transmission end. With the time passing, when the multi-digit states of local code match with latched transmitting states, the comparator will output a matching pulse as receiving "1" pulse to make range counter stop counting. The time interval between two pulses of transmitting "1" and receiving "1" is the time delay of local code relative to transmission end, which represents transmission delay of target echo.

(4) The counting pulse number N of code clock of range register in one Gate ON – OFF period is coarse-range and fine range is determined by code clock phase Φ_0. The range can be expressed as:

$$R = \frac{(N \cdot 360^\circ + \Phi)c}{720F} \tag{2.170}$$

where F is code clock frequency.

The pseudo code ranging can use either single code or compound code.

Compound code is composed of several single codes (also called subcode) of different code length in some logic combination. The code lengths of subcodes are better to be coprime (relatively prime) and closer with each other for the most. For example, in US "Apollo" TT&C system, ranging code was a compound code composed of five subcodes, whose code lengths were: $CL = 2$, $A = 11$, $B = 31$, $C = 63$, and $D = 127$.

As code lengths of subcodes are relatively prime, the code length of the compound code is the least common multiple of subcode code lengths (product of each subcode's code length), i.e.,

$$P = \prod_{i=1}^{n} p_i \tag{2.171}$$

For the "Apollo" example, $P = 2 \times 11 \times 31 \times 63 \times 127 = 5{,}456{,}682$

As can be seen, the code length of such compound code is long enough so that the corresponding maximum unambiguous range is farther than the distance to the Moon.

Capture time of compound code is the sum of individual capture time of each subcode, i.e., if each subcode captures, the whole compound code captures.

$$T_{\text{compound}} = \left(\sum_{i=1}^{n} p_i \right) \tau_0 \tag{2.172}$$

where p_i is code length of subcode; n is the number of subcodes composing the compound code; τ_0 is the capture time of each code element. The capture time of a single pseudo code with the same code length as the compound code is:

$$T_{\text{single}} = P\tau_0 = \left(\prod_{i=1}^{n} p_i \right) \tau_0 \tag{2.173}$$

As $T_{\text{single}} \geq T_{\text{compound}}$, in general pseudo code ranging system the signal form of single pseudo code is rarely used. The ranging accuracy of pseudo code depends on code element width (corresponding with code clock). The smaller the code element width is, the smaller the ranging error will be.

2.4.1.3 Hybrid Ranging System of Code and Tone

Pure tone ranging and pseudo code ranging have their respective advantages and disadvantages. Main advantages of pure tone ranging are that the high-accuracy ranging can be achieved by raising tone frequency, and its occupied bandwidth is rather narrow and capture is fast; but its phase de-ambiguity is complicated. Main advantage of pseudo code ranging is the long non-ambiguity range; but code element width must be decreased to raise accuracy, which increases occupied bandwidth, makes code capture more complicated, and takes longer time. Hybrid system of pseudo code and tone can make them have complementary advantages. Its basic idea is to use high tone for accuracy, use pseudo code for range de-ambiguity, and use pseudo code signal for PSK demodulation on high tone.

There are following three schemes for hybrid system of code and tone.

(1) Pseudo code + high tone

It is a common scheme. It makes the pseudo code be PSK modulated on high tone and selects a higher frequency for high tone to improve ranging accuracy. Its basic principle is the same as above-mentioned pseudo code modulated on carrier, and so it is not necessary for more details here.

(2) Spread spectrum code + high tone

For composing ranging signal, PN code is firstly used to spread the spectrum of the low rate de-ambiguity code (Barker code).The spread-spectrum signal is used to conduct BPSK modulation on 1 MHz high tone and then phase modulation on carrier. Carrier modulation sideband is used for ranging and residual carrier is used for velocity measurement. Spread-spectrum ranging signal has strong anti-interference capability and has CDMA capability in the case of multi-station working at the same time.

The ratio of PN code clock frequency to high tone frequency can be selected to be rather small, so it is easy to implement complete cycle de-ambiguity of the high tone by code clock phase and decrease the accuracy requirements on PN code clock phase.

De-ambiguity code for maximum range is a 13-bit Barker code, which is used for PN code phase de-ambiguity. If the non-ambiguity range is required to be farther, the 16-bit de-ambiguity code can be selected. De-ambiguity code shall have sharp self-correlation peak and low side-lobe.

By contrast, in some TT&C systems which adopt hybrid system of tone and code, when pseudo code uses multiple code, signal power will be occupied, causing signal power dispersion, reducing ranging accuracy and de-ambiguity capability, and capture time of multiple code structure is hard to be shortened.

In such scheme, Barker code is spectrum-spread by pseudo code, reception end accomplishes code tracking and recovery, high tone is directly modulated by pseudo code, which not only improves anti-interference capability, but also makes full use of signal energy. High tone, pseudo code, and Barker code do not separately occupy signal energy. Correlation receiver of de-ambiguity code uses the signal energy in 13 pseudo code cycles, which makes correlation peak for range de-ambiguity have higher output SNR to ensure no mismatch error occurs during ranging.

Meanwhile, single code modulation makes code capture process simple and code capture time can be shortened by semi-parallel processing of multiple correlators. The capture time of N correlators in parallel can be N times shorter than the capture time of N correlator in serial.

After PN code is captured, the arrival time of correlation peaks can be detected by de-ambiguity hardware of matching filters. Ranging process is simple and capture time is composed by two parts of PN code capture and tone loop-locking.

Its composition block diagram is as shown in Fig. 2.41. Ranging terminal is composed of ranging signal generator, ranging signal extractor, range delay measurement unit, ranging computer, etc. Ranging signal extractor is composed of PN code parallel detector, PN code tracking loop, high tone loop, and Barker code matching filter.

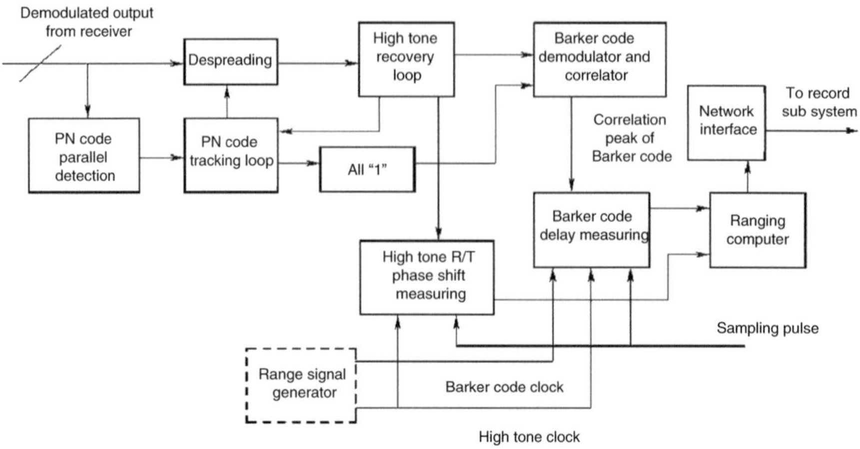

Fig. 2.41 Composition block diagram of ranging terminal

The uplink carrier modulated by ranging signal is returned to ground receiver by transponder. After demodulation there is much noise overlapped in the ranging signal and so filtering is needed to recover the pure ranging signal for measuring the *R*/*T* delay difference, which is sent to ranging computer to calculate target radial range and then sent to data record subsystem through network interface.

In measuring range through delay, both Barker code delay measurement and high tone *R*/*T* phase shift measurement are employed for facilitating measuring and obtaining high resolution. Barker code delay is used to measure coarse range (quantization unit is the cycle number of high tone) and *R*/*T* phase shift of high tone is used to measure fine range. PN code clock is used for phase de-ambiguity of high tone.

(3) Sequence code + high tone

The uplink ranging signal of such scheme includes high frequency tone f_r and sequence codes $r_n(t)$ from its subharmonic waves, which are phase-modulated on the tone. The resulting ranging signal can be expressed as:

$$s_{ur}(t) = \cos\left[2\pi f_r t + \Phi_r r_n(t)\right] \tag{2.174}$$

1) Ranging code

The de-ambiguity signal $r_n(t)$ is produced by subharmonic wave of the ranging tone, its period is:

$$T_n = \frac{2^n}{f_r} \quad 0 \le n \le 20 \tag{2.175}$$

where 2^n is code length and f_r is ranging signal frequency.

Code $r_n(t)$ is a bipolar (±1) periodic rectangular function:

$$r_n(t) = Q_1 \oplus Q_2 \oplus Q_3 \oplus \cdots Q_n \tag{2.176}$$

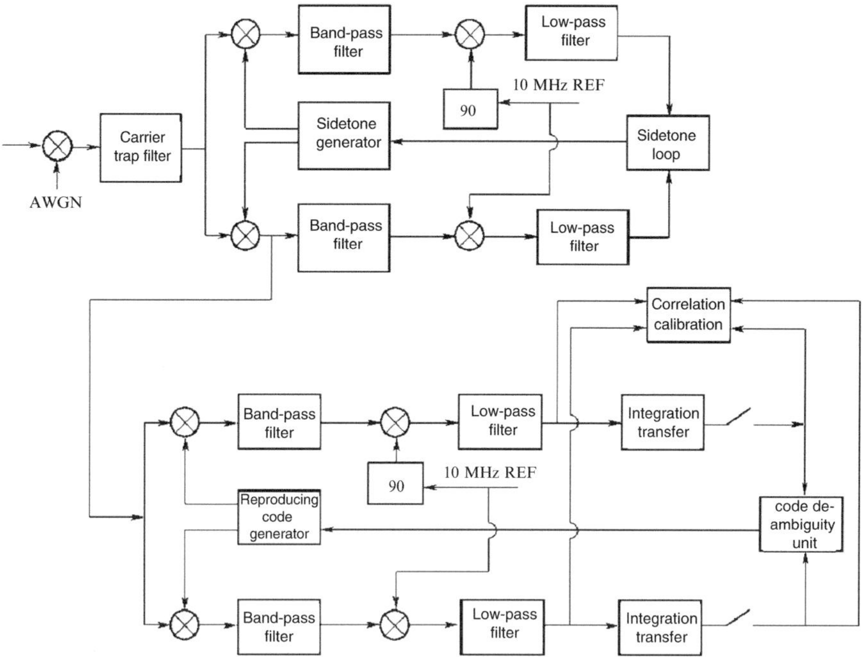

Fig. 2.42 Terminal block diagram

where Q_i can be the output of 1:2 frequency dividing trigger link. Symbol is used to represent sum of "Module 2."

The first trigger of frequency dividing link is driven by the ranging tone. Code signal $r_n(t)$ can also be represented by a recursion expression:

$$r_n(t) = \begin{cases} r_{n-1}(t), & 0 \le t \le T_n/2 \\ -r_{n-1}(t - T_n/2), & T_n/2 < t \le T_n \end{cases} \tag{2.177}$$

2) Working principle

Figure 2.42 shows a block diagram of IF correlation process. The upper branch is a PLL, while the lower branch is a ranging signal reproducer.

10 MHz IF signal from coherent receiver is sent to a narrowband trap filter for reducing the unwanted carrier signal in input end. Following the trap filter is a phase quadrature tone demodulator, which multiples the input modulated IF carrier with in-phase and quadrature phase local tone signal. It is then multiplied by in-phase carrier component recovered by carrier loop and thus is converted to baseband.

The digital tone phase-locking loop filters filter out phase noise, the filtered signal is sent to control tone frequency. A frequency compensation generated by measured carrier Doppler is also sent to tone generator and makes such phase-locking loop work in narrow loop bandwidth. Send signal after quadrature tone component correlator to code correlator. In code correlator, signal is firstly multiplied with local I and Q reproducing codes and then converted to baseband after multiplied with recovered carrier.

3) Capture process

The capture process is similar to above "signal reproducing method" and starts from transmitting and receiving tone only. The phase-locked tone is received in a preset time interval. It is used to reproduce sequence codes and the first code is used to modulate the uplink tone. As the first code is synchronous with tone and its period is 4 times of tone period, correlator can find out 4 phase ambiguities of the 4 frequency divider and solve the ambiguities in-phase or out-off-phase with tone, which are used to adjust the state of frequency divider to make reproducing code in the same phase as the input code. Then the second code is used to replace the first code, which can make alternate judgment on ambiguity again. In order to reduce the necessary correlation times of range de-ambiguity, 4 frequency division step is employed for code cycle. Based on correlation feature of 2 quadrature codes, we can use 2 correlators to calculate the 4 phase ambiguities of the 4-frequency divider. At the end of capture, the longest reproducing code is the same as longest received code, which determines the maximum non-ambiguity range. Sequence code is also transmitted to detect whether there is ranging error such as mismatch error. As it is required only for testing, the needed energy is small. Modulation degree of sequence code on tone can be decreased.

4) Ranging

Uplink code generator produces "start pulse" and reduplication code generator produces "end pulse." The time interval between start pulse and end pulse is measured by a high resolution timer/counter. Ranging accuracy is determined by high tone while non-ambiguity range is determined by longest code.

2.4.2 Vector Analysis Method of Ranging Error

Ranging accuracy is determined by sinusoidal wave signal of the higher frequency (e.g., high tone, code clock, subcarrier signal), so the accuracy of sinusoidal wave signal is the basis of analyzing CW ranging accuracy and it will get analyzed and calculated in this section. For other ranging signal derived from sinusoidal wave signal, such as hybrid signal of code and tone, its ranging spectrum is broadened to a frequency band, whose amplitude-/phase-frequency characteristics will introduce additional ranging system errors. In this case, the total system error can be treated as ranging error of a single sinusoidal wave attached with a loss factor. In case of pure PN code or spread-spectrum signal, the system error resulted from zero change of code loop time discriminator shall be considered.

Based on above reason, this section will focus on analyzing the ranging error of sinusoidal wave ranging signal. Ranging error includes system error and random error.

System error of equipment includes error caused by *R*/*T* channel delay change, range calibration residual error, dynamic error, multipath effect error, and transmission line delay error, among which the error resulted from delay is the fundamental component, especially for the transponder working in space, which may have a considerable delay change during its several years of life cycle. It is the main factor that restricts the increase of equipment accuracy.

As group delay is derived from the differential of phase-frequency characteristic, phase-frequency characteristic is of first importance in network characteristics. So this book will first analyze the effect of frequency characteristics on ranging accuracy.

(1) Frequency characteristics analysis methods of ranging error

Nowadays with the development of digital technology, the low-pass filter has been digitalized and its impact has become ignorable. So the following will only analyze the impact of band-pass frequency response on ranging error.

Tone delay resulted from channel phase-frequency characteristic, amplitude-frequency characteristic, and non-orthogonal product demodulation [11].

Channel phase-frequency characteristic and amplitude-frequency characteristic mentioned in this section refer to the frequency characteristics of any part in the channel (e.g., filter, modulator, AGC attenuation network), or frequency response introduced by any abnormal condition (e.g., effect of bad matching), or the total frequency characteristic, or asymmetry of signal spectrum line or phase shift caused by other factors.

Let the sinusoidal tone signal input to filter is:

$$B_i(t) = B_{im} \sin \Omega t \tag{2.178}$$

Carrier for tone phase modulation:

$$A_{ip}(t) = \sin\left[(\omega_0 t) + m \sin(\Omega t)\right] \tag{2.179}$$

where ω_0 is carrier frequency, Ω is tone frequency, m is phase modulation index.

For convenience sake, suppose the initial phases of carrier and tone both are 0, carrier amplitude $A_m = 1$, then Bessel expansion of Expression (2.179) is:

$$\begin{aligned} A_{ip}(t) = & J_0(m) \sin \omega_0 t + J_1(m) \sin(\omega_0 + \Omega)t - J_1(m) \sin(\omega_0 - \Omega)t \\ & + J_2(m) \sin(\omega_0 + 2\Omega)t + J_2(m) \sin(\omega_0 - 2\Omega)t + \cdots \end{aligned} \tag{2.180}$$

There are $(\omega_o \pm n\Omega)$ side frequencies but we only take the first order side frequency for discuss. The reason is that in actual scheme, narrowband filter with center frequency at Ω is used after product demodulation, so only Ω component is extracted and other 2Ω, 3Ω,..., $n\Omega$ components are all filtered. In Expression (2.180), only first order sideband component is the useful component. In case of small modulation coefficient, components above first order are rather small and their combined components produced after passing through non-linear system is even smaller, so taking first order component only into consideration is allowed by engineering design practice. When only first order component is taken, mathematic processing is simple and Expression (2.180) becomes

$$A_{ip}(t) = J_0(m) \sin \omega_0 t + J_1(m) \sin(\omega_0 + \Omega)t - J_1(m) \sin(\omega_0 - \Omega)t \tag{2.181}$$

After passing through a four-terminal network with amplitude-frequency characteristic $H(\omega)$ and phase-frequency characteristic $\Phi(\omega)$, such phase-modulated wave is to be product demodulated (i.e., sent to a multiplier after digitized). Phase demodulation model is as shown in Fig. 2.43. Corresponding $H(\omega)$ and $\varphi(\omega)$ characteristic curves are as shown in Fig. 2.44.

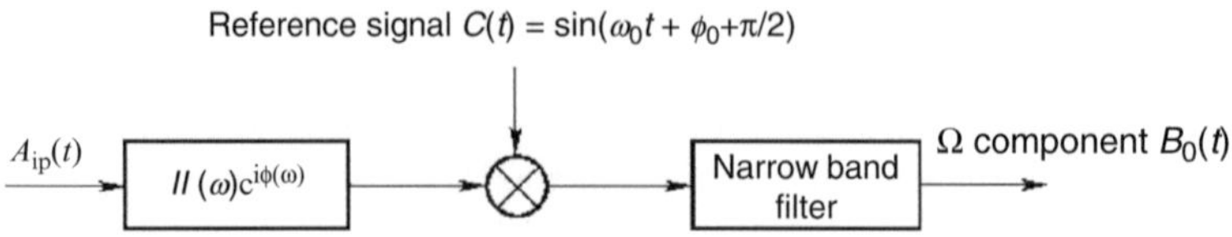

Fig. 2.43 Phase demodulation model

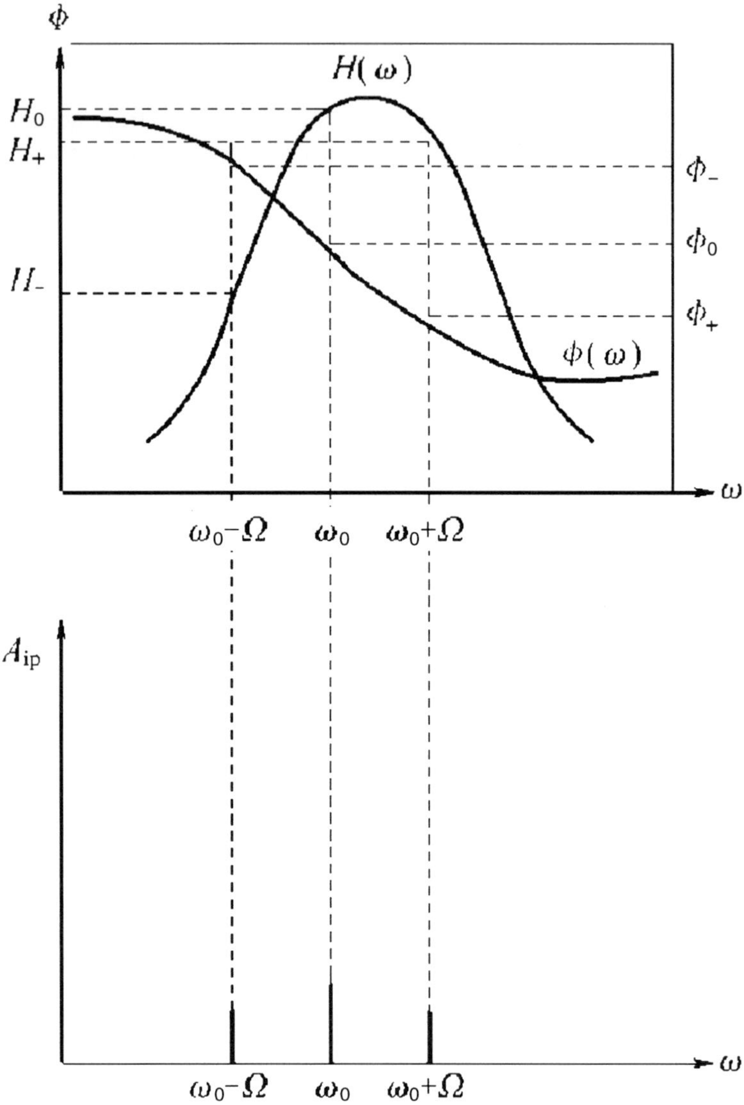

Fig. 2.44 Curves of amplitude-frequency characteristic and phase-frequency characteristic

Obviously, spectrum lines will not be symmetrical after being transmitted through the four-terminal network with amplitude-frequency characteristic and phase-frequency characteristic as shown in Fig. 2.44. Its output signal will be:

$$A_{\mathrm{op}}(t) = H_0 J_0(m)\sin(\omega_0 t + \phi_0) \\ + H_+ J_1(m)\sin\left[(\omega_0 + \Omega)t + \phi_+\right] H_- J_1(m)\sin\left[(\omega_0 - \Omega)t + \phi_-\right] \tag{2.182}$$

Product demodulation is to realize

$$A_{\mathrm{op}}(t) \times C(t)$$

where $C(t)$ is the reference signal that is correlated to and orthogonal with residual carrier $H_0 J_0(m)\sin(\omega_0 t + \Phi_0)$.

In case of no strict orthogonal intersection, there is non-orthogonal phase difference $\triangle\Phi$

$$C(t) = \sin\left[(\omega_0 t + \phi_0) + \frac{\pi}{2} + \Delta\phi\right] \tag{2.183}$$

where $\pi/2$ is orthogonal requirement, $\triangle\Phi$ is non-orthogonal phase difference, which is caused by static phase error, dynamic phase error, phase jitter, circuit phase shift, etc., of residual carrier phase-locked loop.

$$A_{\mathrm{op}}(t) \times C(t) = \frac{1}{2} H_0 J_0(m)\cos\left(-\frac{\pi}{2} - \Delta\phi\right) - \frac{1}{2} H_0 J_0(m) \cdot \\ \cos\left[2\omega_0 t + 2\phi_0 + \frac{\pi}{2} + \Delta\phi\right] + \frac{1}{2} H_+ J_1(m) \cdot \\ \cos\left[\Omega t + (\phi_+ - \phi_0) - \Delta\phi - \frac{\pi}{2}\right] - \frac{1}{2} H_+ J_1(m) \cdot \\ \cos\left[(2\omega_0 + \Omega)t + \phi_+ - \phi_0 + \frac{\pi}{2} + \Delta\phi\right] - \frac{1}{2} H_- J_1(m) \cdot \\ \cos\left[\Omega t + (\phi_0 - \phi_-) + \Delta\phi + \frac{\pi}{2}\right] + \frac{1}{2} H_- J_1(m) \cdot \\ \cos\left[(2\omega_0 + \Omega)t + \phi_0 + \phi_- + \Delta\phi + \frac{\pi}{2}\right] \tag{2.184}$$

Take and arrange Ω component after narrowband filtering:

$$\begin{aligned} B_0(t) = & \frac{1}{2}H_+J_1(m)\sin\left[\Omega t + \left(\phi_+ - \phi_0\right) - \Delta\phi\right] \\ & + \frac{1}{2}H_-J_1(m)\sin\left[\Omega t + (\phi_0 - \phi_-) + \Delta\phi\right] \end{aligned} \tag{2.185}$$

Expression (2.185) represents: two components are created for Ω frequency after demodulation. One component creates $[(\phi_+ - \phi_0) - \Delta\phi]$ phase shift for input tone signal $B_i(t) = B_m \sin \Omega t$, the other component creates $[(\phi_0 - \phi_-) + \Delta\phi]$ phase shift. Output tone $B_0(t)$ is the sum of such two components.

For Fig. 2.44, its vector diagram is as shown in Fig. 2.45.

As shown in Fig. 2.45, Φ is the phase shift of $B_0(t)$ for $B_i(t)$, i.e., tone phase shift. In such vector relationship, we can clearly see physical significance of Φ and reason of Φ change.

According to Fig. 2.45 and by use of simple triangle relation, we can obtain:

$$\phi = \arctan\frac{H_-\sin\left[(\phi_0 - \phi_-) + \Delta\phi\right] + H_+\sin\left[\left(\phi_+ - \phi_0\right) - \Delta\phi\right]}{H_-\cos\left[(\phi_0 - \phi_-) + \Delta\phi\right] + H_+\cos\left[\left(\phi_+ - \phi_0\right) - \Delta\phi\right]} \tag{2.186}$$

$$B_{om} = \frac{J_1(m)}{2}\left\{H_+^2 + H_-^2 + 2H_+H_-\cos\left[(\phi_0 - \phi_-) - \left(\phi_+ - \phi_0\right) + 2\Delta\phi\right]\right\}^{\frac{1}{2}} \tag{2.187}$$

Output tone:

$$B_0(t) = B_{om}\sin\left(\Omega t + \phi\right)$$

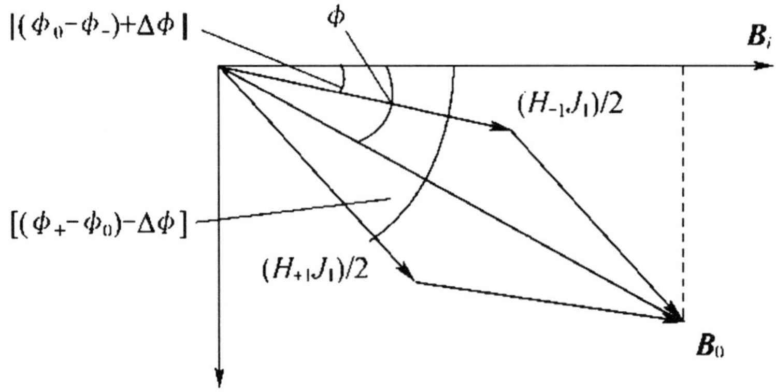

Fig. 2.45 Vector relationship of tone components output from product demodulator

It can be converted into tone delay τ from phase shift:

$$\begin{aligned}\tau &= \frac{\phi}{\Omega} \\ &= \frac{1}{\Omega}\arctan\frac{H_- \sin\left[(\phi_0 - \phi_-) + \Delta\phi\right] + H_+ \sin\left[(\phi_+ - \phi_0) - \Delta\phi\right]}{H_- \cos\left[(\phi_0 - \phi_-) + \Delta\phi\right] + H_+ \cos\left[(\phi_+ - \phi_0) - \Delta\phi\right]}\end{aligned} \tag{2.188}$$

where ϕ unit is rad, Ω unit is rad/s, τ unit is s.

Expression (2.188) is the complete expression of tone delay, which is the basis of tone ranging analysis. On the basis of derivation process of Expression (2.188), phase difference of two side frequency phase shift to carrier phase shift and asymmetry of two side frequency amplitude may impact tone delay, which are impacted by amplitude-/phase-frequency characteristics of filter, as well as spectrum line asymmetry from nonlinear, parasitic amplitude modulation, etc.

Expression (2.188) represents that in addition to phase-frequency characteristics, there are following factors that may impact tone delay.

1) Symmetry (H_+/H_-) of two-side frequency, in case of even symmetry (constant value of amplitude-frequency characteristics is a special case)

Substitute $H_+ = H_-$, $H_+/H_- = 1$ into Expression (2.186) and obtain:

$$\phi = \arctan\frac{\phi_+ - \phi_-}{2} = \frac{\phi_+ - \phi_-}{2}$$

According to the above expression, φ is not related to $\Delta\varphi$. When amplitude-frequency characteristics are symmetrical, non-orthogonal product demodulation does not introduce tone delay. But according to Expression (2.187), output tone amplitude will be reduced, SNR is reduced and random error is increased. If there is amplitude-phase conversion in ranging unit, it may cause ranging error.

2) Odd symmetry ($\varphi_+ - \varphi_0 = \varphi_0 - \varphi_-$) of phase-frequency characteristics and quadrature demodulation ($\Delta\phi = 0$), substitute into Expression (2.186) to obtain

$$\phi = \arctan\frac{\sin(\phi_+ - \phi_0)}{\cos(\phi_+ - \phi_0)} = \frac{\phi_+ - \phi_-}{2} \tag{2.189}$$

In case of quadrature demodulation, amplitude-frequency characteristics does not impact tone delay.

Such (1) and (2) are for most actual cases, under most cases, $\phi = (\phi_+ - \phi_-)/2$, $\tau = (\phi_+ - \phi_-)/2\Omega$

3) In the above cases, if phase-frequency characteristic has an odd symmetry and linearity, refer to Fig. 2.46 for linear phase-frequency characteristics one can get

$$\tau = \frac{\phi_+ - \phi_-}{2\Omega} = \tan\theta = \frac{d\phi}{d\omega}$$

i.e., tone delay is equal to group delay, which is a special case of Expression (2.186).

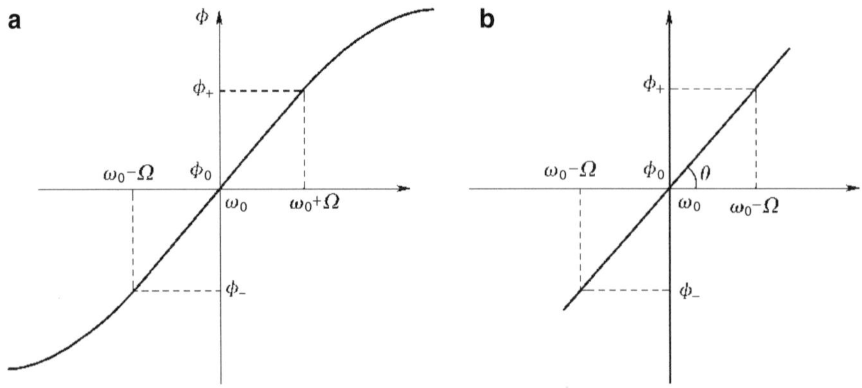

Fig. 2.46 Odd symmetry of phase-frequency characteristics

4) If $H_+ \neq H_-$, according to Expression (2.186), nonorthogonal demodulation still introduces delay, i.e., odd symmetry of phase-frequency characteristics; in case of non-even symmetry of amplitude-frequency characteristics, nonorthogonal demodulation still introduces additional delay. According to Expression (2.187), amplitude is decreased.

In conclusion, such factors can be summarized: ① if phase-frequency characteristics is not constant, it will introduce tone delay; ② if phase-frequency characteristics is not odd symmetry, amplitude-frequency characteristics will impact tone delay; ③ if amplitude-frequency characteristics is not even symmetry, nonorthogonal demodulation will impact tone delay.

To clarify some concepts, it is necessary to make clear following points.

In one four-terminal network, tone delay $\neq$ phase delay $\neq$ group delay (carrier delay is also called phase delay and equal to $[\phi(\omega)/\omega]$, but there are two special cases:

If phase-frequency characteristics are linear, we have tone delay = group delay = constant. If phase-frequency characteristics is linear and through the origin, we have tone delay = group delay = phase delay = tanθ = constant. This is the case of non-dispersive medium.

When amplitude-frequency characteristics is given, phase delay, group delay, and tone delay are solely determined by phase-frequency characteristics, phase-frequency characteristics is of prior importance.

In the non-dispersive medium (phase delay does not vary with frequency), group velocity of wave is equal to phase velocity. For one tone-modulated carrier signal, its modulated tone velocity is equal to carrier velocity. In the non-dispersive medium (similar to the free space of non-dispersive medium), carrier delay and tone delay are the same, with velocity equal to velocity of light. Tone delay can be used for ranging. In the dispersive medium, group velocity is not equal to phase velocity, and group delay is not equal to phase delay. Group velocity is also not tone velocity and calibration or remedy is needed.

(2) Ranging error introduced by phase-frequency characteristics, amplitude-frequency characteristics, and product demodulation orthogonality.

Inherent delay of equipment on tone can be reduced by range calibration, but delay change may cause ranging error. Such delay error is a major component of ranging accuracy.

External factors causing delay change: temperature change, Doppler frequency shift, signal level change (AGC impact), power supply voltage change and aging, etc. Such factors can change tone delay through phase-frequency characteristics, amplitude-frequency characteristics and nonorthogonal product demodulation of channel. The quantitative relation of such impact can be obtained through total differentiation on Expression (2.186), i.e., total error expression:

$$
\begin{aligned}
\Delta\Phi = \mathrm{d}\phi &= \frac{\partial\phi}{\partial(\Delta\phi)}d(\Delta\phi) + \frac{\partial\phi}{\partial\left(\frac{H_+}{H_-}\right)}\mathrm{d}\left(\frac{H_+}{H_-}\right) + \frac{\partial\phi}{\partial(\phi+ - \phi_0)}\mathrm{d}(\phi_+ - \phi_0) + \frac{\partial\phi}{\partial(\phi_0 - \phi_-)}\mathrm{d}(\phi_0 - \phi_-) \\
&= M\left\langle\left\{1 - \left(\frac{H_+}{H_-}\right)^2\right\}\mathrm{d}(\Delta\phi) + \left\{\cos\left[(\phi_0 - \phi_-) + \Delta\phi\right]\sin\left[(\phi_+ - \phi_0) - \Delta\phi\right]\right.\right. \\
&\quad \left. - \sin\left[(\phi_0 - \phi_-) + \Delta\phi\right]\cos\left[(\phi_- - \phi_0) - \Delta\phi\right]\right\} \times \mathrm{d}\left(\frac{H_+}{H_-}\right) \\
&\quad + \left\{\frac{H_+}{H_-}\cos\left[(\phi_0 - \phi_-) + \Delta\phi\right]\sin\left[(\phi_+ - \phi_0) - \Delta\phi\right] + \left(\frac{H_+}{H_-}\right)^2\right\}\mathrm{d}(\phi_+ - \phi_0) \\
&\quad + \left\{\frac{H_+}{H_-}\sin\left[(\phi_0 - \phi_-) + \Delta\phi\right]\sin\left[(\phi_+ - \phi_0) - \Delta\phi\right] + \left(\frac{H_+}{H_-}\right)^2\right\}\mathrm{d}(\phi_+ - \phi_0) \\
&\quad + \left\{1 + \frac{H_+}{H_-}, \cos\left[(\phi_+ - \phi_-) - \Delta\phi\right], \cos\left[(\phi_0 - \phi_-) + \Delta\phi\right]\right. \\
&\quad \left.\left. + \left(\frac{H_+}{H_-}\right), \sin\left[(\phi_+ - \phi_0) - \Delta\phi\right], \sin\left[(\phi_0 - \phi_-) + \Delta\phi\right], \right\}\mathrm{d}(\phi_0 - \phi_-)\right\rangle \\
M &= \left\langle\left\{\cos\left[(\phi_0 - \phi_-) + \Delta\phi\right] + \tfrac{H_+}{H_-}\cos\left[(\phi_+ - \phi_0) - \Delta\phi\right]\right\}^2\right. \\
&\quad \left. + \left\{\sin\left[(\phi_0 - \phi_-) + \Delta\phi\right] + \tfrac{H_+}{H_-}\sin\left[(\phi_0 - \phi_-) - \Delta\phi\right]\right\}^2\right\rangle^{-1}
\end{aligned}
\tag{2.190}
$$

where $\Delta\Phi$ is tone phase shift increment; $\Delta\phi$ is nonorthogonal phase difference.

Thus, reducing M will decrease delay error. Minimum value of M shall be sought out in design.

From discussion on Expression (2.186) we can conclude:

1) Impact of phase-frequency characteristics. The third item in Expression (2.186) is the tone delay change when relative phase difference $(\phi_+ - \phi_0)$ of $(\omega_0 + \Omega)$ side frequency to carrier changes in case of no change for $\Delta\phi$, (H_+/H_-) and $(\phi_0 - \phi_-)$.

For special case: $(H_+/H_-) = 1, \phi_0 - \phi_- = \phi_+ - \phi_0$, substitute into Expression (2.186) to obtain $M = 1/4$, $\mathrm{d}\phi(\phi_0 - \phi_-) = \frac{1}{2}\mathrm{d}(\phi_0 - \phi_-) = \frac{1}{2}\Delta(\phi_0 - \phi_-)$, which is equal to 1/2 of phase difference change of side frequency to carrier.

Similarly

$$\mathrm{d}\phi(\phi_+ - \phi_0) = \frac{1}{2}\mathrm{d}(\phi_+ - \phi_0) = \frac{1}{2}\Delta(\phi_+ - \phi_0)$$

Tone delay variation is:

$$\begin{aligned}\Delta\phi_1 &= \mathrm{d}\phi(\phi_0 - \phi_-) + \mathrm{d}\phi(\phi_+ - \phi_0) = \Delta(\phi_+ - \phi_-)\\ \Delta R_1 &= \frac{c}{720 f_r} \bullet \Delta(\phi_+ - \phi_-)\end{aligned} \tag{2.191}$$

2) Impact of amplitude-frequency characteristics asymmetry. The second item in Expression (1.186) is the delay change $\mathrm{d}\phi(H_+/H_-)$ when amplitude-frequency characteristics non-even symmetry (H_+/H_-) changes in case of no change for $\Delta\phi, (\phi_+ - \phi_0)$ and $(\phi_0 - \phi_-)$.

$$\begin{aligned}\Delta\phi_2 &= \mathrm{d}\phi\left(\frac{H_+}{H_-}\right)\\ &= M\left\{\begin{matrix}\cos[(\phi_0 - \phi_-) + \Delta\phi]\sin\left[(\phi_+ - \phi_-) - \Delta\phi\right] - \\ \sin[(\phi_0 - \phi_-) + \Delta\phi]\cos\left[(\phi_+ - \phi_-) - \Delta\phi\right]\end{matrix}\right\}\Delta\left(\frac{H_+}{H_-}\right)\end{aligned} \tag{2.192}$$

$\Delta\phi_2 = 0$ can be obtained by making $\mathrm{d}_\phi(H_+/H_-) = 0$

By resolve Expression (2.190) we can get:

$$\begin{aligned}&\frac{\sin\left[(\phi_+ - \phi_0) - \Delta\phi\right)}{\cos\left[(\phi_+ - \phi_0) - \Delta\phi\right)} = \frac{\sin\left[(\phi_0 - \phi_-) + \Delta\phi\right)}{\cos\left[(\phi_0 - \phi_-) + \Delta\phi\right)}\\ &\left\{\begin{array}{l}(\phi_+ - \phi_0) - \Delta\phi = (\phi_0 - \phi_-) + \Delta\phi\\ \Delta\phi = \frac{1}{2}[(\phi_+ - \phi_0) - (\phi_0 - \phi_-)]\end{array}\right.\end{aligned} \tag{2.193}$$

Above analysis shows: ① when adjusting product demodulation non-orthogonal phase difference $\Delta\phi$ to equal to Expression (2.193), change of amplitude-frequency characteristics even symmetry does not cause tone delay change; ② if product demodulator is orthogonal, i.e., $\Delta\phi = 0$, substitute into Expression (2.193)

$$\phi_+ - \phi_0 = \phi_0 - \phi_-$$

It is odd symmetry of phase-frequency characteristics.

When product demodulation is orthogonal, if phase-frequency characteristics are odd symmetry, change of amplitude-frequency characteristics asymmetry does not introduce delay change, which matches above discussed results.

3) Error caused by quadrature demodulation. The first item of Expression (2.189) is the delay error when non-orthogonal phase difference $\Delta\phi$ changes in case of no change for (H_+/H_-), $(\phi_+ - \phi_0)$ and $(\phi_0 - \phi_-)$,

$$\Delta\phi_3 = \mathrm{d}\phi(\Delta\phi) = M\left[1 - \left(\frac{H_+}{H_-}\right)^2\right]\Delta(\Delta\phi) \tag{2.194}$$

When $H_+/H_- = 1$, $\mathrm{d}\phi(\Delta\phi) = 0$, i.e., in case of amplitude-frequency characteristics even symmetry, nonorthogonal product demodulation will not introduce tone delay, which is the same as above conclusion.

On the basis of above analysis, phase-frequency characteristics is the most serious impacting factor on tone delay and other factors' impact can be properly controlled for decrease.

2.4.3 *Group Delay Characteristics Analysis Method of Ranging Error [12]*

According to above analysis, phase-frequency characteristics is the most direct and accurate method for ranging accuracy analysis. But group delay index or curve is often given in engineering, it is necessary to analyze ranging error as per group delay characteristics.

2.4.3.1 Group Delay Analysis Method of Tone Ranging Error [12]

(1) Several definition methods of channel delay characteristics

1) Group delay τ_g: The delay of channel on wave group signal is called group delay. According to Fourier analysis, one complex signal is composed of many sinusoidal wave components. Use such signal to modulate (including PM, FM, AM, BPSK, QPSK, etc.) carrier to produce one modulated wave, group delay is the delay of its modulated signal (integrated signal of wave group), its mathematic expression:

$$\tau_g = \frac{\mathrm{d}\Phi(\omega)}{\mathrm{d}\omega}$$

2) Ranging tone delay τ: Ranging tone is one single-frequency sine wave and the delay of channel on such single-frequency sine wave is called ranging tone delay. Reference [1] provides the relationship of ranging tone delay and channel transmission characteristics

$$\tau = \frac{\Delta\Phi}{\omega_m} = \frac{1}{\omega_m}\arctan\frac{H_{-1}\sin(\Phi_0 - \Phi_{-1}) + H_{+1}\sin(\Phi_{+1} - \Phi_0)}{H_{-1}\cos(\Phi_0 - \Phi_{-1}) + H_{+1}\cos(\Phi_{+1} - \Phi_0)} \tag{2.195}$$

where ω_m is ranging tone frequency; Φ_0 is phase shift produced by carrier ω_0 in phase-frequency characteristics; Φ_{+1} and Φ_{-1} are the phase shift corresponding to $\omega_0 + \omega_m$ and $\omega_0 - \omega_m$ in phase-frequency characteristics; H_{+1} and H_{-1} are the amplitude value corresponding to $\omega_0 + \omega_m$ and $\omega_0 - \omega_m$ in amplitude-frequency characteristics.

3) Test delay τ_e: The delay tested by test apparatus on channel is called test delay or envelope delay. In ordinary test apparatus, use one single-frequency sine wave to conduct amplitude modulation on carrier, measure the delay of such amplitude-modulation envelope after passing channel. Its mathematic expression is:

$$\tau_e = \frac{\Phi_{+1} - \Phi_{-1}}{2\Omega} \tag{2.196}$$

Such digital expression is similar to the ranging tone expression. Only when Ω is equal to ranging tone and modulation mode is phase modulation (or frequency modulation), it can represent ranging tone delay.

In Expression (2.196), when $\Omega \to 0$, then

$$\tau_e = \frac{\Delta\Phi}{\Delta\Omega} = \mathrm{d}\Phi(\omega)/\mathrm{d}\omega = \tau_g \tag{2.197}$$

But in actual test, $\Delta\Omega \to 0$ is impossible. Hence, test delay τ_e is similar to group delay.

In addition, "phase delay," "phase truncation delay," etc., are used to describe delay characteristics. New test apparatus can directly measure $\phi(\omega)$; microprocessor in test apparatus directly calculate differentiation $\mathrm{d}\phi(\omega)/\mathrm{d}\omega$ to obtain $\tau(\omega)$. Such various delay characteristics are hidden in and solely determined by frequency characteristics $H(Jw)$.

(2) Expression calculation method of tone ranging error

Expression of group delay $\tau(\omega)$

$$\tau(\omega) = \frac{\mathrm{d}\Phi}{\mathrm{d}\omega} \tag{2.198}$$

Use Expression (2.198) to obtain the phase difference between phase Φ_{+1} of frequency point $(\omega_0+\Omega_1)$ and phase Φ_{-1} of frequency point $(\omega_0-\Omega_1)$

$$(\Phi_{+1}-\Phi_{-1})=\int_{\omega_0-\Omega_1}^{\omega_0+\Omega_1}\frac{\tau(\omega)}{d\omega} \tag{2.199}$$

where $(\omega_0 \pm \Omega_1)$ unit is rad/s; $\tau(\omega)$ unit is s; $(\Phi_{+1}-\Phi_{-1})$ unit is rad.

1) Ranging error from linearity distortion item and parabolic distortion item. Nonlinear item of group delay characteristics can be approximated with Taylor series (above third order can be omitted), group delay fluctuation item can be approximated with Fourier series (above second order can be omitted), total group delay characteristics can be represented:

$$\tau(\omega_\Delta)=a_0+a_1\omega_\Delta+a_2\omega_\Delta^2+\tau_m\sin\left[\frac{2\pi\omega_\Delta}{\Omega_T}+\theta_0\right] \tag{2.200}$$

where ω_Δ is the frequency difference of offset carrier center frequency ω_0, $\omega_\Delta=\omega-\omega_0$, $\tau(\omega_\Delta)$ is the group delay characteristics function with ω_Δ as variable.

Expression (2.200) includes two parts; one part changes with frequency, i.e., second, third and fourth items, in which $a_1\omega_\Delta$ and $a_2\omega_\Delta^2$ are respectively the first order item (linearity distortion item) and second order item (parabolic distortion item) of group delay characteristics provided in engineering design; the fourth item is group delay fluctuation item with fluctuation amplitude of τ_m; the above three items are not respectively provided and called in-band group delay change. The other part is a_0, a_1 and a_2, which is group delay change varying with time change, i.e., group delay stability provided in index (group delay change in how many hours, factors causing change includes temperature, span, power supply, input level, etc.).

Substitute Expression (2.199) into Expression (2.200) to obtain:

$$(\Phi_{+1}-\Phi_{-1})=\int_{-\Omega}^{+\Omega}\frac{\tau(\omega_\Delta)}{d\omega_\Delta}=2a_0\Omega+\frac{2a_2}{3}\Omega^3 \tag{2.201}$$

where Ω is tone frequency.

In most actual cases, ranging tone delay is determined by Expression (2.189).

$$\tau=\frac{(\Phi_{+1}-\Phi_{-1})}{2\Omega}=a_0+\frac{1}{3}a_2\Omega^2 \tag{2.202}$$

By above expression, the change of τ caused by changes of a_0 and a_2 can be calculated.

Ranging error from Doppler frequency shift ω_d can be obtained from change $(\Delta\phi_{+d}-\Delta\phi_{-d})$ of $(\phi_{+1}-\phi_{-1})$ (general ω_d can be extended as other frequency change factors, including signal center frequency drift, filter center frequency drift, etc.). In case of $(\omega_0+\Omega)$, $\Delta\Phi_{+d}$ from change is:

$$\begin{aligned}\Delta\Phi_{+d} &= \int_{+\Omega}^{\Omega+\omega_d}\tau(\omega_\Delta)\mathrm{d}\omega_\Delta = \int_{+\Omega}^{\Omega+\omega_d}\left(a_0+a_1\omega_\Delta+a_2\omega_\Delta^2\right)\mathrm{d}\omega_\Delta \\ &= a_0\omega_d+\frac{a_1}{2}\left[(\Omega+\omega_d)^2-\Omega^2\right]+\frac{a_2}{3}\left[(\Omega+\omega_d)^3-\Omega^3\right]\end{aligned} \tag{2.203}$$

Similarly, at $(\omega_0-\Omega)$, $\Delta\Phi_{-d}$ from change ω_d is represented as follows:

$$\begin{aligned}\Delta\Phi_{-d} &= \int_{-\Omega}^{-\Omega+\omega_d}\tau(\omega_\Delta)\mathrm{d}\omega_\Delta \\ &= a_0\omega_d+\frac{a_1}{2}\left[(-\Omega+\omega_d)^2-\Omega^2\right]+\frac{a_2}{3}\left[(-\Omega+\omega_d)^3+\Omega^3\right]\end{aligned} \tag{2.204}$$

Ranging error from Doppler frequency shift is:

$$\Delta\tau_d = \frac{\Delta\Phi_{+d}-\Delta\Phi_{-d}}{2\Omega} = a_1\omega d + a_2\omega_d^2 \tag{2.205}$$

where a_1 and a_2 are usually provided in specification. If the specification provides group delay characteristics curve, a_1 and a_2 can be calculated from such curve, as shown in Fig. 2.47.

In Fig. 2.47, dotted line is the linearity distortion curve including a_1, solid line is the total curve including a_1 and a_2.

The method for calculating a_1, a_2 with above curves is as follows:

① Linear component a_1 (first order item). Obtain from Fig. 2.47:

$$a_1 = \frac{\tau_A-\tau_B}{B_s} \tag{2.206}$$

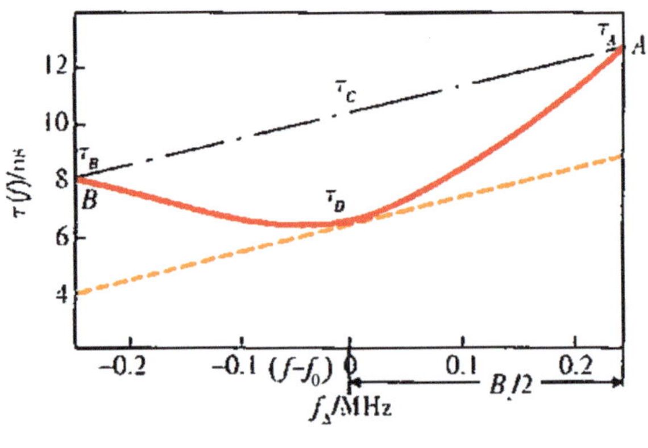

Fig. 2.47 Group delay characteristics curve including a_1 and a_2 distortion

It represents with the time delay of unit band with unit of ns/MHz; B_s is the bandwidth occupied by ranging signal spectrum.

② Parabolic curve component a_2 (second order item): In Fig. 2.47, τc is the group delay value of straight line connecting point A and B at center frequency f_s.

According to Fig. 2.47,

$$\tau_D = a_0 \text{ (corresponding to } \omega_\Delta = 0)$$

$$\tau_A = a_0 + a_1\left(\frac{B_s}{2}\right) + a_2\left(\frac{B_s}{2}\right)^2 \left(\text{corresponding to } \omega_\Delta = \frac{B_s}{2}\right)$$

then

$$\tau_c = \tau_A - a_1\left(\frac{B_s}{2}\right)$$

$$(\tau_c - \tau_D) = a_2\left(\frac{B_s}{2}\right)^2$$

$$a_2 = \frac{(\tau_c - \tau_D)}{(B_s/2)^2} \tag{2.207}$$

2) Ranging error from group delay fluctuation. Group delay fluctuation index is provided in engineering design and its fluctuation characteristics curve can be various. We can use Fourier series to decompose it into the sum of many sinusoidal fluctuation harmonics. When first order item is taken, it is represented:

$$\tau(\omega_\Delta) = \tau_m \sin\left[2\pi\frac{\omega_\Delta}{\Omega_T} + \theta_0\right] \tag{2.208}$$

where Ω_T is the frequency interval of one cycle sinusoidal fluctuation of group delay with unit of rad/s (or Hz); ω_Δ is independent variable with the same unit as Ω_T; τ_m is the maximum group delay peak value with unit of s; θ_0 is the initial phase with unit of rad.

Group delay fluctuation characteristics curve is as shown in Fig. 2.48.

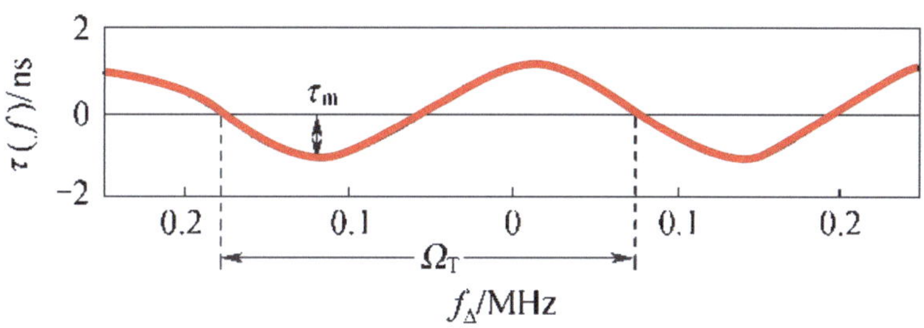

Fig. 2.48 Group delay fluctuation characteristics

From Fig. 2.48 we can obtain

$$\theta_0 = 180^\circ - \left[\frac{(\Omega_1 - \Omega_2)}{\Omega_T}\right] \times 360^\circ$$

Expression of phase difference $\Delta\Phi_{+d}$ and $\Delta\Phi_{-d}$ from Doppler frequency shift ω_d is as follows:

$$\Delta\Phi_{-d} = \int_{-\Omega}^{-\Omega+\omega_d} \tau_{\max} \sin\left[2\pi\frac{\omega_\Delta}{\Omega_\tau} + \theta_0\right] d\omega_\Delta = -\tau_{\max}\frac{\Omega_T}{2\pi}\left[\cos\frac{2\pi}{\Omega_T}(\Omega - \omega_d) - \cos\frac{2\pi}{\Omega_T}\Omega\right]\cos\theta_0$$
$$+\tau_{\max}\frac{\Omega_T}{2\pi}\left[\sin\frac{2\pi}{\Omega_T}(\omega_d - \Omega) + \sin\frac{2\pi}{\Omega_T}\Omega\right]\sin\theta_0$$

Be same as that,

$$\Delta\Phi_{+d} = \int_{+\Omega}^{\Omega+\omega_d} \tau_{\max} \sin\left[2\pi\frac{\omega_\Delta}{\Omega_\tau} + \theta_0\right] d\omega_\Delta$$

Ranging error from Doppler frequency shift ω_d is:

$$\Delta\tau_d = \frac{\Delta\Phi_{+d} - \Delta\Phi_{-d}}{2\Omega}$$
$$= \tau_{\max}\frac{\Omega_T}{2\pi\Omega}\sin\frac{2\pi\Omega}{\Omega_T}\left[\sin\frac{2\pi\omega_d}{\Omega_T}\cos\theta_0 + \left(\cos\frac{2\pi\omega_d}{\Omega_T} - 1\right)\sin\theta_0\right] \quad (2.209)$$

Ranging error can be calculated according to Expression (2.209). It can be seen that the ranging error resulted from group delay fluctuation characteristics is related to fluctuation amplitude $\tau_{\max}$ and initial phase θ_0 of fluctuation. Different ω_d/Ω_T and Ω/Ω_T may cause different ranging errors. When $\Omega \gg \Omega_T$, $\Delta\tau_d \to 0$ and Ω_T gradually decreases to be less than $\Omega/2$, it evolves to be near linearity distortion or square distortion. When $\Omega \ll \Omega_T$ and $\theta_0 = 0^\circ$, $\Delta\tau_d = \tau_{\max}$ is transformed into Δa_0, it does not reach the maximum when ω_d is maximum, but reaches the maximum in case of some ω_d. Under some conditions, it may not introduce ranging error (e.g., when $\theta_0 = 0^\circ$ and $2\pi\omega_d/\Omega_T = n\pi$, its physical significance is that plus-minus half-cycle integration of sine wave counteracts each other). On the basis of such complexity, group delay index of engineering design does not include such index as Ω_T and θ_0. Use group delay characteristics curve and following graphic(al) method for resolution.

3) Ranging error from group delay stability. Definition of group delay stability: assuming with the change of time, environment conditions, AGC circuit, etc., a_0 has a change of Δa_0 and a_2 has a change of Δa_2, they will cause a ranging error of $\Delta\tau_0$ which can be calculated from Expression (2.202), i.e.,

$$\Delta\tau_0 = \Delta a_0 + \frac{\Delta a_2}{3}\Omega^2 \quad (2.210)$$

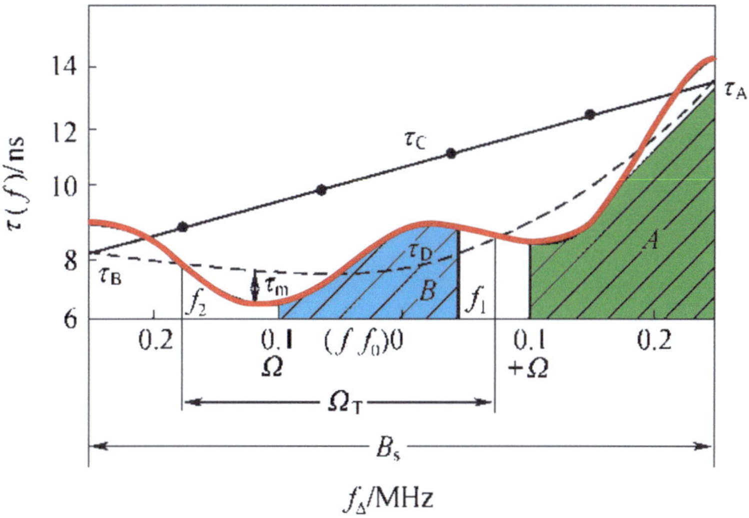

Fig. 2.49 Graphic method of ranging error

$\Delta\tau_0$ is the major component in ranging error and decreased by "range calibration," but change between the two "zero calibration" may cause ranging error.

(3) Graphic calculation method of tone ranging error

If actual $\tau(\omega)$ curve is complex and Taylor series and Fourier series cannot be used for representation, according to the measured curve diagram $\tau(\omega)$, use graphic area solution method for integration of Expressions (2.203) and (2.204), which can also calculate the ranging error. Figure 2.49 is a $\tau(f)$ curve, take it as an example for graphic calculation of ranging error when $\Omega = 100$ kHz, $f_d = 150$ kHz.

In Fig. 2.49, the dotted line is the peak-peak average value curve of the fluctuation.

According to Expressions (2.203) and (2.204), integration of $\Delta\Phi_{+d}$ is the area A under $\tau(f)$ envelope in an interval of f from +0.1 MHz to (+0.1 + 0.15) MHz = 0.25 MHz. Similarly, $\Delta\Phi_{-d}$ is the area B under $\tau(\omega)$ envelope in an interval of f_Δ from −0.1 MHz to (−0.1 + 0.15)MHz = 0.05 MHz. They are represented by the shaded area in Fig. 2.49, from which one can obtain:

$$\Delta\Phi_{+d} = \text{area A} = 4.4 \text{ rad}$$

$$\Delta\Phi_{-d} = \text{area B} = 1.95 \text{ rad}$$

From Expression (2.211) we can also obtain:

$$\Delta\tau_d = \frac{\Delta\Phi_{+d} - \Delta\Phi_{-d}}{2\Omega} = \frac{(4.4 - 1.95) \times 10^{-3} \text{ rad}}{2 \times 0.1 \times 10^6 \times 2\pi \text{ rad/s}} = 1.95 \text{ ns}$$

Below, we will calculate them again by formula method and make a comparison of the two calculation results.

1) Ranging error $\Delta\tau_d^{'}$ from first order item and second order item can be obtained from Fig. 2.49:

$$B_s = 0.5 \text{ MHz},$$
$$\tau_A - \tau_B = 7.4 \text{ ns}, \quad \tau_C - \tau_D = 3 \text{ ns}$$

From Expression (2.207), we get

$$a_1 = 5.2 \text{ ns}/0.5 \text{ MHz} = 10.4 \text{ ns/MHz}$$

From Expression (2.208), we get

$$a_2 = \frac{3 \text{ ns}}{(0.25 \text{ MHz})^2} = 48 \text{ ns/MHz}^2$$

From Expression (2.206), we get

$$\Delta\tau_d^{'} = a_1\omega_d + a_2\omega_d^2$$

Substitute into a_1, a_2 and ω_d, etc., the calculation result is:

$$\Delta\tau_d^{'} = 2.64 \text{ ns}$$

2) Ranging error $\Delta\tau_d^{''}$ from group delay fluctuation. From Fig. 2.49, we get:

$$\Omega_T = (f_1 - f_2) = 250 \text{ kHz}, \quad \tau_m = 1.2 \text{ ns},$$
$$\theta_0 = 180^\circ - [(f_1 - f_0/\Omega_T)] \times 360^\circ = 72^\circ$$

$\Delta\tau_d^{''}$ can be obtained by substituting these values into Expression (2.211) and results are as listed in Table 2.9.

The total ranging error $\Delta\tau_d = \Delta\tau'_d + \Delta\tau''_d = 2.1$ ns when $\omega_d = 0.15$ MHz is close to 1.95 ns which is obtained by graphic method. Moreover, we can conclude that ranging error from group delay fluctuation does not increase when ω_d increases.

Table 2.9 Calculation results of ranging error

ω_d(MHz)	0	0.03	0.06	0.09	0.12	0.13	0.14	0.15
$\Delta\tau_d''$(ns)	0	−0.015	−0.12	−0.375	−0.525	−0.547	−0.55	−0.54

2.4.3.2 Group Delay Characteristics Analysis Method of PN Code Ranging Error and Spread Spectrum Ranging Error [13]

(1) Principle of generation

PN code ranging and spread spectrum ranging is usually implemented by delay locking loop, and the null point of error discriminator in the loop is symbol measurement point of the PN code ranging delay. The change of such null point on the time axis reflects that of the ranging delay. Its physical mechanism is: the output voltage of the time error discriminator in the loop is zero in case of stable tracking of PN code. If influenced by other factors, the original time discrimination characteristics become a new time discrimination characteristics, and its output voltage changes by ΔE accordingly. Such ΔE will make the frequency and phase of the local PN code clock change and, in the meantime, make phase (i.e., delay) of PN code change simultaneously. After it changes by the delay $\Delta\delta_0$, the output voltage becomes zero on the new time discrimination characteristics; at this time, the delay locking loop will become stable again. $\Delta\delta_0$ is the ranging error arising from change of time discrimination characteristics, and is corresponding to zero point of output voltage on new time discrimination characteristics. Therefore, if the value of $\delta_0^{'}$ corresponding to the output zero point of new (changed) time discrimination characteristics is solved and then minus by δ_0 of the original time discriminator, the ranging delay error $\Delta\delta_0 = \delta_0^{'} - \delta_0$ can be obtained, and then solve the range error through delay error.

Change of channel group delay will result in change in time discrimination characteristics. The ranging error arising out of change in group delay characteristics can be solved according to the foregoing mechanism. However, it is very difficult to get its mathematical expression via analytical method, so simulation method is recommended. The section below will set out its impact upon ranging error through simulation.

(2) Simulation block diagram

The simulation block diagram is as shown in Fig. 2.50:

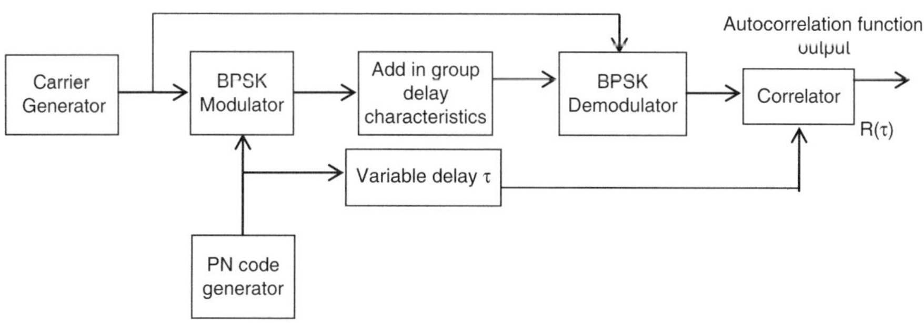

Fig. 2.50 Simulation block diagram of PN code ranging error

In Fig. 2.50 “add in group delay characteristics” include first degree, quadratic term, sinusoidal variation term, and constant-value term of group delay characteristics set out in Sect. 2.4.3.1. It simulates the group delay characteristics in the following expression:

$$\tau(\omega_\Delta) = a_0 + a_1\omega_\Delta + a_2\omega_\Delta^2 + \tau_m \sin\left[\frac{2\pi\omega_\Delta}{\Omega T} + \theta_0\right]$$

where ω_Δ is the frequency difference deviating from the central frequency ω_0.

Group delay characteristics can be simulated by full-pass filter to solve the auto-correlated function $R(t)$ while t is changing, and then calculate the time discrimination characteristics $D(t)$ of advanced code and lagging code at an interval of a code element width T_c: $R(\tau) - R(\tau + T_c) = D(\tau)$.

Then, further solve the value of τ_0' corresponding to $D(\tau) = 0$ in case of different group delay characteristics and the value of τ_0 in absence of group delay characteristics (namely, the group delay is zero). Calculate $\delta_0 = (\tau_0' - \tau_0)$ to obtain the ranging signal delay δ_0 (i.e., the delay in absence of group delay) arising out from different group delay characteristics; the change $\Delta\delta_0$ caused by change in such δ_0 along with the group delay characteristics is the ranging error introduced by the group delay characteristics.

(3) Simulation results [13]

When the rate of PN code is 3 and 10 Mbps, the simulation result is as shown in Fig. 2.51.

In Fig. 2.51, the subfigure (a) is PN code ranging delay caused by sinusoidal group delay variation, B_I means the bandwidth of filter (unit: radians/second), usually as $B_I = 1/T_c$. The simulation result reveals that the ranging delay $\delta_0 = \overline{\tau}_{B_I}$, namely,

$$\delta_0 = \overline{\tau}_{B_I} = \frac{1}{B_I}\int_{-B_I/2}^{+B_I/2} \tau_m \sin\left[\frac{2\pi\omega_\Delta}{\Omega T} + \theta_0\right] \mathrm{d}\omega_\Delta \qquad (2.211)$$

Where, $\overline{\tau}_{B_I}$ is the ratio of average value of group delay to the value of B_I within B_I According to the expression above, it is shown that δ_0 is relevant to amplitude of the group delay variation, bandwidth of B_I, starting phase of variation within the bandwidth (similarly, it is also relevant to the ending phase). But there is a special circumstance, under which, the positive semicircle number of the variation is equal to the negative semicircle number when both starting phase and ending phase is zero; therefore, in case of $\overline{\tau}_{B_I} = 0$, whatever is amplitude and times of the variation, no PN code delay is caused.

Any group delay variation curve can be decomposed to be the sum of n times of harmonic curve; respectively solve δ_{0n} and then solve δ_Σ.

In Fig. 2.51, the subfigure (b) is PN code ranging delay caused by the linear group delay, a_1 is the slope of group delay characteristics (unit: ns/MHz), and t_B means the value of group delay at the passband edge (namely, *at* $\omega_0 \pm B_I/2$, $B_I = 1/T_c$)

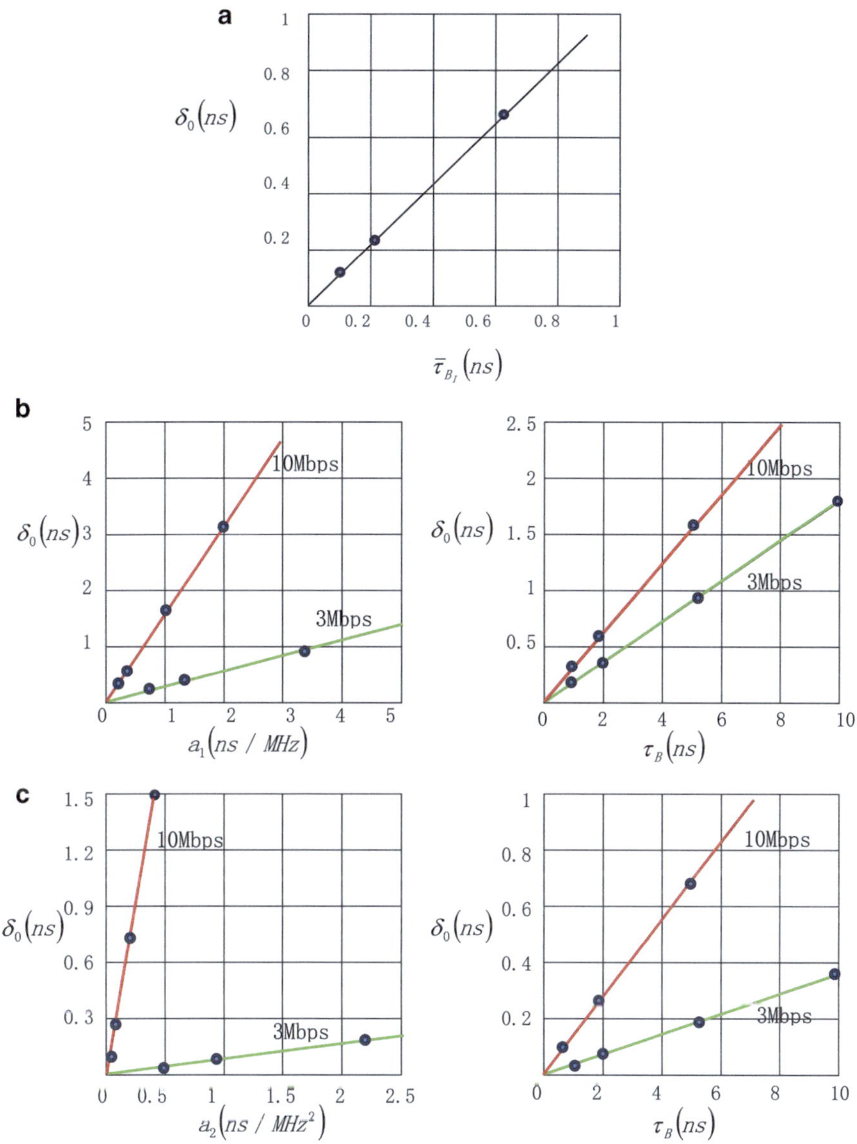

Fig. 2.51 Simulation results when PN code delay is caused by group delay. (**a**) Relationship between PN code delay and sinusoidal variation group delay characteristics. (**b**) Relationship between PN code delay and linear group delay characteristics. (**c**) Relationship between PN code delay and parabolic group delay characteristics

(called as "boundary delay"). The simulation result shows that δ_0 is in linear relationship with a_1 which can be fit to a mathematical expression:

$$\begin{aligned} &\delta_0 = 0.25a_1, \delta_0 = 0.17\tau_B \text{ (in case of 3 Mbps code rate)} \\ &\delta_0 = 1.5a_1, \delta_0 = 0.3\tau_B \text{ (in case of 10 Mbps code rate)} \end{aligned} \quad (2.212)$$

In Fig. 2.51, the subfigure (c) is PN code ranging delay caused by the parabolic group delay, a_2 is the coefficient of the parabola (unit: ns/MHz^2), τ_B is the value of group delay at the passband edge (namely, at $\omega_0 \pm B_l/2$, $B_l = 1/T_c$). The simulation result shows that the greater a_2 is, the greater τ_B becomes; as a result, the greater δ_0 so caused. δ_0 is approximately in linear relationship with a_2, which can be fit into a mathematical expression:

$$\begin{aligned} &\delta_0 = 0.085a_2, \quad \delta_0 = 0.038\tau_B \text{ (in case of 3 Mbps code rate)} \\ &\delta_0 = 3.3a_2, \quad \delta_0 = 0.15\tau_B \text{ (in case of 10 Mbps code rate)} \end{aligned} \quad (2.213)$$

The PN code ranging signal delay introduced by constant-value group delay is equal to the constant-value coefficient of group delay characteristics, namely, $\delta_0 = a_0$.

Expressions (2.211), (2.212), and (2.213) reveal that a_1, a_2, a_0, τ_B, and $\overline{\tau}_{B_l}$ will introduce PN code ranging signal delay. It introduces range zero value which can be deducted by range calibration. After being calibrated, the change in a_1, a_2, a_0, τ_B, and $\overline{\tau}_{B_l}$ will be introduced into the ranging error.

2.4.4 *Random Ranging Error*

Thermal noise and short-term stability of ranging signal will cause the random ranging error. More details are discussed below.

2.4.4.1 Ranging Error Introduced by Thermal Noise

According to CW ranging principle $R = \frac{c\Phi}{4\pi f_R}$, we can obtain that

$$\sigma_{R_1} = \frac{C}{4\pi} \frac{\delta_\Phi}{f_R} \quad (2.214)$$

where δ_Φ is the phase measurement error of the equipment, f_R is the frequency of high tone, and σ_{R_1} is the required ranging accuracy.

The phase measurement error of the equipment δ_Φ is limited by the SNR of the tone signal to be modulated. For the purpose of improving the phase measurement accuracy for the high tone, the phase-locked loop (PLL) has to be used. Because the closed-loop narrowband filtering of the phase-locked loop is able to improve the

SNR of tone, under the function of AWGN $n(t)$, when the SNR of phase-locked loop $(S/N)_L$ is greater than the threshold value (≥ 7 dB), the relation between the phase jitter output by the main tone loop and the loop SNR may be expressed that

$$\delta_\Phi = \sqrt{\delta_\Phi^2} = \sqrt{\delta_\Phi^2(t)} = \frac{1}{\sqrt{2\left(\frac{S}{N}\right)_L}} = \frac{1}{\sqrt{\frac{S}{\Phi B_L}}} \tag{2.215}$$

If $(S/N)_L < 7$ dB, the losing lock probability of the phase-locked loop increases, and the pulse noise will be found in the phase noise δ_Φ, consequently making the result greater.

Substitute the δ_Φ in Expression (2.215) into Expression (2.214), then we obtain

$$\sigma_{R_1} = \frac{c}{4\pi f_R} \frac{1}{\sqrt{2}} \frac{1}{\sqrt{\left(\frac{S}{N}\right)_L}} \approx \frac{c}{18 f_R \sqrt{\left(\frac{S}{N}\right)_L}} = \frac{c}{18 f_R \sqrt{\left(\frac{S}{\Phi}\right)\frac{1}{2B_L}}} \tag{2.216}$$

where $(S/N)_L < 7$ dB in general, $N = 2\Phi B_L$, Φ is noise power spectral density, B_L is the equivalent noise bandwidth for a single side of loop, $B_L = \int_0^\infty |H(j2\pi f)|^2 df$, which is related to the type and exponent number of the phase-locked loop. See Table 2.7 for calculation.

2.4.4.2 Ranging Error Caused by Short-Term Stability of Ranging Signal

The phase noise of the ranging signal (shown as short-term stability in time domain) may introduce additional ranging error. The reason is that the phase noise has frequency modulation flicker noise and frequency drift noise, and then these noises may be radiated as time extended. Therefore, for the deep-space two-way ranging with long time delay, if the short-term stability of a signal is not required strictly, the non-ignorable ranging error will appear. However, due to a short R/T time delay in CC&T for the orbit satellites of the Earth, such error may be neglected.

The ranging signal is expressed that

$$u_R(t) = A\cos\left[\omega_s t + \varphi(t)\right]$$

where ω_s is the nominal frequency, and $\varphi(t)$ is the phase noise.

In a basic model, the ranging distance is the phase of *R/T* CW ranging signal, which has a space time delay T. Assuming that the additional phase noises may not be introduced by the uplink/downlink channel for ranging and the transponder, that is, the frequency fluctuation (short-term stability) should only be caused by the main vibration source, then the two-way ranging model is simplified as the model shown in Fig. 2.52.

Fig. 2.52 Ranging model

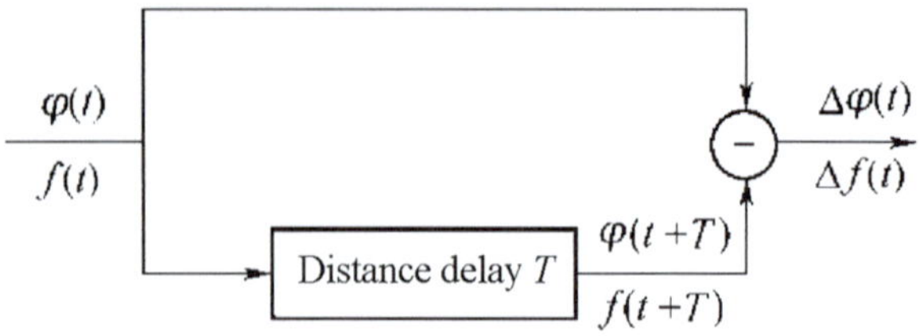

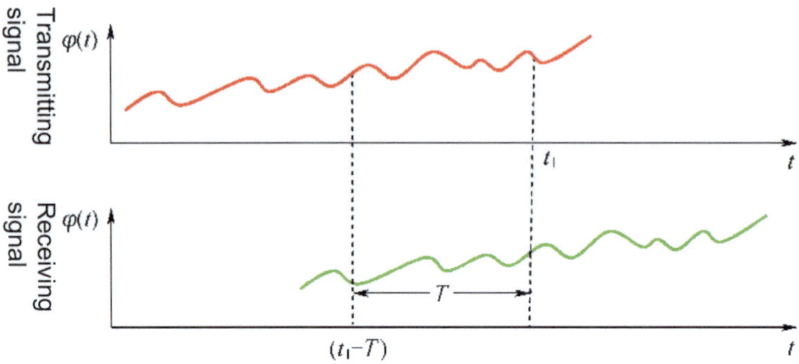

Fig. 2.53 Phase and frequency relationship of *R*/*T* signals

In Fig. 2.52, the phase difference $\Delta\varphi(t)$ relative to ranging distance is equal to the difference between the phase $\varphi(t)$ and the phase $\varphi(t+T)$ of transmitter, which have T time difference then

$$\Delta\varphi(t) = \varphi(t+T) - \varphi(t) \tag{2.217}$$

Correspondingly, the difference of frequency fluctuation is expressed that

$$\Delta f(t) = f(t+T) - f(t) \tag{2.218}$$

In Expression (2.217), the first term and the second term are the two measured values of two times of transmitting signals respectively at T time interval; the corresponding time relationship is as shown in Fig. 2.53.

In Fig. 2.53, the measurement starts at t_1, and the time delay of measured *R*/*T* signal is T. The measurement shown in Fig. 2.53 is equivalent to the measurement for the transmitting signal shown in Fig. 2.54.

In Fig. 2.54, for the *R*/*T* signal with delay time T, the measurements of phase difference $\Delta\varphi$ and frequency difference Δf are equivalent to those of phase difference and frequency difference between $(t_1 - T)$ and t_1.

Distance ranging is implemented by measurement of phase difference. Therefore, the relationship of the short-term stability with the phase fluctuation difference $\Delta\varphi$ shall be derived to obtain the result. The relationship between the phase and the frequency is expressed that

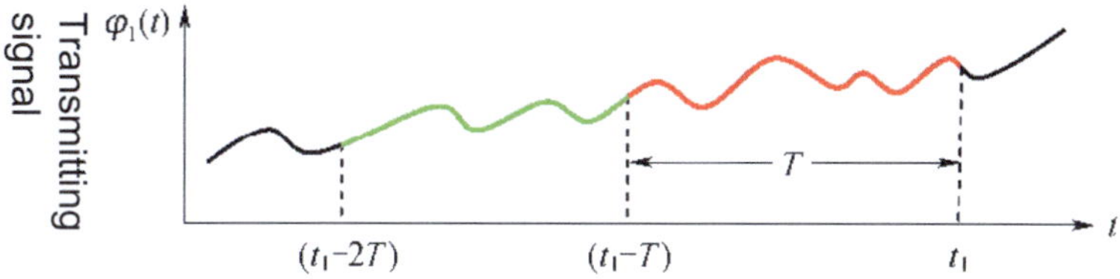

Fig. 2.54 Equivalent measurement for transmitting signal

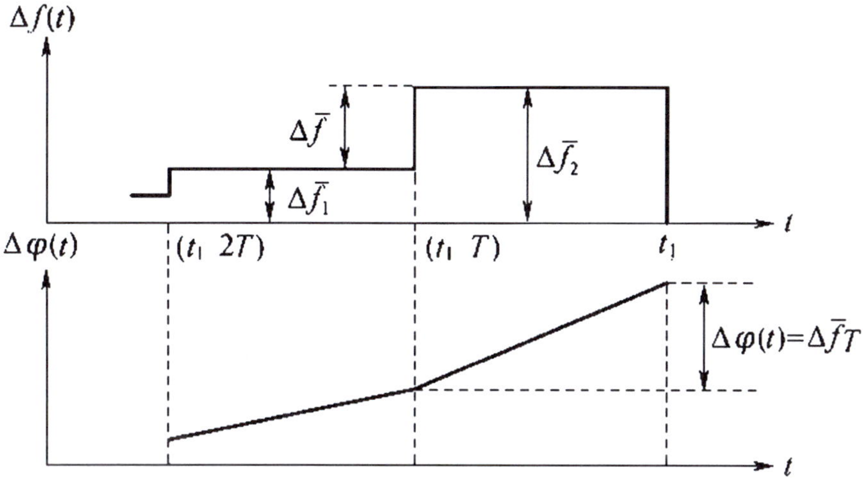

Fig. 2.55 Relationship of phase and frequency fluctuation

$$\Delta\varphi(T) = 2\pi\langle\Delta\bar{f}\rangle T \tag{2.219}$$

where $\Delta\varphi(T)$ is the phase difference of *R*/*T* signal caused by frequency fluctuation at delay time T, $\Delta\bar{f}$ is the change of frequency fluctuation averages within T time against the previous time, and <...> refers to average value of multiple times for improving measurement accuracy.

It is assumed that the previous time is T (so it can be related to Allan variance in this way), as shown in Fig. 2.55.

In Fig. 2.55, $\Delta\bar{f}_2$ is the average value of the difference $\Delta f_2(t) = f(t) - f_s$ of the frequency fluctuation $f(t)$ and the frequency nominal value f_s within $(t_1 - T) - t_1$, $\Delta\bar{f}_1$ is the average value of the difference $\Delta f_1(t)$ of the frequency fluctuation $f(t)$ and f_s within $(t_1 - 2T)$, and the $\Delta\bar{f}_2 - \Delta\bar{f}_1 = \Delta\bar{f}$ expresses the change of average values for adjacent two intervals that occurred at $(t_1 - T)$ and t_1, so that the phase change $\Delta\varphi(T)$ shall evolve as $\Delta\varphi(T) = \Delta\bar{\omega}T = 2\pi\langle\Delta\bar{f}\rangle T$, which is the ranging error relative to the frequency fluctuation (short-term stability). Following two physical concepts shall be noted in this analytical method.

(1) T is not the integration time of the circuit but the delay between T/R signals which exists objectively in a physical relationship when the phase difference is derived from the frequency fluctuation (Expression (2.219)).

(2) And take the previous time segment also as T, for the purpose of measurement with Allan variance, then the relationship between the ranging error introduced by short-term stability and the Allan variance may be obtained (or using other methods to gain that, however, the results obtained with different methods shall be the same).

According to Expression (2.55), we can obtain that:

$$\Delta\bar{f} = \Delta\bar{f}_2 - \Delta\bar{f}_1$$

$$\left\langle \left(\Delta\bar{f}\right)^2 \right\rangle = \left\langle \left(\Delta\bar{f}_2 - \Delta\bar{f}_1\right)^2 \right\rangle$$

$$\frac{\left\langle \left(\Delta\bar{f}\right)^2 \right\rangle}{f_s^2} = \frac{\left\langle \left(\Delta\bar{f}_2 - \Delta\bar{f}_1\right)^2 \right\rangle}{f_s^2}$$

Substitute the relative frequency instability $y = \Delta f/f_s$ into the expression above; then

$$\frac{\left\langle \left(\Delta\bar{f}\right)^2 \right\rangle}{f_s^2} = \left\langle \left(\bar{y}_2 - \bar{y}_1\right)^2 \right\rangle$$

Because the Allan variance for zero clearance is

$$\sigma_y^2(T) = \frac{1}{2}\left\langle \left(\bar{y}_2 - \bar{y}_1\right)^2 \right\rangle$$

Therefore,

$$\left\langle \left(\Delta\bar{f}\right)^2 \right\rangle = 2f_s^2\sigma_y^2(T) \tag{2.220}$$

Substitute Expression (2.220) into Expression (2.219) to obtain:

$$\Delta\varphi(T) = 2\pi\sqrt{2}f_s\sigma_y(T)T$$

The one-way ranging error caused by short-term stability shall be:

$$\sigma_{R_2} = \frac{c\cdot\Delta\varphi}{2\cdot 2\pi f_s} = \frac{CT}{\sqrt{2}}\sigma_y(T) \tag{2.221}$$

The two-way ranging error caused by short-term stability is $2\sigma_{R_2}$.

Normally $\sigma_y(T)$ does not decrease linearly with increasing of T when T value is large. Therefore, the ranging error caused by it will be larger because of longer ranging distance. The result calculated from the expression above shows that such error is negligible [2], if the distance is not so far (e.g., to the Mars), but for extreme deep-space exploration, in case of Pluto along the solar system or entering into the Galaxy outside solar system, if the short-term stability of a signal is good enough, the non-negligible ranging error shall be introduced. Let's take an example of Pluto exploration. The Pluto is the farthest planet away from the Earth among the nine planets in solar system, and the time delay of electric wave between the Pluto and the Earth is about 27 h, that is,

$$T \approx 27 \times 3,600 = 9.72 \times 10^4 (s)$$

If hydrogen clock is used for measurement, by substituting $\sigma_y(9.72 \times 10^4 \mathrm{s}) \approx 8 \times 10^{-15}$ into Expression (2.221), we can obtain $\Delta R = 0.16$ m, which is not the major component in the total error.

If cesium clock is used for measurement, by substituting $\sigma_y(9.27 \times 10^4 s) \approx 8 \times 10^{-14}$ into Expression (2.221), we can obtain $\Delta R = 1.6$ m, which is the major component in the total error. Therefore, in extreme deep-space ranging, the effect of short-term stability of a signal on ranging accuracy shall be considered if the distance is extremely far, and a high-quality atomic clock is preferred.

The analysis above is an objective process in physics without effects of additional circuit, and integration filter is usually used in the practical circuit. The frequency domain method is recommended to analyze the effect of that filter. The transmission function diagram is as shown in Fig. 2.56.

As shown in Fig. 2.56, $\phi_i(\omega)$ is the power spectrum of input phase noise, and $\phi_0(\omega)$ is the power spectrum of output phase noise. $|H_1(\omega)|^2$ is the power transmission function of delay canceling.

$$|H_1(\omega)|^2 = 4 \sin^2 \frac{\omega T}{2}$$

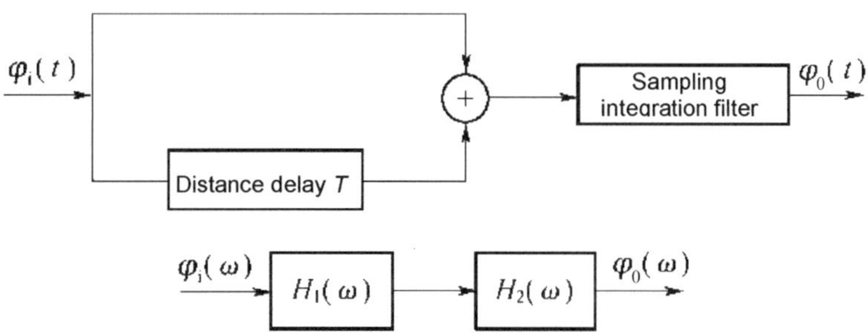

Fig. 2.56 Ranging transmission function block diagram

$|H_2(\omega)|^2$ is the power transmission function of integration filter when the sampling smoothing time is τ.

$$|H_2(\omega)|^2 = \frac{[\sin^2(\omega\tau/2)]}{(\omega\tau/2)^2}$$

The variance of the output phase noise is:

$$\sigma_\varphi^2 = \frac{1}{2\pi}\int_0^\infty \left\{\varphi_i(\omega)\cdot 4\sin^2\left(\frac{\omega T}{2}\right)\cdot\left[\sin^2\left(\frac{\omega\tau}{2}\right)\right]/\left(\frac{\omega\tau}{2}\right)^2\right\}\mathrm{d}\omega \qquad (2.222)$$

The frequency characteristics of $\varphi_i(f)$, $|H_1(f)|^2$ and $|H_2(f)|^2$ are as shown in Fig. 2.57.

As shown in Fig. 2.57, in case of good short-term stability (i.e., the $\varphi_i(f)$ in the figure is narrow), for deep-space TT&C with long time T, the integration filter works only when τ is close to T or τ is greater than T. If the sampling integration time τ is short, so the integration filter may not takes a well effect. In a practical circuit, the integration time is usually less than the delay time T of deep-space TT&C.

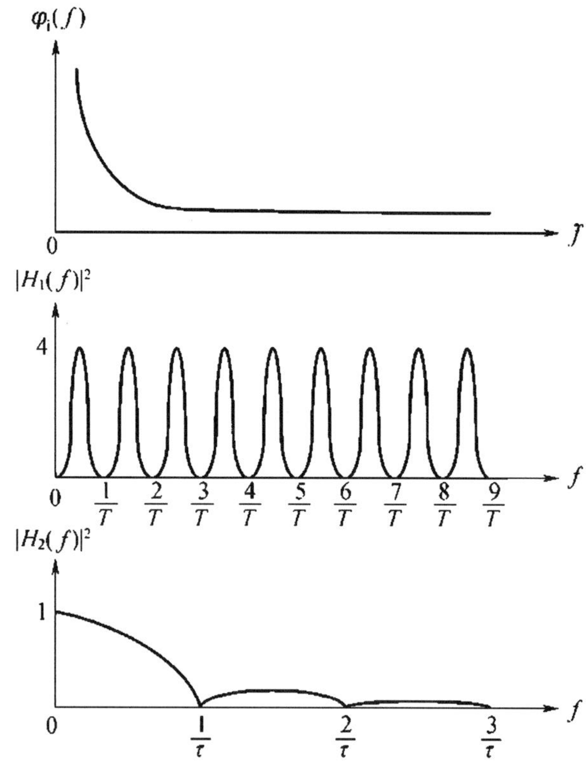

Fig. 2.57 Frequency domain analysis of ranging error

2.4.5 Theoretical Calculation of Ranging Accuracy

The ranging accuracy is mainly restricted by the random and system measurement errors which appear as random variable or system variable in measuring round-trip time of the ranging signal. Error calculation expressions are given below.

(1) Random error

The one-way random error that affects the equipment ranging accuracy has following error sources.

1) The error caused by thermal noise (see Sect. 2.4.5).

$$\sigma_{R_1} = \frac{c}{18f_r}\left[\frac{S}{2\varphi B_n}\right]^{-\frac{1}{2}} \tag{2.223}$$

where c is the velocity of light, f_r is the frequency of ranging principal tone, S/φ is the ratio of the power spectral density of principal tone signal and the noise spectral density, and B_n is the equivalent noise bandwidth of single-side principal tone loop.

2) One way error caused by instability of short-term frequency of the ranging principal tone (see Sect. 2.4.5).

$$\sigma_{R_2} = \frac{cT}{\sqrt{2}}\sigma_y(T) \tag{2.224}$$

where $\sigma_y(T)$ is the Allan variance of the instability of principal tone short-term frequency and T is the time delay in transmitting/receiving signal.

3) Error caused by multiple communication paths (see Sect. 5.4.4).

$$\sigma_{R_3} = \frac{\left(\frac{U_{m_2}}{2\pi U_{m_1}}\right)}{2f_R} \tag{2.225}$$

where U_{m_2} is the amplitude of multipath reflection signal, and U_{m_1} is the amplitude of direct signal.

4) Error caused by the phase noise of principal tone ring VCO.

$$\sigma_{R_4} - \frac{\sigma_\varphi c}{4\pi f_R} \tag{2.226}$$

where σ_φ is the phase noise (rad) of principal tone ring VCO.

5) Error caused by pulse jitter in case of range counter ON/OFF.

$$\sigma_{R_5} = \frac{\sqrt{2}}{2}c\sigma_T \tag{2.227}$$

where σ_T is the RMS value of pulse time jitter in case of counter ON/OFF.

6) Quantization error. The maximum error of the pulse measured time between range counter ON and OFF is ±1 count clock, then

$$\sigma_{R_6} = \frac{c}{2}\sigma_T = \frac{c}{2}\frac{1}{\sqrt{6}f_R} \tag{2.228}$$

where f_R is the frequency of count clock.

7) Total random error.

$$\sigma_R = \sqrt{\sigma_{R_1}^2 + \sigma_{R_2}^2 + \sigma_{R_3}^2 + \sigma_{R_4}^2 + \cdots} \tag{2.229}$$

σ_R is determined by σ_{R_1}. If f_r and S/φ are given, the bandwidth of principal tone loop B_n may be reduced to satisfy the requirement of system random error.

(2) System error

The one-way system error that affects the equipment ranging accuracy has following error sources.

1) Ranging error introduced by variation of system frequency characteristics.

① In case of calculating with amplitude-/phase-frequency characteristic: In engineering design, if the variations of phase-frequency characteristic and amplitude-frequency characteristic are given, the expression below may be used to calculate:

The ranging error introduced by the variation of phase-frequency characteristic (see Sect. 2.4.2), without considering effect of amplitude-frequency characteristic:

$$\Delta R_1 = \frac{c}{720 f_R}\Delta(\Phi_+ - \Phi_-) \tag{2.230}$$

where Φ_+ is the phase shift of the phase-frequency characteristic corresponding to $(f_O + f_R)$, Φ_- is phase shift of the phase-frequency characteristic corresponding to $(f_O - f_R)$, and $\Delta(\Phi_+ - \Phi_-)$ is the variation of $(\Phi_+ - \Phi_-)$.

The ranging error caused by the change of amplitude-/phase-frequency characteristics is:

$$\tau = \frac{\phi}{\Omega} = \frac{1}{\Omega}\text{tg}^{-1}\frac{H_- \sin[(\phi_0 - \phi_-) + \Delta\phi] + H_+ \sin[(\phi_0 - \phi_-) - \Delta\phi]}{H_- \cos[(\phi_0 - \phi_-) + \Delta\phi] + H_+ \cos[(\phi_0 - \phi_-) - \Delta\phi]} \tag{2.231}$$

where τ is the time delay of ranging tone, Ω is the frequency of ranging tone, $\Delta\phi$ is the non-orthogonal phase difference by orthogonal demodulation, ϕ_0 is the phase-frequency characteristic value relative to the carrier frequency ω_0, ϕ_+ is the phase-frequency characteristic value relative to the upper sideband of $(\omega_0 + \Omega)$, ϕ_- is the phase-frequency characteristic value relative to the lower sideband of $(\omega_0 - \Omega)$, H_+ is the amplitude-frequency characteristic value relative to the upper sideband of $(\omega_0 + \Omega)$, and H_- is the amplitude-frequency characteristic value relative to the lower sideband of $(\omega_0 - \Omega)$.

Use Expression (2.231) to calculate τ_0 and τ_1 respectively before/after change of $(H_+, H_-, \phi_+, \phi_-, \phi_0, \Delta\phi)$; then the variation of one-way distance caused by change of time delay $\Delta\tau = (\tau_1 - \tau_0)$ is that

$$\Delta R_1 = \frac{c(\tau_1 - \tau_0)}{2} \tag{2.232}$$

② In case of calculating with group time-delay characteristic (see Sect. 2.4.3): In engineering design, if the variation of group time-delay characteristic is given, then the following expressions are used to calculate:

The ranging error caused by stability of group time delay is:

$$\Delta R_1 = \frac{C}{2}\left(\Delta a_0 + \frac{\Delta a_2}{3}\Omega^2\right) \tag{2.233}$$

where a_0 is the zero-order term of group time-delay characteristic; a_1 and a_2 are the first-order term (linear distortion term) and the second-order term (parabola distortion term), respectively; Δa_0 and Δa_2 are the variations of a_0 and a_2; and Ω is the frequency of fine tone.

The ranging error caused by the change of Doppler frequency ω_d is:

$$\Delta R_1 = \frac{\left(a_1\omega_d + a_2\omega_d^2\right)c}{2} \tag{2.234}$$

where ω_d is the Doppler frequency shift (or the change of signal frequency and filter center frequency caused by other factors).

2) Dynamic error ΔR_2. The dynamic error is the error that affected by the target velocity, acceleration, etc., which is mainly caused by the phase difference $\Delta\theta_e$ of the principal tone loop, and the expression is shown below:

$$\Delta R_2 = \frac{c}{4\pi f_R}\Delta\theta_e \tag{2.235}$$

For each type and exponent of phase-locked loop, $\Delta\theta_e$ is calculated with the expressions in Table 2.10 [14].

As shown in Table 2.10, B_L is the noise equivalent loop bandwidth of the (single-side) principal tone ring, α is the Doppler rate (Hz/S), β is the Doppler acceleration (Hz/S^2), t is the time since Doppler acceleration starts (if the Doppler acceleration starts before the principal tone is locked, then t is the time since the loop is acquired and locked). If the Doppler acceleration appears continuously, then the periodic cycle skipping will be found in the Type II loop.

For the Type II loop shown in the table, the phase transmission function is:

$$H(s) = \frac{sK_1 + K_2}{s^2 + sK_1 + K_2} \tag{2.236}$$

Table 2.10 Phase error of principal tone ring $\Delta\theta_e$ (rad)

Loop	Constant range rate Constant Doppler frequency shift	Derivative of constant range rate Constant Doppler rate	Second derivative of constant range rate Constant Doppler acceleration
Type II loop with standard underdamping	0	$\frac{9\pi}{16B_L^2}\cdot a$	$\left\{\frac{9\pi\beta}{16B_L^2}\right\}t-\frac{27\pi\beta}{64B_L^3}$
Type II loop with supercritical damping	0	$\frac{25\pi}{32B_L^2}\cdot a$	$\left\{\frac{25\pi\beta}{32B_L^2}\right\}t-\frac{125\pi\beta}{128B_L}$
Type III loop with standard underdamping	0	0	$\frac{12{,}167\pi}{8{,}000B_L^3}\cdot\beta$
Type III loop with supercritical damping	0	0	$\frac{35{,}937\pi}{16{,}384B_L^3}\cdot\beta$

where $K_1=(8/3)B_L$, $K_2=(1/2)K_1^2$ (standard underdamping); $K_1=(16/5)B_L$, $K_2=(1/4)K_1^2$ (supercritical damping).

For the Type III loop shown in the table, the phase transmission function is:

$$H(s)=\frac{s^2K_1+sK_2+K_3}{s^3+s^2K_1+sK_2+K_3} \tag{2.237}$$

where $K_1=(60/23)B_L$, $K_2=(4/9)K_1^2$, $K_3=(2/27)K_1^3$ (standard underdamping); $K_1=(32/11)B_L$, $K_2=(1/3)K_1^2$, $K_3=(1/27)K_1^3$ (supercritical damping).

In the Type II loop, because $\sigma_{R_1}\propto\sqrt{B_n}$ and $\Delta_{R_1}\propto\left(1/B_n^2\right)$, the requirements of the random error and the system error cannot be satisfied simultaneously under the limit conditions of threshold level, maximum speed $\dot{R}_{\max}$, and maximum acceleration $\ddot{R}_{\max}$. The method below is recommended.

① The variable bandwidth method is applied to the principal tone loop. The flight vehicle has a large acceleration in low Earth orbit, but the s/φ value is much greater than the threshold value, so the bandwidth of the principal tone loop is widened to satisfy the requirements of σR_1 and ΔR_1. When the flight vehicle is operated in high Earth orbit, the s/φ is low, and the acceleration is small, so the bandwidth of principal tone loop is narrowed to satisfy the requirements of σR_1 and ΔR_1.

② If the Type III loop is taken as the principal tone loop, then the error caused by the Doppler rate of change becomes zero.

③ The group time delay of the principal input filter varies with the change of temperature and time, and then drift error Δ_{R_5} is introduced.

Table 2.11 lists the ranging errors caused by different phase changes of the principal tone.

Table 2.11 Effects on ranging errors by the input filter of principal tone loop

Principal tone (kHz)	Phase drift (°)	Ranging error (m)
100	1	4.17
27	1	15

Following methods may be used to minimize the error.

- A thermostat is added on the principal tone input filter.
- When the bandwidth of input filter if the SNR of principal tone input is satisfied, so as to even the phase characteristic of the filter.
- A digital filter is used. The error is fundamentally eliminated after using the digital ranging terminal.
- Zero residual Δ_{R_4}. Because the pattern of each factor that affects the system error is hard to obtained, as a result, the zero value of the equipment is hard to adjust, which consequently leads to ranging error.
- Ranging error Δ_{R_5} introduced by up-link. The phase jitter and drift caused by the phase modulator may change the phase of the baseline signal, and the amplitude-/phase-frequency characteristics (or group time delay) of the filter next to the modulator may cause the phase drift in case of being affected by temperature, so that the ranging error may be introduced. Its calculation has been described above.

The power amplifier is a wideband component that hardly affects the ranging error, however, the change of time delay shall be minimized in design.

- Ranging error Δ_{R_6} introduced by down-link. The ranging error may be caused by following reasons: phase jitter of carrier loop caused by thermal noise and combination intervene, change of receiver input level, the change of amplitude-/phase-frequency characteristics (or group time delay) of the down-link channel caused by changes of frequency and temperature and aging. Its calculation has been described above.
- Ranging error Δ_{R_8} caused by the change of antenna phase center. Radio ranging uses the center of antenna phase as reference; if the center is changed, it will cause ranging error.
- The total system error is calculated by square and extraction of a root, that is,

$$\Delta R_{\Sigma} = \sqrt{\Delta R_1^2 + \Delta R_2^2 + \Delta R_3^2 + \Delta R_4^2 + \Delta R_5^2 + \Delta R_6^2 + \Delta R_7^2 + \Delta R_8^2} \qquad (2.238)$$

2.4.6 *One-Way Ranging Technique*

For the two-way ranging technique introduced above, the flight vehicle requires a transponder to cooperate with the ground station and to form an uplink/downlink loop to measure the delay difference between uplink and downlink signals, thus obtaining the range. In that method, the transponder increases the load on flight vehicle, and the range error brought about by it is a main factor limiting the improvement of range accuracy in the current ranging system. The one-way ranging technique, however, needs no transponder, and so there's no such shortcoming to overcome.

2.4.6.1 One-Way Ranging by Interferometer

One-way ranging by interferometer refers to ranging by utilizing downlink signals. In this method, the flight vehicle transmits a random waveform signal, and the range then can be worked out by using the two range differences measured by three signal-receiving-only ground stations located in a straight line. This method is also called "Ranging by Interferometer" [15]. "One-way ranging" can be applied provided that the target is not too far from ground stations, and the radio waves which are allowed to be received by ground stations are either plane waves not spheric waves. That is to say, from the perspective of geometric optics, the three rays received by the three ground stations are not parallel to each other but form a triangle ΔA_1SA_2, as shown in Fig. 2.58. Angle measuring by interferometer, however, is usually for targets at a far distance and the three rays received are assumed to be parallel to each other. This is exactly the main distinction between "ranging by interferometer" and "angle measuring by interferometer."

In the left of Fig. 2.58, the space flight vehicle S is within the 3D coordinates of $OXYZ$ and the three ground stations A_0, A_1, and A_2 are on OY axis. The length of baseline between stations is B. A_0 is the master station and A_1 and A_2 are slave stations. The distances from S to A_0, A_1, and A_2 are R_0, R_1, and R_2, respectively. Parameter to be measured is the distance R_0 between master station A_0 and target S. Take out the plane formed by S and OY axis from the 3D coordinates in Fig. 2.58 and a 2D coordinate system A_1SY is got, as shown in the right figure above. L is the vertical distance between S and OY axis. From the geometric relationship in the figure above, we can get:

$$R_0^2 = (B+b)^2 + L^2 = (B+b)^2 + \left(R_2^0 - b^2\right) = B^2 + 2Bb + R_2^2 \tag{2.239}$$

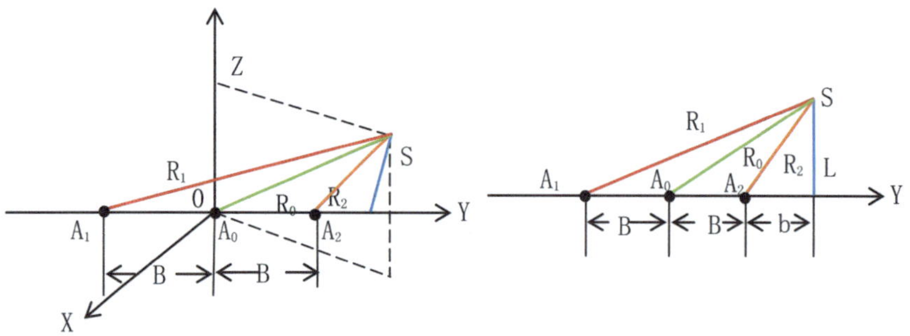

Fig. 2.58 Geometric relationship of rays for ranging by interferometer

Since

$$L^2 = R_1^2 - (2B + b)^2 = R_2^2 - b^2$$

Therefore,

$$2Bb = \frac{R_1^2 - R_2^2}{2} - 2B^2 \tag{2.240}$$

Substitute Expression (2.240) into Expression (2.239) to obtain:

$$R_0^2 = \frac{R_1^2 + R_2^2}{2} - B^2 \tag{2.241}$$

Expression (2.241) shows the relationship between distances R_0 and (R_1+R_2). However, distance (R_1+R_2) is hard to measure in engineering, so it is necessary to transform Expression (2.241) to a range difference equation, making it easy for measurement in engineering. From Expression (2.241), we get:

$$\begin{aligned}
2R_0^2 &= R_1^2 + R_2^2 - 2B_2^2 = \left[\left(R_1^2 - 2R_0R_1 + R_0^2\right) - \left(-2R_0R_1 + R_0^2\right)\right] \\
&\quad + \left[\left(R_2^2 - 2R_0R_2 + R_0^2\right) - \left(-2R_0R_2 + R_0^2\right)\right] - 2B^2 \\
2R_0^2 &= (R_1 - R_0)^2 + (R_2 - R_0)^2 + 2R_0R_1 + 2R_0R_2 - 2R_0^2 - 2B^2 \\
& 2R_0(2R_0 - R_1 - R_2) = (R_1 - R_0)^2 + (R_2 - R_0)^2 - 2B^2 \\
R_0 &= \frac{(R_1 - R_0)^2 + (R_2 - R_0)^2 - 2B^2}{2[(R_0 - R_1) + (R_0 - R_2)]} = \frac{2B^2 - \left[(R_1 - R_0)^2 + (R_2 - R_0)^2\right]}{2[(R_1 - R_0) + (R_2 - R_0)]}
\end{aligned} \tag{2.242}$$

Expression (2.242) is the one for solving range R_0 by using range difference. There are two methods for measuring range difference:

(1) Delay difference measurement method

Convert range difference into delay difference:

$$(R_1 - R_0) = (\tau_1 - \tau_0)C = \Delta\tau_{10}C, (R_2 - R_0) = (\tau_2 - \tau_0)C = \Delta\tau_{20}C$$

Substitute the equation above into Expression (2.242) to obtain:

$$R_0 = \frac{2B^2 - \left[\Delta\tau_{10}^2 + \Delta\tau_{20}^2\right]C^2}{2[\Delta\tau_{10} + \tau_{20}]C} \tag{2.243}$$

The delay difference can be measured by using the marking signals of downlink signals, such as the frame synchronization signal of telemetry signal, "all 1" signal of PN code signal, etc.

(2) Phase difference measurement method

To have a higher measurement accuracy, the delay difference can be measured by utilizing the phase difference of carriers. The relationship between delay difference and phase difference is as below:

$$\Delta\tau_{10} = \frac{\Delta\varphi_{10}}{\omega_0} \quad \Delta\tau_{20} = \frac{\Delta\varphi_{20}}{\omega_0}$$

Substitute the equation above into Expression (2.243) to obtain:

$$R_0 = \frac{2B^2 - \left[\left(\Delta\varphi_{10}^2 + \Delta\varphi_{20}^2\right)/\omega_0^2\right]C^2}{2[(\Delta\varphi_{10} + \Delta\varphi_{20})/\omega_0]C} \tag{2.244}$$

Although carrier phase difference measurement method can get a higher measurement accuracy, it can easily cause phase ambiguity. In that case, the value measured by "delay difference" can be used for resolving such ambiguity. So it is suggested that "delay difference" be used for coarse ranging and "carrier phase difference" for fine ranging.

The measurement accuracy of one-way ranging method relates to the length and accuracy of baseline, delay and phase-shift stability of measuring instruments, wave length and signal form of measuring signal, as well as propagation characteristics of radio waves. Delay, phase calibration, and wired baseline transmission (connected end interferometry) are usually the methods for improving ranging accuracy. With the same delay difference measuring accuracy, the ranging accuracy decreases with the increase of range and deviation angle of target.

2.4.6.2 One-Way Ranging Technique by the Time Synchronization Between Satellite and Earth

The two-way ranging technique aforesaid is to generate the ranging signal to be transmitted based on the clock of ground station, and to measure the two-way delay of the received/transmitted signals based on the same clock to obtain higher ranging accuracy. If the satellite and Earth have high-accuracy synchronization in time, the ranging signal to be transmitted then can be generated by the satellite and the ground station can utilize the highly synchronized clock signal to measure the one-way delay, thus obtaining the one-way range. However, error in the time synchronization between satellite and ground station may directly lead to range error, so its engineering application will significantly influence the development of satellite-Earth high-accuracy synchronization technology. This method is really a very promising ranging technique.

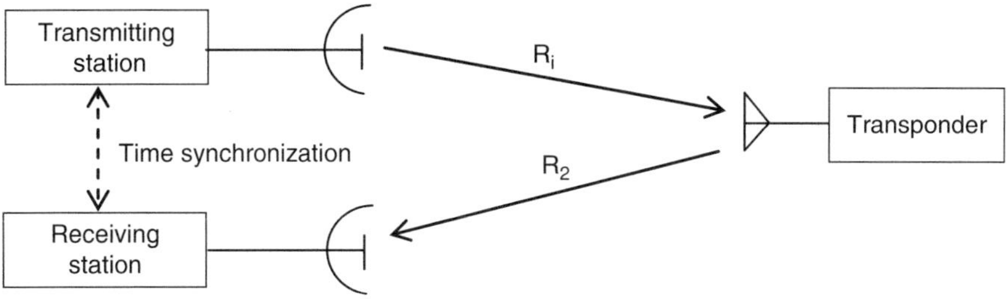

Fig. 2.59 Schematic diagram of a three-way ranging

2.4.7 *Three-Way Ranging Technique*

For the two-way ranging technique described above, ranging signals are transmitted and received by the same ground station. The technique allowing ranging signals to be transmitted by one ground station and received by another is called three-way ranging. Its working principle block diagram is as shown in Fig. 2.59.

As shown in Fig. 2.59, high-accuracy inter-station synchronization technology is utilized to synchronize the time between receiving and transmitting stations. Assuming that the synchronization error is $\Delta\tau$ and if the time synchronized in the receiving station is used as the reference time for measuring the delay of downlink signal, the delay τ measured by receiving station will be as below:

$$\begin{gathered}\tau = \frac{R_1}{C} + \frac{R_2}{C} + \Delta\tau \\ (R_1 + R_2) = \tau C - (\Delta\tau)C = C(\tau - \Delta\tau)\end{gathered} \tag{2.245}$$

As shown in Expression (2.245), the value measured is the sum of ranges.

The accuracy of three-way ranging is lower than that of two-way ranging for the following reasons:

(1) The time synchronization error $\Delta\tau$ between stations. usually solved by utilizing high-accuracy inter-station synchronization. The main solutions include time synchronization by common-view GPS satellite and time synchronization by two-way transmission comparison.

(2) The ranging error caused by the uplink delay of transmitting station equipment and the downlink delay of the receiving station equipment: for two-way ranging, the uplink and downlink equipment delays are generated in the same ground station and can be deducted through range calibration; while for three-way ranging, however, the equipment delays exist in the one-way delay of two stations respectively, which makes three-way range calibration impossible. One solution is that the two stations calibrate their own two-way equipment delay $\Delta\tau_1$

and $\Delta\tau_2$ respectively and then work out the average uplink delay and downlink delay of the two station equipment: $\left(\frac{\Delta\tau_1+\Delta\tau_2}{2}\right)$, that is,

$$\begin{aligned} \Delta\tau_1 &= \left(\Delta\tau_{\text{uplink}}\right)_1 + \left(\Delta\tau_{\text{downlink}}\right)_1 \\ \Delta\tau_2 &= \left(\Delta\tau_{\text{uplink}}\right)_2 + \left(\Delta\tau_{\text{downlink}}\right)_2 \\ \frac{\Delta\tau_1 + \Delta\tau_2}{2} &= \frac{\left(\Delta\tau_{\text{uplink}}\right)_1 + \left(\Delta\tau_{\text{uplink}}\right)_2}{2} + \frac{\left(\Delta\tau_{\text{downlink}}\right)_1 + \left(\Delta\tau_{\text{downlink}}\right)_2}{2} \end{aligned} \quad (2.246)$$

Obviously, this can only reach the approximate range calibration. If the consistency between the two station equipment is good, the calibration error will be small accordingly.

Three-way ranging technique is usually used in deep-space TT&C. Under that circumstance, the receiving/transmitting signal delay is very long, and due to the Earth's rotation, the antenna beam of a ground station cannot receive its return signal after it is transmitted, and when the signal returns to the Earth's surface, it is needed to have another beam to receive it, thus realizing three-way ranging.

2.4.8 *Deep-Space Ranging System*

Currently, three main systems are used for deep-space ranging, that is, sequential tone ranging, PN code ranging, and tone and code mixed ranging. The analytical results for each system are discussed below.

(1) Sequential tone ranging [12, 16, 17]. The signal used for sequential tone ranging is a correlated tone sequence with frequency ratio of 2. It has one ranging clock and multiple ambiguity-resolving tones, of which the previous one decides the accuracy, and the latter determines the distance without ambiguity. When a tone sequence is transmitted, every tone may last for a while so that the correlated integration is conducted to improve the SNR. Under operating condition, a ranging data can be obtained by transmitting a tone sequence. Therefore, *n* ranging data will be obtained by transmitting periodically for n times. Strictly speaking, sequential tone ranging is discrete ranging. Due to that the ranging data is determined by the code clock; thus, the method aforesaid may be used to calculate the ranging error by substituting the code clock for the tone.

(2) PN code ranging [12, 16]. The signal used for deep-space PN code ranging is a compound code ranging signal logically combining a ranging clock and several PN codes ("and" and "or" logic), wherein the ranging clock decides the accuracy, and the compound code resolves the distance ambiguity. Because such signal is a continuous signal, so that PN code ranging is continuous ranging. For such PN code signal only used for ranging, the compound code ranging signal is designed to obtain higher accuracy and longer distance ambiguity, that is, the major part of power of a signal concentrates on the ranging clock for improving the ranging

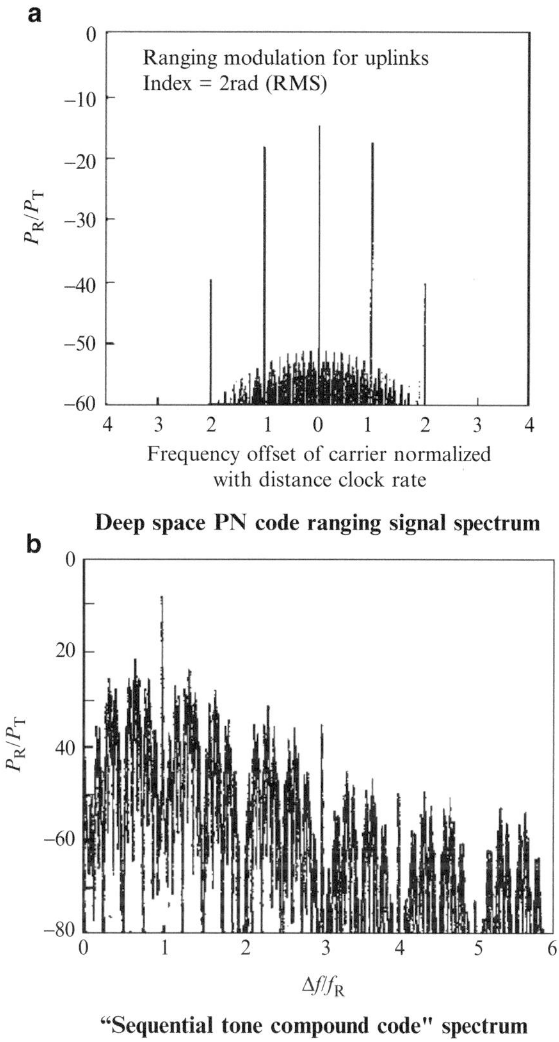

Fig. 2.60 Deep-space ranging signal spectrum. (**a**) Deep-space PN code ranging signal spectrum. (**b**) "Sequential tone compound code" spectrum

accuracy and acquiring signal easily, therefore, such signal basically is a periodic sequence with a period of 2 chips. In case of resolving ambiguity, other PN codes make the ranging clock to reverse in a half-period occasionally, but such reversion is not happened frequently. One specific example for such compound PN code is a regenerative compound code for ranging used in the Deep-Space Network of the United States. The frequency spectrum of the code is as shown in Fig. 2.60a [12].

As shown in Fig. 2.60a, P_T is the total power of signal, P_R is the power of ranging signal, f_R is the frequency of ranging clock, and the Δf is the frequency offset relative to the center frequency of carrier.

From Fig. 2.60a, the ranging clock has a great first-order sideband component, and the spectrum line between the carrier frequency and the second-order component is small, the reason is that the component for resolving ambiguity is produced by the semi-cycle conversion occurred occasionally with the compound PN code.

The main spectrum component of ranging signal still is the ranging clock component, and other components rarely affect the ranging clock. Therefore, the method aforesaid can be remained for estimating the ranging error.

The important thing is that the method aforesaid is not applicable to spread spectrum TT&C system. Because the spread spectrum PN code is not only used for ranging but also used for anti-jamming, anti-interception, code division multiple access (CDMA), etc., so that the spectrum of the ranging signal is similar with that of "white noise."

(3) Tone and code mixed ranging. Here is a typical example presented in *ESA Standards* (ESA PSS-04-104). For tone and code mixed ranging, the ranging signal is such signal that obtained by 45° phase modulation with one Sequential tone code against the ranging clock (i.e., high tone), wherein the ranging clock determines the ranging accuracy, and the Sequential tone code is used for resolving the distance ambiguity, it can be expressed as

$$C_n = Q_1 \oplus Q_2 \oplus Q_3 \oplus \cdots \oplus Q_n$$

where C_n is the nth code, $\oplus$ represents for XOR logic, and Q_i stands for the two-divided frequency sequence square wave generated by ranging clock.

From this expression, the code is compounded by the Sequential tones, not the PN codes. The difference between such code and the Sequential tone is that, since the ranging clock always exists, the code is not acquired by searching using relevant functions but acquired in sequence by modulation of different tones; therefore, this kind of code acquisition is more simple and reliable, which is the reason why such a code is better than PN compound code. The spectrum of such ranging signal after distance acquisition is shown in Fig. 2.60a.

Observed from Fig. 2.60b, because the ranging clock only undertakes 45° phase modulation, therefore, residual clock component is great, so that other components hardly affect the ranging clock. Consequently, the method aforesaid can be used for estimating the ranging error.

The analytical results obtained above may be used to calculate the ranging error caused by group delay. If a_1, a_2, τ_m, Ω_T, and θ_0 are given, the ranging error may be calculated with equation; if the group delay characteristic curve is given, the ranging error may be calculated in graphic way. The latter is more accurate. For PN code ranging signal, the method aforesaid also can be used to calculate the ranging error only when the component of code clock is the main component of ranging signal spectrum.

2.5 Angle Measurement Theory and Technology

The angle measurement using antenna-tracking and the interferometer angle measurement are usually adopted for TT&C system. For angle measurement using antenna-tracking, the antenna-feed-servo subsystem is used to automatically track the target, so that the electronic boresight of antenna aims to the target correctly, then the displayed value of the antenna pedestal position, after compensation, is the angle measured value. It is a common method used for TT&C system. However, this method has limited measurement accuracy due to several factors including the processing accuracy of antenna-feed-servo system, antenna rotating and tracking receiver. In this case, the interferer system shall be adopted for further improving the accuracy of the height and angle measurement. The reason is that the aperture of a single antenna is limited, but the interferer may improve the accuracy of the height and angle measurement by extending the distance (or referred to as "baseline") between the two measurement units (or referred to as "terminal station"). Moreover, with the development of electrical measuring technology, the measuring element of the phase interferer is the phase or time delay of a signal, consequently, the angle measurement in high accuracy and long distance recently adopts the Very Long Baseline Interferometry (VIBI) technology.

Angle tracking is a technology combining auto control and wireless radio, in which a lot of edge technical problems exist. We will discuss the specific problems below.

2.5.1 Angle Measurement Using Antenna Tracking – Three-Channel, Dual-Channel, and Single-Channel Monopulse Technologies

Since the conical scanning radar was invented in 1942, its reliability and accuracy always were the major issues for further improvement. The rotating joint and the conical scanning motor of the radar are the main weak links for its reliability. In the 1950s, a three-channel monopulse system is presented. It significantly improved the accuracy and reliability, making the precision tracking radar entered into a new era. This system is still in use today but complicated. After that, a dual-channel monopulse (DCM) system which was easier than the three-channel monopulse system was presented and often been used in recent TT&C system. However, it has some problems in cross coupling and angle error caused by phase shift inconsistence of the sum and difference channels, and still complicated. Then a single-channel monopulse (SCM) system was presented in 1960s, which simplified the system and was more reliable than conical scanning but had a poorer accuracy than the two systems aforesaid. Therefore, most medium-accuracy systems (e.g., telemetry receiver, guidance system, etc.) adopted this system from the 1960s to 1970s, but it also had some issues such as conversion loss caused by the single-channel converter, lower tracking accuracy and higher sidelobe. Then, the radiator scanning radar system (RADSCAN), a new conical scanning system, was appeared in 1982. It uses one waveguide radiator (or bevel waveguide), which is deviated from the rotating axis, to move around the rotating axis, so as to produce a conical scanning beam. It has not the defects that the single-channel converter in SCM has. Due to the transmitter weighs light (or even can be made of carbon fiber), so the rotating motor only requires a small driving power. The brushless DC motor is recommended with some advantages like high reliability, high G/T value, lower sidelobe and light weight. Since then, several angle tracking methods similar with such "sequence lobe" system had been presented, for example, electronic scanning (ESCAN) antenna, etc.

The amplitude comparison and the phase comparison are included in simultaneous lobe angle measurement system. For amplitude comparison, it has three-channel monopulse, dual-channel monopulase, single-channel monopulse, conical scanning, and amplitude extreme tracking. For phase comparison, it includes phase monopulse and interferer system.

Some angle tracking systems that are commonly used in TT&C system are discussed below.

2.5.1.1 Three-Channel Monopulse (TCM) System

The working principles of amplitude comparison and equisignal method are normally used for this system, as shown in Fig. 2.61.

For measuring the angle of the two planes in azimuth and elevation, the four antennas (or feed horn) shown in Fig. 2.61a, b may be used to measure it with amplitude measurement. Theoretically, the sum signal E_{Σ} produced by the comparator, the error signal E_A of azimuth angle and the error signal E_E of the elevation angle form the amplitude monopulse diagram and the sum-difference directional diagram as shown in Fig. 2.61d, e respectively. Because there are three signals, E_{Σ}, E_A, and E_E, shall be processed, it is so called TCM system. In Fig. 2.61c, beam 1 and beam 2 are same but overlapped in part, and the line from the intersecting point of the two beams to the origin O is the equisignal line (or referred to as electronic boresight). When the target places on the OA line, the two beams receive the signals with same amplitude and phase, it gets zero by subtracting, which is the origin point of angle tracking. The equisignal method has a higher accuracy of angle measurement, but the antenna feeds are more complicated. Two beams may be alternatively produced by sequential rotating of one lobe (sequential lobing), or simultaneously produced by two lobes (simultaneously lobing).

The block diagram of tracking servo system is as shown in Fig. 2.61f. The angle error signal obtained using the difference directional pattern is converted into the DC error signal through processing and modulation in receiver, and its voltage amplitude is in direct proportional to the azimuth (or elevation angle) that the antenna deviates from the target. Then such signal is transmitted to the servo system for amplification and correction, driving the execution mechanism consisting of

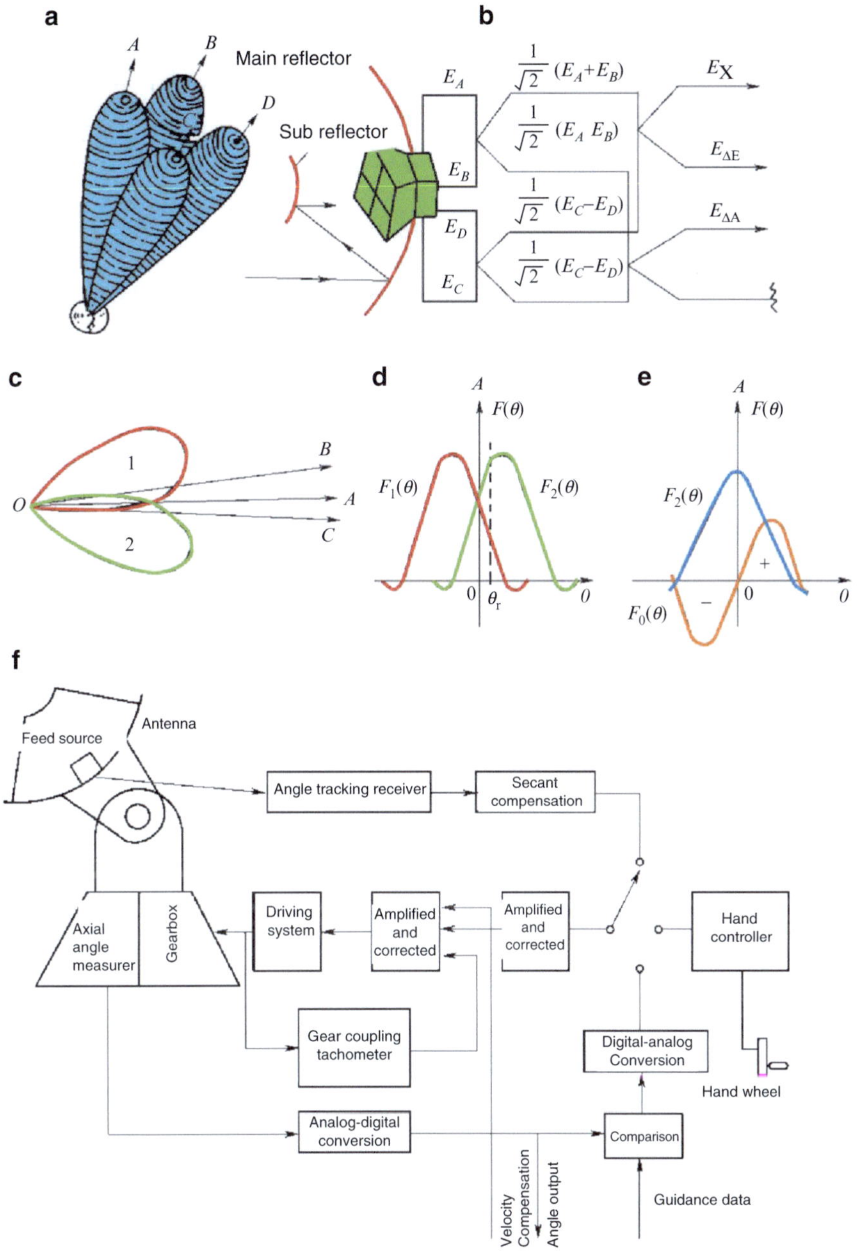

Fig. 2.61 Three-channel monopulse system concept diagram

motor and reduction gearbox to make the antenna to move along the direction of reducing sighting error, so that the electronic boresight of antenna points the target. For angle tracking, the error that the target deviates from the electronic boresight is referred to as tracking error. The intersection point of the antenna azimuth and elevation rotating shafts is considered as the reference point, then the rotating position of the antenna azimuth shaft and the elevation shaft reflects the angle of target. Therefore, for angle measurement, the shaft angle measurer installed on the rotating shaft of antenna is used to measure the target angle in the coordinate system, and the measuring error is referred to as angle measurement error.

2.5.1.2 Dual-Channel Monopulse System (DCM)

For simplifying system, the azimuth error signal and the elevation error signal in TCM are usually combined as one signal with the orthogonal phase multiplexing method, which becomes the dual-channel monopulse system. As shown in Fig. 2.62, $E_{\Delta E}$ and $E_{\Delta A}$ are combined to E_{Δ}, which can be realized by multiple horn feeds and 90° phase shifter, or by multimode feed with circular waveguide TE_{11} and TE_{21}.

As shown in Fig. 2.62, K_{Σ} and Φ_{Σ}, respectively, are the gain and phase shift of the sum channel, and K_{Δ} and Φ_{Δ} are the gain and phase shift of the sum channel, which are normalized to the amplitude with AGC in the sum channel and the orthogonal phase are divided by two coherent multipliers with phase difference of 90° to modulate the azimuth error signal and the elevation error signal. The reference signal of the multiplier is coherent to the sum-channel signal with a phase-locked loop. If the amplitude of the sum-channel signal is set to 1:

$$E_{\Delta} = \mu\theta_A \cos\omega_0 t + \mu\theta_E \cos\left(\omega_0 t + \frac{\pi}{2}\right) = \mu\theta_A \cos(\omega_0 t - \varphi) \tag{2.247}$$

$$E_{\sum} = \cos\omega_0 t$$

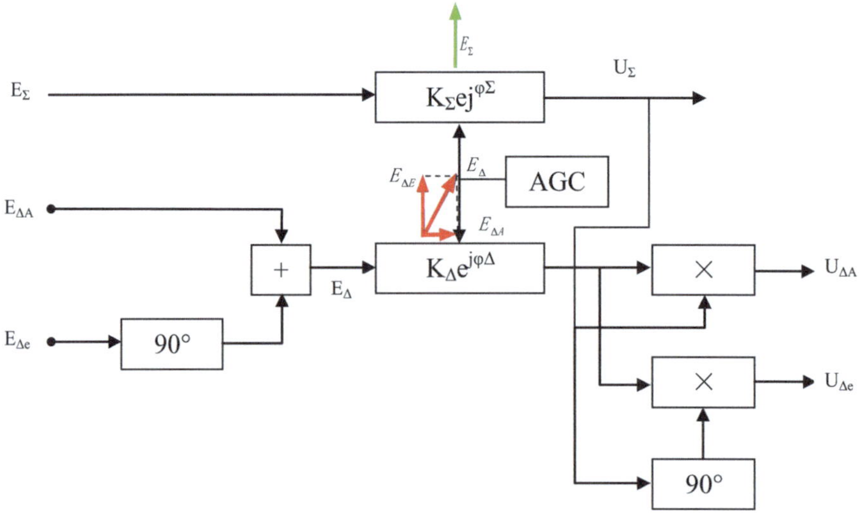

Fig. 2.62 Schematic diagram of dual-channel monopulse

where θ_A and θ_E are the azimuth deviation angle and the elevation deviation angle, respectively, and μ is the slope of relative angle error of the antenna.

$$\theta = \sqrt{\theta_A^2 + \theta_E^2}, \quad \Phi = \arctan\frac{\theta_E}{\theta_A}$$

where Φ is the phase of combined carrier and θ is the included angle between the target and the electronic boresight of antenna.

After magnification and AGC normalization through respective channel,

$$U_\Delta = \frac{K_\Delta}{K_\Sigma}\mu\theta_A \cos\omega_0 t + \frac{K_\Delta}{K_\Sigma}\mu\theta_E \cos\left(\omega_0 t + \frac{\pi}{2}\right) \tag{2.248}$$
$$U_\Sigma = \cos(\omega_0 t + \Delta\varphi)$$

where $\Delta\varphi = \varphi_\Sigma - \varphi_\Delta$ is the phase inconsistency of the sum channel and the difference channel when the difference channel is taken as a reference channel. The reference signal is coherent with the sum-channel signal, i.e., $\cos\left(\omega_0 t + \Delta\varphi + \frac{\pi}{2}\right)$, and when it is multiplied with the difference-channel signal $U\Delta$ by LPF, we can obtain that

$$U_{\Delta A} = \frac{K_\Delta}{K_\Sigma}\mu\theta_A \cos\Delta\varphi + \frac{K_\Delta}{K_\Sigma}\mu\theta_E \sin\Delta\varphi \tag{2.249}$$

According to the expression above, the first term is the effective azimuth error output signal, and there is a second term if $\Delta\varphi$ exists, which is related to the elevation deviation angle, making the azimuth error signal has the elevation deviation signal; consequently, a cross coupling will be formed for elevation against azimuth. Therefore, when the antenna is tracking a target, the spiral tracking will be produced with slow velocity of convergence and poor dynamic tracking performance. After 90° phase shift of reference signal, $\cos\left(\omega_0 t + \Delta\varphi + \frac{\pi}{2}\right)$ is multiplied with the difference channel signal U_Δ through LPF, then we can obtain that

$$U_{\Delta E} = \frac{K_\Delta}{K_\Sigma}\mu\theta_E \cos\Delta\varphi + (K_\Delta/K_\Sigma)\mu\theta_A \sin\Delta\varphi \tag{2.250}$$

According to the expression above, the first term is the effective elevation error output signal, and when the phase shift of the sum channel and the difference channel are different, the cross coupling signal for azimuth against elevation will be produced. The cross coupling for azimuth against elevation K_{cE} is defined as: when $\theta_E = 0$ but θ_A exists, then the ratio of the cross-coupling interfering signal output from the elevation channel and the output signal of the azimuth channel is obtained from Expressions (2.249) and (2.250):

$$K_{cE} = tg\,\Delta\varphi \tag{2.251}$$

And the cross coupling for elevation against is also be solved in a same way with the identical expressions.

Even though the dual-channel monopulse system has the cross coupling, but it simplifies the receiver, with the simple sum-difference signals produced by multimode feeds, so this system is widely applied in TT&C system.

If the sum phase and the difference phase have a good consistency, then the accuracy of angle measurement in DCM is similar with that in TCM, where the tracking error caused by thermal noise is:

$$\sigma_1 = \frac{\theta_{0.5}}{K_m\sqrt{\frac{S}{N}\frac{B_{1F}}{B_n}}} = \frac{1}{\mu\sqrt{\frac{S}{N}\frac{B_W}{B_n}}} \tag{2.252}$$

where $\theta_{0.5}$ is the beam width of half power point, *S/N* is the SNR of receiver (bandwidth is B_{1F}), B_{1F} represents the equivalent noise bandwidth of medium-frequency amplifier of the receiver, B_n is the equivalent noise bandwidth of the servo, K_m represents the normalized difference slope (generally ranging between 1.2 and 1.7), $K_m = K\left(\theta_{0.5}/\sqrt{G_\Sigma}\right)$, $K = (\partial F(\theta)/\partial\theta)\big|_{\theta=0^0}$ is the absolute slope of difference beam voltage gain of the antenna (when $\theta = 0°$), which is determined by the antenna aperture, wherein, $F(\theta)$ is the difference voltage pattern, G_Σ is the gain of sum beam, $\sqrt{G_\Sigma}$ is the sum beam voltage gain, so that K_m is defined as: the slope of the difference beam voltage gain is normalized by the sum beam voltage gain, with measurement unit of 3 dB beam width. (Note: Another normalization difference slope $\mu = \left[\frac{\partial F(\theta)}{\partial\theta}/\sqrt{G_\Sigma}\right]$ $= K_m/\theta_{0.5}$ is often used in antenna. Do not confuse the former one with the latter).

$K = K_m\sqrt{G_\Sigma}/\theta_{0.5}$ is obtained from K_m expression, then the voltage gain of the difference beam deviated from the electronic boresight of $\Delta\theta$ is $\sqrt{G_\Sigma} = \Delta\theta K$, and the voltage gain ratio of the sum beam and the difference beam is $\sqrt{G_\Delta}/\sqrt{G_\Sigma} = K_m(\Delta\theta/\theta_{0.5})$, and then the voltage of the receiver's sum/difference input signal may be simulated base on these expressions.

2.5.1.3 Single-Channel Monopulse System (SCM)

For further simplifying the system, the dual-channel or the three-channel described above are combined as single-channel monopulse respectively. The combination of the azimuth error signal and the elevation error signal is based on time division multiplexing, while the combination of the angle error signal and the sum-channel signal is normally implemented by conducting ($0/\pi$) phase modulation and then adding sum-channel signal to generate the amplitude modulation signal.

More and more equipment provided with single-channel monopulse tracking system are used in the system that has low accuracy requirement for tracking system, such as business TT&C station, RSGS, satellite reconnaissance signal receiving station, and telemetering ground station.

(1) Dual-channel monopulse combines as single-channel monopulse: On the basis of dual-channel monopulse shown in Fig. 2.62, the difference signal is processed by quadri-phase modulation with square wave modulation signal and then combines

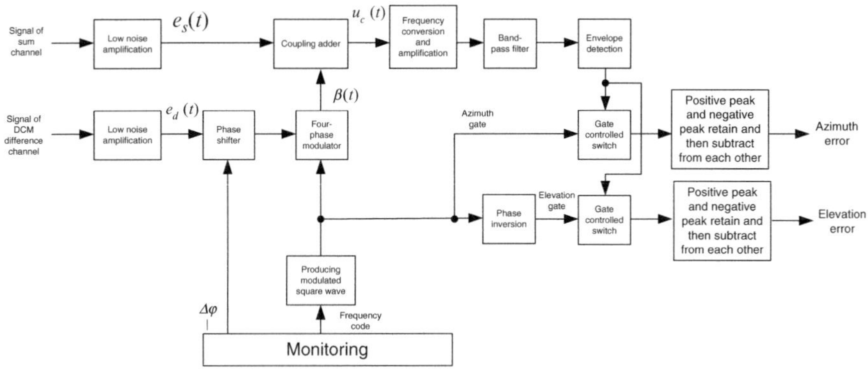

Fig. 2.63 Schematic block diagram of combination of DCM into SCM

with sum signal, thereby becoming one channel output signal, that is, a combined signal. The error voltage for such signal may be obtained only through envelope detection [18]. Since the channels are combined, the phase and gain after combination of sum channel and difference channel will not be inconsistent for this system. All equipment except LNA are allowed to be installed at the generator room, so as to improve the reliability, usability, and maintainability of the equipment; meanwhile, the cost is greatly decreased with reduction of equipment numbers.

Refer to Fig. 2.63 for the theory of combining dual-channel monopulse into single-channel monopulse. For ensuring the *G*/*T* value of the system, the combination of the sum signal and difference signal shall be implemented after LNA.

Suppose the signal from the antenna feed is single-frequency signal, after the feed output and the signal instantaneous value is amplified for K_Σ through LNA, then we can obtain that

$$e_s = K_\Sigma Am \cos \omega t$$

The difference-channel signal after combination of the two phase-quadrature channels is amplified for K_Δ through LNA, and then we can obtain that:

$$e_d = (\mu\theta) K_\Delta Am \cos(\omega t - \Phi)$$

$$\varphi = \arctan \frac{A}{E}$$

After four-phase modulation, the difference signal is:

$$e_d = (\mu\theta) K_\Delta A_m \cos(\omega t - \Phi + \beta(t) + \Delta\varphi) \tag{2.253}$$

where *E* is the elevation deviation angle, *A* is the azimuth deviation angle, $\Delta\varphi$ is the relative phase difference of the sum signal and the difference signal, and $\beta(t)$ is the time-varying function of the phase modulation angle, which is controlled by the square wave modulation signal $\alpha(t)$ to produce the change of phase $\beta(t)$; it is expressed as below in four-phase modulation:

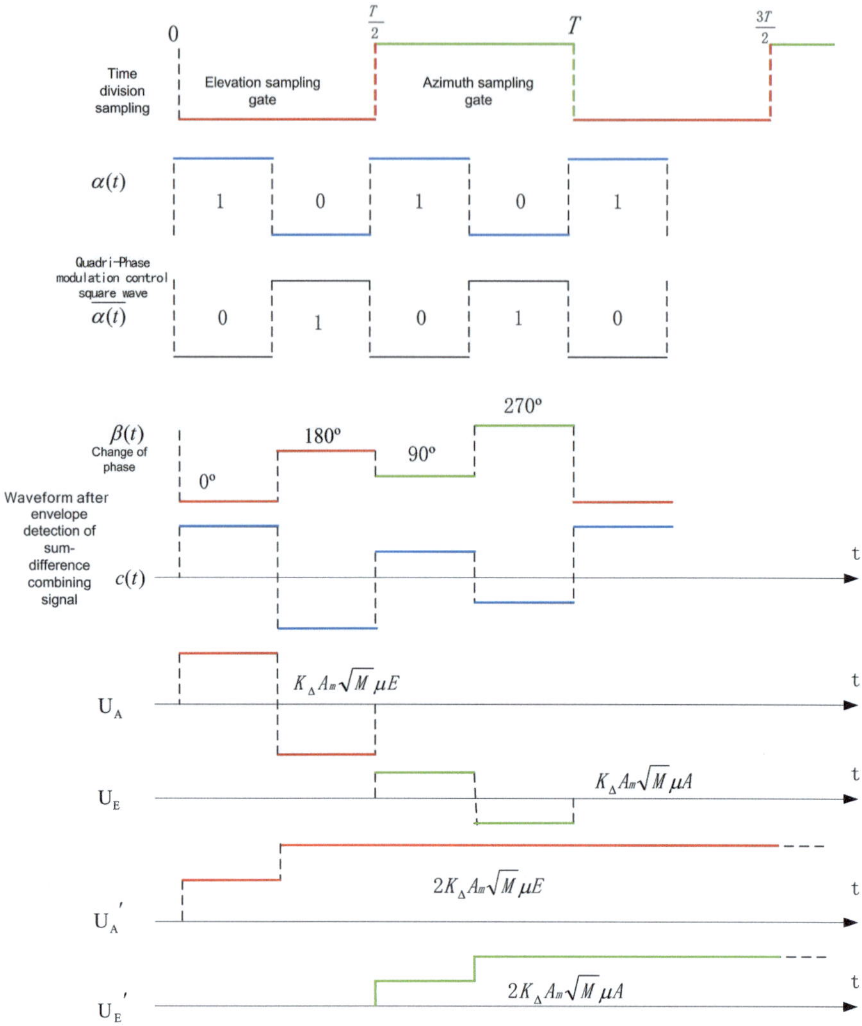

Fig. 2.64 Synchronous demodulation waveform of single-channel monopulse angle error signal

$$\beta(t) = \begin{cases} 0 & t = 0 - \dfrac{T}{4} \\ \pi & t = \dfrac{T}{4} - \dfrac{T}{2} \\ \dfrac{\pi}{2} & t = \dfrac{T}{2} - \dfrac{3T}{4} \\ \dfrac{3\pi}{2} & t = \dfrac{3T}{4} - T \end{cases}$$

The change of phase is as shown in Fig. 2.64.

After modulation, the difference signal is combined with the sum signal by the directional coupler with power coupling coefficient of M, then the combined signal is expressed as:

$$\begin{aligned} u_c &= K_\Sigma Am\cos\omega t + K_\Delta Am\sqrt{M}(\mu\theta)\cos\left[\omega t - \Phi + \beta(t) + \Delta\varphi\right] \\ &= c(t)\cos\left[\omega t + \psi(t)\right] \end{aligned} \tag{2.254}$$

The voltage of angle error that is in relation to the amplitude $C(t)$ of $u_c(t)$ is extracted by envelope detector. Using the formula $a\cos\omega t + b\cos(\omega t + \alpha) = a[1 + (b^2/a^2) - (2b/a)\cos(180° - \alpha)]\cos[\omega t + \psi(t)]$ to expand Expression (2.254), and as θ is a small value under tracking situation, we can obtain the following expression through derivation:

$$C(t) = K_\Delta Am\sqrt{M}(\mu\theta)\cos\left[\beta(t) - \Phi + \Delta\varphi\right] \tag{2.255}$$

Within one period of four-phase modulation, according to different value of $\beta(t)$, then we can obtain that:

$$C(t) = \begin{cases} K_\Delta Am\sqrt{M}\mu\theta\cos\left[\Phi - \Delta\varphi\right] & t = 0 - \dfrac{T}{4} \\ -K_\Delta Am\sqrt{M}\mu\theta\cos\left[\Phi - \Delta\varphi\right] & t = \dfrac{T}{4} - \dfrac{T}{2} \\ K_\Delta Am\sqrt{M}\mu\theta\sin\left[\Phi - \Delta\varphi\right] & t = \dfrac{T}{2} - \dfrac{3T}{4} \\ -K_\Delta Am\sqrt{M}\mu\theta\cos\left[\Phi - \Delta\varphi\right] & t = \dfrac{3T}{4} - T \end{cases} \tag{2.256}$$

In equipment calibration, the relative phase difference before combination of the sum signal and the difference signal shall be adjusted to close to 0, i.e., $\Delta\varphi \approx 0$. As $E = \theta\cos\Phi$, $A = \theta\sin\Phi$, then Expression (2.251) becomes:

$$\begin{aligned} &t = 0 - \frac{T}{4} && U_{E1} = K_\Delta A_m\sqrt{M}\mu E \\ &t = \frac{T}{4} - \frac{T}{2} && U_{E2} = -K_\Delta A_m\sqrt{M}\mu E \\ &t = \frac{T}{2} - \frac{3T}{4} && U_{A1} = K_\Delta A_m\sqrt{M}\mu A \\ &t = \frac{3T}{4} - T && U_{A2} = -K_\Delta A_m\sqrt{M}\mu A \end{aligned} \tag{2.257}$$

As shown in Expression (2.252), the elevation error signal E is transmitted by segmenting within $(0 - T/2)$; similarly, the azimuth error signal A is transmitted by segmenting within ($T/2-T$), so as to implement time division multiplexing of dual-channel. Meanwhile, the servo system for angle tracking is becoming a discrete self-regulation system, which approximates a continuous self-regulation system only if $1/T$ is greater than (5–10) times of self-regulation loop bandwidth and the peak-holding circuit is used to retain the error voltage within $T/2$ interval. As shown in Fig. 2.63, the "gate controlled switch" is used for time division demultiplexing, and the "peak-holding circuit" is used to convert the discrete signal into continuous signal, and subtract the positive peak voltage from the negative peak voltage, so as to double the error voltage. After processing, the output of elevation error angle is:

$$U_E = 2K_\Delta A_m \sqrt{M}\mu E \tag{2.258}$$

Similarly, the output of azimuth error signal is:

$$U_A = 2K_\Delta A_m \sqrt{M}\mu A \tag{2.259}$$

If there has a time-delay error Δt between the modulating square wave and the demodulated envelop's square wave U_A, the video cross coupling will be produced. Consequently, Δt must be aligned to 0 at factory acceptance test. Because the value of Δt changes very little, there is no need to calibrate again on site.

The other kind of cross coupling is "RF cross coupling," which is caused by phase shift inconsistence $\Delta\varphi$ of DCM sum/difference RF signals before combination in SCM, and the result is shown as in Expression (2.251).

Meanwhile, $\Delta\varphi$ causes the loss of error signal, i.e., $L_\Delta = \cos\Delta\varphi$; as a result, the SNR of error signal is decreased, thus making the error of angle tracking increase.

For mitigating the effect on angle tracking of these two factors, the RF phase calibration shall be performed, to make $\Delta\varphi = 0$. If the index can be controlled to enable that $\Delta\varphi \leq \pm 14°$ is true under all conditions, then $K_\Delta = \text{tg}14° = 0.25$, $L_\Delta = \cos 14° = 0.9703$, so that RF calibration shall not be performed in case of lower tracking accuracy. As shown in the result, the cross coupling is the main affecting factor.

For both DCM and SCM, the phase inconsistence between sum channel and difference channel will cause cross coupling and static tracking error. The reason is: when the phase difference between the sum and difference signals is $\Delta\varphi = 90°$, there will be no DC error signal output represented by first term of Expression (2.249) even when there is input of IF signal in the difference channel, which will bring in a static tracking error corresponding to the IF input in difference channel; when the phase difference between the sum and difference signals is $\Delta\varphi \neq 0°$, there is a DC error signal output represented by first term of Expression (2.249), which will cause a cross coupling. For DCM, its error detector is a phase discriminator with discrimination characteristic of cos ($\Delta\varphi$). If the phase difference between the sum and difference signals is $\Delta\varphi = 90°$, the output is 0; if $\Delta\varphi \neq 0°$, the output is not equal to 0. For SCM, the error signal detector is the envelope detector for amplitude modulation. If the phase difference between the sum and difference signals is $\Delta\varphi = 90°$, through $0/\pi$ modulation and addition, the difference signal will not form amplitude modulation to the sum signal, so the output of the envelope detector is 0; if $\Delta\varphi \neq 0$, through $0/\pi$ modulation and addition, the difference signal will form amplitude modulation to the sum signal, so the output of the envelope detector is not equal to 0. Since SCM and DCM have the same principles of cross coupling and static tracking error caused by phase inconsistence of the sum channel and difference channel (refer to Sect. 2.5.5.1 for principle of static tracking error, and refer to Sect. 2.5.1.2 for principle of cross coupling), so the calculation expressions of them are the same (see Expression (2.251)).

Due to the SCM aforesaid is produced from the channel combination for DCM, so the expression of tracking error caused by thermal noise in SCM may be derived from the DCM Expression (2.252), then from Expression (2.252), we obtained that:

$$\delta_1 = \frac{\theta_{0.5}}{K_m\sqrt{\frac{S}{N}\frac{B_{IE}}{B_n}}} = \frac{\sqrt{G_\Sigma}}{K\frac{U_I}{U_n}\sqrt{\frac{B_{IF}}{B_n}}} \tag{2.260}$$

After combined into SCM, the expression above then has following effects:

1) The absolute slope of difference lobe gain is K: it affects the output voltage of the difference channel $U_\Delta = K\theta$ (θ is the angle deviated from the electronic boresight). If other conditions remain the same, the change of U_Δ can be equivalent to that caused by K changed to K':

Because: $U_\Delta{}' = K'\theta$, $U_\Delta = K\theta$

then, $K' = K\frac{U_\Delta{}'}{U_\Delta}$

Factors affecting the signal amplitude of the difference channel are the power coupling coefficient of summit or M ($M < 1$) and the phase inconsistence between sum and difference channels $\Delta\varphi$, that is,

$$U_\Delta' = U_\Delta\sqrt{M}\cos\Delta\varphi$$

Then,

$$K' = K\sqrt{M}\cos\Delta\varphi \tag{2.261}$$

2) Noise after channel combination

The RMS value of the noise voltage of sum and difference channels are both set to U_n. Whereas the noise voltage of difference channel is coupled $\sqrt{M}$ times and then adds the noise of sum channel, so the noise power after addition increases up to $U_n^2 + MU_n^2 = (1+M)U_n^2$, correspondingly, the noise voltage increases to $\sqrt{1+M}U_n$.

3) Aliasing noise

The thermal noise of the difference channel will cause the random error of angle tracking. Refer to Fig. 2.65 for noise spectrum of the difference channel in SCM and DCM.

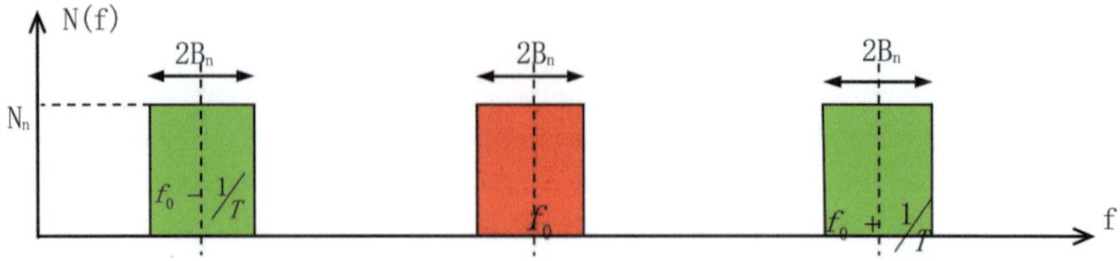

Fig. 2.65 Noise spectrum of difference channel

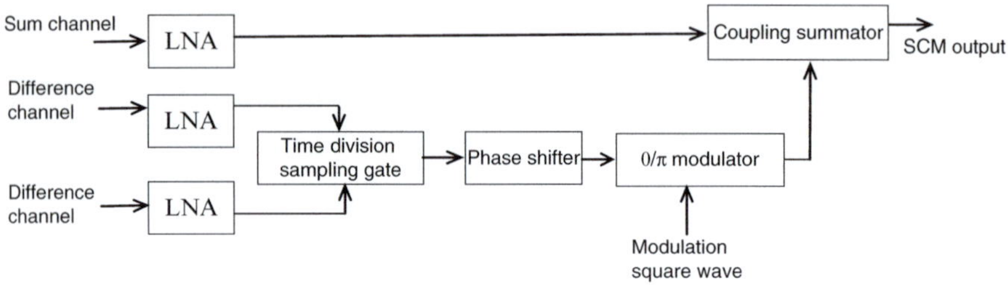

Fig. 2.66 Principle block diagram of combining TCM into SCM

As shown in Fig. 2.65, the spectrum of angle tracking random error in DCM is within $f_0 \pm B_n$ band, and its noise power is $2B_nN_0$, where N_0 is the noise power spectral density. The spectrum of angle tracking random error in SCM is within $2B_n$ band at two symmetrical frequency points $f_0 \pm 1/T$ ($1/T$ is the modulation frequency for forming SCM amplitude modulation), then the two symmetrical noise spectrums are overlapped through envelop detection, and its noise power is $4B_nf_0$ as twice as that of DCM, correspondingly, the random error of angle tracking increases $\sqrt{2}$ times.

Substitute the effects of those three factors into Expression (2.260), then the random error σ_n caused by the thermal noise in SCM is obtained:

$$\sigma_n = \frac{\sqrt{G_\Sigma}}{K\sqrt{M}(\cos\Delta\varphi)\frac{U_\Sigma}{\sqrt{1+MU_n}}\sqrt{\frac{B_{IF}}{B_n}}} = \frac{\theta_{0.5}}{K_m\sqrt{M}(\cos\Delta\varphi)\sqrt{\frac{1}{(1+M)}\frac{S}{N}\frac{B_{IF}}{B_n}}} \tag{2.262}$$

(2) Three-channel monopulse combines into single-channel monopulse:

Then the antenna outputs three signals of the sum channel, azimuth difference channel, and elevation difference channel. Refer to Fig. 2.66 for one combination method in common use.

As shown in Fig. 2.66, the azimuth difference channel and the elevation difference channel are combined into one channel using TDM method, meanwhile, part of the signals from the two channels are coupled via the same one $0/\pi$ modulator and then adds the signal from the sum channel to become an amplitude-modulated signal, after that, its working principle and the random error analysis are the same as that of combining DCM into SCM. It is easier since that no cross coupling exists, and the phase modulator changes from the four-phase modulation to $0/\pi$ two-phase modulation with period of $T/2$. Its analysis is the same as that of combination of DCM into SCM.

2.5.2 *Angle Measurement with Interferometer*

The interferometer angle measurement is a method that uses the phase difference or time-delay difference of the signal transmitted from the target received by the two receiving antennas to measure the angle with coherent principles. When it receives single-frequency signal, the interferometer is used to measure the phase, so we called it the phase difference interferometer; when it receives broadband signal (e.g., broadband noise of radio star), the time-delay difference is usually used to measure since the phase measurement is more complicated, and we called it the time-delay-difference interferometer. The target signal can be response signal, beacon signal, or others.

2.5.2.1 Phase Difference Interferometer

Its principle is as shown in Fig. 2.67.

The origin of coordinate is the midpoint of the line between the two antennas (referred to as baseline) O, and α is the angle of target which defines the included angle of the baseline and the target direction. The phase difference of the target signals received from the two antennas is $\Delta\phi$. When the target is far away from the two antennas, the intensities of the signals received from the two antennas are approximately equal and can be considered as a plane wave (i.e., parallel ray); therefore, the phase difference caused by the distance difference between the two antennas is expressed as:

$$\Delta\phi = \left(\frac{2\pi}{\lambda}\right)(R_A - R_B) = \left(\frac{2\pi}{\lambda}\right) D\cos\alpha$$

where D is the distance between the two antennas of receivers, λ is the wavelength of the target signal, and $\cos\alpha$ is the direction cosine.

As long as the phase difference $\Delta\phi$ is measured, then the direction cosine $L(L = \cos\alpha)$ can be solved, and then the angle coordinate of the target α can be

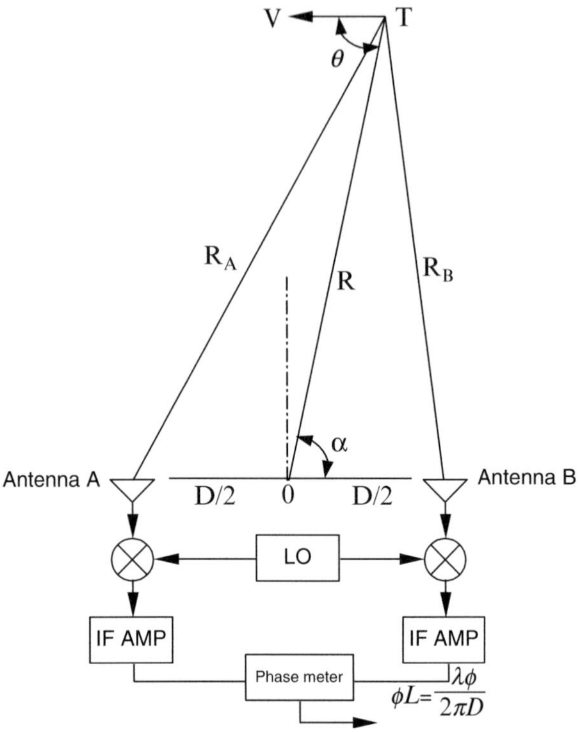

Fig. 2.67 Angle measurement principle of phase difference interferometer

solved. Since increasing the baseline length D is helpful to improve the measurement accuracy of α, it is obvious that the improvement of $\Delta\phi$ measurement accuracy is also useful to improve the angle measurement accuracy.

$\Delta\phi$ measurement method: mix the RF signals received from the two antennas to IF signals, and then, transmit them after amplification to the phase meter to measure $\Delta\phi$. Since the period of phase is 2π, and $\Delta\phi$ has multiple measured values, leading to angle ambiguity (similar to the tone ranging), for which we should take appropriate measures. In addition, in the course of frequency conversion and amplification of the two signals aforesaid, the phase inconsistence of the two channels will introduce the phase error, which also requires appropriate technical means, thus leading to complicated implementation, higher technical specification, and the necessity of requiring the equipment to be stable and fixed, which is inconvenient for mobility. However, the angle measurement accuracy of it is the highest in radio TT&C system.

In the event that the baseline of the interferometer is extreme short, the interferometer becomes phase-comparison monopulse radar. Under this situation, the antenna is fixed on an antenna pedestal and uses the direction cosine as the error signal to track the angle. Four antennas are required for 2D tracking in azimuth and elevation.

Key technologies for interferometer are as follows.

(1) Phase balance of the interferometer phase measurement system. The error of phase measurement shall be minimized in order to provide higher angle measurement accuracy for the interferometer under the entire operation environment. The phase error normally is caused by the phase inconsistence of the interferometer receiving system and system SNR. Following error sources may cause inconsistence of the phases between the channels:

- Polarization error and antenna pattern error
- Mechanical error
- Change of phase center of the antenna
- Phase inconsistence of antenna element and feed waveguide
- Phase inconsistence of each local oscillator and receiving channel
- Phase inconsistence of each IF amplifier and baseband

In order to minimize these errors, the antenna, reference signal, front end, and the local oscillator of balanced mixer shall be consistent as far as possible so as to minimize the phase inconsistence in case of change of environment. The waveguide transmission line shall be as short as possible. The phase matching antenna and balanced mixed shall be used. The IF amplifier with consistent phase shift and the transmission lines are used to minimize the inconsistence of the phases between IF channels. The automatic phase calibration unit is adopted in the system to solve the phase inconsistence caused by change of environment, using frequency division, time division, or code division method.

(2) The length error of the baseline between the end stations shall be measured accurately.

(3) Digitalization: to minimize the drift error caused by phase drift of analog circuit on the phase interferometer.

Refer to Fig. 2.68 for typical 2D deployment of the interferometer antenna [19].

As shown in Fig. 2.68, the design of antenna deployment shows: on elevation plane: $b_1 = 0.577\lambda$, $b_2 = 15b_1$, and antenna A_1, A_2 and A_0 are used for azimuth angle measurement, where antenna A_3, A_4 and A_0 are perpendicular to antenna A_1, A_2 and A_0. Since there are five antennas in total, with which five channels shall be connected correspondingly. A_1–A_2 is used to accurately measure the azimuth angle, A_1–A_0 is used to solve the azimuth ambiguity, A_3-A_4 is used to accurately measure the elevation angle, and A_3-A_0 is used to solve the elevation ambiguity. In each DF channel as shown in Fig. 2.68, the input signal is converted into a 70 MHz IF signal through LNR and one down-conversion, and then transforms by *A/D* conversion at 70 MHz to form a digital signal. The digitalized phase measurement is used to improve the measurement accuracy.

Phase calibration technology is adopted to calibrate the long-term drift error so as to improve DF accuracy. The frequency of calibration signal source that is close to the received frequency (frequency division) is generated by the frequency synthesizer which uses high-stability crystal oscillator as frequency reference. The frequency synthesizer also generates *Tx*/*Rx* frequency of local oscillator, *Tx* signal carrier frequency, and sampling frequency of *A/D*.

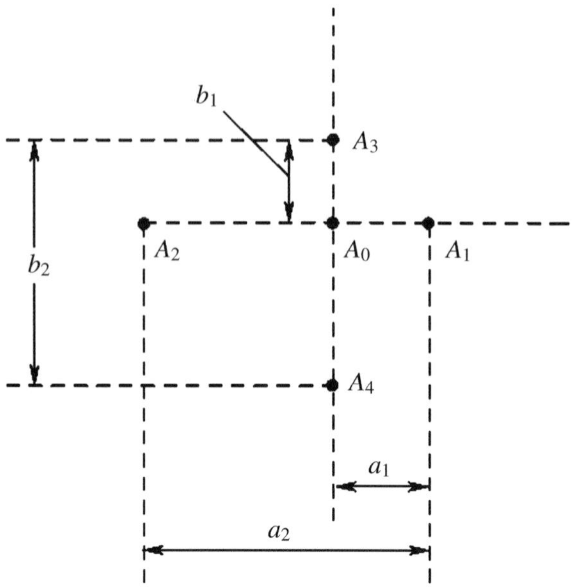

Fig. 2.68 Schematic diagram of antenna 2D deployment

The phase error required for angle measurement accuracy may be calculated by the expression below, that is,

$$\begin{aligned}\sigma_\theta &= \frac{1}{2\pi d}\,\frac{1}{\cos\theta}\sigma_\varphi \\ \sigma_\varphi &= \sigma_\theta \frac{2\pi d\cos\theta}{\lambda}\end{aligned} \tag{2.263}$$

where σ_θ is the RMS index of angular accuracy, σ_φ is the RMS value of inconsistence of the phases between the channels, d is the longest distance of one baseline, that is, $d = 15b_1 = 8.355\lambda$, λ is the working wave length, and θ is the angle in the greatest coverage direction.

1) The measurement error of phase difference required for ambiguity solution is expressed as:

$$\varphi_d < \frac{\pi}{(n_i/n_{i-1}) + 1} \tag{2.264}$$

where n_i/n_{i-1} is the ratio of baselines.

$\varphi_\mathrm{d} > \sigma_\varphi$ is required for solving ambiguity.

2) Allocation of system DF errors: Table 2.12 lists all the factors causing DF error and its allocated values.

Table 2.12 System phase errors

Error source	Phase inconsistency error (1σ)/(°)	Error source	Phase inconsistency error (1σ)/(°)
Polarization mode	1	Mixer	2
Center of phase	0.5	Calibration error	1
Mechanical calibration	2	Phase error caused by integrated baseband	3
Waveguide circuit	3	Phase error caused by thermal noise	2
IF amplifier	3	Total error	6.5

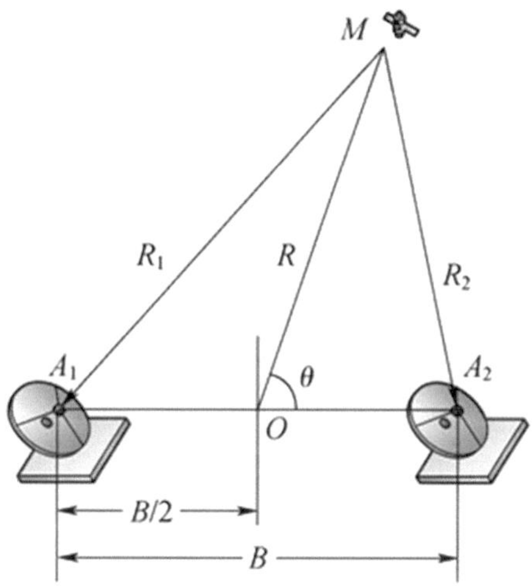

Fig. 2.69 Angle measurement principle of time-delay difference interferometer

2.5.2.2 Time-Delay Difference Interferometer

The principle of the time-delay difference interferometer is as shown in Fig. 2.69, in which M is the instantaneous position of the target, A_1 and A_2 are the positions where the two antennas locate in the same coordinate axis, B is the distance between the two antennas, the length of baseline, O is the center of baseline, θ is the included angle from the target direction OM and the baseline, R_1 and R_2 respectively are the distance from the target to each of two antennas, and R is the distance from the target to the center of baseline.

According to cosine theorem, there is:

$$R_1^2 = R^2 + \left(\frac{B}{2}\right)^2 + 2R\left(\frac{B}{2}\right)\cos\theta$$

$$R_2^2 = R^2 + \left(\frac{B}{2}\right)^2 - 2R\left(\frac{B}{2}\right)\cos\theta$$

then,

$$R_1^2 - R_2^2 = 4R\left(\frac{B}{2}\right)\cos\theta$$

The direction cosine is:

$$l = (R_1 + R_2)(R_1 - R_2)/(2RB)$$

If $R \gg B$, then $2R = R_1 + R_2$, and

$$l = \cos\theta \approx \frac{(R_1 - R_2)}{R} = \frac{r}{B} \tag{2.265}$$

where r is the range difference.

According to the time-delay difference τ of the same signal received from the two stations (measured with correlation method), we obtain:

$$r = \tau c$$

where c is the light velocity.

$$l = \cos\theta \approx \frac{\tau c}{B} \tag{2.266}$$

By differentiating Expression (2.244), we can obtain that

$$\begin{aligned} \delta l &= (\delta\theta)\sin\theta = \frac{(\delta\tau)c}{B} \\ \delta\theta &= \frac{(\delta\tau)c}{(B\sin\theta)} \end{aligned} \tag{2.267}$$

where δ represents error increment. This expression shows the measurement error caused by time-delay difference interferometer.

2.5.2.3 Deep-Space TT&C Application with Interferometer

Actually, the angle measurement with interferometer is a very old concept. In application to deep-space TT&C, the baseline is required to be extended to improve the measurement accuracy, which forms the Very Long Baseline Interferometry (VLBI). Currently, the baseline has extended to the edges of continents and reached to the ends of the Earth, which is tending to the limits. Although there are points of view that the baseline shall be extended to the outer space, it hasn't performed yet. Lengthening baseline brings some disadvantages as follows:

- Measurement error of "time difference" caused by different sources between the two terminal stations
- Error caused by different radio wave propagation paths during receiving the signals of the two terminal stations
- Error caused by different equipment used in the two terminal stations
- Errors caused by incorrect baseline measurement
- Loss of some real-time capability due to the post processing of range difference data between the two distanced terminal stations
- Smaller overlapped area covered with one spacecraft by the two terminal stations because of long baseline

Therefore, some technologies like Difference Very Long Baseline Interferometry (ΔVLBI), Same-Beam Interferometry (SBI), and Connected Element Interferometry (CEI) are presented to minimize these errors. ΔVLBI and SBI are designed to put the reference point for measurement into the outer space, and change the measurement of long range to that of relatively short range. If the location of the reference point is known, the position of deep-space probe shall be determined by measuring the relative positions of the deep-space probe and the reference point, with an improved measurement accuracy by mitigating relevant system errors with differential technology. The principle of ΔVLBI is that: Firstly, the two long-distanced DSNs simultaneously observe the deep spacecraft and then generate a time difference by processing of measured data with accuracy of 20 ps, and the time-delay difference at that time is converted to an angle. Then, these two deep-space stations simultaneously take measurements of the same radio star in a known location near the spacecraft, which yields two values, the time difference and angle, from which the equipment errors can be calculated by subtracting the known radio star position, thus improving the accuracy by differentiating this value with the measured data of spacecraft. Spacecraft CEI technology uses the optical fiber for baseline transmission so as to improve measurement accuracy of the time difference between the two end stations and acquire real-time ability. For SBI technology, two antennas in the deep-space station are used to simultaneously measure the two targets within the same beam, thus to obtain the measured data of difference interferometry. The accurate relative positions of the two targets can be obtained with SBI. Moreover, the deep-space interferometer also adopts following methods:

(1) Angle measurement system with Delta differential one-way ranging (ΔDOR)

The time-delay difference measurement may be implemented with one-way range difference; therefore, VLBI will evolve to differential one-way ranging (DOR), and further evolves to ΔDOR plus differential technology.

ΔDOR is used to differentiate the time-delay difference of the target signals received from the two TT&C stations at the ends of baseline with that of the radio star signals received from the calibration receiver, where the position of radio star is as a known reference point (obtained from astronomy). Because all ranging signals are the CW signals, the range-ambiguity-solving method is required (see Sect. 2.4.1

for its basic principle). Thus, the range-ambiguity-solving method and its "big hop counts" algorithm are still the key technologies. For ΔDOR ambiguity-solving method, except for the residual carrier after modulation, the transmit signal of the target shall carry the strong subcarrier (or side tone) and harmonic components of each sub-harmonic, that is, the transmitting spectrum of the target receiver is wide enough, and the ground receiver is able to receive the residual carrier of the transmitting signal and the even harmonic component of the modulated subcarrier of the forwarding signal, thus to solve the rang difference with higher harmonic as well as the ambiguity with other harmonics.

(2) CEI angle measurement

The Connected Element Interferometry (CEI) based on the same reference frequency, can implement accurate measurement to the received time-delay signals from two end stations through the fiber communication link between the two tracking-measurement stations distanced from tens to 100 km. With this high precision data the included angle of the baselines between the two stations relative to the radio emission source can be determined. The equivalent measurement accuracy may be implemented on the relatively short baseline without any long-term and continuous arc segment data. The fiber communication link is used to transmit the acquired data to the same correlator for processing in real time.

Its angular accuracy $\delta\theta$ is inversely proportional to the length of baseline B and is proportional to the measurement accuracy $\delta\tau$ of time-delay difference τ. In the event of achieving high accuracy of angle measurement, the longer baseline shall be used (by lengthening baseline B) or the measurement accuracy of time delay for interferometer shall be improved (by reducing $\delta\tau$). In addition, the factor causing a lot of errors is a relevant factor on the short baseline and reduced by cancellation. Since the short baseline prior model has a small error, the high accurate phase delay data can be obtained.

CEI measurement has such advantages as follows:

1) Real-time processing
2) Coherent processing between the two stations owing to the use of the same clock
3) Obviously reduced errors of transmission medium by cancellation of short baseline
4) A higher elevation angle and a longer observation time achieved with a shorter baseline

This method has more advantages in deep-space orbit determination, especially for orbit measurement when approaching/falling of the Moon and the planets. Also, this technology has a special advantage in real-time high-accuracy orbit determination of LEO and MEO, implementing measurement with multiple stations jointly without transmission of uplink signals.

The error sources affecting the CEI phase delay measurement mainly include the length of baseline, position of radio source, troposphere, ionosphere, stability of clock, and measurement accuracy. Many error sources may be corrected by simultaneously observing the radio sources (spaceship and radio star) adjunct to the two angle positions and then conducting the differential calculation, for example, the unknown delay and phase jitter of the clock and local oscillator between the two stations, unknown delay of measuring equipment, equipment phase jitter in the course of measurement, delay caused by antenna deformation, transmission medium delay, and baseline error.

The prior model is used to solve the phase ambiguity in measurement of time delay. For dual mode of S-band and X-band, the prior model is used to solve the integer ambiguity of phase measurements of S-band and then X-band signals. The prior model is developed on the basis of available information, such as station address, source position and transmission medium delay, etc. This model shall be a feature of accuracy for solving phase ambiguity.

The random jitter of troposphere is the main error source affecting the measurement accuracy of CEI phase relay, where the statics error is inversely proportional to the elevation angle while observing the spaceship.

(3) Angle measurement with same beam interferometer

In case two spaceships' angle location, are very approximate they can be in the same beam of the ground antenna, and two antennas in the two deep-space stations are used to observe these spaceships simultaneously and then generate differential interferometry data, which is called as Same Beam Interferometry (SBI) technology that provides the high accurate measurement for relative angle positions. SBI data supplements the radial direction information from the traditional Doppler data and further improves the orbit determination accuracy. Observing from the Earth, the angle between the two spacecraft is far smaller than that between the spacecraft and the radio star. Since measurement errors are proportional to the angle, hence the interferometry of spacecraft vehicle may be more accurate than the traditional way. When the two spacecraft locate in the same antenna beam, the carrier phase of them can be tracked at the same time, so that the carrier phase is used for further improving the measurement accuracy instead of group time delay. SBI technology is better than the traditional interferometry of spacecraft-radio star on operation. It's not required to compare with the radio star, and the antenna is able to point at the spacecraft without shift, so that real-time phase measurements may be implemented at the same time. The determination of orbit plan combining the same beam interferometry data with dual-way Doppler data or one-way/dual-way Doppler data has many advantages and multiple spacecraft may be tracked with this method.

2.5.3 Theoretical Calculation of Angle Tracking Accuracy [20]

(1) Random error

1) Error caused by thermal noise of the receiver is:

$$\sigma_1 = \frac{\theta_{0.5}}{K_m\sqrt{\frac{S}{N}\frac{B_{IF}}{B_n}}} = \frac{1}{\mu\sqrt{\frac{S}{N}\frac{B_{IF}}{B_n}}} \tag{2.268}$$

where $\theta_{0.5}$ is the beam width of half power point, *S/N* is signal-to-noise ratio of receiver (bandwidth is B_{IF}), B_{IF} represents IF amplification equivalent noise bandwidth of the receiver, B_n is servo equivalent noise bandwidth, K_m represents normalized difference slope (generally ranging between 1.2 and 1.7), and μ is relative difference slope $\mu = \left[\frac{\partial F(\theta)/\partial\theta}{\sqrt{G_\Sigma}}\right] = \frac{K_m}{\theta_{0.5}}$.

2) Multipath reflection. When the elevation is relative low, the target signal passes through the antenna side lobe and enters the receiver after being reflected by ground. The elevation measurement error generated by the jamming servo system is:

$$\sigma_2 = \frac{\rho\theta_{0.5}}{\sqrt{8A_3}} \tag{2.269}$$

where ρ represents reflection coefficient; A_3 is gain ratio of main lobe to side lobe.

3) Variation error caused by dynamic delay of the target. The variation error of angle delay caused by the target angular acceleration and angular jerk is the largest when dynamic variation occurs at low orbit and will reduce along with increase in distance. Variation error of the target is:

$$\sigma_3 = \left(\frac{\dot{\omega}}{K_v} + \frac{\ddot{\omega}}{K_d}\right)\frac{1}{B_n} \tag{2.270}$$

where $\dot{\omega}$ represents angular acceleration of the target, $\ddot{\omega}$ is angular jerk of the target, K_v represents speed error constant of the servo system, and K_d is acceleration error constant of the servo system.

4) When azimuth axis is not perpendicular to horizontal plane, the introduced error is:

$$\begin{aligned} \sigma_4 &= \gamma \mathrm{tg}\, E \sin A \ \ (\text{azimuth angle}) \\ \sigma_4 &= \gamma \cos E \ \ (\text{elevation angle}) \end{aligned} \tag{2.271}$$

where *r* represents random tilt of the azimuth axis to horizontal plane.

5) When azimuth axis is not perpendicular to elevation, the introduced error is:

$$\sigma_5 = \delta \mathrm{tg}\, E \ \ (\text{azimuth angle}) \tag{2.272}$$

where δ is random quantity when two axles are not perpendicular.

6) When optical axis is not perpendicular to elevation axis, the introduced error is:

$$\sigma_6 = \Delta \sec E (\text{azimuth angle}) \tag{2.273}$$

where $\triangle$ is random quantity of optical axis deviating from elevation axis.

7) Torque in-equilibrium error generated by gusty wind is:

$$\sigma_7 = 2K_W\bar{v}\left[\int_0^{f_{\max}} \frac{\varphi_W(f)}{K_T^2(f)}\mathrm{d}f\right]^{1/2} \tag{2.274}$$

where K_w represents wind torque constant, $\bar{v}$ is mean wind speed (m/s), $f_{\max}$ represents the highest frequency of gusty wind, $K_T^2(f)$ is system torque transmission function, and $\varphi_W(f)$ represents gusty wind spectrum.

8) Quantization error of data sensor is

$$\sigma_8 = \frac{g}{\sqrt{12}} \tag{2.275}$$

where σ is the minimum-digit quantization unit.

Additionally, there are angle fluctuation error σ_9 caused by irregular troposphere, angle measurement error σ_{11} introduced by zero drift of phase discriminator of receiver, error σ_{12} introduced by zero drift of servo amplifier, random error σ_{13} introduced by servo noise and blind zone, random error σ_{14} caused by pedestal unlevelness, and angular conversion error $\sigma_{15.}$

9) Total random error is

$$\sigma = \sqrt{\sigma_1^2 + \sigma_2^2 + \cdots + \sigma_{15}^2} \tag{2.276}$$

(2) System error

1) Dynamic delay error is

$$\Delta_1 = \frac{\omega}{K_v} + \frac{\dot{\omega}}{K_d} \tag{2.277}$$

where ω is angular velocity of the target and $\dot{\omega}$ is angular acceleration of the target.

2) Polarization error is

$$\Lambda_2 = \frac{\theta_{0.5} \cdot \delta}{2K_m}\sqrt{\frac{3}{8}\left[(1-K)^2 + \frac{4}{3} \cdot K\sin^2\Delta\varphi\right]} \tag{2.278}$$

where δ is the length difference between major axis and minor axis, which is normalized with respect to the minor axis, K_m is normalized slope of difference lobe, K is degree of inconsistency of axial ratio of two polarizers, and is difference between polar tilt angles of two polarizers, and $\Delta\varphi$ is difference between polar tilt angles of two polarizers.

3) Amplitude unbalance before the comparator is

$$\Delta_3 = \frac{\Delta G/G}{2\mu} \tag{2.279}$$

where $\Delta G/G$ is relative value of antenna gain unbalance.

Inconsistency between the sum and difference gain of tracking receiver will cause open-loop gain change of the servo system, thus causing dynamic error variation (see Expression (2.267)).

4) Phase unbalance before and after the comparator is

$$\Delta_4 = \frac{\tau \cdot tg\varphi}{2K_m}\theta_{0.5} = \frac{\theta_{0.5}tg\varphi}{K_m\sqrt{G_n}} \tag{2.280}$$

where τ is phase unbalance before the comparator; φ is phase unbalance after the comparator (inconsistency between sum and difference phases of the tracking receiver); $\sqrt{G_n} = 2/\tau$ is null depth of differential beam, representing the ratio of G_Σ to the minimum gain $G_{\min}$ at differential beam zero, namely, $G_\Sigma/G_{\min}$.

5) Coupling error of sum-difference channel is

$$\Delta_5 = \frac{\theta_{0.5} \cdot \cos\varphi}{K_m \cdot \sqrt{F_1}} \tag{2.281}$$

where φ is signal coupling phase shift angle of sum-difference channel; F_1 is isolation degree of sum-difference channel.

6) Unbalance and zero drift of servo amplifier, namely,

$$\Delta_6 = \frac{\Delta V}{K_\varphi} \tag{2.282}$$

where $\triangle V$ is unbalance voltage value of all servo systems converted to the output terminal of the detector; K_φ is directional sensitivity of the antenna to output terminal channel of error discriminator.

7) Due to unbalance in antenna structure, the error between antenna deformation resulting from the torque generated by stable wind and the antenna pointing is

$$\Delta_7 = \frac{\overline{T}_W}{K_T} \tag{2.283}$$

where $\overline{T}_W$ is mean wind torque:

$$\overline{T}_w = K_w \times V^2$$

8) When azimuth axis is not perpendicular to horizontal plane, the error is:

$$\Delta_8 = \gamma' \tan E \sin A(\text{azimuth angle}) \tag{2.284}$$

$$\Delta_8 = \gamma' \cos A(\text{elevation angle}) \tag{2.285}$$

where γ' represents inclination of the azimuth axis to horizontal plane.

9) When azimuth axis is not perpendicular to elevation axis, the error is:

$$\Delta_9 = \delta' \tan E(\text{azimuth angle}) \tag{2.286}$$

where δ' refers to non-perpendicularity between the two axles.

10) When optical axis is not perpendicular to elevation axis, the error is:

$$\Delta_9 = \Delta' \sec E(\text{azimuth angle}) \tag{2.287}$$

where Δ' represents non-perpendicularity between the two axles.

11) Optical-electrical axis non-parallelism error is

$$\Delta_{11} = \psi \sec E(\text{azimuth angle}) \tag{2.288}$$

$$\Delta_{11} = \psi(\text{elevation angle}) \tag{2.289}$$

where ψ is adjustment surplus of optical-electrical axis parallelism.

Total system error is

$$\Delta = \sqrt{\Delta_1^2 + \Delta_2^2 + \cdots + \Delta_{11}^2} \tag{2.290}$$

2.5.4 Theoretical Analysis of Angle Tracking of Wideband Signal – Cross-Correlation Function Method

With exponential increase in high-speed data transfer rate, more and more attention arises to angle tracking of wideband signals. In the past, during angle tracking of FM wideband signal, a small part of signal bandwidth is segmented to perform angle tracking. Namely, the wideband signal becomes narrowband signal. Such method simplifies the proposal but causes loss of signal energy, thereby enlarging angle error and decreasing angle tracking sensitivity. Especially, there are obvious defects if partial bandwidth exists in *C/N* wideband angle tracking. The better method is to solve cross-correlation function of monopulse sum and difference signal, and demodulate wideband difference signal from low *C/N* in the case of equal time delay and large integral constant. Additionally, it is required to normal ize angle error signal to enable angle servo loop to operate normally. Angle acquisition shall be made before the servo loop is closed, which is usually implemented through detection of sum channel signals (e.g., sum AGC). However, in case of low *C/N*, the common detection method fails to detect sum channel signal, which involving key technology of cross-correlation method. The updated angle tracking technology is given in the section below.

(1) Amplitude comparison pulse tracking principle of wideband signal

Common angle tracking system is a narrowband system and generally analyzed by single-frequency signal, with a phase discriminator used as the angle error detector. The impact of sum and difference channels being discussed is the

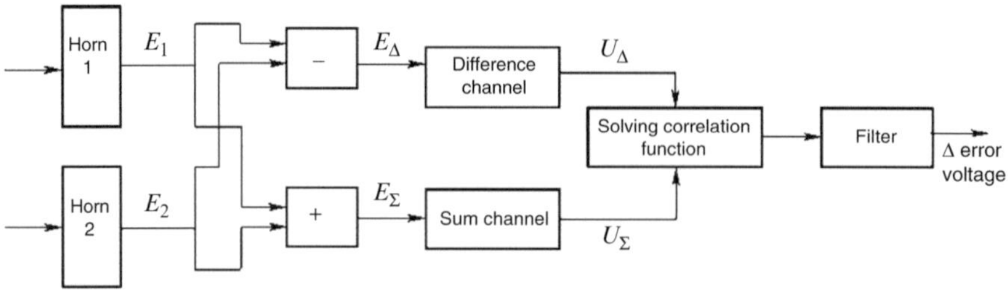

Fig. 2.70 Amplitude comparison monopulse receiver for receiving wideband signal

consistency of phase and gain; for quadrature dual-channel combination only those with 90° phase difference are considered. However, angle tracking system of wideband signal involves tracking to the signal full spectrum, so there are many special problems.

Take typical four-horn amplitude comparison monopulse for instance; refer to Fig. 2.70.

Figure 2.70 shows a pair of horns at elevation direction (another pair of horns are the same in terms of operating principle). Signal E_1 of horn 1 is taken as reference to discuss correlation of all signals in the figure. Since horn 2 and horn 1 receive the same wideband signal, E_1 is correlated with E_2, and the sum signal E_Σ and difference signal $E_\Delta = E_1 - E_2$ are both cross-correlated with E_1. When $E_1 > E_2$ (upward deviate corresponding to the target), its cross-correlation value is positive, when $E_1 < E_2$, its cross-correlation value is negative (downward deviate corresponding to the target). Positiveness and negativeness of cross-correlation value can be used to determine direction to which of the target is deviated. The cross-correlation value is determined by the following cross-correlation function expression [21], namely,

$$C_{\Sigma\Delta}(\tau) = U_\Sigma U_\Delta K \cos\left[(2\pi f_0 + \pi B)\tau\right] \frac{\sin \pi B\tau}{\pi B\tau} \tag{2.291}$$

Expression (2.291) is corresponding to wideband signals of flat spectrum with reference significance. τ is delay difference between sum channel and difference channel, B is signal bandwidth, U_Σ and U_Δ are amplitude of sum and difference signal, and K is transmission coefficient of correlator. To achieve peak value of $C_{\Sigma\Delta}(\tau)$, an additional time delay $(-\tau)$ shall be added between sum channel and difference channel to achieve $\tau + (-\tau) = 0$, at this time, its peak value is:

$$C_{\Sigma\Delta}\left(O_g\right) = U_\Sigma U_\Delta K \tag{2.292}$$

$KE^2(U_\Delta/U_\Sigma)$ (E is sum channel AGC comparison level) can be obtained after the correlation value is normalized. It is angle error signal, which is in direction proportion to the angle deviation but irrelevant to distance. Namely, the angle error slope is normalized, and normalization of its sum and difference signals can be implemented by conventional AGC. Also, normalization can be achieved through sampling U_Σ and U_Δ simultaneously, and calculating U_Δ/U_Σ.

As a result, differences between wideband signal tracking and single-carrier signal (or narrowband signal) tracking are as follows:

1) Since there is not residual carrier signal in wideband signal, the conventional carrier phase-locked loop method cannot be used. The angle error extraction method, in which the sum signal and difference signal are in direct cross-correlation, is employed.
2) Time delay between sum channel and difference channel is required to be consistent to make difference between time be, thus obtaining relevant peak value.
3) Amplitude-/phase-frequency features of sum channel and difference channel are required to be consistent, in order to make sum and difference signals not in distortion or maintain consistent distortion, thereby obtaining large relevant value.

Therefore, time delay of the channel shall be controlled to be in consistence during angle tracking of wideband signals. Additionally, phase shift of each sum and difference LO signal may be inconsistent, and that required to be adjusted. As LO signal is single frequency signal, a phase shifter can be used for adjustment.

According to above, high accuracy of angle measurement may be achieved for wideband signals as long as design method is appropriate. When power spectrum density of signals maintains unchanged, the wider the bandwidth is in the signal spectrum range, the stronger the signal power becomes, and the stronger its cross-correlation value is.

See Fig. 2.71 for block diagram of wideband angle tracking receiver.

In Fig. 2.71, delay adjustment is used to calibrate inconsistency between sum channel and difference channel, phase shifter used to calibrate inconsistency between phase shifts of LOs. Wideband filtering requires consistency between amplitude-/phase-frequency features. Figure 2.71 shows "3-channel" monopulse

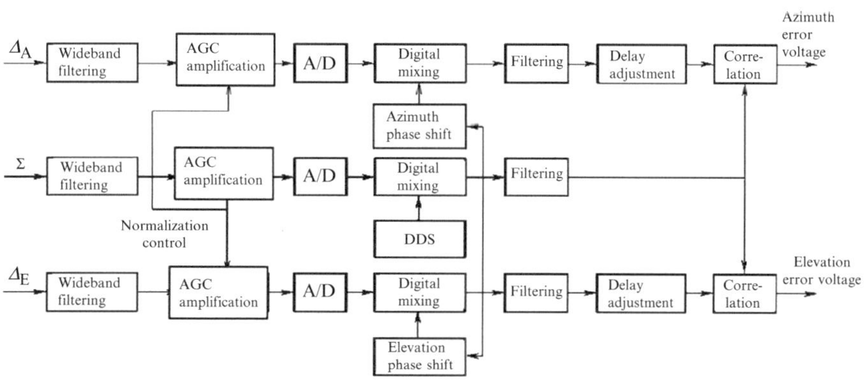

Fig. 2.71 Block diagram of a 3-channel monopulse wideband receiver

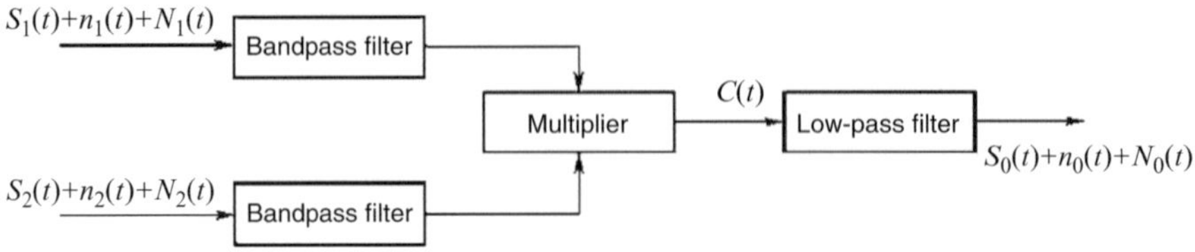

Fig. 2.72 Block diagram of cross-correlation detection

proposal. It is revealed from Expression (2.291) that an additional time delay t_0 must be added between elevation difference signal and azimuth difference signal to implement "dual-channel monopulse" making that the correlation value of error signal from dual-channel to the elevation difference is maximum while that to the azimuth difference is relatively small (vice versa). Application of "single-channel monopulse" proposal will result in more than 6 dB signal-noise ratio loss, which conflicts with low *C*/*N* receiving.

(2) Output signal-noise ratio of angle error cross-correlation detection

Block diagram of cross-correlation detection is as shown in Fig. 2.72.

In Fig. 2.72, $S_1(t)$ and $S_2(t)$ are the two-channel signals being correlated (in angle detection, respectively being sum channel and difference channel). $n_1(t)$ and $n_2(t)$ are uncorrelated part of 2-channel noise, and $N_1(t)$ and $N_2(t)$ are correlated part of 2-channel noise. Band-pass filter allows signal spectrum to pass but limit white noise, with bandwidth being *B*. Low-pass filter filters the relevant output signal, with bandwidth being B_L.

Cross correlator can solve cross-correlation function $R_{xy}(0)$ when $t=0$, multiplier can multiply 2-channel signal, its output is:

$$\begin{aligned} C(t) &= [S_1(t)+n_1(t)+N_1(t)]\cdot[S_2(t)+n_2(t)+N_2(t)] \\ &= S_1(t)S_2(t)+S_1(t)n_2(t)+S_1(t)N_2(t)+n_1(t)S_2(t) \\ &\quad + n_1(t)n_2(t)+n_1(t)N_2(t)+N_1(t)S_2(t)+N_1(t)n_2(t)+N_1(t)N_2(t) \end{aligned} \tag{2.293}$$

1) In case of low carrier-noise ratio

In case of low carrier-noise ratio, as $N_1(t) \ll n_1(t)$, $S_2(t) \ll n_2(t)$, $N_1(t) \ll n_1(t)$, $N_2(t) \ll n_2(t)$ and the amplitude of $n_1(t)$ and $n_2(t)$ is equal approximately; therefore,

$$C(t) \approx S_1(t)S_2(t)+n_1(t)n_2(t)+N_1(t)N_2(t) \tag{2.294}$$

where $S_1(t)S_2(t)$ is signal part; $n_1(t)n_2(t)$ is non-correlation noise part; $N_1(t)N_2(t)$ is correlated noise part.

① Signal output. Output signal is the cross-correlation function of $S_1(t)$ and $S_2(t)$ in bandwidth B when $= 0$:

$$R_{S_1S_2}(o) = \lim_{T\to\infty} \frac{1}{1T} S_1(t)S_2(t)\mathrm{d}t \tag{2.295}$$

$S_1(t)$ is correlated with $S_2(t)$ but they have different amplitude. Let $\frac{S_1(t)}{S_2(t)} = m$, when $E(n) = 0$, then

$$R_{S_1S_2}(o) = m \cdot D\left[S_1^2\right] = m \cdot \sigma_{S_1}^2 \tag{2.296}$$

where $D\left[S_1^2\right] = \sigma_{S_1}^2$ is variance of $S_1(t)$.

② Cross-correlation output of correlated noise $R_{N_1N_2}(o)$. Since $N_1(t)$ is correlated with $N_2(t)$, according to above, the following expression can be obtained:

$$R_{N_1N_2}(o) = M \cdot D\left[N_1^2\right] = M \cdot \sigma_{N_1}^2 \tag{2.297}$$

where,

$$M = \frac{N_1(t)}{N_2(t)}$$

where $\sigma_{N_1}^2$ is variance of correlated noise $N_1(t)$.

③ Output of non-correlated noise. Typical analytic method of its cross-correlation function $R_{n_1n_2}(o)$ is that the cross-correlation function, $R_{xy}(t)$ is derived from inverse Fourier transformation of cross-spectral density $R_{xy}(\omega)$ of random function $x(t)$ and $y(t)$. Here, we employ frequency domain method to make analysis, in order to obtain clear understanding on the project development.

Strictly speaking, random signals cannot be decomposed in Fourier method, because random signals are usually not periodic or square integrable. However, Wiener put forward "broad sense harmonic" concept, which points out that: Fourier analysis of random signals can be discussed in limitation significance, namely, take out random signals $x(t)$ within limited time period $(-T \sim T)$, but the time is limited; therefore, there is Fourier transformation in $x(t)$. When $T \to \infty$, it approaches actual condition. Therefore, Fourier series can approximately express frequency domain of $x(t)$.

To implement digital signal processing, continuous random signals are usually sampled to form discrete random signals which are separated evenly, on account of its periodicity, can be dealt with Fourier series decomposition.

Therefore, $n_1(t)$ and $n_2(t)$ can be approximately expressed as:

$$n_1(t) = \sum_{n=0}^{\infty} \sigma_{1n} \cos\left(n\Omega t + \varphi_{1n}\right) \tag{2.298}$$

$$n_2(t) = \sum_{n=0}^{\infty} \sigma_{2n} \cos\left(n\Omega t + \varphi_{2n}\right) \tag{2.299}$$

where σ_{1n} and σ_{2n} are root-mean-square value; $\Omega = 2\pi/T_0$.

In Expressions (2.298) and (2.299), all spectral components are uncorrelated with each other and, when $T \to \infty$, become continuous spectrum.

For flat noise spectrum with $\sigma_{1n} = \text{constant} = \sigma_{10}$ and $\sigma_{2n} = \text{constant} = \sigma_{20}$, then

$$n_1(t) = \sigma_{10}\sum_{n=0}^{\infty} \cos(n\Omega t + \varphi_{1n}) \tag{2.300}$$

$$n_2(t) = \sigma_{20}\sum_{n=0}^{\infty} \cos(n\Omega t + \varphi_{2n}) \tag{2.301}$$

After being bandlimited by the band-pass filter BPF with bandwidth $2_\pi B$ (unit is rad/s, unit of B is Hz) and center frequency at $\omega_{0,}$ they become band-pass noise, namely,

$$\text{BPF}[n_1(t)] = \sigma_{10}\sum_{n=x}^{x+b} \cos(n\Omega t + \varphi_{1n}) \tag{2.302}$$

$$\text{BPF}[n_2(t)] = \sigma_{20}\sum_{n=x}^{x+b} \cos(n\Omega t + \varphi_{2n}) \tag{2.303}$$

where $x\Omega = \omega_0 - (1/2)2\pi B$ is $x = (\omega_0/\Omega) - (\pi B/\Omega)$, $(x+b)\Omega = \omega_0 + (1/2)\pi B$, that is $b = 2\pi B/\Omega$, we can obtain:

$$\begin{aligned} \text{BPF}[n_1(t)n_2(t)] = {} & \sigma_{10}\cos(n\Omega t + \varphi_{1x})\cdot\sigma_{20}\sum_{n=x}^{x+b}\cos(n\Omega t + \varphi_{2n}) + \cdots \\ & \sigma_{10}\cos\left[(x+b)\Omega t + \varphi_{x+b}\right]\cdot\sigma_{20}\sum_{n=x}^{x+b}\cos(n\Omega t + \varphi_{2n}) \end{aligned} \tag{2.304}$$

There are b terms to be added in Expression (2.304). After the calculation result passes through the low-pass filter LPF with bandwidth being $B_{L,}$ we can obtain that the spectral component included in $n_0(t)$ is

$$\begin{aligned} n_0(t) = \text{LPF}\{\text{BPF}[n_1(t)n_2(t)]\} = {} & \sigma_{10}\sigma_{20}\cos(x\Omega t + \varphi_{1x})\cdot\sum_{n=x}^{x+b}\cos(n\Omega t + \varphi_{2n}) + \cdots \\ & \sigma_{10}\sigma_{20}\cos\left[(x+b)\Omega t + \varphi_{1(x+b)}\right]\cdot\sum_{n=(x+b)}^{x+b+L}\cos(n\Omega t + \varphi_{2n}) \end{aligned} \tag{2.305}$$

where $L = 2\pi B_L/\Omega$

There are $\left[\left(\frac{2\pi B}{\Omega}\cdot\frac{2\pi B_L}{\Omega}\right) - \sum_{m=1}^{L} m = 1\right]$ terms in expression (2.305). As $B \gg B_L$, it is approximately $\frac{2\pi B}{\Omega}\cdot\frac{2\pi B_L}{\Omega}$ terms. Considering image frequency beat, the terms will be doubled, namely, there are $\sigma_{n_1 n_2} = \sqrt{\frac{8\pi^2 BB_L}{\Omega^2}\frac{\sigma_{10}^2\sigma_{20}^2}{4}} = \sqrt{\frac{2\pi^2 BB_L}{\Omega^2}}\sigma_{10}\sigma_{20}$ terms.

According to this, when the bandwidth B of the bandpass filter grows, the beat terms falling in B_L will increase accordingly, that is physical reason why output noise will increase while B increasing. Therefore, the noise increment can be called as "beat loss."

Amplitude of each beat term is $1/2\sigma_{10}\sigma_{20}$ and is not correlated mutually. Perform operation of mean square and root extraction to $(8\pi^2 BB_L/\Omega^2)$ terms to get output noise root-mean-square value $\sigma_{n_1n_2}$ of the correlator, namely,

$$\sigma_{n_1n_2} = \sqrt{\frac{8\pi^2 BB_L}{\Omega^2}\frac{\sigma_{10}^2\sigma_{20}^2}{4}} = \sqrt{\frac{2\pi^2 BB_L}{\Omega^2}}\sigma_{10}\sigma_{20} \tag{2.306}$$

After $n_1(t)$ and $n_2(t)$ being bandlimited by bandwidth B, the variance σ_{n1}^2 and σ_{n2}^2 respectively are:

$$\sigma_{n1}^2 = \frac{2\pi B}{\Omega}\sigma_{10}^2, \quad \sigma_{n2}^2 = \frac{2\pi B}{\Omega}\sigma_{20}^2$$

therefore,

$$\sigma_{n_1n_2} = \sigma_{n_1}\sigma_{n_2}\sqrt{\frac{B_L}{2B}} \tag{2.307}$$

④ Signal-noise ratio of output voltage in case of low carrier-noise ratio
(a) Output voltage signal-noise ratio $(S/N)_n$ of non-correlated noise

$$\left(\frac{S}{N}\right)_n = \frac{R_{S_1S_2}(o)}{\sigma_{n_1n_2}(o)} = \sqrt{\frac{2B}{B_L}\frac{m\sigma_{S_1}^2}{\sigma_{n_1}\sigma_{n_2}}} = \frac{\sigma_{S_1}}{\sigma_{n_1}}\cdot\frac{\sigma_{S_2}}{\sigma_{n_2}}\sqrt{\frac{2B}{B_L}} = \sqrt{\frac{P_{S_1}}{P_{n_1}}}\sqrt{\frac{P_{S_2}}{P_{n_2}}}\sqrt{\frac{2B}{B_L}} \tag{2.308}$$

(b) Output voltage signal-noise ratio $(S/N)_N$ of correlated noise

$$\left(\frac{S}{N}\right)_N = \frac{R_{S_1S_2}(o)}{\sigma_{N_1N_2}(o)} = \frac{\sigma_{S_1}^2}{\sigma_{N_1}}\frac{m}{M} = \frac{\sigma_{S_1}}{\sigma_{N_1}}\cdot\frac{\sigma_{S_2}}{\sigma_{N_2}} = \sqrt{\frac{P_{S_1}}{P_{N_1}}}\sqrt{\frac{P_{S_2}}{P_{N_2}}} \tag{2.309}$$

Quantization loss shall be considered in actual project. It is 0.5 in case of 1 bit quantization, 0.7 in case of 2 bit quantization, and 0.8 in case of the higher quantization.

According to Expression (2.308), in case of white noise, the bigger the bandwidth B of the band-pass filter is, the more the beat component of noise becomes, in which, output noise voltage in B_L increases in directly proportional to $\sqrt{B}$. However, the two signals are correlated and their beat component are added arithmetically within B_L, namely, they increase in direct proportion to B (in strict sense, only uniform signal spectrum is strictly direct proportional to B). Therefore, the output signal-noise ratio is in direct proportion to $\sqrt{B}$ within main energy range of the signal spectrum. $(B/B_{\mathrm{L}}) = K_B$ is known as bandwidth gain, therefore, a higher K_B can be obtained for wideband signal of B, which shall increase output S/N, and this is the reason why wideband angle tracking system can implement low C/N tracking. Therefore, in case of low C/N, the signal bandwidth shall be used as possible to implement angle tracking. If only a section of signal bandwidth is segmented, the output S/N within B_{L} will decrease. However, when the bandwidth B increases and exceeds signal spectrum, increment of B will not improve output signal voltage. But, the output noise voltage still increases according to $\sqrt{B}$, at this time, the output signal-noise ratio will decrease. Therefore, there is an optimum bandpass bandwidth $B_{\mathrm{on}T}$ to ensure a maximum output S/N. The value of $B_{\mathrm{on}T}$ is relevant to the spectral shape of the signal. Namely, there is different $B_{\mathrm{on}T}$ for different spectral shape. Such value can be obtained by simulation or experiment.

2) In case of high carrier-noise ratio

At this time, a section of signal bandwidth is segmented to perform angle tracking and thus simplifying the proposal. Within the intercepted narrower bandwidth B, the signal spectrum shall be deemed to be uniform. Since the carrier-noise ratio is extremely high, the Expression (2.293) is simplified as:

$$C(t) = S_1(t)S_2(t) + S_1(t)n_2(t) + S_2(t)n_1(t) + N_1(t)N_2(t) \tag{2.310}$$

where $S_1(t)$ and $S_2(t)$ are correlated, and $N_1(t)$ and $N_2(t)$ are correlated. Their correlated output expression has been solved as shown in expressions (2.296) and (2.297). $S_1(t)$ and $n_2(t)$, $S_2(t)$ and $n_1(t)$ in Expression (2.310) are non-correlated. Because of uniform spectrum, the derivation principle of correlated output is the same as that of non-correlated noise. Its formula is similar to the one output by non-correlated noise expressed by Expression (2.307).

① Correlated output of $S_1(t)$ and $n_2(t)$:

$$\sigma_{s_1}\sigma_{n_2}\sqrt{B_{\mathrm{L}}/2B} \tag{2.311}$$

② Correlated output of $S_2(t)$ and $n_1(t)$:

$$\sigma_{s_2}\sigma_{n_1}\sqrt{\frac{B_{\mathrm{L}}}{2B}} \tag{2.312}$$

③ Total non-correlated noise:

$$\sqrt{\sigma_{s_1}^2\sigma_{n_2}^2 + \sigma_{s_2}^2\sigma_{n_1}^2}\sqrt{\frac{B_{\mathrm{L}}}{2B}} \tag{2.313}$$

When tracking is conducted, the sum signal $S_1(t)$ is larger than difference signal $S_2(t)$; the Expression (2.303) is approximately expressed as:

$$\sigma_{s_1}\sigma_{n_2}\sqrt{\frac{B_L}{2B}} \tag{2.314}$$

④ Output voltage signal-noise ratio in case of high carrier-noise ratio:
(a) Output voltage signal-noise ratio of non-correlated noise

$$\left(\frac{S}{N}\right)_n = \frac{\sigma_{S_1}\sigma_{S_2}}{\sigma_{S_1}\sigma_{n_2}} \cdot \sqrt{\frac{2B}{B_\mathrm{L}}} = \frac{\sigma_{S_1}}{\sigma_{n_1}}\sqrt{\frac{2B}{B_\mathrm{L}}} = \sqrt{\frac{P_{S_2}}{P_{n_2}}}\sqrt{\frac{2B}{B_\mathrm{L}}} \tag{2.315}$$

where $\sqrt{P_{s_2}/P_{n_2}}$ is the signal-noise ratio of the segmented bandwidth B.
(b) Output voltage signal-noise ratio of correlated noise

$$\left(\frac{S}{N}\right)_n = \sqrt{\frac{P_{S_1}}{P_{N_1}}}\sqrt{\frac{P_{S_2}}{P_{N_2}}} \tag{2.316}$$

The foresaid noise includes two parts: ① antenna feed noise and LNA noise, ② sky noise (cosmic noise and atmospheric loss noise). The LNA is not shared, so LNA noise is non-correlated. But, sky noise is correlated with some antenna feed noise. In most applications, correlated noise is generated by antenna noise (except for deep-space detection) and determined by correlation of the patterns of sum and difference sidelobes touching to ground. Finally, it will be the minimum value of sum and difference cross-correlated output noise, namely, it determines the threshold value. This correlated noise cannot be deducted by prolonging integral time. However, non-correlated noise can be greatly weakened by prolonging integral time, so the output signal-noise ratio is finally determined by correlated noise.

If correlated noise $\sigma_{N_1}^2$ is $1/L$ of non-correlated noise σ_n^2, the output noise after long integral time can be obtained through Expression (2.297): $R_{N_1N_2}(o) = \frac{M}{L}\sigma_{n_1}^2$.

(3) Angle tracking error in case of cross-correlated angle error detection

The precondition of still and balance of tracking antenna is that the servo motor has no drive voltage; in this case, the output of angle error detector shall be 0, namely,

$$E_\Delta - E_n = 0 \tag{2.317}$$

where E_Δ is the signal voltage output by angle detection, E_n is the value of noise voltage output by angle detection; and the negative sign in the expression above indicates the servo system is negative feedback.

According to Expression (2.317), there is:

$$\frac{E_\Delta}{E_n} = 1 \tag{2.318}$$

The meaning of Expression (2.318): if there is noise E_n, the servo system will make the antenna deviate from $\Delta\theta$ to generate error voltage E_Δ, which offsets E_n to make output of angle error detector be 0, so as to enable the antenna achieve a new balance and stable point. When the instantaneous value of the angle detector

$E_\Delta/En = 1$, its r.m.s value equals to 1, which is determined by Expression (2.308), namely,

$$\frac{S}{N} = \sqrt{\frac{P_\Sigma}{P_{n\Sigma}}}\sqrt{\frac{P_\Delta}{P_{n\Delta}}}\sqrt{\frac{2B}{B_\mathrm{L}}} - \frac{U_\Sigma}{U_{n\Sigma}} \cdot \frac{U_\Delta}{U_{n\Delta}} \cdot \sqrt{\frac{2B}{B_L}} = 1$$

then,

$$U_\Delta = \frac{U_{n\Sigma}}{U_\Sigma} \cdot U_{n\Delta} \cdot \sqrt{\frac{B_\mathrm{L}}{2B}} \tag{2.319}$$

The angle error $\Delta\theta$ corresponding to U_Δ is determined by the following expression in angle measurement analysis (converted to feed outlet), namely,

$$\Delta\theta = \frac{\sqrt{G_\Delta}}{K} = \frac{\theta_{0.5}\sqrt{G_\Delta}}{K_m\sqrt{G_\Sigma}} = \frac{\theta_{0.5}}{1.4} \cdot \frac{U_\Delta}{U_\Sigma} = 0.7 \times \theta_{0.5}\frac{U_\Delta}{U_\Sigma}$$

where G_Σ is gain of sum beam; G_Δ, gain of difference beam; K, absolute difference slope; $K = K_m\left(\sqrt{G_\Sigma}/\theta_{0.5}\right)$, $\theta_{0.5}$ is 3 dB width of sum beam; normalization difference slope $K_m = 1.2–1.7$, 1.4 is applied.

Substitute it into U_Δ in Expression (2.319)

$$\Delta\theta = 0.7\theta_{0.5}\left(\frac{U_\Delta}{U_\Sigma}\right) \cdot \left(\frac{U_{n\Sigma}}{U_\Sigma}\right)\sqrt{\frac{B_L}{2B}} \tag{2.320}$$

When the sum noise temperature is same as the difference noise temperature, i.e., $T_\Sigma = T_\Delta$, then $U_{n\Sigma} = U_{n\Delta}$ and

$$\Delta\theta = 0.7\theta_{0.5}\left(\frac{U_{n\Sigma}}{U_\Sigma}\right)^2\sqrt{\frac{B_L}{2B}} = 0.7\theta_{0.5}\frac{P_{n\Sigma}}{P_\Sigma}\sqrt{\frac{B_L}{2B}} \tag{2.321}$$

(4) Four-channel monopulse proposal

The foregoing angle error signals will not serve as drive signal of angle tracking system until being normalized by sum signal. Additionally, when angle acquisition is conducted, it is required to detect amplitude of sum channel signal to serve as normalized reference signal and angle acquisition identification signal, which is usually extracted by sum channel signal. When the signal-noise ratio of sum channel signal is quite low, it will be a difficult problem. In that case, the advanced modern signal processing technology can be used to extract weak signal against the background of strong noise, such as cyclic spectrum technology and spectral line enhancement technology. In combination with monopulse angle tracking system, four-channel monopulse proposal is applied to extract such signal; namely, on the basis of typical "five-horn" 3-channel monopulse feed, the original four horns are employed to form an additional "reference" signal channel, as shown in Fig. 2.73.

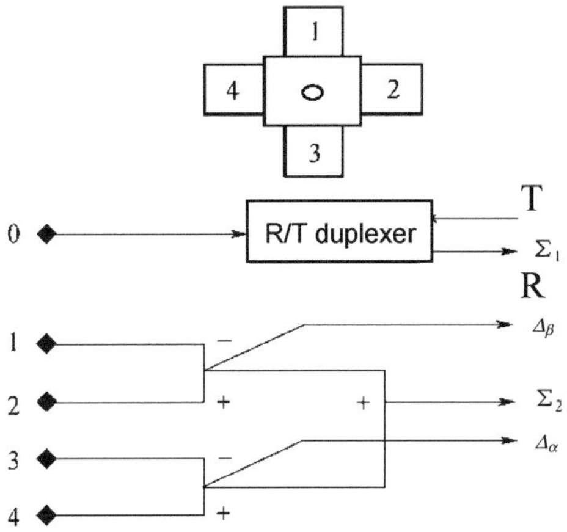

Fig. 2.73 Four-channel monopulse feed

Figure 2.73 shows a five-horn feed, Σ_1 is sum channel, employing Σ_1, Δ_α, Δ_β to form monopulse; Σ_2 is reference channel, Σ_1 and Σ_2 are used to perform cross-correlation detection of normalized signal and angle acquisition signal; Σ_1 serves as transmitting reference (like safety control signal). Σ_1 outputs P_Σ and P_Δ, $P_{n\Delta}$ outputs P_0 and $P_{n\Sigma 2}$, $P_{n\Sigma 1}$ and $P_{n\Sigma 2}$ are noise of sum channel and reference channel, which contains uncorrelated and correlated parts. Since they are generated by different LNA, horns, and feeders, therefore, the correlated part is relatively low, the SNR solved by cross-correlation are high extremely, and finally limited by relevant part of antenna noise.

When cross-correlation method is applied to extract normalized signals, the output signal-noise ratio expression may replace P_Δ and $P_{n\Delta}$ with P_0 and P_{n0}, we can get

$$\mathrm{SNR} = \sqrt{\frac{P_\Sigma}{P_{n\Sigma}}}\sqrt{\frac{P_0}{P_{n0}}}\sqrt{\frac{2B}{B_L}} \tag{2.322}$$

Additionally, "square method" can be used to recover carrier (such as PSK signal). Square loss may exist when square method is used, however, when P_s/N_0 of the signal is relatively large, the single-carrier C/N_0 recovered by the squarer is still large and can be applied as normalized reference signal after being extracted by the carrier tracking ring, even serve as tracking signal of signal-channel monopulse. Since P_s/N_0 of wideband signal is very large, such method is likely to be used.

Furthermore, the possibility of non-normalization will be discussed in case of low C/N: with decrease in C, the angle error slope will decline and the dynamic performance of angle tracking will be deteriorated. When C is small, the distance is far. Therefore, C changes mildly with change in distance, so it can be coordinated with the servo system to make the servo system in normal operation within the specified variation range of the angle error slope. When C improves to achieve proper normalization, enter normalized operating state.

According to analysis above, for low *C*/*N*, angle acquisition and tracking to can be implemented through the following proposal:

1) For non-periodic wideband signals: when there is no signal beacon (e.g., there are FM non-periodic signals only), normalization is key point, in this case, "four-channel monopulse" proposal can be employed. The threshold of the minimum receivable signal is limited by the correlated antenna feed noise between sum and difference channels. The improvement approach is to reduce relevance between the sidelobes of the antenna, i.e., the sidelobes of the two antennas. In addition, the sum and difference channel correlated noise cancellation method may be discussed. The other method is not to modulate FM signals at acquisition state, in order to make it be single-frequency carrier signals which can be acquired easily. Such signals will be modulated after being acquired and tracked.

2) For PSK wideband signals: "square method" may be applied to recover "single-frequency carrier," and then angle acquisition and tracking may be performed for single-frequency carrier. Since there is square loss, therefore, it is required to recover P_Σ/N_0. The acquisition and tracking requirements can be met. In case of the specified carrier-noise ratio, the wider the signal bandwidth is, the higher the corresponding P_Σ/N_0 is, so it may be applicable to rapid data transmission signals.

3) For S/Ka-band signals: where there are S-band signals, S-band wide beam may be used to guide Ka-beam narrow beam.

4) For periodic wideband signals: Theoretically speaking, "auto-correlation method" may be used to extract normalized reference signals. "Auto-correlation" required in-phase in carrier, but the frequency at Ka-band is high and there is Doppler frequency shift, so it is difficult to implement such extraction.

5) For single-frequency narrowband signal. Angle error detection of single-frequency signal is actually cross-correlation detection. However, due to tracking and filtering of phase-locked loop, $P_\Sigma/N > 1$, therefore, in case of the same P_Σ/N_0, its random error is relative small compared with wideband signal and the acquisition probability is high. Although there is locking time, the technology is matured in terms of USB and deep-space detection, therefore, this method is the best choice in presence of signal beacons. In order to perform angle tracking and improve the system reliability in absence of beacon signal, the system may be provided with two tracking operation modes, namely, "beacon signal" and "wideband signal" modes.

2.5.5 *Angle Measurement by Phased Array Tracking and Angle Measurement Accuracy – Space Window Sliding and Projective Plane Methods*

2.5.5.1 Angle Measurement Principle by Phased Array Tracking

Angle measurement method used in phased array CW radar, compared with that employed in mechanical scanning radar, has some features due to the fact that the phased array receiving antenna is of multichannel system and the antenna beam has phased array characteristics.

In terms of principle, the monopulse angle measurement method used in phased array CW radar is the same as monopulse precise tracking radar with mechanical scanning feature. There are three methods: amplitude contrast method, phase contrast method, amplitude and phase contrast method. These methods are used to conduct comparison and inner-angle insertion to the phase or amplitude of receiving signal from different beams, so as to improve angle measurement accuracy. Differently, mechanical tracking monopulse highly-accurate measurement radar commonly uses amplitude sum-difference monopulse system, the phased array CW radar generally employs phase sum-difference monopulse system. Amplitude sum-difference monopulse method requires that distance between two feeds shall meet $d < 2\lambda$, which is difficult to be achieved in constrained feed phased antenna. However, when monopulse phase contrast method to measure angel in phased array, it is easy to divide the entire antenna array into two or four parts, and to employ the entire receiving array to receive beam. In phased array antenna as shown in Fig. 2.74, two sub-antennas consisting of two sub-arrays can conduct phase sum-difference monopulse angle measurement.

In Fig. 2.74, the antenna array is equally divided into two sub arrays, with the distance between the two array phase centers of D, and the electrical boresight direction of θ. There is phase difference φ between the signals received by the two sub-arrays from θ direction, then,

$$\varphi = \frac{2\pi}{\lambda} D \sin\theta \tag{2.323}$$

θ can be obtained from measured φ.

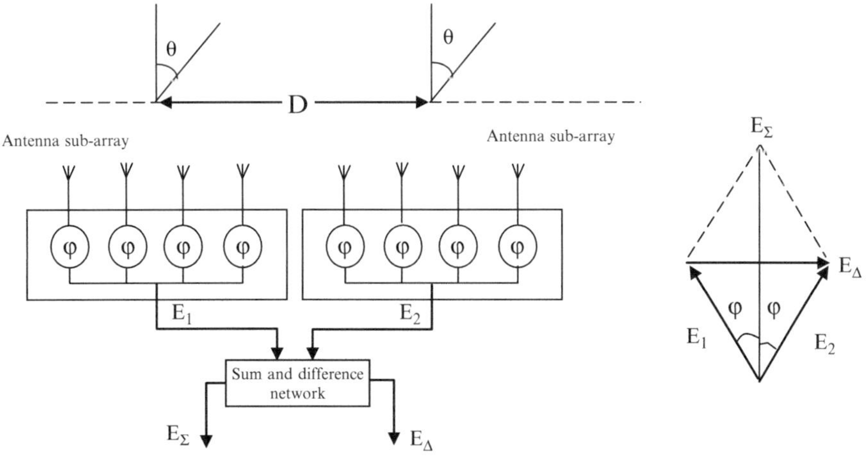

Fig. 2.74 Block diagram of phase sum-difference monopulse

In the phase sum-difference monopulse scheme as shown in Fig. 2.74, phase contrast is conducted between left and right sub-arrays in the sum-difference channel. The output differential signal $E_{\triangle}$ is relative to φ and θ. Meanwhile, the output E_{Σ} from the sum-difference channel is relative to θ. After the differential signal is normalized with $E_{\triangle}/E_{\Sigma}=K(\theta)$, the angle-sensitive function of the phase sum-difference monopulse can be obtained:

$$\Delta\theta = \frac{\lambda}{\pi D\cos\theta_0}\mathrm{tg}^{-1}k(\theta) \tag{2.324}$$

where $k(\theta)=\frac{E_{\Delta}}{E_{\Sigma}}$ is known as normalized differential signal.

When phased array antenna performs targets continuous tracking, the targets will be in location nearby θ_0 during most of tracking period; therefore, when $\Delta\theta \ll \Delta\theta_{3\mathrm{dB}}/2$, Expression (2.324) is simplified as:

$$K(\theta) = \frac{E_{\Delta}}{E_{\sum}} \approx \frac{\pi}{\lambda}D\cos\theta_0 \times \Delta\theta \tag{2.325}$$

$$\Delta\theta = \frac{\lambda}{\pi D\cos\theta_0} \times K(\theta) \tag{2.326}$$

However, when the phased array antenna conducts discrete tracking to the targets on a time-sharing multi-target basis, especially, when the tracking accuracy stays low and the angle of the target deviating from θ_0 is approximately $\pm\Delta\theta_{3\ \mathrm{dB}}/2$, Expression (2.324) shall also be used to get deviation value of the target $\Delta\theta$. Accordingly, the target position θ is $\theta_0 \pm \Delta\theta$, where θ_0 is coarse measurement and obtained from phase control code of the phased array. $\Delta\theta$ is fine measurement and obtained from the angle-sensitive function Expression (2.324).The expressions above are the basis of phased array open-loop angle measurement.

In fact, angle measurement is carried out by antenna and receiver. The block diagram of the whole angle measurement system is as shown in Fig. 2.75.

In Fig. 2.75, the angle error output signal of the receiver is $\Delta u = K(\theta) = E_{\Delta}/E_{\sum}$, of which the normalization is achieved by sum-channel AGC, and phase change of $E\Delta$ is detected by phase discriminator. Phase shift of $\pi/2$ is to make $E_{\triangle}$ and E_{Σ} be

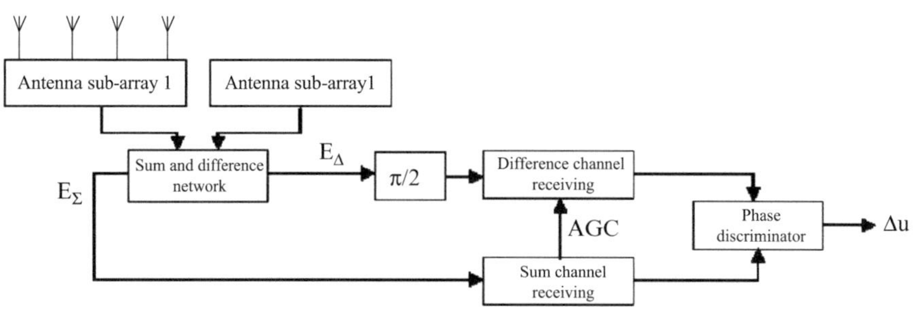

Fig. 2.75 Block diagram of phase sum-difference monopulse angle tracking receiving system

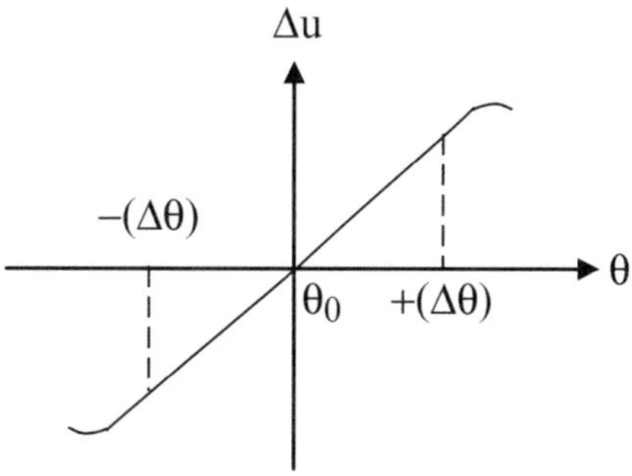

Fig. 2.76 Angle-sensitive feature of tracking receiver's output

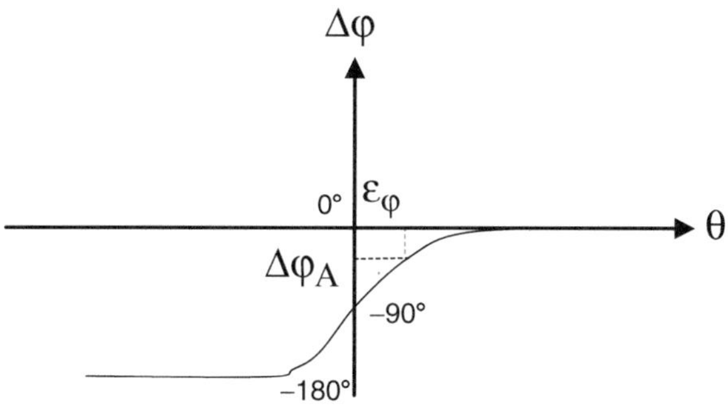

Fig. 2.77 Phase change between sum channel and difference channel

in-phase, so as the phase discriminator can be used to test angle error voltage. When the target is in the direction $\Delta\theta$ deviating from the electrical axis, the differential channel will output E_Δ, and the corresponding sum channel will output E_Σ; when the target moves from the left side of the electrical axis to right side, phase of E_Δ will change 180°. The angle-sensitive function curve containing receiver is as shown in Fig. 2.76.

At this time, the angle-sensitive function is:

$$\Delta\theta = \frac{\lambda}{\pi D \cos\theta_0} \arctan(\Delta u), \tag{2.327}$$

In this angle measurement method, the phase difference of the signals received by two antennas spatially separated is used to determine deviation of the target from the beam pointing θ, and sum and differential beams are formed to achieve differential signal normalization and to judge the direction of the target deviating from θ, therefore, such method is also known as amplitude-phase monopulse angle measurement method. In such method, after phase-shifting of E_Δ, and when θ is changing from $-(\Delta\theta)$ to $+(\Delta\theta)$, the phase is changing from $-180°$ to 0°, which will be $-90°$ nearby θ_0. Therefore, even if there is E_Δ, the output from the discriminator is zero. It is concluded that even if the antenna has null depth, output from phase discriminator still be zero to stabilize angle tracking. Phase reversion process is shown in Fig. 2.77, the deeper the null depth is, the smaller the transitions scope from $-180°$ to 0° becomes.

The value of null depth affects position of $\Delta u = 0$ on θ axis, this is because the null depth is generated from inconsistence between sum/difference phase before comparator. The smaller the null depth is, the larger the inconsistence of the phase before comparator becomes. Additionally, the phases of the sum/differential channel of the receiver are also inconsistent. When the sum attains to $\pi/2$, zero point of Δu will exist (namely, angle tracking static stable tracking position). In case the sum-difference phase of the receiver is a give value, $\Delta\varphi R$, the zero value of Δu will appear when the antenna sum-difference phase is $\Delta\varphi_A = \left(-\frac{\pi}{2} - \Delta\phi_R\right)$. It is revealed from Fig. 2.77 that it is corresponding to ε_φ value, such value is angel tracking static error. $\Delta\varphi_A$ varies from different $\Delta\varphi_R$, ε_φ will change accordingly. The above section illustrates the physical concept of generating angle tracking static error by null depth. In case of amplitude sum-difference monopulse, it is expressed as following:

$$\varepsilon_\varphi = \frac{\theta_{3\ \mathrm{dB}} \tan\left(\Delta\varphi_R\right)}{K_m\sqrt{G_n}} \tag{2.328}$$

where $\theta_{3\mathrm{dB}}$ *is* 3 dB beam width of the antenna, K_m represents normalized difference slope, G_n is null depth of difference beam.

Based on the foresaid angle error open-loop measurement principle, the principle block diagram of phased array monopulse closed-loop angle tracking can be obtained as shown in Fig. 2.78. In this figure, the negative feedback signals in angle tracking loop is phase control value to each array element output from a beam controller. They control the phase discriminator of each array element to generate phase shift ϕ_1, ϕ_2,... ϕ_N, and the phase difference between adjacent array elements is $\Delta\phi_x$. When the target is on the direction of the antenna array electrical boresight, the spatial phase difference between adjacent array elements $\Delta\phi_x = 0$, at this time, the beam controller makes $\Delta\phi_x = 0$, so plus the same phases output from

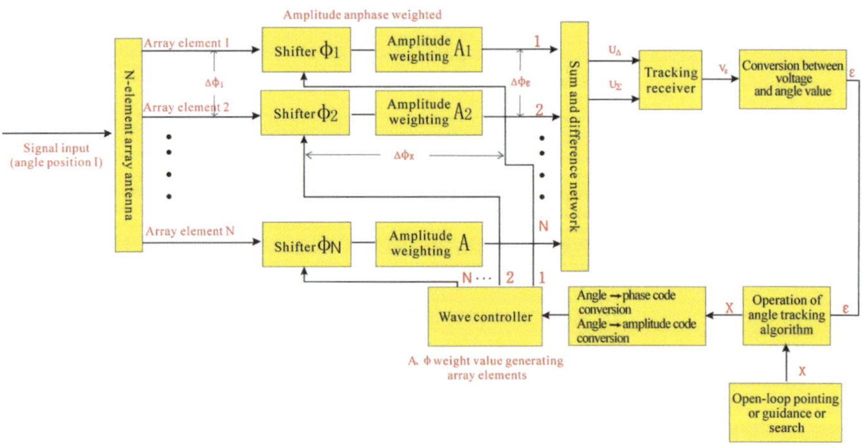

Fig. 2.78 Principle of phased array monopulse tracking

discriminators to synthesize a maximum gain. When the target deviates from the electrical axis, $\Delta\phi_i \neq 0$, the phase difference $\Delta\phi_\varepsilon$ between the phase shifter output of adjacent array elements is $\Delta\phi_\varepsilon \neq 0$. Applying the principle of angle error signals generation, the error signal V_Δ will be generated by the sum and difference device, which is amplified, processed and filtered by the tracking receiver to output angle error voltage V_ε, which is converted into angle value ε by using angle-sensitive functions, and a new antenna pointing angle X shall be calculated with the angle tracking algorithm. X is converted to the new weight value φ and A of each array element of the phased array by beam controller. Such recursion algorithm shall be performed until $\Delta\phi_i \neq -\Delta\phi_x$, so $\varepsilon = 0$, thereby angle tracking implemented. In fact, it is phase-shift tracking, and the phase shift value is corresponding to the angle measurement data.

2.5.5.2 Monopulse Tracking of Planar Phased Array – Projective Plane Method

"Open-loop angle measurement" or "closed-loop angle measurement" can be used for phased array angle measurement. The former, comparatively simpler, is based on angle-sensitive function as shown in Expression (2.327) and implement angle deviation measurement by measuring the output voltage of angle error, which obviously has a relation to accuracy of angle-sensitive characteristics. The latter has no relation to accuracy of angle-sensitive characteristics, because the output of angle error is zero after tracking is stable, and implements angle measurement with the phase value from the phase shifter, which has a high accuracy. For different target angle, the projective plane of planar array on radio wave equiphase surface will roll with the changing of the target angle. It is so called Projective Plane method, wherein gain loss of scan angle exists. The method shall be mainly described in this section. The analysis can be carried out in time domain or frequency domain.

(1) Time domain analysis

It is well known that the time domain analysis adopts difference equation, and if the current output of the equation is related not only to the current and past inputs, but also to the past output, it is the recursive equation. The phased array angle tracking is the case. The phased array angle tracking is usually implemented by a set of recursive equations calculation. The values of recursive functions are calculated based on the previous one or several calculation results, so this is a recursive step tracking method. Different recursive functions can compose different type of systems, such as Type I system (including one integrating element, Type I filter), Type II system (including two integrating elements, Type II filter), Type III system (including three integrating elements, Type III filter) and uses the Kalman filter, etc. There are different dynamic errors according to different types of loops. Such tracking are realized by optimal estimation through following parameters, including: ① error of observation, ② estimated beam position, ③ passing through smooth target location, rate, acceleration, etc. According to design requirements, such tracking algorithm may have different principles, e.g., minimum mean square error criterion. The following introduces a common tracking algorithm based on recursive function:

1) Recursive tracking equation of Type I system

$$X_{n+1} = X_n + a_0 \varepsilon_n \tag{2.329}$$

where α_0 position smoothing constant ($0 < \alpha_0 < 1$), and may slowly converge if it is too small and may oscillate if it is big; X_n is the angle value calculated from the nth sampling point; ε_n is the angle error measured from the nth sampling point; X_{n+1} is the resulted forecast angle value calculated from the $(n+1)$-th sampling point.

The above expression is a recursive equation. To perform one recursive calculation within t, $1/t$ is used as sampling clock frequency. Continuous summing up (equal to integral) shall result in an angle error closing to zero, which realizes the angle tracking of no position static state error and is equal to one integration element. The higher the calculation frequency, the better the tracking dynamics is.

2) Recursive tracking equation of Type II system

$$X_{n+1} = X_n + \alpha_0 \varepsilon_n + t\overline{\dot{X}}_n \tag{2.330}$$

where $\overline{\dot{X}}_n$ smooth target angular velocity of the nth sampling point; t is sampling interval. The above expression adds velocity estimation component and increases velocity tracking accuracy, equal to 2 integration elements. One calculation method of $\overline{\dot{X}}_n$:

$$\overline{\dot{X}}_n = \overline{\dot{X}}_{n-1} + \frac{a_1}{t} \varepsilon_n \tag{2.331}$$

where $\overline{\dot{X}}_{n-1}$ smooth target angular velocity of the $(n-1)$-th sampling point; α_1 velocity smoothing constant, $0 < \alpha_1 < 4$, $0 < \alpha_0 < 1$; $\overline{\dot{X}}_n$ also can be estimated according to other methods.

3) Recursive tracking equation of Type III system

$$X_{n+1} = X_n + \alpha_0 \varepsilon_n + t\overline{\dot{X}}_n + \frac{1}{2} t^2 \overline{\ddot{X}}_n \tag{2.332}$$

where $\overline{\ddot{X}}_n$ smooth target angular velocity of nth sampling point. The above expression adds acceleration estimation component and increases acceleration tracking accuracy, equal to three integration elements. One calculation method of $\overline{\ddot{X}}_n$:

$$\overline{\ddot{X}}_n = \overline{\ddot{X}}_{n-1} + \frac{a_2}{t_2} \varepsilon_n \tag{2.333}$$

where $\overline{\ddot{X}}_{n-1}$ smooth target acceleration of $(n-1)$-th sampling point; α_2 acceleration smoothing constant, $0 < \alpha_2 < 8$, $0 < \alpha_1 < 16/9$; $\overline{\ddot{X}}_n$ also can be estimated according to other methods.

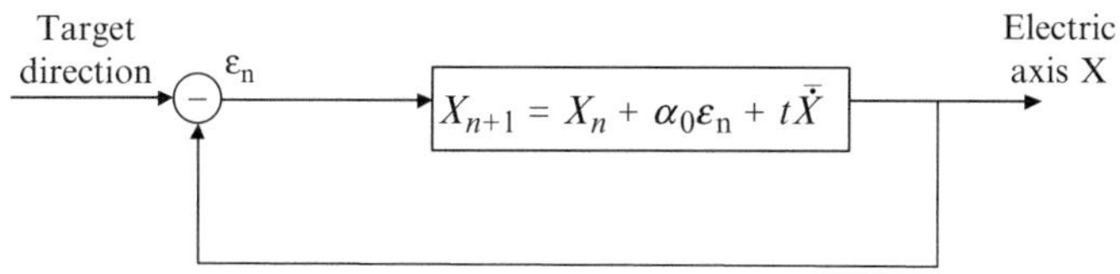

Fig. 2.79 Block diagram of phased array servo system

According to above tracking algorithm, block diagram of phased array angle tracking closed-loop system is as shown in Fig. 2.79.

In Fig. 2.79, ○ is accomplished by difference beam of phased array antenna. Difference signal, after passing through the filter, shall be processed with the recursive tracking algorithm, which is recursive calculation process (the digital transfer function is available by algorithm as PI, PID, etc.). X_n is the sum-beam electric axis direction θ_n, which is the same as zero direction of difference-beam and shifts according to Expression $\theta_n = \sin^{-1}\left(\frac{\lambda}{2\pi d}\Delta\phi\right)$ under the control of phase control code $\Delta\phi$ of phased array. ε_n is the angle difference signal. $\overline{\dot{X}}_n$ is the smooth target velocity of the nth sampling point. t is sampling interval and α_0 is calculation step size. X_0 initial value of angle at the time of tracking start. When enabling the target to fall into the major beam with guidance, etc., there is X_0 and ε_0 and continuous recursive calculation can be done until tracking is stable.

(2) Frequency domain analysis

As everyone knows, frequency domain analysis of digital loop shall be conducted with Z transform and Z-transfer function. Such frequency domain analysis is more frequently applied because the technical specifications of angle tracking loop are generally reflected in its frequency response. However, time domain analysis is connected with frequency domain analysis by Z transform.

The following is the recursive tracking equation of first-order loop of time domain provided by Expression (2.329):

$$X_{n+1} = X_n + \alpha_0\varepsilon_n$$

This is a difference equation, and is a digital integrator physically. Under control of digital error signal, digital-control beam is generated. In the equation, X_{n+1} is the $(n+1)$-th output sampling point of integrator, ε_n is the nth input sampling point, and α_0 is proportion coefficient. In recursive operation, X_{n+1} is stored in integration register. Z transform for integrator with an unit delay is:

$$X(Z) = \frac{\alpha_0 Z^{-1}\varepsilon(Z)}{1 - Z^{-1}}$$

where, $X(Z)$ is the output, $\varepsilon(Z)$ is the input, and transfer function $K_A(Z)$ of corresponding digital integrator is

$$K_A(Z) = \frac{X(Z)}{\varepsilon(Z)} = \frac{\alpha_0 Z^{-1}}{1 - Z^{-1}} \tag{2.334}$$

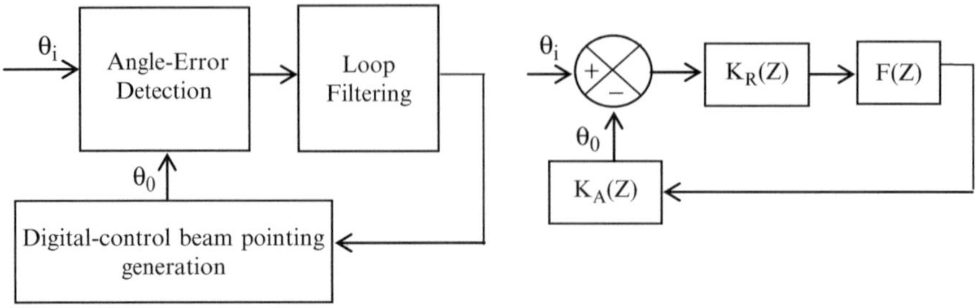

Fig. 2.80 Block diagram of phased array angle tracking loop

High-order loop should be applied to obtain better loop dynamic performance. The following shows a simple block diagram of such loop:

In Fig. 2.80, angle-error detection is conducted by difference beam and tracking receiver; its transform function is represented by $K_R(Z)$. Therefore, when receiver bandwidth is larger than loop bandwidth, it is a proportional element $K_R(z) = k_r =$ constant. Digital-Control Beam Pointing Generation is conducted by the phase shifter that controls such phased array. It is a digital integrator. Its transform function is represented by $K_A(Z)$, while $F(Z)$ is the transform function of loop filter. Closed-loop transfer function $H(Z)$ of angle tracking loop is

$$H(Z) = \frac{K_R(Z)F(Z)K_A(Z)}{1 + K_R(Z)F(Z)K_A(Z)} = \frac{K_r\alpha_0 Z^{-1}F(Z)}{\left(1 - Z^{-1}\right) + K_r\alpha_0 Z^{-1}F(Z)} \tag{2.335}$$

In Eq. (2.235), $K_A(Z)$ is the transform function of digital-control pointing angle generator. Multiple integral elements can be included in $F(Z)$. If only one integral element is included, the loop is called Type II loop. If two integral elements are included in $F(Z)$, it is called Type III loop. If no integral element is included in $F(Z)$, it is called Type I loop. The $F(Z)$ design method is already mature technology in PLL design. The design of phased array angle tracking loop can use many design methods in digital PLL.

2.5.5.3 Angle Tracking of Curved Surface Phased Array – Space Window Sliding Method

There are three defects for planar phased array. Firstly there is limited coverage of scanning angle with theoretical value of ±60°; secondly, no conformity to carrier, which impacts the aerodynamic characteristics of aircraft and limits increase of phased array area; thirdly, no special beamforming and curved surface phased array is applied. Common curved surface phased arrays include: hemispheric array, paraboloid array, hyperboloid array, cone array, etc. As is well known, since phased array's price is high, cost effectiveness is one of the key factors for options. Cost is

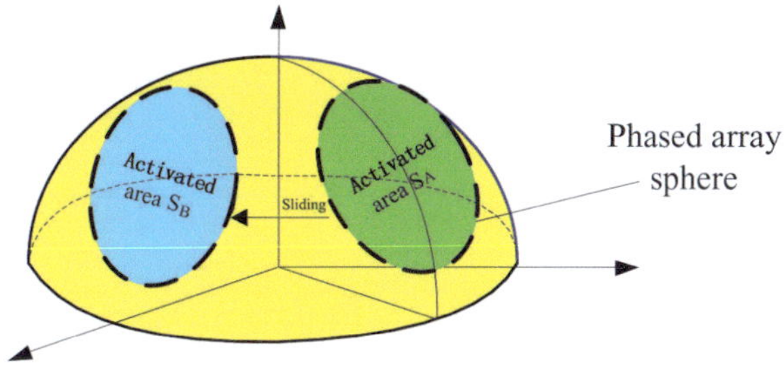

Fig. 2.81 Schematic diagram of empty window sliding of sphere-phased array

mainly determined by array element number, which is related to surface area of antenna array. There are various performance specifications. If we make a comparison with the specifications of continuous azimuth coverage of 360°, elevation of 0° to +90° and the same gain, it shall be found that in the above schemes, when elevation changes from 0° to +90°, the performance/price ratio will be: hemispheric array 0.19–0.38, paraboloid 0.19–0.48, hyperboloid array 0.19–0.62, cone 0.15–0.72. For elevation is 0°, the performance/price ratio decreases because of the only use of some area of curved surface, which makes gain decrease. One of the solutions is to expand phased array area downward and keep the effective area unchanged, e.g., on the basis of hemisphere surface, expand one section of cylinder downward so as to keep gain unchanged at low elevation. But for the triangular pyramid composed of three planar arrays, when scanning angle is 0°–45°–90°, performance-price ratio 0.15–0.22–0.15 and is below curved surface array. Generally speaking, curved surface array scheme is much better than planar array scheme.

The actual work area (area producing antenna gain) of curved surface array is the area occupied by that array element activated by electromagnetic wave in beam direction, which is called activated area or irradiated area. As shown in Fig. 2.81, when activated surface moves from S_A to S_B (for processing of amplitude/phase signal at rear end of array element, that means a way of window sliding), beam pointing moves from θ_A to θ_B.

The activated area in Fig. 2.81 is equal to one paraboloid area. When it moves from S_A to S_B, it is looked like a parabolic movement, its sum beam, difference beam, and angle error characteristics are sliding correspondingly. So the angle tracking of sphere-phased array is "activated area sliding," and the activated area is similar to one of the windows of the sphere-phased array. Therefore, it can also be called "space window sliding." As the value of activated area wouldn't change when sliding, there won't be any scanning angle gain loss. So the scan, acquisition, and tracking antenna of the spheric array and parabolic antenna are similar. During spheric array design, we can refer to the parabolic antenna in its scanning, capturing, and tracking.

(1) Monopulse simultaneous lobe method

Different from planar array, in the curved surface array, even if the DOA direction is the same as that of normal line of irradiated side, the signal delay received by array element in irradiated area is different, which causes different output phase value of array elements, which shall be made a phase compensation to ensure in-phase summation of output signal of array elements, as shown in Fig. 2.82.

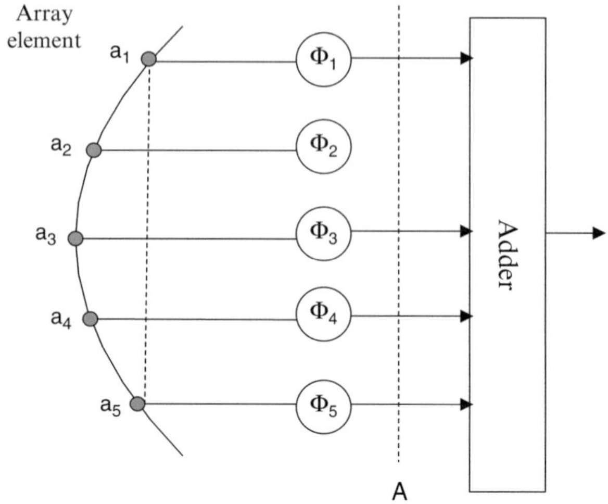

Fig. 2.82 Phase compensation of spheric array

Spheric array example is as shown in Fig. 2.82, after spatial delay difference of array element a_1, a_2, a_3, a_4, a_5 is compensated by phase shifter Φ_1, Φ_2, Φ_3, Φ_4, Φ_5, it forms the in-phase signal at cable A and becomes the same as planar array, the subsequent sum and difference beam forms, which is the same as monopulse tracking of planar array in Sect. 2.5.5.2.

For the spheric array, spheric equation is known and according to geometry relationship of bow and string in circle, array element delay difference can be obtained and mathematical function relationship of array element phase shift and beam direction in spheric phased array is achieved.

For better illustrating of physical concept of phased array angle tracking of spheric surface, the following table describes the function relationship explaining the physical process of angle tracking. We use Figs. 2.81 and 2.82 to obtain the relationship of beam electric axis pointing and array element phase shift as shown in the Table 2.13.

The S_A and S_B in Table 2.13 are as shown in Fig. 2.81. When θ_A points, amplitude weighted value of all array elements except S_A is 0 and phase shift value of array elements $a_1...a_5$ in S_A is $\varphi_1...\varphi_5$. When the beam pointing is required to change to θ_B, set the amplitude weighted value of all array elements except S_B to be 0, phase shift value of array elements $b_1...b_5$ in S_B is $\varphi_1...\varphi_5$ (as spheric surface is symmetrical, its phase value is the same as that of S_A). Other pointing can be obtained in a similar way to realize phased beam scanning. For the purpose of monopulse angle tracking, it can form four monopulse subarrays in irradiated area S_A to form sum-difference phase monopulse angle. Then the angle tracking can be implemented with the phased array angle tracking system in Fig. 2.83, i.e., when signal is on the normal line X_{n-1} (on the direction of electric axis) of irradiated area, difference signal is zero; when it deviates from the direction of electric axis, it outputs one difference signal ε_{n-1}, from which we can calculate a new electric axis pointing angle X_n according to expression in Fig. 2.79, which corresponds to θ_n in

Table 2.13 Relationship between spheric beam electric axis pointing and array element phase shift

Electric axis direction	θ_A					θ_B					...	θ_n					...
Illumination area S	S_A					S_B					...	S_n					...
Array element included in illuminated area	a_1	a_2	a_3	a_4	a_5	b_1	b_2	b_3	b_4	b_5	...	n_1	n_2	n_3	n_4	n_5	...
Compensation phase of array elements to planar array	φ_1	φ_2	φ_3	φ_4	φ_5	φ_1	φ_2	φ_3	φ_4	φ_5	...	φ_1	φ_2	φ_3	φ_4	φ_5	...

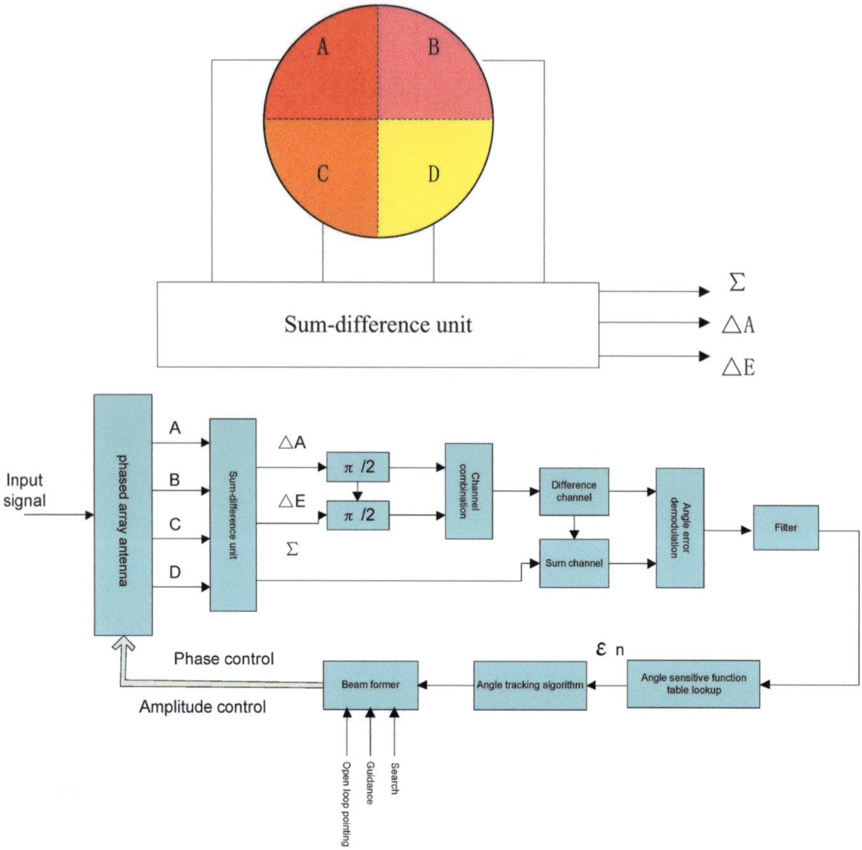

Fig. 2.83 Block diagram of phased array monopulse tracking system

Table 2.13. With the phase shift value and amplitude weighted value of array elements referred to the table, the first recursive tracking calculation can be performed and sum and difference beam can also be formed. Then, the recursive computation continues to make the sum and difference beam slide constantly until the tracking is stable.

In Fig. 2.83, A, B, C, and D are four monopulse subarrays in planar array and composite signal of subarrays form phase sum-difference monopulse, which is different from electromechanical scanning antenna.

The method given above, based on a table to find out the corresponding array element phase shift with the beam pointing θ_n, is clear in physical concept and is feasible, except for a requirement of mass operation. In order to reduce the workload, $\phi_n = f(\theta_A)$ function expression of different curved surfaces can be used. If it is applicable, phase control value ϕ_n can be obtained through real-time calculation of computer, which is the same as planar array. As for the signal amplitude value received by array element, as array element is on the curved surface, the gain difference of array element antenna pointing makes signal amplitude difference between array elements bigger in comparison with planer array, which can be obtained from pattern. It also needs additional amplitude weighted value to obtain required sidelobe diagram or maximum *S/N*. Projected area slides with the sliding of electrical boresight of antenna and is realized by movement of amplitude weighted value, which continuously changes in 0–1. Through a method of maximum signal to noise ratio (SNR) synthesis weighting, the best synthesis gain will be available, which can refer to maximum ratio combination in Expression (2.354). When it adopts digital beam forming scheme, such amplitude-phase weight of continuous sliding is easy to be realized, and the phase jerk and amplitude jerk are small, so the velocity measurement, distance measurement, and the angle measurement accuracy are high.

(2) Phase scan sequential lobe method

The advantage of the above monopulse simultaneous lobe method is: for monopulse, using simultaneous lobe method can obtain the real-time, high-precision angle information, so as to obtain a high angle tracking accuracy. Its disadvantage is that its realization is complicated and it has high requirements on the amplitude consistency and phase consistency of four subarrays. The amplitude inconsistency or phase inconsistency before its sum-difference will cause angle tracking error. For an asymmetric curved surface array, such influence is more severe and it requires amplitude and phase compensation which will increase the circuit complexity. In order to simplify the circuit scheme, we can only use the amplitude of the antenna lobe and the sequential lobe scan to obtain the angle error signal if certain sacrifice of thermal noise angle tracking error is allowed. This scheme is similar to the conical scanning scheme of the mechanical scanning antenna. However, it is the electronic phased array scanning that is used in phased array antenna, and in lobe scanning scheme, binary scanning is implemented only in 2D of elevation and azimuth. Its realization is much simpler, and its angle tracking accuracy is higher than that of conical scanning scheme. Its working principle is shown in Fig. 2.84.

What is shown in Fig. 2.84 is the elevation one-dimensional electronic scanning. OA axis in the lobe diagram is the pointing axis, and it is also the measuring axis of angle measurement as well as the tracking axis of angle tracking. When the antenna is not scanning, the electronic axis of the beam will point at A, while in scanning, the beam will deviate upward from the electronic axis to an angle $(+\Delta\theta)$ (indicated by the beam location in full line in the diagram) within the time interval of ΔT_1, and the beam will deviate downward from the electronic axis to an angle $(-\Delta\theta)$ (indicated by the beam location in dotted line in the diagram) within the time interval of ΔT_2. Periodic binary $(+\Delta\theta, -\Delta\theta)$ scanning will be conducted according to this time interval. If the target is located at direction A of the pointing axis, it can be seen from the diagram that what is received by the antenna will be continuous wave U_0 within ΔT_1 and ΔT_2. If the target is located at direction *B* above the pointing axis, the signal received by the antenna will be U_{1B} within the time interval of ΔT_1 and U_{2B} within ΔT_2. From the diagram, we can know $U_{1B} > +U_{2B}$, so an amplitude-modulated wave will be formed, with a modulation degree of

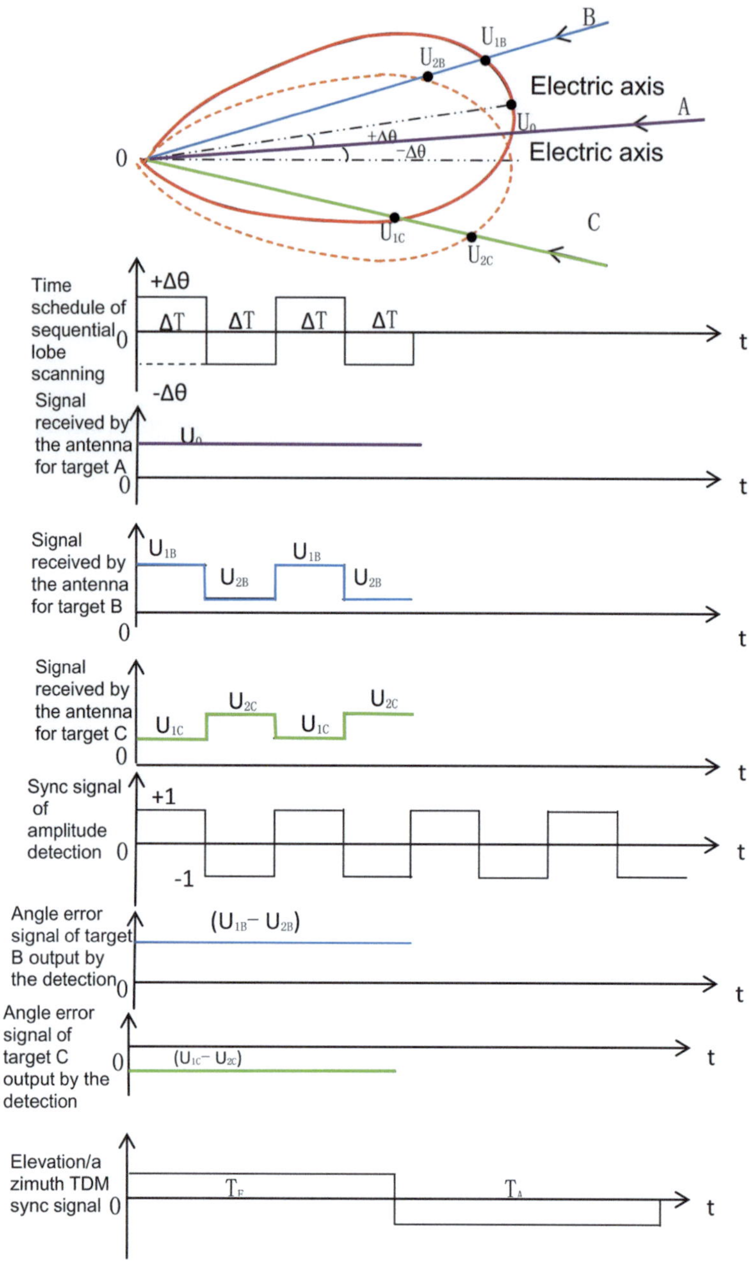

Fig. 2.84 "Phase scan sequential lobe method" principle diagram

$M_B = (U_{1B} - U_{2B}/2U_0)$ which is in direct proportion to the angle of deviation of the target deviating from the pointing axis. Therefore, the error signal can be obtained by detecting the amplitude-modulated wave. Similarly, when the target is located at the direction C below the pointing axis, an amplitude-modulated wave with a modulation degree of $M_c = (U_{1C} - U_{2c}/2U_0)$ will be received by the antenna. The value of M_C is also is in proportion to the angle of deviation of the target deviating from the pointing axis, but $U_{1C} < U_{2C}$, so the phase of M_C modulated signal is opposite to that of M_A modulated signal. This means target A and target B are respectively located at the opposite direction of the pointing axis. Thus, the phase indicates the direction of the target deviating from the pointing axis. Therefore, we can obtain the angle error signal with positive and negative polarity by using the sync signal indicated in the diagram as the phase reference to detect the modulated signal synchronously.

The above-mentioned is the elevation one-dimensional situation. When implementing elevation and azimuth 2D angle tracking, the time division method can be used, i.e., implementing electronic scanning at elevation within the time interval of T_E, and taking out the elevation angle error signal from the tracking receiver synchronously, and keeping a sync time interval. While implementing electronic scanning at azimuth within the time interval of T_A, we can use the same method to take out the azimuth angle error signal. Refer to Sect. 2.5.1.3 for the principle of this time division multiplexing.

This scheme is similar to conical scanning scheme. Refer to relevant materials of conical scanning radar and the derivation method of Expression (2.262) for its angle tracking accuracy calculation.

(3) Phase weighting of curved array

In the above list method, it uses angle of spread (θ_n) to check corresponding array element phase shift (ϕ_n), which has a clear physical concept and makes it easy to understand. So next, let's further derive its mathematical expression based on the physical conception.

First, we will derive a mathematical expression of a curved array. Here, we put a curved array in any shape into a coordinate system OYZ that is plane and two-dimensional. In theory, the origin of such coordinate can be chosen by random. However, in consideration of the ease of engineering fulfillment, the choice is better to concentrate in the region which makes the curve seem to be symmetrical. Figure 2.85 shows a sketch diagram of any curved array placed in a coordinate system OYZ.

Then, we will derive, with geometric optics, the phase shift values that should be additionally added to each element in the random curved array. In Fig. 2.85, the equiphase surface (O_2OO_1) is formed by radio wave when propagates freely (i.e., with no sheltering from linear array antenna) in the space. Here, we take such equiphase surface as reference. Since such surface is perpendicular to the direction of radio wave propagation (in coordinate system OYZ, the angle of direction of radio wave is represented by θ), when elements O, O_1, and O_2 are deployed in such surface, in-phase addition can be conducted for each of them receives the same signal phase information. However, if these elements are moved to positions (a_0, a_1,

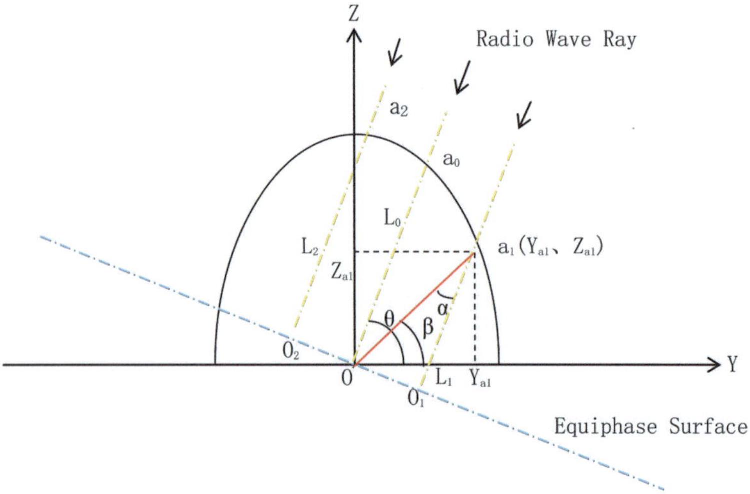

Fig. 2.85 Sketch diagram of geometric position of curved array in a two-dimensional coordinate

and a_2) on the curved array shown in figure above, such in-phase addition cannot be conducted, for there exist signal phase differences in space ($\Delta\phi_0$, $\Delta\phi_1$, $\Delta\phi_2$, corresponding to position a_0, a_1, and a_2), because geometric path length $(a_0o) \neq (a_1o_1) \neq (a_2o_2)$. Therefore, in-phase addition for signal at each element can be conducted only when phase shift value for each element is added. Next, let's take $\Delta\phi_1$ and $\Delta\phi_0$ and as examples to derive the phase shift value.

From the geometric relationship in Fig. 2.85, we can get:

$$\begin{aligned}(a_1o_1) &= (a_1o)\cos\alpha = (a_1o)\cos(\theta-\beta)\\ &= (a_1o)(\cos\theta\cos\beta + \sin\theta\sin\beta) = Y_{a1}\cos\theta + Z_{a1}\sin\theta\end{aligned} \tag{2.336}$$

Phase difference in space ($\Delta\phi_{a1}$) from distance is (a_1o):

$$\Delta\phi_{a1} = (a_1o)\left(\frac{2\pi}{\lambda}\right) = \frac{2\pi}{\lambda}(Y_{a1}\cos\theta + Z_{a1}\sin\theta) \tag{2.337}$$

So phase shift value $(-\Delta\phi_{a1})$ must be added in the array to compensate such difference.

Similarly, we can derive:

$$\Delta\phi_{an} = \frac{2\pi}{\lambda}(Y_{an}\cos\theta + Z_{an}\cos\theta) \tag{2.338}$$

where Y_{an} and Z_{an} are the coordinate values of element a_n, and θ is the angle of spread of the beam. Therefore, the phase shift value needed by each element can be calculated with Expression (2.338).

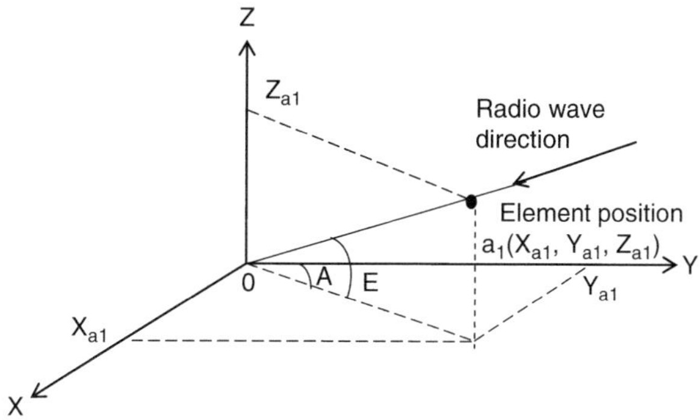

Fig. 2.86 Sketch diagram of curved surface array in a three-dimensional coordinate

For a random curved-surface array, we can establish a three-dimensional coordinate system, and use the same theory to derive [22]:

$$\Delta\phi_{an} = \frac{2\pi}{\lambda}(X_{an}\cos E\cos A + Y_{an}\cos E\sin A + z_{an}\sin E) \tag{2.339}$$

where, $\Delta\phi_{an}$ is the phase shift value for element a_n, X_{an}, Y_{an}, and Z_{an} are the three-dimensional coordinate values of such element. If a random curved surface array contains n elements, there will be n sets of coordinate values in total for these elements. Such values can be determined at design and manufacturing phase, and listed and stored in a memory, from which they can be read during phased array operation and calculated with Eq. (2.339) to get the phase weight value for each element. The three-dimensional relationship of curved surface array is shown in Fig. 2.86.

In Fig. 2.86, (X_{a1}, Y_{a1}, Z_{a1}) are the coordinate values of element, A and E are, respectively, azimuth angle and elevation angle of beam pointing.

2.5.5.4 Time-Sharing Multi-target Tracking, Simultaneous Multi-target Tracking, and Adaptive Tracking

Time-sharing multi-target tracking of phased array uses TDMA principle and points antenna beam to different targets at different time slots, each target is distributed with fixed address code and time slots. In such time slot, phased array antenna beam dwells to accomplish angle measurement, distance measurement, remote measurement, remote control and multi-target TT&C. In the period other than beam dwelling target, it uses tracking filtering to predict the track of target in such period as the orbit prediction of next tracking. Beam directly jumps to prediction position for next tracking. When target velocity is V and interval of two beam irradiation is T, for the second tracking starting point, target coordinate will be limited in one spheric space with radius of TV. If such spheric space does not exceed beam width, the second tracking is very simple; if it exceeds, location correlation and track correlation are needed. It can also use such time slot distribution to realize TWS, i.e., after discovery of first target, it distributes some time slots to track such target and after tracking is stable it distributes some time slots to search the whole tracking airspace. During the search, it uses tracking filter to extrapolate the first target. After searching the second target, it immediately tracks the second target; after tracking is stable, it starts searching on the next target. It does not stop searching and enter full tracking state until searching and tracking of all targets. As phased array TT&C system coordinates the target and target signal is known, orbit is known for orbit aircraft, technical difficulty is smaller. Its main technical problems are tracking filter and orbit prediction. Tracking filter algorithm is key point of multi-target tracking and ordinary contents include: two-point extrapolated filter, Wiener filtering, Kalman filtering, generalized filtering, etc. Simple tracking filter algorithm can be used for low velocity target. When TWS state changes into comprehensive tracking state, work becomes simpler.

Its main technical problems are: ① low sampling rate: low tracking sampling rate is feature of time-shared multi-beam phased array system. It adopts the minimum sampling rate adaptive to tracking filter design and specific target dynamic characteristics to save time. ② "Interpolative-zero value" tracking design: "interpolative-zero value" tracking is the feature of phase scanning radar system. It uses the movement mode of discrete step and low sampling rate to irradiate tracked target. Differential beam zero point hardly points to target on the preset beam position. It needs to insert one corrected increment for zero point to provide accurate angle data. Even if limited interpolation is needed, error slope (by angle error) shall be calibrated, then such corrected increment shall be plus (or minus) discrete beam pointing center. Monopulse is the most accurate technique of beam separation angle measurement and monopulse angle measurement is adopted for interpolation correction. ③ Initial tracking problem: initial tracking after capturing target requires the design of loop filter that makes tracking loop best match. During the initial tracking of target, it moves only with specified coordinate but without target, the specified coordinate is one step input of filter. It requires wider bandwidth and quicker sampling rate than stable work to finish transient process as soon as possible and ensure continuous tracking. ④ Design of loop filter: basis of tracking loop filter: dynamic characteristics of target, accuracy, power, utilization ratio of time (sampling rate required by continuous tracking), and calculation complexity under initial tracking and stable tracking. In various engineering designs, Type II, Type III, and Kalman filter are the commonest design of phased array system.

But for digital transmission of some high speed code (e.g., reconnaissance image), time division method may cause some limitation. Transponder storage and transmission can be used for time sharing transmission, which transmission bit rate is increased by N time (N is the number of multi-targets). If bit rate increase is limited, image compression can be used to decrease bit rate. If it still cannot meet technical requirements, simultaneous multiple beam can be used.

Simultaneous multiple beam tracking: each beam tracks one target and use multiple simultaneous beams for multi-target tracking; there are many tracking loops and each tracking loop design is the same as single target, which has been introduced in Sects. 5.5.5.2 and 5.5.5.3. Its difference from time-sharing multiple beam tracking: there is continuous sampling for its angle error signal and sampling rate is high (1–100 kHz), time-sharing multiple beam tracking provides time-sharing tracking on multiple targets, but angle error sampling and processing of each target is time-sharing and discrete. Its sampling interval is the time interval on different sampling and sampling rate is low (0.2–10Hz). As time interval is long, “extrapolation,” “interposition,” and other tracking filter of moving targets are required to predict target position. For phased array C&T system, as remote metering, remote control, remote sensing, etc., is of continuous signal, time-sharing multiple beam is not used, but simultaneous multiple beam is used. In simultaneous multiple beam, there may be multiple targets in one beam, CDMA and FDMA can be adopted to differentiate different target signals.

Adaptive angle tracking is mostly used in digital transmission. Minimum mean square, CMA, recursion least square, and other adaptive algorithms are adopted for adaptive angle tracking or spatial filtering [23], according to principle of highest signal-to-noise ratio (or best signal/jamming ratio, etc.).

Adaptive antenna technology includes adaptive multi-beam forming, adaptive beam zero alignment, DOA, spatial spectrum analysis, early automatic alignment and integration of phase, etc. Automatic alignment of phase and adaptive beam forming can automatically accomplish beam forming and target capturing, which is closed loop tracking mode; but phase alignment mode cannot adaptively form spatial filter, which is its disadvantage. Contrarily, adaptive beam forming can adaptively form spatial filter and strengthen anti-jamming capability of equipment, which can obtain the maximum ratio of signal to noise. But calculation is complex, equipment capacity is too big, and a reference signal same as signal to be received is needed; convergence procedure of recursive calculation is too long with possibility of divergence. There are many principles for adaptive beam forming, including highest signal-to-noise ratio principle, maximum signal-to-interference noise ratio principle, LMS, etc. Its angle tracking accuracy is not high and is not used for angle measurement.

Space spectrum means that one time varying signal is composed of incoming wave components (e.g., incoming wave of mainlobe and sidelobe) of different spatial directions. Space spectrum analysis can analyze the components in all spatial directions, which can be used for measurement. The phase of array elements in phased array represents direction and phased array is good for spatial spectrum analysis.

Such above modes can be integrated for application; e.g., DOA or spatial spectrum analysis is used for direction guidance, multi-beam for tracking, LMS for anti-jamming.

2.5.5.5 Angle Measurement Accuracy of Phased Array Tracking

Tracking angle measurement accuracy is an important index in phased array TT&C system. Compared with normal monopulse tracking radar, it has no essential difference, but under the condition of multi-target tracking, it has higher relative measurement accuracy for the common system errors and time differences are eliminated. As for absolute angle measurement accuracy, phased array decreases the error introduced in electromechanical link, and can quickly and flexibly change the servo loop parameters, so as to improve the dynamic performance. However, it also increases some error factors. In a planar array, the increased angle error factors are as follows.

(1) Scan angle factor

Beam width θ is inversely proportional to the cosine of scan angle Φ of the beam deviating from the array, i.e.,

$$\theta = \frac{\theta_0}{\cos\Phi} \tag{2.340}$$

where θ_0 is the width of the beam at the direction of array normal. When the beam deviates from the direction of array normal, beam width will increase and the measurement error item related to beam width (e.g., angle thermal noise) will directly increase. With the increase of beam width, its corresponding antenna gain will decrease, which will decrease the receiving echo SNR and increase the thermal noise error.

(2) Beam pointing random error

Due to such factors as array deformation, position error of array element, phase shift error of phase shifter, the fixed insertion phase shift error of phase shifter, the amplitude and phase of antenna aperture will have error, which will cause beam deformation and changes of phase center. Such errors are caused by fabrication error, changes of environmental conditions, etc., and will generate beam pointing error.

(3) Quantization error of phase shifter

Digital phase shifter is generally used in a phase shifter unit. It will make the phase distribution of the antenna aperture in a discrete form and generate aperture error which will cause quantization error to the beam pointing.

(4) Changes of phase center

Phase center is the reference point for ranging, angle measurement and velocity measurement, and its changes will induce orbit-measuring error. In a phased array antenna, the phase center will change with the changes of the arrangement of array elements, array deformation, beam direction, array element mutual coupling, fabrication tolerance, etc. Besides reducing the changes of such factors to the maximum extent in design and fabrication to improve the stability of phase center, measurement of phase center is also required. Angle and range calibration can also be applied to eliminate this error.

(5) Consistency and calibration of time delay and amplitude of each channel

When an active phased array scheme is applied, the consistency of time delay and amplitude of the *T*/*R* assembly of each array element will have great effect on the synthesis gain, beam pointing, beam shape of the phased array. The stability of time delay can induce ranging errors. Therefore, measures must be taken to reduce their inconsistencies. For a narrowband phased array system, changes of time delay can be converted into changes of phase, so the amplitude and phase calibration technology is usually applied to ensure the amplitude and phase consistency of each channel. The angle measurement of a phased array angle tracking system is realized by the phase measurement of a phase shifter; therefore, phase calibration is very important in high-precision angle measurement. Generally, phase calibration inside the array is applied for the calibration of phase shift consistency of each array element channel. A better method is phase calibration outside the array. Its calibration includes the array and array element channel, so it is more accurate. Because the amplitude of each array element channel will influence the synthesis result, amplitude calibration is also required, to fully realize vector calibration. In adaptive tracking, *S*/*N* is required to be up to its maximum value, while this requirement can be complied with for amplitude and phase calibration can be automatically done under the adaptive function. Thus, phase calibration (or simplified phase calibration) is not required in theory.

For other angle measurement error, formulas related to electromechanical angle tracking system may be used for calculation (Table 2.14).

2.5.6 *Independent Guidance, Self-Guidance, and Multi-beam Guidance*

The first step of target acquisition by C&T system is to realize the acquisition of flight vehicles by ground station at angles. With the improvement of operating frequency band and the increase of antenna aperture, antenna beam becomes narrower and narrower, which makes the target acquisition by narrow beam a key technical issue. Angle guidance technology is required to be used. Some typical schemes will be introduced in the followings.

2.5.6.1 Independent Guidance

Independent guidance is a scheme which uses special guidance equipment (also called synchronous guidance) or other prior target position data (also called number guidance) to guide the antenna of the main C&T equipment (equipment to be guided) to point at the target spatial coverage, and its block diagram is as shown in Fig. 2.87.

In Fig. 2.87, the wide beam "guiding equipment" will first track the target, and the target position data acquired by it will be sent to the servo system of the "equipment to be guided." It will be continuously used to guide "the equipment to be guided" to follow. The following duration is called continuous guidance time *T*. The guidance probability of this scheme is described by continuous guidance

Table 2.14 Angle error sources of phased array tracking

Error source	Error formula and affecting factors
1. Thermal noise	$\sigma_1 = \frac{\theta_{0.5}}{K_{\mathrm{m}}\sqrt{25/N}}$
	where σ_1 is angle thermal noise error (mrad); S/N is SNR; K_m is normalized monopulse difference slope, $K_m = 1.89$ (typical value), $K_m = 1.2\text{–}2.0$ (general scope)
2. Multipath reflection	$\sigma_2 = \frac{\theta_{0.5}}{\sqrt{8A_3}}$
	where σ_2 is angle multipath error (mrad); ρ is surface reflectivity; A_3 is the ratio of main lobe to side lobe at the direction of reflected ray
3. Dynamic lag	In case of type II tracking loop with critical damping, employ minimum mean square error criterion and exponential weighting:
	$\alpha_3 = \frac{\ddot{\omega} T_S^2}{a_1}(1 - a_0)$
	where σ_3 is angle dynamic lag error (rad); $\ddot{\omega}$ is the maximum angular acceleration (rad/s^2); T_s is sampling period (s); α_0 is position smoothing constant; α_1 is velocity smoothing constant
4. Errors caused by changes of amplitude and phase of monopulse feed source and receiver	In case of four-horn monopulse feed source:
	$\sigma_4 = \left[1 - G_1 \frac{G_2 \cos(a_1 + a_2 + a_3) + \sin(a_1 + a_3)}{G_2 \cos(a_2 + a_3) - \sin a_3}\right]$
	where σ_4 is the error (mrad sine) caused by amplitude and phase inconsistency, G_1 is the amplitude inconsistency before comparator (gain ratio), α_1 is the phase inconsistency before comparator (°), G_2 is the amplitude inconsistency after comparator (gain ratio), α_2 is the phase inconsistency after comparator (°), α_3 is the phase inconsistency caused by other factors (°)
5. Error of electrical boresight direction	$\sigma_5 = \frac{\varepsilon \theta_{0.5}}{2\sqrt{N}}$
	where σ_5 is pointing error (milliradian), ε is the value of root extraction of the phase error quadratic sum of each antenna unit (rad, which is caused by unit spacing, axis pointing shift, tolerance of phase shifter, cable, and feed), $\theta_{0.5}$ is 3 dB beam width (mrad), and N is the total number of array elements
	Caused by fabrication tolerance (mechanical and electrical), temperature, wind, sun, attraction, base deviation, and array deviation
6. Quantization of phase shifter	$\sigma_6 = \frac{2.6\theta_0}{N_L 2^P}$
	where σ_6 is beam pointing error caused by phase quantization (m rad sine), P is the bit of phase shifter, N_L is the unit number of linear array

(continued)

Table 2.14 (continued)

Error source	Error formula and affecting factors
7. Digital angle quantization	$\sigma_7 = \frac{L_\Delta}{\sqrt{12}}$
	where σ_7 is angle quantization error (rad) and L_Δ is the value of minimum effective bit (rad)
8. Scanning factor	Error caused by the changes of monopulse slope leaving the normal with scanning angle and the loss of scanning angle (based on the minimum mean square fitting)

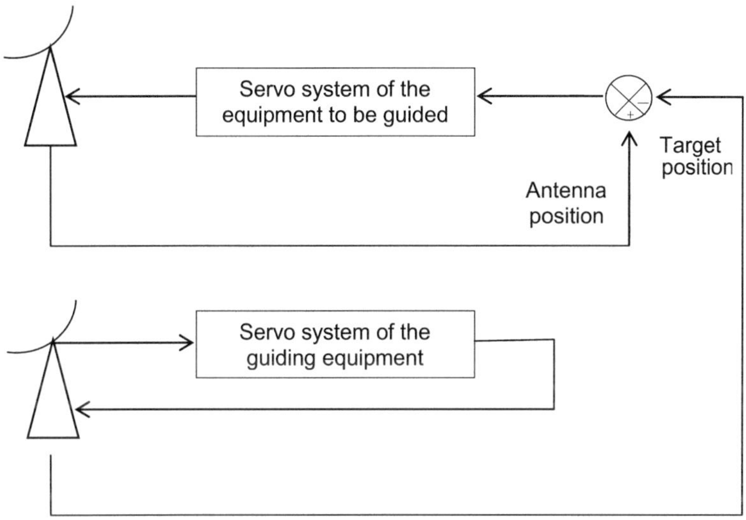

Fig. 2.87 Principle block diagram of independent guidance

probability [16], which can be analyzed according to guiding antenna tracking error along with the following assumptions:

(1) The error of guidance system is random and stable.
(2) Guidance probability is close to 1.
(3) The average number of times that the target exceeds the air space within unit time is small.

When such assumptions are satisfied, the guidance probability that allows exceeding air space K times within guidance time T is approximately subject to Poisson distribution:

$$P_{K,T} = \frac{(\overline{N}T)^K}{K!} e^{-\overline{N}T} \tag{2.341}$$

Expression (2.341) indicates: the probability of an event occurring K times in the time interval of $[0, T)$ is $\frac{(\overline{N}T)^K}{K!}\mathrm{e}^{-\overline{N}T}$, where $\overline{N}$ is the mathematical expectation of event occurrence times within unit time.

Therefore, the guidance probability when the times that target exceeds the air space is less than j in continuous guidance time T is:

$$P_{j/T} = \sum_{K=0}^{j} P_{K,T} = \sum_{K=0}^{j} \frac{(\overline{N}T)^K}{K!}\mathrm{e}^{-\overline{N}T} \tag{2.342}$$

It can be concluded from the above expression that in guidance time T, the guidance probability when not even one time that the target exceeds the specified air space is:

$$P_{0/T} = \mathrm{e}^{-\overline{N}T}$$

$$\overline{N}_{(R>R_0)} = \frac{1}{\sqrt{2\pi}}\frac{\dot{\sigma}}{\sigma}\frac{R_0}{\sigma}I_0\left(\frac{R_0\sqrt{\Delta_A^2+\Delta_E^2}}{\sigma^2}\right)e^{-\frac{1}{2\sigma^2}\left(R_0^2+\Delta_A^2+\Delta_E^2\right)} \tag{2.343}$$

where:

R_0 is the radius of specified spatial coverage which actually depends on the antenna beam width of the equipment to be guided;
Δ_A and Δ_E are, respectively, the azimuth systematic error and elevation systematic error of the angle of guiding equipment;
σ is random angle error of the guiding antenna;
$\dot{\sigma}$ is the root-mean-square value $\dot{\sigma} = \frac{2\pi\beta}{\sqrt{3}}\sigma$ of random error time change rate of Δ_A or Δ_E.
β is random error equivalent bandwidth (related to the bandwidth of the servo system of guiding equipment);
I_0 is the modified Bessel function of the first kind.

Independent guidance includes digital guidance, data guidance, program tracking, and so on. The Δ_A, Δ_E, σ, $\dot{\sigma}$ in Expression (2.343) are the errors between the angle guidance data and real angle of the target.

2.5.6.2 Self-Guidance

Self-guidance refers to the use of the wide beam antenna on the same antenna pedestal of the main C&T equipment to complete its narrow beam guidance. When the signal is guided into the narrow beam, the guidance mission is fulfilled. Then the next mission is handed over to the servo tracking system of the narrow beam in

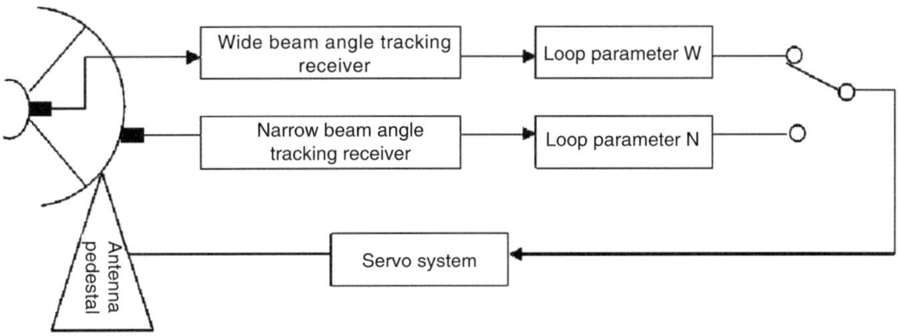

Fig. 2.88 Principle block diagram of self-guidance

"waiting for acquisition" state, by which the narrow beam acquisition and tracking are performed. This "waiting for acquisition" capability is provided by most narrow beam in modern TT&C equipment. The guidance probability of the self-guidance scheme is calculated according to the "hit probability" when the signal hits into the narrow beam. Such wide beam antenna includes: small aperture antenna of the same frequency, low frequency band wide beam of coplanar dual-band antenna, etc. Under ideal conditions, the electrical boresights of wide beam antenna (guiding antenna) and narrow beam antenna (antenna to be guided) are coincident. When wide beam antenna tracks and aims at a target, the angle tracking system will convert into narrow beam from wide beam to realize the self-guidance. However, there are random errors and systematic errors when wide beam antenna tracks a target, and these errors may lead to the result that the signal fails to drop into the narrow beam, which specifies the existence of the probability of guidance success. Take sub-reflector antenna as an example for the analysis of probability of guidance success.

In Fig. 2.88, a small aperture wide beam antenna of the same frequency is installed on the sub-reflector. Good electrical boresight consistency between the wide beam and narrow beam is required in this scheme. When the system starts to work, the target is made to enter the range of wide beam by using waiting for acquisition, scanning for acquisition, digital guidance and other means. The starting position of the switch in Fig. 2.88 is *W*, so the target will first enter the self-tracking range of wide beam antenna. At the same time, the narrow beam antenna will move along with the movement of the pedestal and stays in "waiting for acquisition" state. When the wide beam antenna tracks and aims at the target, the target is located at the range of narrow beam. The narrow beam tracking receiver will give an indication signal of "signal has been received and normalized" represented by the voltage of the angle error and AGC voltage. According to this signal, the switch in the Figure will be switched to *N*. The input angle error signal of servo system will be converted to the output of narrow beam tracking receiver from the output of wide beam tracking receiver. At the same time, the servo loop parameter will be changed. In this way, the angle self-tracking of narrow beam antenna is completed.

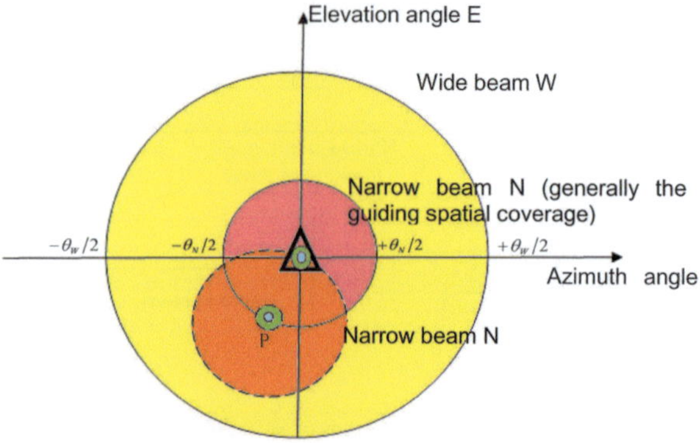

Fig. 2.89 Diagram of the cross section of the wide and narrow beam when there is no systematic error

Table 2.15 Self-guidance probability

θ_N/σ	P
0.5	0.1
1	0.2
2	0.8
3	0.88
4	0.95

(1) Self-guidance probability if there is no relative systematic error or only random error existed between electrical boresights of the wide and narrow beam

At this moment, the electrical boresights of the wide and narrow beam is completely coincident, as shown in Fig. 2.89.

In Fig. 2.89, △ is target position, width of wide beam is θ_w, position of electrical boresight is represented by a big circle "○," width of narrow beam is θ_N, position of electrical boresight is the center of the narrow beam, represented by a small circle "○." The required guiding spatial coverage is θ_N. When there is no relative systematic error, the big circle "○" and the small circle "○" are coincident and at the center of the circle. When there is random error of wide beam tracking, the relative position between the target and the electrical boresight will have a relative shaking. If this shaking is still within a circle with a radius of $\theta_N/2$, the lobe width of the narrow beam (marked by the dotted line in the figure) can still cover the target, which means the guidance succeeds, otherwise the guidance fails. The shaking is generally subject to normal distribution, the probability of the error less than $(-\theta_N/2 \sim +\theta_N/2)$ can be calculated based on statistical theory and it is:

$$P = 1 - \exp\left[-\frac{x^2}{2}\right] \tag{2.344}$$

Where P is called "hit probability," or the probability of self-guidance, $x = \frac{\theta_N}{\sigma}$, σ is root mean square value of the angle tracking random error under the wide beam guidance. Partial calculated results of the above expression are shown in the following Table 2.15:

(2) The guidance probability with systematic error in guidance subsystem

The systematic errors here mainly include:

1) Static relative error θ_1 between the electrical boresight pointing of the narrow beam and the electrical boresight of the wide beam.

2) Sum θ_2 of dynamic lag error and other relevant systematic errors during wide beam tracking. The former is the main part of θ_2, and the latter includes sum and difference phase inconsistency, amplitude unbalance before comparator, coupling error among channels, zero drift of phase discriminator, dead zone of servo system, unbalance and zero drift of servo amplifier, inconsistency of electrical boresight pointing change of wide and narrow beam caused by gust, etc. However, the conversion error and radio wave propagation error in monopulse angle measurement error will not affect the guidance probability of same-pedestal wide beam, and the systematic error which will maintain the same in a long time can be decreased by calibration.

3) In the conversion period T_s of wide beam guidance to narrow beam tracking, before the conversion of conversion dynamic error θ_3 caused by target movement, the PLL locking and AGC of the main tracking receiver have been completed, but in the period of T_s, the self-tracking loop of the wide beam is off, and the narrow beam loop is not connected yet, so θ_3 will be imported into the movement velocity of the target $\dot{\theta}_0$.

$$\theta_3 = T_s\dot{\theta}_0 \tag{2.345}$$

When there is velocity memory in wide beam angle tracking, and if the memory velocity is $\dot{\theta}_A$, then:

$$\theta_3 = T_s\left(\dot{\theta}_0 - \dot{\theta}_A\right) \tag{2.346}$$

For the current angle tracking equipment, the conversion can be quickly completed in computer, so T_s is very small. Besides, there is velocity memory, so θ_3 is a minor component.

When there are θ_1, θ_2, θ_3, the relationship among all kinds of errors in guiding spatial coverage θ_N is as shown in Fig. 2.90.

It can be seen from Fig. 2.90 that the electrical boresight of narrow beam will be located at O_N point, and the beam will shake around O_N point. In order to ensure 100 % success of guidance, the minimum value for which the shaking cannot exceed O_N is $\left(\frac{\theta_N}{2} - \theta_1 - \theta_2 - \theta_3\right)$, and the probability P' for not exceeding it is:

$$P' = 1 - \exp\left(-\frac{x_N^2}{2}\right) \tag{2.347}$$

$$x_N = \left(\frac{\theta_N}{2} - \theta_1 - \theta_2 - \theta_3\right)$$

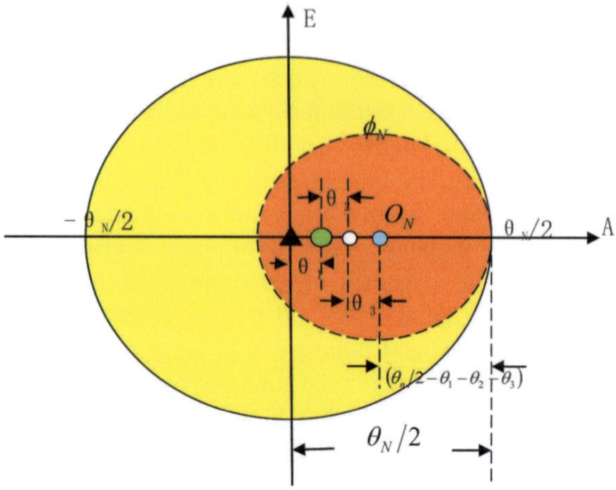

Fig. 2.90 Diagram of the cross section of the wide and narrow beam when there is systematic error

In the above expression, the physical meaning of p' is the probability that the shaking of O_N point in Fig. 2.90 will not exceed the dotted circle ϕ_N. However, the shaking between the ϕ_N circle and θ_N circle also cannot exceed the guidance range θ_N. Set the probability as P'', then the total hit probability is:

$$P_s = 1 - \left(1 - P'\right)\left(1 - P''\right) \tag{2.348}$$

Obviously, $Ps > P'$, $P_s < P$ (refer to (2.344) for the expression concerning P). In order to improve Ps, it is required to take measures to decrease θ_1, θ_2, θ_3.

The description above is the qualitative physical concept of P_s decrement. The quantitative computational formula of P_s is: [16].

$$P_s(R \leq R_0) = 1 - \Phi\left(\frac{R_0}{\sigma} \bullet \frac{\sqrt{\Delta_A^2 + \Delta_E^2}}{\sigma}\right) \tag{2.349}$$

where P_s is the hit probability for a single guidance, R is the angle of guiding antenna deviating from the target, R_0 is the guiding spatial coverage required, here it is narrow beam width θ_N, σ is random error of guiding subsystem, and Δ_A and Δ_E are the systematic error of guiding subsystem. $\Phi\left(\frac{R_0}{\sigma} \bullet \frac{\sqrt{\Delta_A^2+\Delta_E^2}}{\sigma}\right)$ is an integral formula, which can be worked out by the general curve shown in Fig. 2.91.

When N independent guidances are repeated, the guidance probability P_N is

$$P_N = 1 - (1 - P)^N \tag{2.350}$$

From the above, N guidances can increase the guidance probability, but it is required to meet the time interval T_G ($T_G = 2/B_L$, and B_L is the bandwidth of servo

Fig. 2.91 $\Phi\left(\frac{R_0}{\sigma} \bullet \frac{\sqrt{2}\Delta}{\sigma}\right)$ Curve (assumed $\Delta_A = \Delta_E = \Delta$)

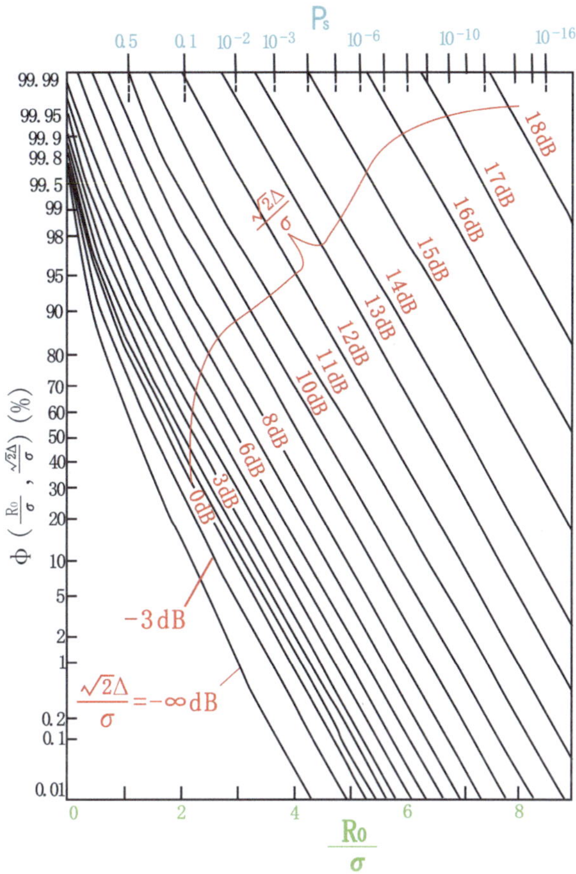

loop) required by N independent guidances. The time spent for N independent guidances is NT_G, which means that the guidance probability is increased at the cost of extending time. But the target cannot fly outside the space domain within NT_G time.

Under certain exceptional circumstances, if the narrow-beam tracking system requires the wide-beam guiding subsystem to provide a relatively long continuous guiding time T_Σ to ensure the narrow-beam tracking shifts to auto-tracking reliably, Expression (2.343) for calculating continuous guidance probability will then be used. The exceptional circumstances include circumstances where the ability of narrow-beam angle tracking system to wait for capture is poor and the system needs to be guided for a period of time T_Σ, or where the delay time of the effective angle error voltage output of the narrow-beam tracking system is relatively long or even longer than the acquisition time of angle servo system.

According to the analysis above, the system error will lower the guidance probability and must, therefore, be reduced as far as possible. The measures for decreasing are: ① in case of long range, using narrow bandwidth for servo to

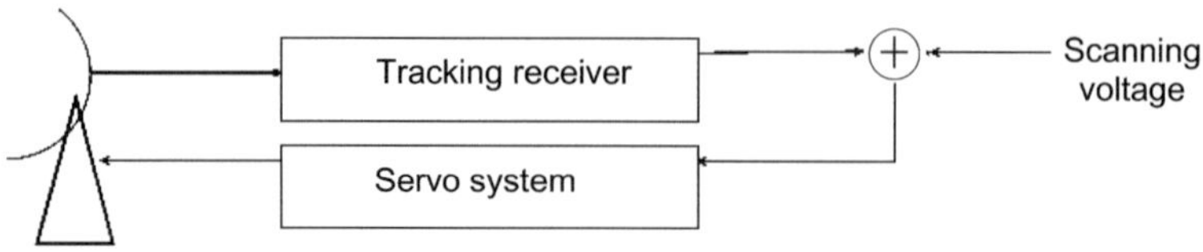

Fig. 2.92 Principle block diagram of "scan while tracking"

decrease random error for angle velocity is small and C/N is low; in case of short range, using wide bandwidth for servo bandwidth to decrease dynamic error for angle velocity is large and C/N is high; ② decreasing T_s and improving the velocity memory performance of the servo system to decrease the dynamic error in conversion period; ③ using fast servo system to decrease dynamic error; ④ improving the processing and installation accuracy of small antenna to ensure their electrical boresight consistency; and ⑤ adopting "calibrate while tracking," "scan while tracking," and other schemes to decrease systematic error. The principle block diagram of "scan while tracking" is as shown in Fig. 2.92.

In Fig. 2.92, the servo system of wide beam is in closed-loop tracking target state. It will overlap a small scanning voltage on the error voltage output by the tracking receiver while tracking the target, to make the antenna have a small scanning around the direction of tracking target. When the narrow beam scanning at the same time scans the target, it will convert into narrow beam tracking.

2.5.6.3 Multi-beam Guidance

(1) Overview of multi-beam guidance

Multi-beam guidance refers to using several adjacent narrow beams to realize the wide angle coverage and obtain angle error information, which is represented as the relative angle difference between each beam pointing and the main beam pointing.

The following are schemes for multi-beam guidance:

1) Planar phased array multi-beam: use phased array to form multi-beam for angle measurement and then guide into the main beam. It includes feed source phased array and face phased array. Its advantage is that it can realize a wide guiding spatial coverage, while its disadvantages are that the system is complicated and the cost is high.
2) Offset-focus array feed source multi-beam: use the principle of feed source offset-focus inducing beam deviation and use the offset-focus arranged array feed source to form multi-beam, i.e., each horn in the array feed source will form a beam. Its pattern is as shown in Fig. 2.93:

The following are options for using offset-focus array feed multi-beam antenna to realize multi-beam guidance:

① Open loop angle measurement re-guidance scheme: it uses the angle value of multi-beam open loop measurement based on parameter estimation theory to

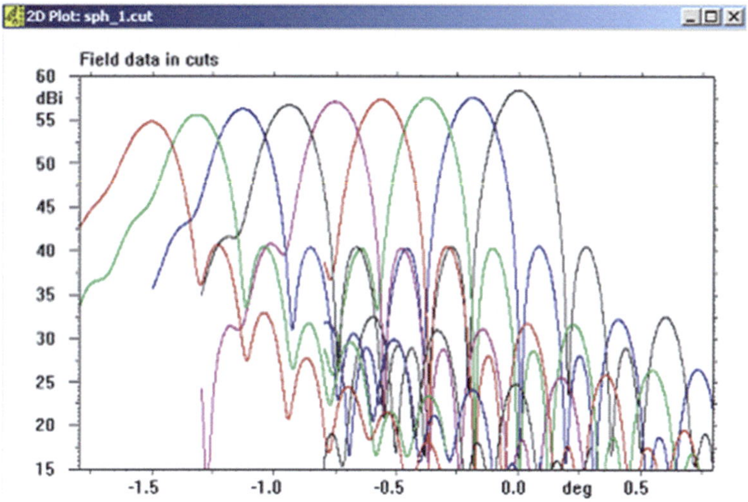

Fig. 2.93 An example of multi-beam pattern

predict the angle value of the target relative to main beam. This value will be regarded as the guidance data and sent to the follow-up servo system which is under guiding state. The system mechanically drives the antenna to make the main beam point at the target.

② Closed loop tracking convergence scheme: use "multi-beam closed loop tracking" to replace "multi-beam open loop measurement," and the acquisition process of dynamic prediction will be changed into dynamic tracking process. It uses: (1) multi-beam feed source and its antenna; (2) high-speed sampling switch and multichannel Ka-band integrated channel; (3) signal processing unit with wide angle error characteristics, to collectively form a wide angle error characteristics corresponding to multi-beam angle coverage range (equivalent to a wide beam). Its output information is sent to the closed loop of servo system and the wide angle coverage tracking is realized based on discrete automatic regulatory theory. When the tracking converges to the coverage of main beam width, it converts to main beam tracking. Its principle is as shown in Fig. 2.94.

If there is any error of the target not being able to be aimed at when the convergence ends, the above-mentioned "scan while tracking" can be used as a solution.

This method can overcome the disadvantages of open loop measurement which require high accuracy and speed. During tracking process, the requirement on the accuracy of angle error characteristics is not high. As long as there is angle error voltage, the servo system can be driven to converge to the coverage of main beam width. And the sampling rate of signal is determined by the bandwidth of servo system. Due to the narrow servo system bandwidth, the sampling rate is low. Thus, time division multiplexing and equipment simplification can be realized. The scheme is simple and reliable.

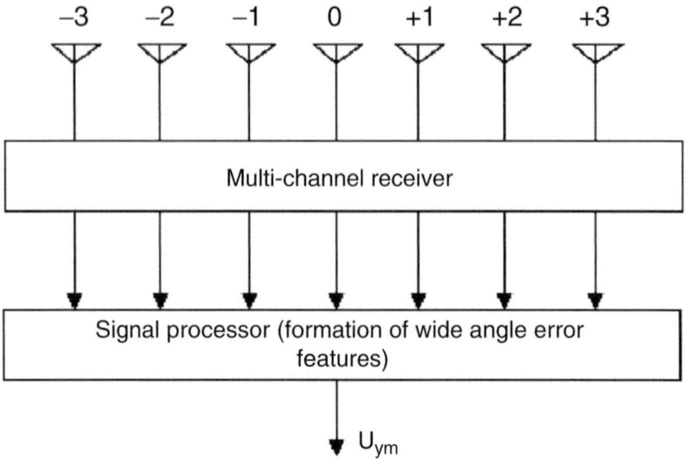

Fig. 2.94 Principle block diagram of multi-beam wide angle tracking

③ Multi-beam solution of phased array feeder: multi-beam can be formed by the parabolic feeder focus-deviated solution described above, and each beam corresponds to one angle information. Besides, angle information that is more precise can be further obtained by comparison of adjacent beams. However, such feeder focus-deviated will cause distortion of phase distribution, which will affect the directivity pattern of parabolic antenna, decrease the gain, widen the lobe and increase the sidelobe. Such negative influence will be decreased as the focus distance (f) gets longer, and relation between the deviation angle of beam ($\Delta\theta$) and focus (f), and relation between distance of horizontal feeder deviation (L) and focus (f) can be determined by the following equation:

$$\Delta\theta = \frac{L/f}{1 + (L/f)^2}\,\text{rad} \tag{2.351}$$

For energy collection by parabolic focusing, parallel radiations can be focused on the same focus only when their direction parallels with the focus axis; parallel radiations in other directions will be focused on difference focuses and will face defocusing. Therefore, there exists a focus area for acquisition of most energy, and such area can be planar or curved. If a phased array feeder is deployed in such focus area for phase and amplitude weight to match the field distribution of the focus plane, then the distortion of directivity pattern and decrease of gain can be effectively eliminated. Such phased array feeder has the following advantages:

(a) Having a higher gain and supporting a larger scan angle under requirement of certain scan consumption.
(b) Supporting continuous beam scan without beam loss at crossover caused by discrete beam in the solution of multi-beam of feeder focus-deviated.

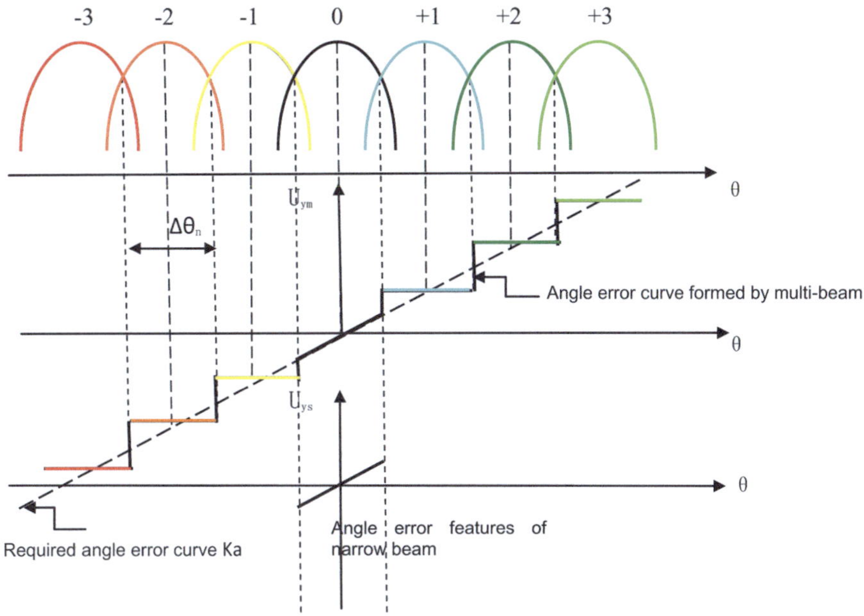

Fig. 2.95 Comparison between the wide angle error characteristics formed by multi-beam and narrow beam error characteristics

(c) Supporting larger element interval when given a smaller scan angle, thereby decreasing its feeder array element.

However, its disadvantage is a more complex multi-beam forming.

(2) The principle of using multi-beam to generate wide angle error characteristics

Its principle is as shown in Fig. 2.95.

Figure 2.95 is one-dimensional. In the figure, −3, −2, −1, 0, +1, +2, and +3 are seven multi-beams and 0 is the main beam, which is one element of the multi-beams. The space pointing of each beam includes angle information. The difference between the nth beam pointing θ_n and the electrical boresight of main beam θ_0 is angle error θ. When target signal is detected within the nth beam, the angle error θ of the target relative to the electrical boresight of the antenna will be known. And then, according to requirements of servo system for the slope of angle error characteristics curve, signal processor will give the corresponding angle error voltage U_{ym}. Different beam corresponds to different U_{ym}, so as to form the wide angle error characteristics corresponding to multi-beam. For example, if a signal is received in the +3rd beam (or + n beam), judge whether there is a signal or not in the signal detection units according to amplitude adjustment criterion. If "there is a signal," it indicates the target is in No. +3 beam, and the signal processing circuit will make it output a U_{+3} voltage. When there is a signal in No. +2 beam, the signal processing circuit will make it output a $U_{+2} = \frac{2}{3}U_{+3}$ voltage. When there is a signal in No. +1 beam, the signal processing circuit will make it output a $U_{+1} = \frac{1}{3}U_{+3}$ voltage. When there is a signal in No. 0 beam, the output is error voltages within the main beam. Similarly, U_{-3}, U_{-2}, U_{-1} can be obtained. Because the multi-beam does not continuously cover the airspace but be quantized by each beam width, there is quantization error so as to form the angle error characteristics of stepped appearance as shown in Fig. 2.95. It brings two kinds of influences. One is to cause angle quantization random error but it has little effect on the acquisition. "Amplitude comparison angle measurement" may be used as the supplement of decrease of quantization error, i.e., using "beam position number" as large number and the value of amplitude comparison angle measurement as decimal. The other one is nonlinear, which is equivalent to the phenomenon when there is bias between the angle error slope output by angle tracking receiver and the slope required by servo system. When there is such nonlinearity, the tracking system can still converge the tracking, but may affect the dynamic quality.

What are shown in Fig. 2.95 are the angle error characteristics of multi-beam and the angle error characteristics of the main beam. It can be seen that multi-beam largely increases the scope of angle error characteristics so as to increase the angle acquisition scope. When multi-beam antenna angle tracking converges to the main beam, the main beam channel will receive signals. Signal processing unit will judge "there are signals," and the multi-beam closed loop will be converted into main beam closed loop to realize narrow beam self-tracking. The tracking state of the nth multi-beam and 2D antenna can be obtained in a similar way.

(3) Use sampling discrete servo system to realize wide angle acquisition and tracking—closed loop discrete automatic tracking method

The explanation of the above operating principle is based on that each beam horn can output continuous signals to form a continuous automatic regulatory system. This system requires that each horn shall have 1-channel receiver and its signal detector, which makes the equipment very complicated. A simplified method is to use error signal discrete sampling to form discrete automatic regulatory system, as shown in Fig. 2.96.

In Fig. 2.96, the sampling switch is a fast electronic active switch. Low gain LNA_1 can be used to decrease the effect of insertion loss of sampling switch on G/T value. The HF amplification of reception channel is mainly provided by high gain LNA_2 after the sampling switch. As there's only one LNA_2, the total stages and price of LNAs are largely reduced. Voltage holding time is designed longer than the sampling period T design of each beam. If there is voltage input in the memory time, update the voltage holding value. If there is still no signal at the end of memory time, zero set the holding voltage and make the angle servo system open loop to restart the acquisition.

If signals are detected in a beam when sampling switch is used for the discrete sampling of multi-beam, a pulse signal will be outputted. The width of the pulse signal is the sampling residence time τ at that beam and pulse period is the sampling period T of that beam. In this way, a discrete automatic regulatory system is formed. There is no error signal in the time interval of $(T - \tau)$, and it is required to use a voltage holding circuit to make the continuous output of error signals. If the voltage remains unchanged in $(T - \tau)$, it is called "zero-order hold." If it presents linear slope, it is called "first-order hold." The performance of the latter is better that of the former. Discrete automatic regulatory theory points that, if sampling frequency $1/T$

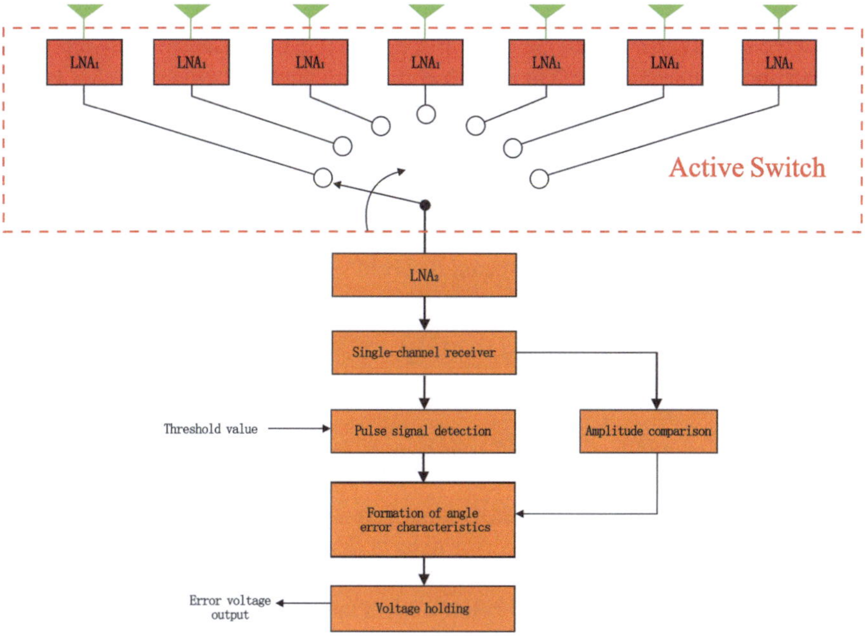

Fig. 2.96 Principle diagram of sampling discrete angle tracking

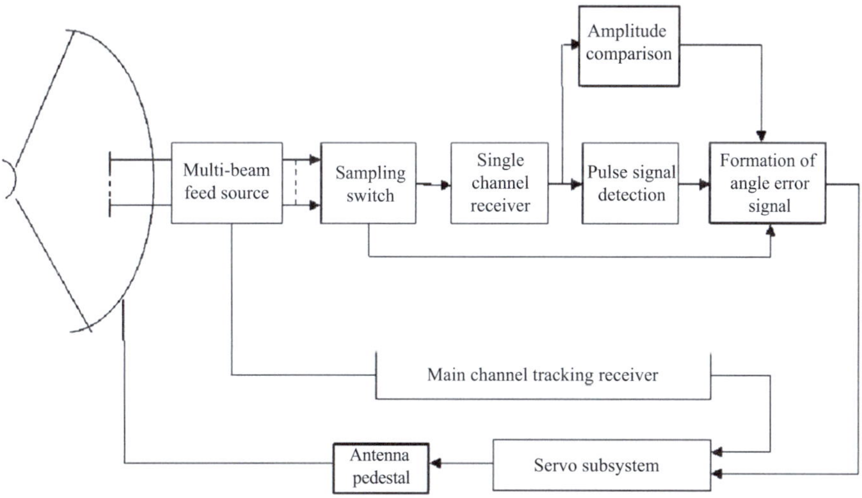

Fig. 2.97 Simplified block diagram of sampling discrete angle tracking system

is too low, large tracking error and poor dynamic quality will be caused. When sampling frequency $1/T$ is (5–10) times larger than servo bandwidth, the phase lag of "zero-order holder" is small and can be omitted. At this moment, the discrete automatic regulatory system can be handled approximately as the continuous automatic regulatory system. Block diagram of this scheme is as shown in Fig. 2.97.

When the signals received are too strong, the adjacent beams have the possibility to receive signals. An amplitude comparison circuit can be used for the amplitude comparison of such signals. The strongest one is determined as within that beam. Meanwhile, the amplitude ratio of adjacent beams can also be used in angle measurement by comparing amplitude, which can be a supplement to acquire more accurate angle error characteristics.

The detection of sampling signal is the detection of a single pulse signal whose width is τ and period is T. Different from radar waveform design, it is required to complete the detection of n multi-beam received signals within T. When switching time of the sampling switch is too short, the residence time of each beam

$$\tau = \frac{T}{n} \tag{2.352}$$

where τ is pulse width. Generally, the signals in τ can be a single carrier or a modulated carrier (e.g., FM, QPSK).

As shown in Fig. 2.97, angle error signal corresponding to a sub-beam pointing is acquired when pulse signal detector detects that sub-beam has an pulse signal output (under the action of sampling by switch, the output is an pulse signal) for the first time. If loop of servo system is closed when that pulse signal output is detected for the first time, the guiding system will then enter into the process of closed-loop tracking convergence. At that moment, under the action of that angle error signal, the loop will start automatic adjustment and tracking will gradually converge to the main beam at the center of multi-beams to finally complete angle guidance. As long as there is an error signal, automatic adjustment is possible. In this scheme, error signal can be acquired and loop can be closed as long as a single pulse signal is detected. So the guidance probability is mainly determined by the detecting probability of single pulse.

When loop of servo system is closed, a series of pulses (n pulses) will act on the closed-loop system subsequently. $n = T_r f_c$, where T_r is the time for convergence, and f_c is the sampling frequency. The n pulses can generate n angle error voltages and the voltage that is finally applied on the servo system is the mean voltage E of those n error voltages. However, some pulses may be lost due to the impact of single pulse detection probability P_D. The number of lost pulses is $n(1 - P_D)$. The losing of pulses will make E decreased to $\frac{n-n(1-P_D)}{n}E = P_D E$, which will then make the angle error slope K_R decreased to $P_D K_R$. Since open-loop gain of loop is $K_a = K_R K_0$ (K_0 is a known design value), K_a will also be decreased to $P_D K_a$. The decrease of K_a will have the following influences:

1) Servo loop bandwidth B_n becomes narrower: this is due to [19]. For type II servo system, $B_n = \sqrt{K_a/3.5}$.

2) $\sigma\,\% \leq 10\,\%$ The time for convergence becomes longer: the fast convergence time of servo system is the rise time t_r (the time for reaching 90 % of the stable value) during its transient process. When overshoot $\sigma\%$ in the transient process of servo system is equal to or less than 10 %, i.e., $\sigma\,\% \leq 10\,\%$, $t_r \approx \frac{1}{2B_n} = \sqrt{\frac{3.5}{4K_a}}$; when $\sigma\,\% \geq 10\,\%$, $t_r \approx \sqrt{\frac{2.5}{4K_a}}$

Table 2.16 Relationship between detection probability and IF SNR

P_f / $(S/N)_I$ (dB) / P_d	10^{-3}	10^{-4}	10^{-5}	10^{-8}	10^{-10}	10^{-12}
0.8	10.4	11.5	12	13.5	14.5	15.2
0.9	11	12	12.5	14.2	15	15.8
0.95	11.5	12.3	13	14.5	15.5	16.1
0.98	12.2	13	13.5	15.5	16	17
0.999	13.8	14.5	15	16.2	17	17.5

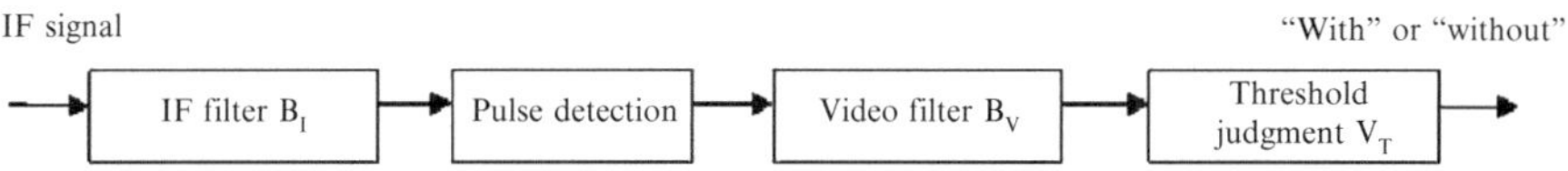

Fig. 2.98 Pulse signal detection model

3) Dynamic error Δ_a becomes greater: for type II servo system, $\Delta_a = \ddot{\theta}/K_a$, where $\ddot{\theta}$ represents the angular acceleration of target. When Δ_a is less than half of the width of main beam, servo system can shift to main-beam tracking, in which case the guidance is successful; when is less than half of the multi-beam coverage angle, servo system works under the continuous guiding state of tracking convergence process θ_w; when $\Delta_a > \theta_w/2$, the target is beyond the guiding range, which will lead to guidance failure.

Generally, value of P_D is designed to be a great one and approximate to 1. It has minor impact on convergence and will not cause failure of guidance.

For the same reasons as above, the stepped appearance of angle error as shown in Fig. 2.95 will also make the angle error slope deviate from its optimum value K_a as designed. And the decrease of K_a will have influences similar to those mentioned above. The difference lies in that the stepped error voltage is formed due to the fact that the error characteristics of linear K_a is quantized by sub-beam width $\Delta\theta_n$. For each $\Delta\theta_n$, however, the mean error voltage generated by linear K_a is equal to the stepped error voltage corresponding to that $\Delta\theta_n$. During target tracking and convergence process, the angle of deviation of target is variable, and its error voltage also varies; so, the main component that is finally applied on narrowband servo system is the mean error voltage smoothed by narrowband filter. Therefore, the impact of stepped error on guidance will not cause the failure of guidance.

The acquisition probability of this scheme can be given based on the detection theory of pulse radar signal. The detection theory of pulse radar signal [24] gives the following data of detection probability P_d, false alarm P_f, and IF SNR $(S/N)_I$.

The data given in Table 2.16 is based on the detection model of Fig. 2.98.

In Fig. 2.98, the input IF signal is a single-carrier pulse signal, B_I is the bandwidth of IF filter, B_V is the bandwidth of video filter after detection, and the optimal bandwidth is $B_I = \frac{1}{\tau}, B_V = \frac{0.5}{\tau}$. In radar signal detection analysis, assume $B_V \geq B_I/2$, i.e., mainly decided by IF signal envelope. *S/N* also corresponds to IF SNR $(S/N)_I$.

When IF signal is a modulated signal, its bandwidth B_I is usually very large, and narrow band filter cannot be added at IF, or signal energy will be lost. The optimal bandwidth of video signal is decided by pulse width τ, $B_v = 1/2\tau$, because $1/2\tau < B_I/2$, the envelope of video signal is mainly decided by $1/2\tau$. Because the narrowing of video bandwidth, the video SNR $(S/N)_V$ is improved. But in the case of low $(S/N)_I$, amplitude detection will cause square loss. The expression of detection output $(S/N)_V$ is:

$$\left(\frac{S}{N}\right)_V = \left(\frac{S}{N}\right)_I^2 \frac{B_I}{2B_V} = \left(\frac{S}{N}\right)_I^2 B_I\tau \tag{2.353}$$

P_d and P_f are determined by video signal envelope. After $(S/N)_V$ is calculated using the above expression, replace $(S/N)_I$ with $(S/N)_V$, then the approximate value of P_d and P_f will also be calculated. In real project design, computer simulation or hardware test are adopted to calculate P_d and P_f.

To sum up, the design method of this scheme is: when servo bandwidth B_n, IF bandwidth B_I, IF SNR $(S/N)_I$ and multi-beam data n are known: ① calculate T by $1/T \geq (5 \sim 10)B_n$; ② calculate τ by $\tau = T/n$; ③ calculate $(S/N)_V$ under the required acquisition probability P_d and false-alarm probability according to Table 2.16; ④ calculate the required $(S/N)_I$ by Expression (2.352).

If the calculated $(S/N)_I$ cannot meet the requirement, increase τ to improve $(S/N)_I$ so as to decrease *n*. In this way, array feed source will be divided into several sub-arrays for parallel processing and *n* of each sub-array will be decreased.

(4) Use "sequential sub-lobe" and "double time division" to realize wide angle acquisition

Based on "discrete automatic regulatory theory," the above scheme uses time division sampling to decrease the number of channels. At the border of acquisition spatial coverage, the sampling interval cannot be longer than the dwell time of the target at narrow beam, or the target will fly out of that beam at the second sampling. There is another thought which uses "sequential lobe method" theory to divide the multi-beam array feed source into several sub-arrays, and each sub-array correspond to one sub-lobe. Conduct time division sequential scan to the sub-lobe to cover a wide acquisition spatial coverage. The time division sampling interval cannot be longer than the dwell time of the target at the sub-beam, so its sampling interval may be long. There are a lot of channels in the sub-beam required, and parallel processing is required.

In system design, comprehensive comparison may be conducted according to the features of the two kinds of "time division multiplexing" schemes, or integrate the advantages of the two by using "double time division multiplexing" method, i.e., using "discrete automatic regulatory theory" for the first time division and the sequential lobe theory for the second time division. One principle block diagram of "double time division multiplexing" scheme is shown in Fig. 2.99.

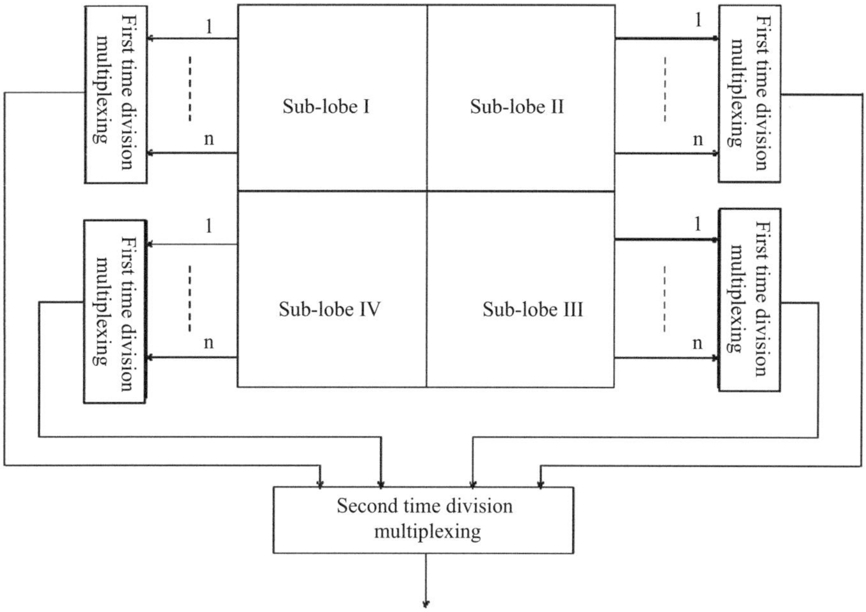

Fig. 2.99 Principle block diagram of "second time division multiplexing"

In Fig. 2.99, divide the $4n$ horn feed source arrays into four sub-arrays, and each sub-array includes n horns. Each sub-array uses "discrete automatic regulatory theory" for the first time division multiplexing to multiplex n-channel into 1-channel output. It corresponds to one sub-lobe. The output of four sub-arrays will go through a 4-channel switch to complete the second time division multiplexing. Then it will multiplex 4-channel into 1-channel output. The 4-channel switch will be switched on, which is equivalent to the sequential scan of four sub-lobes so as to complete wide angle coverage.

2.5.7 Polarization Diversity-Synthesized Technology of Angle Tracking

2.5.7.1 Function of Polarization Diversity-Synthesized Technology

In flight vehicle TT&C and information transmission system, polarization diversity-synthesized technology is generally applied, which functions are as follows:

(1) Anti-multipath interference, to gain a stable tracking.
(2) To gain two kinds of orthogonal polarization wave for diversity synthesis to obtain synthesis gain, so as to improve SNR and tracking accuracy.
(3) The two orthogonal polarization receiving channels back up each other to improve the reliability of the system.

Reasons why there are two kinds of polarization waves in the radio wave received are:

(1) When the spacecraft enters at a low altitude and low elevation, the antenna elevation of the TT&C station is very low, and multipath interference will be caused due to refraction by the ground (or sea surface). Thus, signal fluctuation and fadeout will be caused, and there will be two kinds of orthogonal polarization components.

(2) When the antenna of the spacecraft is polarized by a line, it will be decomposed into two kinds of polarization components, i.e., the left-hand rotation and right-hand rotation, which can be received by the ground station through its dual-polarization antenna for polarization synthesis. 3 dB S/N gain will be obtained in theory.

(3) In order to obtain an omnidirectional antenna free of "dead spot" on a revolving flight vehicle, the space synthesis of the pattern can be changed into polarization synthesis of circuit. Adopt "Four array elements orthogonal polarization synthesis" scheme, i.e., use 4 antennas 90° apart from each other. The two symmetrical antennas 180° apart from each other are co-polarized and the other two antennas are orthogonal polarized. Thus, the polarization synthesis technology can be used to obtain an omnidirectional synthetic beam free of dead spot.

(4) If the thunder, rain, etc., make the atmospheric composition change and form plenty of small refractive bodies and reflective bodies during the process of radio wave propagation, the generated cross polarization (e.g., the influence of rain drops) will suffer irregular random fadeout.

(5) The antenna on the spacecraft and the shell of spacecraft will affect each other. When the antenna is close to the shell, it will change the polarization index (e.g. the axial ratio). In fact, the change of axial ratio is the change of proportion between the left-hand rotation and right-hand rotation circular polarized waves. Figure 2.100 shows the polarization change of the circular polarization satellite antenna.

In ideal conditions, there is only circular polarized wave, while in fact it will change into elliptic polarized wave and even linear polarized wave when the antenna deviates from the direction of the electrical boresight of the antenna. Therefore, polarization attenuation will be caused if a fixed polarization antenna is used for receiving.

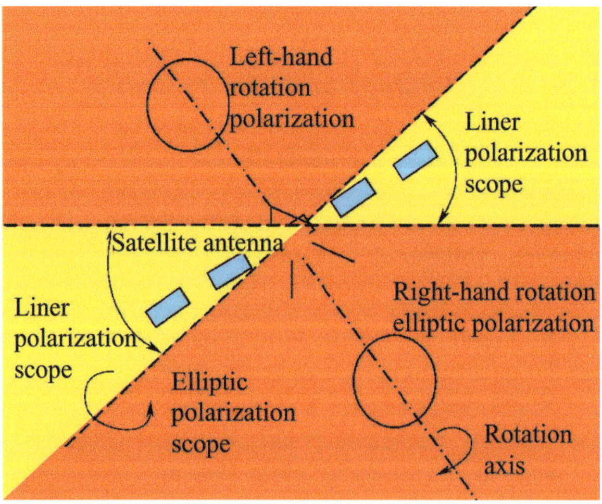

Fig. 2.100 Circular polarization satellite antenna

In addition, during the process of TT&C satellite positioning or when satellite control system fails, rolling phenomenon will occur. If both left-hand and right-hand rotation polarization antennas are installed on the satellite, the signal received by the ground station will be left-hand rotation signal for a moment and then the right-hand rotation signal for another moment.

(6) When the spacecraft is in high-speed movement or rotation, changes of rotating state will cause the rotation of spacecraft antenna polarization direction and zero point of the pattern, which will cause a long-term fadeout of the signal received by the antenna. Such fadeout includes cross polarization fadeout which is different from the fadeout of multipath effect and generally occurs in high SNR conditions. Cross polarization can not only weaken the signal but also have great effect on self-tracking performance.

(7) The flame jetted from the engine of the flight vehicle (during the working of the engine or stage separation) makes the atmosphere ionize to form a plasma area. Therefore, the signal passed through by it will have a rapid random change with respect to its amplitude and polarization.

Since there are two kinds of orthogonal polarized waves in the radio wave from the spacecraft, the ground station shall have the corresponding polarization diversity receiving capability to weaken the effect of signal fading so as to obtain a stable tracking and to improve *S*/*N*.

2.5.7.2 Mode and Principle of Signal Synthesis

In diversity synthesis receiving, signals obtained by *N* different independent signal branches can be used to obtain diversity gain according to different combination schemes. From the point of view of the synthesis position, the synthesis before a detector can be called "synthesis before detection," i.e., the synthesis at IF and RF and almost the synthesis at IF. The synthesis can also be performed after a detector, called "synthesis after detection," i.e., the synthesis at video.

(1) Maximum ratio synthesis

N branches will be formed after frequency diversity at the receiving terminal, and the branches will have in-phase addition and be sent to the detector for detection after phase adjustment and multiplied by a proper weighed coefficient. The principle diagram of maximum ratio synthesis is as shown in Fig. 2.101.

It can be proved based on Chebyshev's inequality when the variable gain weighting coefficient is $G_i = \frac{A_i}{\sigma^2}$, the SNR after diversity synthesis can be up to the maximum value, where A_i indicates the signal amplitude of the *i*th diversity branch and σ^2 is the noise power of that branch, $i = 1,2,\ldots,n$.

Output after synthesis is:

$$A = \sum_{i=1}^{N} G_i A_i = \sum_{i=1}^{N} \frac{A_i}{\sigma^2} \times A_i = \frac{1}{\sigma^2} \sum_{i=1}^{N} A_i^2 \tag{2.354}$$

It can be seen that a branch with a higher SNR has a higher contribution to the signal after synthesis. When the statistics of each branch is independent and all the

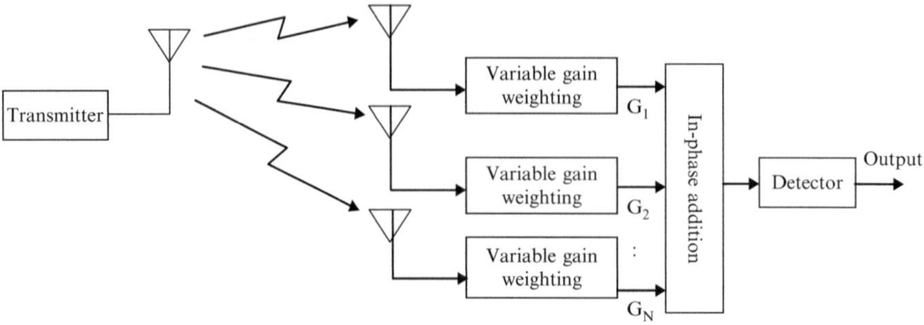

Fig. 2.101 Principle diagram of maximum ratio synthesis receiving of space diversity

branches have a same average SNR $\overline{\text{SNR}}$, the average output SNR after maximum ratio synthesis is:

$$\overline{SNR_M} = N \bullet \overline{SNR} \tag{2.355}$$

where $\overline{\text{SNR}}_M$ indicates the average output SNR after maximum ratio synthesis, and N indicates the number of diversity branches, i.e. diversity multiplicity.

Synthesis gain is:

$$K_M = \frac{\overline{SNR_M}}{\overline{SNR}} = N \tag{2.356}$$

Therefore, synthesis gain is directly proportional to the number of diversity branches N.

(2) Equal gain synthesis

In the above maximum ratio synthesis, making $A_i = 1, \quad i = 1, 2, \ldots, n$ will obtain equal gain synthesis, and the average output SNR after equal gain synthesis is:

$$\overline{\text{SNR}_E} = \overline{\text{SNR}}\left[1 + (N-1)\frac{\pi}{4}\right] \tag{2.357}$$

Synthesis gain of equal gain synthesis is:

$$K_M = \frac{\overline{SNR_M}}{\overline{SNR}} = 1 + (N-1)\frac{\pi}{4} \tag{2.358}$$

When N (diversity multiplicity) is large, equal gain synthesis is almost the same as maximum ratio synthesis, only about 1 dB difference. The realization of equal gain synthesis is easy and the required equipment is simple.

(3) Selective synthesis

Principle diagram of selective synthesis is shown in Fig. 2.102.

In Fig. 2.102, the receiving terminals are receivers of N diversity branches $i = 1, 2, \ldots, N$. In R_i, use the selection logic to select a channel with the highest SNR $\overline{\text{SNR}}_{i=1}$ as the output. The average output SNR of selective synthesis is:

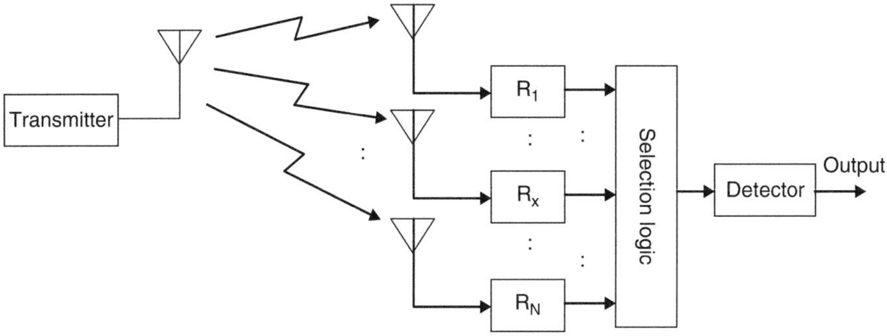

Fig. 2.102 Block diagram of selective synthesis receiver

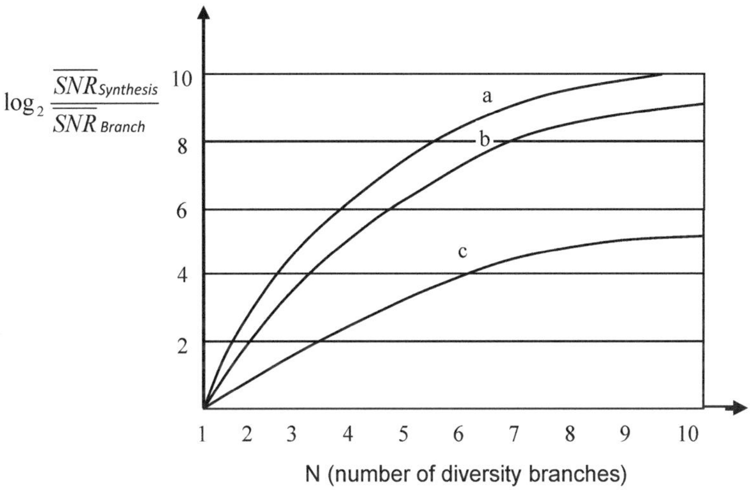

Fig. 2.103 Performance comparisons of three kinds of diversity synthesis. (**a**) Maximum ratio synthesis, (**b**) equal gain synthesis, and (**c**) selective synthesis

$$\overline{\text{SNR}}_S = \overline{\text{SNR}}_{i=1} \sum_{i=1}^{N} \frac{1}{i} \tag{2.359}$$

It can be seen that the contribution to selective diversity output SNR by adding one diversity branch is only the reciprocal of the total number of diversity branches. Synthesis gain of selective synthesis is:

$$K_s = \frac{\text{SNR}_S}{\text{SNR}_{i=1}} = \sum_{i=1}^{N} \frac{1}{i} \tag{2.360}$$

(4) Performance comparison of three main synthesis modes

What is shown in Fig. 2.103 is the improvement degree of the average SNR of the three synthesis modes.

When $\overline{SNR}$ of branches are different, the performance difference among them will be greater.

In addition, selective synthesis scheme is greatly influenced by the correlation of the synthesis signal fading. Only when their fading correlation coefficient is less than 0.3, the influence will be less. The equal gain synthesis scheme is greatly influenced by the fading distribution characteristics.

Therefore, the maximum ratio synthesis scheme is the best and it is frequently adopted in actual engineering.

2.5.7.3 Polarization Synthesis Scheme of Angle Tracking Channel

When there are two kinds of components in the radio wave received, it will cause the signal fluctuation and fading and have great effect on angle tracking performance, which makes the antenna unable to aim at the target and cause jitter. Therefore, polarization diversity receiving is generally adopted for angle measurement channel to improve tracking performance.

(1) IF polarization synthesis

For the monopulse system of three-channel and two-channel, diversity phase lock is unable to be done for there is no signal in difference channel (or S/Φ is very small) during tracking, thus it is difficult to realize it. One IF polarization synthesis scheme is to realize left-hand rotation and right-hand rotation diversity phase lock and maximum ratio synthesis in a sum channel with available signal, and use sum channel VCO as the reference signal of difference channel error detection. The phase of the sum channel and difference channel shall be consistent to ensure the same phase of the left-hand signal and right-hand signal in difference channel. The difference channel shall adopt the same weighted value as that of the sum channel; thus, the difference channel can also realize maximum ratio synthesis. AGC voltage can be used to realize diversity weighting. This scheme uses a diversity synthesis circuit whose sum and difference channels are completely the same to realize the polarization diversity synthesis of difference channel. Since the sum channel can automatically and self-adaptively realize maximum ratio synthesis, and the difference channel is consistent with the sum channel, the polarization diversity synthesis of difference channel can also be realized.

(2) Video polarization synthesis

Monopulse angle error detection is a coherent detection, and it won't cause output SNR loss in theory; therefore, synthesis technology after detection can be used. The angle error signal has a low frequency and is easy for digitization, so this scheme is called video digital polarization synthesis. Its principle diagram is shown in Fig. 2.104. The reference signal of PD in the Figure can be provided by the reference signal of PLL which is included in IF_{Σ}.

The advantages of this scheme are:

1) Do without diversity phase locked loop so as to simplify the circuit.
2) Digitalized maximum ratio synthesis, simple scheme, good performance, close to theoretical results.

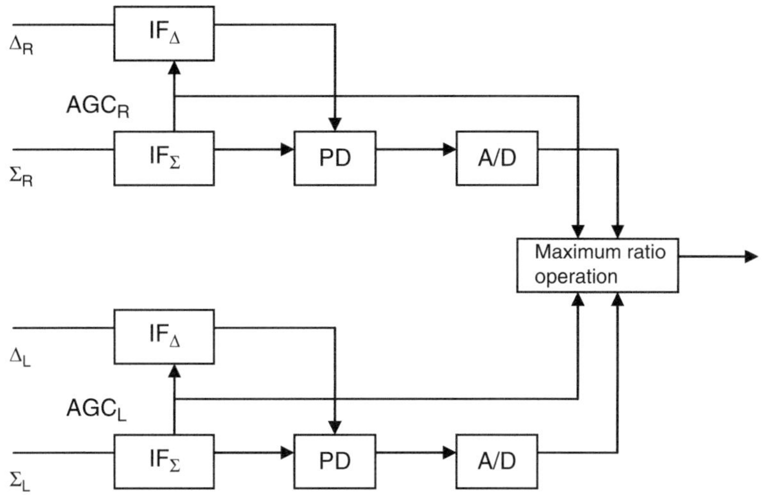

Fig. 2.104 Principle diagram of video digital polarization synthesis scheme of angle tracking channel

3) Fully inherit the previous monopulse channel technology and only add the part of maximum ratio synthesis.
4) The debugging is simple, stable and reliable.

(3) Polarization combination scheme of circular waveguide TE_{11}, TE_{21} module monopulse—"diversity-combination method"

In C&T system, circular waveguide TE_{11}, TE_{21} mode feed source is generally used to compose a DCM monopulse antenna [25]. It has a high efficiency but also some special problems in polarization diversity synthesis.

The principle of conical multimode corrugated horn is to use the pattern distribution and radiation characteristic of fundamental mode and high order mode in circular waveguide, to form the sum pattern by fundamental mode and the difference pattern by high order mode. When the antenna beam axis aims at a target, the incoming wave signal only excites the fundamental mode (TE_{11} mode) in the conical multimode corrugated horn. When the target deviates from the antenna beam axis, the incoming wave signal can excite the fundamental mode (TE_{11} mode) and the high order mode (TE_{21} mode) in the conical multimode corrugated horn at the same time. Based on the design of smooth-walled circular waveguide mode-selective coupler (tracker), only TE_{11} mode is allowed to pass its direct-through arm which is the sum *T*/*R* channel. The side arm can couple two orthogonal TE_{21} modes through eight rectangular waveguide coupling aperture arrays and form the output port for circular polarized different-mode signal through synthesis network. The amplitude of high order TE_{21} mode is directly proportional to the angle of deviation, and forms circular polarized different signal output through synthesis network.

This feed source is different from multi-horn, square horn multimode feed source. The relative phase relationship among its sum/difference signals is shown in Fig. 2.105.

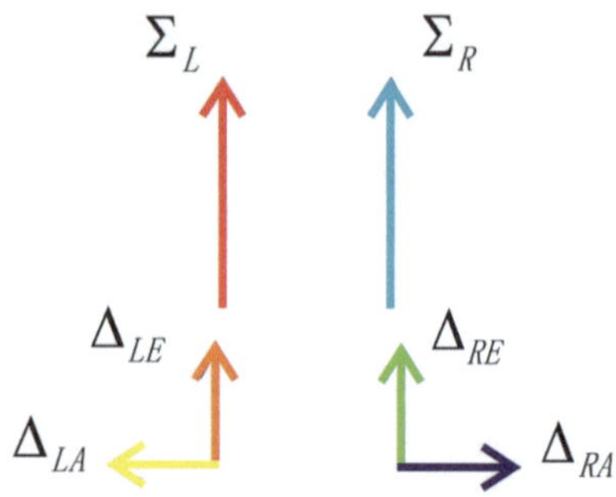

Fig. 2.105 Phase relationship among the sum/difference signals formed by circular waveguide TE11 and TE21 mode feed sources

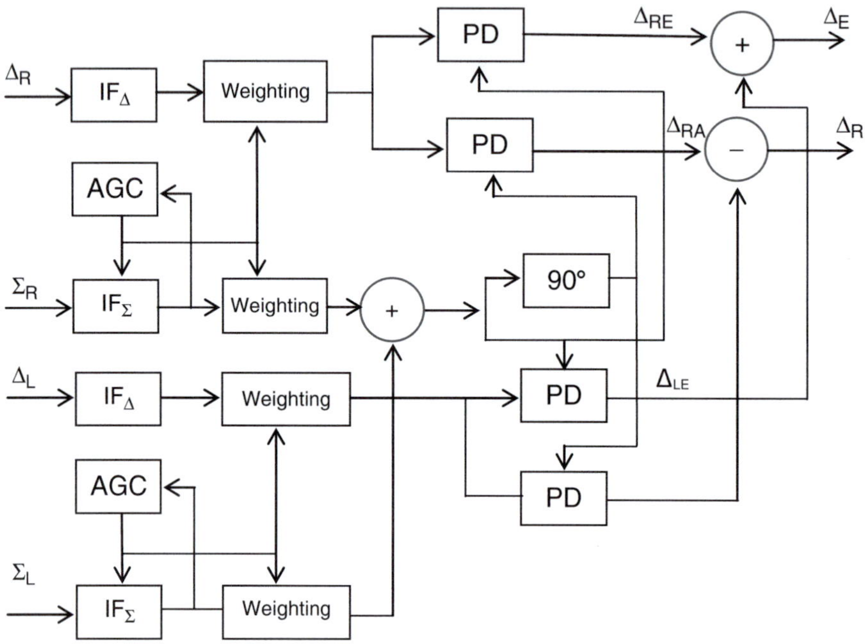

Fig. 2.106 Block diagram of polarization synthesis of circular waveguide TE11 and TE21 mode monopulse

In Fig. 2.105, Σ_L, Δ_{LE}, Δ_{LA} is respectively the sum channel, elevation difference, azimuth difference signal of left-hand rotation signal. The phase difference between Δ_{LE} and Δ_{LA} is 90°, so as to realize DCM channel combination of left-hand polarization in DCM. In a similar way, Δ_{RE}, Δ_{RA} will realize the DCM combination of right-hand rotation polarization channel. It can be seen from that figure that, if the polarization combination is conducted at IF, only Δ_{LE} and Δ_{RE} can have in-phase addition, while Δ_{LA} and Δ_{RA} will have reverse phase subtraction. Therefore, polarization synthesis cannot be realized. In order to solve this problem, it is required to split Δ_{LE} and Δ_{LA} and make the phase of Δ_{LA} component reversed; thus Δ_{LA} and Δ_{RA} can have in-phase addition now. This idea of "synthesis after split" can be realized by video polarization diversity synthesis, as shown in Fig. 2.106.

In Fig. 2.106, difference channel signal is synthesized by video polarization. The subtraction between azimuth right-hand rotation video polarization component and azimuth left-hand rotation video polarization component is used to realize their reverse phase addition. IF polarization synthesis is adopted for sum channel. Because AGC voltage is directly proportional to the voltage of sum channel signal, AGC_R voltage of right-hand sum channel is used for the amplitude weighting of right-hand sum channel signal and its difference channel signal, and the AGC_L voltage of left-hand sum channel is used for the amplitude weighting of left-hand sum channel signal and its difference channel signal, to realize the maximum ratio synthesis.

References

1. Sun Zhongkang et al (1996) Active and passive localization technology with single or multiple base. National Defense Industry Press, Beijing
2. Wu Ping (1996) Optimal station arrangement of typical TT&C system. J Astronaut 17(1):15–23
3. Liu Jiaxing (2007) Frequency domain analysis of velocity measure error. Telecommun Eng 2007 Suppl
4. Huang Xuede et al (2000) Missile TT&C system. National Defense Industry Press, Beijing
5. Liu Jiaxing (2010) Characterization of short term stability and its transform between and $S_1(f)$ [J]. J Spacecr TT&C Technol 6:29–33
6. Bames JA, Chi AR (1971) Characterization of frequency stability. IEEE Trans M-20 No. 1971:10–13
7. Qian Yunmin (1978) Relationship between sampling variance and CW radar velocity measurement accuracy. Telecommun Eng 1978(6)
8. Liu Jiaxing (2005) Effect of short term frequency stability of oscillators on the velocity in a TT&C system. Telecommun Eng 2005(1):1–7
9. Gardner FM (2007) Phase-locked loop technology. Post & Telecom Press, Beijing
10. Huang Chengfang (2012) Full coherent frequency folding multistation transponder. Telecommun Eng 2012(3)
11. Liu Jiaxing (1980) Tone ranging time delay in linear four terminal network and product demodulator. J CETC 10, 1980.2
12. Liu Jiaxing (2009) Group delay characteristics analysis method for range error and its applications in deep space TT&C systems. J Spacecr TT&C Technol 2009.2:22–26
13. Liu Jiaxing, Zhou Wenlan (2013) Effect of group delay on distance-measuring error. Telecommun Eng 2013(5)
14. JPL (2007) DSMS telecommunications link design handbook. http://deepspace.jpl.nasa.gov/dsndocs/810-005/
15. Cai Tingsong (2013) Short-baseline positioning technology based on phase difference. 10th Institute of China Electronics Technology Group Corporation, Chengdu
16. Chen Fangyun (1992) Satellite TT&C manual. Science Press, Beijing
17. Wang Yuanling (2008) Analysis of deep space ranging signal systems. Telecommun Eng 48(5):1–6
18. Wang Xiaoyan (2005) Study on single channel monopulse angle tracking systems. Telecommun Eng 45(3):117–120
19. Liu Jiaxing (2001) A tentative scheme of interferometer MMW rendezvous radar. Telecommun Eng 41(1):13–18

20. Zhao Yefu, Liu Jiaxing et al (2001) Radio tracking system. National Defense Industry Press, Beijing
21. Yu Zhijian (2009) Deep space TT&C system. National Defense Industry Press, Beijing
22. Zhang Guangyi (2006) Phased array radar technology. Publishing House of Electronics Industry, Beijing
23. Yi Runtang (1998) Study on the adaptive processing of TDRSS phased array ground multi-beam formation. Telecommun Eng 1998(4)
24. Skolnik M (2012) Introduction to radar systems (trans: Zuo Qunsheng et al.). Publishing House of Electronics Industry, Beijing
25. Ke Shuren (2012) Discussion on polarization diversity receiving of circular waveguide TE11 and TE21 mode true and false monopulse system, C&T, 2012(2)

Chapter 3
Information Transmission Technologies

Distortion and noise are two major subjects in information transmission research. Essential requirements for a transmission system are low distortion caused by signal transmission and low additional noise as much as possible, which are the major measurement factors for channel performance.

Signal and distortion can be divided into two types, linear distortion and non-linear distortion. Linear distortion is caused by the linear network. It is well known that linear network can be described by its transmission function $\dot{K}(\omega)$, that is, it depends upon the amplitude-frequency characteristic and phase-frequency characteristic. When we study the linear distortion based on $\dot{K}(\omega)$, this is called "frequency domain method" which is associated with time domain method through Fourier Transform.

Non-linear distortion of signal is related to the amplitude of input signal, and the transfer function which is the function of both frequency and signal amplitude, i.e., $u_0(\omega) = \dot{K}(\omega, u)u_i(\omega)$ where $\dot{K}(\omega, u)$ is the transmission function, which relates to signal amplitude and frequency.

Non-linear distortion is characterized by a new frequency component which is generated by the non-linear network.

There are two types of noise in transmission channel. One is thermal and phase noise; the other is interference noise. Generally, the former is overriding noise, which is generated by the front end device of the receiving system (e.g., low-noise pre amplifier and antenna feeder) and conversion, modulation/demodulation device in the channel (e.g., phase noise, intermodulation noise, and threshold noise from the oscillator).

The information transmitted by the information system is various and widely meaningful. It can be summarized into two types, continuous information and discrete information. Before a message arrives at the receiving terminal, there is much uncertainty for the recipient about cognition of such message. "Information", as the fundamental objective of transmission, refers to the "core" of message which can be used to eliminated such uncertainty upon communication establishment.

J. Liu, *Spacecraft TT&C and Information Transmission Theory and Technologies*,
Springer Aerospace Technology, DOI 10.1007/978-3-662-43865-7_3

Information is usually loaded on electric signals. Continuous-valued electric signal is called analog signal and discrete-valued signals can be converted into digital signals. Analog signals can also be converted into digital signals through a sampling and A/D converter. If the channel transmits analog signals, it is called analog transmission system; if it transmits digital signals, it is called digital transmission system. The following gives an introduction to the two aspects of information transmission technology in "analog" and "digital".

3.1 Analog Transmission Technology in C&T

In an analog signal transmission system, the analog signal generator outputs continuous-amplitude signals. For the TT&C system, a typical representative of the analog signal transmission system is the unified S-band system (USB). Its characteristics include:

(1) The frequency-division approach of multiple subcarriers and multiple carriers is used to implement multi-channel information transmission and multi-function integration. The phase modulation of multi-subcarrier analog signal is used at the transmitting end.
(2) At receiving end, the phase-locked coherent receiving and demodulation of phase-modulated signals is used.

3.1.1 Analog Signal Modulation

Analog modulation includes linear modulation and nonlinear modulation. Linear modulation means that the spectrum of modulated carrier is the parallel translation of baseband signal spectrum (modulated signal). Except this parallel translation, there is no new spectrum component, so linear transformation features are provided. Thus it is called linear modulation (such as AM). Nonlinear modulation means that new frequency components (such as baseband high-order component and combined component in the phase-modulated carrier) are generated in the modulated carrier spectrum, and nonlinear transformation features are provided. Thus, it is called nonlinear modulation. In addition, linearity and nonlinearity of modulation characteristics are different from the above meanings. Modulation characteristic refers to the characteristic of correspondence between baseband parameters (e.g., amplitude) and modulated parameters of the modulated wave (e.g., carrier phase in phase modulation). When the correspondence shows a non-linear relationship, it is called nonlinear modulation characteristic; when the correspondence shows a linear relationship, it is called linear modulation characteristic. However, any modulation mode only can be implemented using nonlinear devices. The above-mentioned three "nonlinearities" refer to three different concepts in modulation technology. Additionally, pulse digital modulation refers to a method of converting analog signals into digital signals (e.g., PCM). From the perspective of modulation, it can be seen as modulating a binary pulse sequence using an analog baseband signal, and the modulation result is that pulse "existence" or "nonexistence" (i.e., code) is changed. So it is called pulse digital modulation. Carrier digital modulation refers to a technique that uses a digital signal (e.g., PCM digital code) to modulate sine-wave (e.g., PSK and FSK). It is called digital modulation for short.

3.1.1.1 Mode of Analog Phase Angle Modulation

Phase modulation and frequency modulation are two RF modulation systems commonly used in the united carrier TT&C system. Due to $\omega = \mathrm{d}\phi/\mathrm{d}t$, it can be considered that both phase modulation and frequency modulation are modulation of phase angle, so they are collectively called phase angle modulation. Modulated carrier generated by phase angle modulation of a sinusoidal carrier can be expressed as (given initial phase $\phi_0 = 0°$).

$$s(t) = A\cos\left[\omega_c t + \Delta\Phi(t)\right], \tag{3.1}$$

where A is carrier amplitude; ω_c is carrier frequency; $\Delta\Phi(t)$ is instantaneous carrier phase change caused by the modulated signal $f(t)$.

When amplitude A and carrier frequency ω_c remain constant and $\Delta\Phi(t)$ is a function of $f(t)$, the modulation is called phase modulation (PM), and:

$$\Delta\Phi(t) = k_{PM} f(t), \tag{3.2}$$

where k_{PM} is the phase-shift constant.

In ideal condition, k_{PM} is a constant and the PM characteristic is linear. However, in practical engineering, it is impossible to satisfy the requirement that K_{PM} is a constant. Then the PM characteristic (characteristic curve) is non-linear.

In phase modulation, corresponding modulated signals are called phase-modulated carriers or phase-modulated signals. The mathematical expression is:

$$S_{PM}(t) = A\cos\left[\omega_c + k_{PM} f(t)\right]. \tag{3.3}$$

When amplitude A keeps constant and instantaneous carrier frequency is the linear function of $f(t)$, the modulation is called frequency modulation (or FM). Then the instantaneous frequency of the carrier is:

$$\omega = \omega_c + \Delta\omega = \omega_c + k_{FM} f(t), \tag{3.4}$$

where k_{FM} is the frequency-shift constant.

In ideal condition, K_{FM} is a constant and the FM characteristic is linear. However, in practical engineering, it is impossible to satisfy the requirement that K_{FM} is a constant. Then the FM characteristic is non-linear. Similarly, in frequency modulation, corresponding modulated signals are called frequency-modulated carriers or frequency-modulated signals. The mathematical expression is:

$$S_{FM}(t) = A\cos\left[\omega_c + k_{PM}f(t)\right]t. \tag{3.5}$$

Since instantaneous angular frequency has the following relationship:

$$\omega = \omega_c + k_{FM}f(t) = \frac{d\Phi(t)}{dt}$$

then,

$$\Phi(t) = \int \omega dt = \omega_c t + k_{FM}\int f(t)dt$$

therefore,

$$S_{FM}(t) = A\cos\Phi(t) = A\cos\left[\omega_c t + k_{FM}\int f(t)dt\right]. \tag{3.6}$$

Through comparison, Expression (3.6) and Expression (3.3) can be combined to a uniform formula to express the PM and FM signal, i.e.,

$$s(t) = A\cos\left[\omega_c t + \Delta\Phi(t)\right]. \tag{3.7}$$

In PM, there is:

$$\Delta\Phi(t) = k_{PM}f(t). \tag{3.8}$$

In FM, there is:

$$\Delta\Phi(t) = k_{FM} = \int f(t)dt. \tag{3.9}$$

Expression (3.7) is the uniform expression of angular modulation wave.

Expression (3.9) shows that in order to obtain FM carrier, we can first integrate the modulated signal $f(t)$ and then perform phase modulation of the carrier. This is indirect FM. Similarly, in order to obtain PM carrier, we can first differentiate $f(t)$ and then perform frequency modulation of the carrier. This is indirect PM.

When $f(t)$ is the sine wave, the integration and differentiation processing which will make it become the cosine wave, i.e., a phase change of 90°. Ranging tone and subcarrier in TT&C are both sine waves. Either direct or indirect angular modulation method can be used.

When $f(t)$ is a non-sinusoidal periodic function, it can be expressed by the Fourier series as:

$$f(t) = C_0 + \sum_{i=1}^{k} c_i \cos(\omega_i t + \varphi_i), \tag{3.10}$$

where ω_i is the ith frequency component, or the ith subcarrier component in C&T; C_0 is direct current component.

In the modulator circuit, C_o is generally isolated by the blocking capacitor and doesn't carry information. So the phase modulation can be expressed as:

$$\Delta\Phi(t) = k_{\text{PM}} \sum_{i=1}^{k} c_i \cos[\omega_i t + \varphi_i] = \sum_{i=1}^{k} k_{\text{PM}} c_i \cos[\omega_i t + \varphi_i], \tag{3.11}$$

where $k_{\text{PM}} c_i$ is the peak phase deviation caused by the ith frequency component ω_i; c_i is the amplitude of the ith frequency component; φ_i is the phase of the ith frequency component, which can be modulated, that is, the phase can be the function of the time.

In C&T, φ_i of ranging tone won't be modulated, while φ_i of telemetering subcarrier, command subcarrier, voice subcarrier, etc. will be modulated.

For FM wave, according to Expression (3.9), there is:

$$\begin{aligned} \Delta\Phi(t) &= k_{\text{PM}} \int f(t)\mathrm{d}t \\ &= \sum_{i=1}^{k} \frac{c_i k_{\text{FM}}}{\omega_i} \sin[\omega_i + \theta_i] = \sum_{i=1}^{k} \frac{c_i k_{\text{FM}}}{\omega_i} \cos\left[\omega_i t + \theta_i + \frac{\pi}{2}\right], \end{aligned} \tag{3.12}$$

where $c_i k_{\text{FM}}$ is peak frequency offset caused by the ith frequency component; θ_i is the constant, unmodulated.

By comparing Expression (3.12) and Expression (3.11), it can be seen that FM wave and PM wave can be mutually equivalent. The difference between their phases is 90°, and their amplitudes are $\frac{c_i k_{\text{FM}}}{\omega_i}$ and $k_{\text{PM}} c_i$, respectively. Therefore, the modulator and demodulator of PM wave and FM wave are interchangeable, if only $f(t)$ performs relevant integration and differentiation processing.

In C&T, when ω_i is a sine wave, phase demodulation can be replaced by the frequency demodulation. In this case, the phase difference is a constant $\pi/2$ and the amplitude increases $\frac{K_{\text{FM}}}{K_{\text{PM}}\omega_i}$ times (some modulation degree analyzers measure phase modulation degree based on this principle, so does the zero ranging transponder of the TT&C equipment). If the frequency demodulation is replaced by the phase demodulation, phase difference is a constant 90° and the amplitude increases $\frac{K_{\text{PM}}\omega_i}{K_{\text{FM}}}$ times. When ω_i is the FM subcarrier, θ_i is a function of time. In this case, it is difficult to analyze and calculate an equivalent PM $\Delta\Phi(t)$ expression, because it is difficult to calculate the integral of Expression (3.9). However, the final objective for frequency demodulation is to obtain $f(t)$, and $\Delta\Phi(t)$ is just an intermediate variable, so it is not necessary to know its mathematical expression. The $f(t)$ can be obtained by sending $\Delta\Phi(t)$ to the differentiator. In practical engineering, it is easy to implement the differentiator. So it is of practical purpose for engineering.

3.1.1.2 Frequency Spectrum of Phase Angle Modulation Wave

(1) Frequency spectrum in case of phase angle modulation of single frequency signal

As said above, the uniform expression of $\Delta\Phi(t)$ in phase angle modulation is:

$$\Delta\Phi(t) = KA_m \sin \omega_m t = \beta \sin \omega_m t. \tag{3.13}$$

For the PM wave, $\Delta\Phi(t)$ is a sinusoidal signal; K is the phase-shift constant K_{PM}, which represents a phase change per unit amplitude of the signal; $\beta = A_m \cdot K_{\mathrm{PM}}$ is phase-modulation index, also called angle-modulation index. For the FM wave, $\Delta\Phi(t)$ is a cosine signal; K is frequency-shift constant K_{FM}; $\beta = A_m \cdot K_{\mathrm{PM}} = \frac{\Delta\omega_{\max}}{\omega_m}$ refers to angle-modulation index, where $\Delta\omega_{\max}$ is peak frequency offset caused by the modulated signal. Then, Expression (3.7) can be uniformly used to analyze the frequency spectrum of single-frequency phase angle modulation. Here, the angle-modulation wave expressed by β angle-modulation index can be uniformly expressed as:

$$s(t) = A \cos\left[\omega_{\mathrm{c}} t + \beta \sin \omega_m t\right]. \tag{3.14}$$

Expression (3.14) can be expressed by the Bessel function as the series, namely,

$$s(t) = A \sum_{n=-\infty}^{\infty} J_n(\beta) \cos\left(\omega_{\mathrm{c}} + n\omega_m\right)t. \tag{3.15}$$

Expression (3.15) is the series expression of the angle-modulation wave. It can be seen that the frequency spectrum of the angle-modulated signal contains an infinite number of frequency components. The amplitude of carrier frequency components is in direct proportion to $J_0(\beta)$, and the amplitude of side frequency components symmetric to both sides of ω_{c} is in direct proportion to $J_n(\beta)$.

According to Expression (3.15), the frequency spectrum of angle-modulation wave contains an infinite number of frequency components. However, in practical engineering, only finite bandwidth is used for the angle-modulation wave. In general, there are four methods to take the bandwidth:

1) Carson bandwidth. Here, bandwidth B (considering $\beta = \frac{\Delta\omega_{\max}}{\omega_m} = \frac{\Delta f_{\max}}{f_m}$) is:

$$B = 2\Delta f_{\max}\left(1 + \frac{1}{\beta}\right) = 2(\Delta f_{\max} + f_m) = 2(\beta + 1)f_m, \tag{3.16}$$

where $\Delta f_{\max}$ is maximum frequency offset.

According to Expression (3.16), bandwidth B is not more than $(\beta + 1)$-order sideband. The calculation under Carson bandwidth shows that the amplitude of out-of-band high-order components (i.e., $>(\beta + 1)$ order) is less than 15 % of unmodulated carrier.

2) Bandwidth for narrowband phase angle modulation. Here, $\beta \ll 1$, therefore:

$$B \approx 2f_m. \tag{3.17}$$

It is also called first-order sideband bandwidth. For the modulation of single sinusoidal signals, the sine baseband signals can be recovered only by taking out the first-order sideband, so bandwidth $2f_m$ is often used. This often occurs for the extraction of ranging tones and subcarriers in TT&C.

3) Bandwidth for wide-band phase angle modulation. Here, $\beta \gg 1$, therefore:

$$B \approx 2\Delta f_{\max}. \tag{3.18}$$

4) 1/100 limited bandwidth. Selection condition for order n is:

$$J_n(\beta) \geq 0.01.$$

In this case, out-of-band sideband amplitude is less than 1 % of unmodulated carrier, so most of the energy will fall in the in-band. Here the bandwidth is $2n\omega_m$, which is related to the modulation index β. The calculation results are listed in Table 3.1.

(2) Phase angle modulation of double-frequency and multi-frequency signals

For the double-frequency phase angle modulation,

$$f(t) = f_1(t) + f_2(t) = A_{m_1} \sin \omega_{m_1} t + A_{m_2} \sin \omega_{m_2} t.$$

Table 3.1 Relationship between 1/100 limited bandwidth and β

β	0.1	0.3	0.5	1	2	5	10	20	30
$2n\omega_m$	$2\omega_m$	$4\omega_m$	$4\omega_m$	$6\omega_m$	$8\omega_m$	$16\omega_m$	$28\omega_m$	$50\omega_m$	$70\omega_m$

Then the angle-modulation wave is;

$$s(t) = A\cos\left[\omega_c t + \beta_1 \sin\omega_{m_1} t + \beta_2 \sin\omega_{m_2} t\right].$$

Expression (3.15) can be expressed by the Bessel function as the series, namely,

$$s(t) = R_e[s(t)] = A\sum_{n=-\infty}^{+\infty}\sum_{k=-\infty}^{+\infty} J_n(\beta_1)J_k(\beta_2)\cos\left(\omega_c + n\omega_{m_1} + k\omega_{m_2}\right)t. \quad (3.19)$$

It can be seen from Expression (3.19) that, in addition to $(\omega_c \pm n\omega_{m_1})$ and $(\omega_c \pm k\omega_{m_2})$ high-order components generated by the frequencies, the frequency spectrum of double-frequency phase angle modulation also contains a combined component $(\omega_c \pm n\omega_{m_1} \pm k\omega_{m_2})$ generated by ω_{m_1} and ω_{m_2}, and both of them are symmetrically distributed at both sides of the carrier frequency. Double-frequency modulated spectrum is not only the superposition of each single-frequency modulation, but also produces some new combined components. It shows non-linear characteristics, so the phase angle modulation is a type of nonlinear modulation.

In case of multi-frequency phase angle modulation,

$$f(t) = \sum_{i=1}^{p} A_i \cos\omega_{m_i} t \quad (3.20)$$

then,

$$s(t) = A\cos\left[\omega_c t + \sum_{i=1}^{p}\beta_i \sin\omega_{m_i} t\right].$$

By similar mathematical derivations, one can obtain

$$s(t) = A\sum_{n=-\infty}^{+\infty}\sum_{k=-\infty}^{+\infty}\cdots\sum_{j=-\infty}^{+\infty} J_n(\beta_1)J_k(\beta_2)\cdots J_j(\beta_p)\cos\left(\omega_c + n\omega_{m_1} + k\omega_{m_2} + \cdots + j\omega_{m_p}\right)t. \quad (3.21)$$

3.1.1.3 Nonlinear PM Characteristics

The above discussion is made under the condition that the phase modulator is ideally linear. When the PM characteristic is nonlinear, the phase-shift constant K is not equal to a constant, but a function of the modulating voltage u, namely, $K(u)$. Then the PM index is

$$m = K(u)A_m, \Phi = K(u)u.$$

When $u = A\cos\omega_m t$ and Φ is a periodic function, the expression can be expanded by the Fourier series

$$\begin{aligned}\Phi(t) &= a_0 + a_1 u + a_2 u^2 + \cdots \\ &= \frac{a_2A^2 + 2a_0}{2} + \left(a_1 A + \frac{3a_3A^3}{4}\right)\cos\omega_m t + \left(\frac{a_2A^2}{2}\right)\cos 2\omega_m t + \cdots\end{aligned}$$

then,

$$\begin{aligned}S(t) &= \cos\left[\omega_0 t + \Phi(t)\right] \\ &= \cos\left[\omega_0 t + \Phi_0 + m_1\cos\omega_m t + m_2\cos 2\omega_m t + m_3\cos 3\omega_m t + \cdots\right]\end{aligned} \tag{3.22}$$

where,

$$\Phi_0 = \frac{a_2A^2 + 2a_0}{2}, \quad m_1 = \left(a_1A + \frac{3a_3A^3}{4}\right), \quad m_2 = \frac{a_2A^2}{2}, \quad m_3 = \frac{a_3A^3}{2}.$$

Expression (3.22) shows that due to the impact of nonlinear modulation characteristic, the phase modulation changes from single-frequency signal PM to multi-frequency signal PM. According to Expression (3.22), the PM model with nonlinear PM characteristic can be obtained (see Fig. 3.1).

In Fig. 3.1, the output end of the nonlinear network generates ω_{m_1} and its higher harmonics. Therefore, one multi-frequency signal is added to the input end of the ideal phase modulator. When the input baseband amplifier is of nonlinear characteristics, one model the same as Fig. 3.1 can be obtained. So the analytical methods for them are the same. Since higher harmonics is smaller, it is meaningful to discuss the condition in which the second and third harmonics occur.

(1) In case of $2\omega_m$ distortion component,

$$S(t) = A\sum_{-\infty}^{+\infty}\sum_{n=-\infty}^{+\infty} J_n(m_1)J_k(m_2)\cos(\omega_0 + n\omega_{m_1} + 2k\omega_{m_1})t. \tag{3.23}$$

For $n = +1$, $K = 0$ and $n - -1$, $K = +1$, $(\omega_0 + \omega_{m_1})$ component can be obtained, namely,

$$[AJ_1(m_1) - AJ_1(m_1)J_1(m_2)]\cos(\omega_0 + \omega_{m_1})t. \tag{3.24}$$

Fig. 3.1 Phase modulation model for phase modulator with nonlinear characteristics

For $n = -1, K = 0$ and $n = +1, K = -1$, $(\omega_0 - \omega_{m_1})$ component can be obtained, namely,

$$-[AJ_1(m_1) - AJ_1(m_1)J_1(m_2)]\cos(\omega_0 - \omega_{m_1})t. \tag{3.25}$$

Expression (3.24) and Expression (3.25) show that the asymmetric spectrum occurs, which can be measured by the spectrum analyzer. When the spectrum is asymmetric, it means that the modulation characteristic is nonlinear.

(2) In case of $3\omega_m$ distortion component, a similar analytical method can be used to obtain $(\omega_0 + \omega_{m_1})$ and $(\omega_0 - \omega_{m_1})$ spectrum components, namely,

$$A[J_1(m_1) + J_2(m_1)J_1(m_3) + J_4(m_1)J_1(m_3)]\cos(\omega_0 + \omega_{m_1})t$$
$$-A[J_1(m_1) + J_2(m_1)J_1(m_3) + J_4(m_1)J_1(m_3)]\cos(\omega_0 - \omega_{m_1})t.$$

It is obvious that frequency spectrum is symmetric, but deviates from the theoretical value under linear modulation characteristic.

3.1.2 Demodulation of Analog Signal

TT&C channel is a special transmission channel with the following characteristics:

(1) Signals are flooded by strong noise. In the TT&C system, the signal is very weak due to a great distance between the satellite and the Earth, and *C*/*N* ratio is small to −30 dB, or even lower.
(2) Coherent detection is used. Due to very low *C*/*N* ratio, coherent detection is generally used.
(3) Narrow-band phase-locked loop (PLL) is used to realize tracking & filtering. The reply signal from the spacecraft has Doppler frequency shift and *C*/*N* ratio is very low, so PLL is used for tracking & filtering and to measure velocity and to provide reference signals for coherent detection. Therefore PLL is the core technology of the TT&C system and the short-term stability that will effect velocity-measuring accuracy is a special issue for the TT&C.
(4) It has very high requirements for phase & amplitude stability of the ranging tone. TT&C is a high-accuracy ranging system, so the phase instability of the ranging tone will introduce ranging drift error, amplitude change will introduce ranging error through amplitude/phase conversion, and the distortion of the ranging signal will also cause ranging error.
(5) TT&C signals are usually multi-subcarrier PM signals and the PM and phase demodulation are both nonlinear, so multi-subcarrier intermodulation is a special issue.

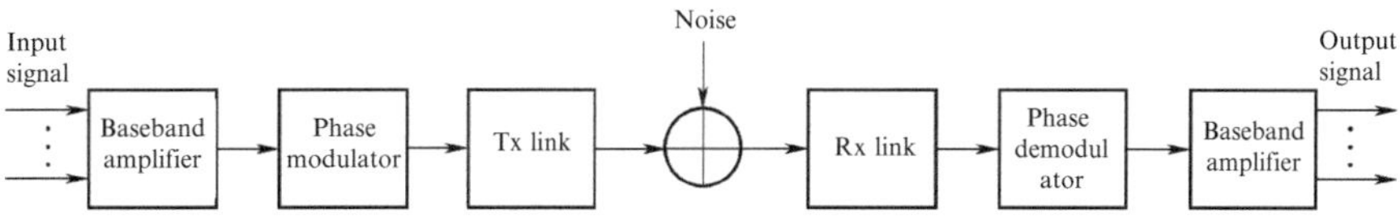

Fig. 3.2 Block diagram of TT&C channel

3.1.2.1 Demodulation of Noise Plus Signal Through an Ideal Multiplier

TT&C channel includes equipment from the baseband amplifier before the phase modulator to the baseband amplifier after the demodulator, as shown in Fig. 3.2.

In Fig. 3.2, input signals are multi-subcarrier signals, and noise is thermal noise produced at the receiver front-end and antenna feeder.

In the TT&C channel, phase angle modulation signal will be interfered with during transmission. Generally, this interference is additive, namely, an additive and ergodic stable white Gaussian noise. Since the bandwidth is always limited in actual transmission channels, the noise with spectrum density being uniform within certain passband and being zero out of the band is called band-limited white noise. If this bandwidth is significantly lower than center frequency ω_0, this noise is called smooth narrowband white Gaussian noise that is a sine wave with a random variation in amplitude and phase. It can be expressed as:

$$n(t) = r_n(t)\sin\left[\omega_0(t) + \Phi_n(t)\right], \tag{3.26}$$

where $r_n(t)$ is the randomly varing envelope; $\Phi_n(t)$ is the randomly varying phase within $\pm\pi$.

Expression (3.26) can also be expanded to be the difference between two components whose phases are in quadrature, namely,

$$n(t) = n_1(t)\cos\omega_0 t - n_\theta \sin\omega_0 t, \tag{3.27}$$

where,

$$\sqrt{n_I^2(t) + n_\theta^2(t)} = r_n(t); \quad \text{arctg}\frac{n_\theta(t)}{n_I(T)} = \Phi_n(t),$$

where $n_I(t)$ and $n_\theta(t)$ are equivalent low-frequency noise components, which are independent mutually and have the same statistic characteristics. $n(t)$, $n_I(t)$ and $n_\theta(t)$ have the same variance, namely,

$$r_n^2 = \sigma_{nI}^2 = \sigma_{n\theta}^2 = N, \tag{3.28}$$

where N is the power of noise.

One-dimensional probability density function of the random variable $r_n(t)$ is of Rayleigh distribution and that of the random variable $\Phi_n(t)$ is of uniform distribution within $\pm\pi$.

The power spectrum density (PSD) of the narrowband Gaussian noise is $\Phi_n(\omega)$. Considering the spectrum folding before and after demodulation, the spectrum density of components $n_1(t)$ and $n_\Theta(t)$ is:

$$\Phi_{nI}(\omega) = \Phi_{n\theta} = \Phi_n(\omega_c + \omega) + \Phi_n(\omega_c - \omega). \tag{3.29}$$

When $\Phi_n(\omega)$ is a constant in $\Delta\omega$ band, namely,

$$\Phi_n(\omega) = \begin{cases} \Phi_n & \left|\omega - \omega_c\right| \leq \dfrac{\Delta\omega}{2} \\ 0 & \text{Other} \quad \omega \end{cases} \tag{3.30}$$

then,

$$\begin{aligned} \Phi_{nI}(\omega) &= \Phi_{n\theta}(\omega) = 2\Phi_n(\omega) = 2\Phi_n & \left|\omega\right| \leq \frac{\Delta\omega}{2}. \\ \Phi_{nI}(\omega) &= \Phi_{n\theta} = 0 & \text{Other } \omega \end{aligned} \tag{3.31}$$

An ideal product demodulator model is shown in Fig. 3.3.

$$U_i(t) = A \cos\left[\omega_c t + \Phi_s(t)\right], \tag{3.32}$$

where $U_i(t)$ is angle-modulation carrier signal; $\Phi_s(t)$ is the phase modulation of baseband signal.

$n_i(t)$ refers to the narrow-band Gaussian noise after bandpass filtering. According to Expression (3.32), it can be expressed as:

$$n_i(t) = n_I(t) \cos\omega_0 t - n_\theta(t) \sin\omega_0 t. \tag{3.33}$$

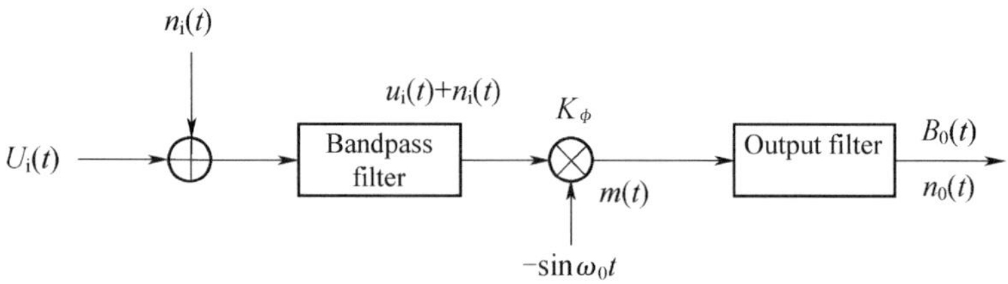

Fig. 3.3 Ideal product demodulator model

The product demodulator performs $(U_i(t)+n_i(t))\times(-\sin\omega_0 t)$. The gain is K_φ and the output is

$$\begin{aligned} m(t) &= -K_\phi[sin\,\omega_0 t][U_i(t)+n_i(t)] \\ &= -K_\phi \sin\omega_0 t\{A\cos[\omega_0 t+\Phi_s(t)]+n_I(t)cos\,\omega_0 t - n_\theta(t)sin\,\omega_0 t\}. \end{aligned} \tag{3.34}$$

Expand Expression (3.34) and take its low-frequency output item:

$$m(t) = K_\Phi \frac{A}{2}\sin\Phi_s(t) + K_\Phi \frac{n_\theta(t)}{2} = B_o(t) + n_o(t), \tag{3.35}$$

where $B_o(t) = K_\Phi \frac{A}{2}\sin\Phi_s(t)$ is the signal component in the output from product demodulator; $n_o(t) = K_\Phi \frac{n_\theta(t)}{2}$ is the noise component.

PSD of $n_o(t)$ is $\Phi_o(\omega)$.

$$\Phi_o(\omega) = \frac{K_\phi^2}{4}\Phi_{nI}(\omega) = \frac{K_\phi^2}{4}\Phi_{n\theta}(\omega) = \frac{K_\phi^2}{2}\Phi_n(\omega), \tag{3.36}$$

where $\Phi_n(\omega)$ is PSD of input noise.

Instantaneous power of $B_o(t)$ is:

$$S_o = [B_o(t)]^2 = K_\phi^2 \frac{A^2}{4}\sin^2\Phi_s(t).$$

The output power in equivalent low-frequency output bandwidth W is:

$$N_o = \Phi_o(\omega)W = \frac{K_\phi^2}{2}\Phi_n W.$$

Output SNR is:

$$\frac{S_o}{N_o} = \frac{A^2}{2}\frac{\sin^2\Phi_s(t)}{\Phi_n W}. \tag{3.37}$$

Because

$C_i/N_i = \frac{A^2/2}{2W\Phi_n}$ (C_i is the carrier power; $2W$ is equivalent intermediate frequency bandwidth)

$C_i/\Phi_n = A^2/2\Phi_n$ (signal noise power spectrum density ratio of input carriers) therefore,

$$\frac{S_o}{N_o} = 2\left(\frac{C_i}{N_i}\right)\sin^2\Phi_s(t). \tag{3.38}$$

Output and input SNR is:

$$G = \frac{(S_o/N_o)}{(C_i/N_i)} = 2\sin^2\Phi_s(t), \tag{3.39}$$

where C_i/N_i is input carrier to noise ratio (CNR).

Under typical single sine wave (single tone) PM, there is:

$$\Phi_s(t) = \beta \sin \omega_m t. \tag{3.40}$$

where β is the phase-modulation index.

For small PM index, $J_1(\beta) \approx \beta/2$, $J_n(\beta) \approx 0 (n > 1)$. According to Expression (3.15), it obtains:

$$U_i(t) = A_0 J_0(\beta) \cos \omega_0 t + A_0 J_1(\beta) \cos(\omega_0 + \omega_m)t \\ - A_0 J_1(\beta) \cos(\omega_0 - \omega_m)t. \tag{3.41}$$

According to Expressions (3.30) and (3.41), the single side frequency signal noise spectrum density ratio of input tone can be written as:

$$\left(\frac{S_{\beta i}}{\Phi}\right)_i = \frac{A_0^2 J_1^2(\beta)}{2\Phi_n}, \tag{3.42}$$

where $S_{\beta i} = \frac{A_0^2}{2} J_1{}^2(\beta)$ is the single side frequency power of tone PM; Φ_n is the input noise spectrum density.

The signals at the output end of product demodulator depend on Expression (3.35). For single tone PM, it can be expressed as:

$$B_o(t) = K_\Phi \frac{A}{2} \sin \Phi_s(t) = K_\Phi A \sum_{n=1}^{\infty} J_{2n-1}(\beta) \sin(2n-1)\omega_m t. \tag{3.43}$$

For small PM index, it obtains:

$$B_o(t) = K_\phi^2 A J_1(\beta) \sin \omega_m(t).$$

The power of output tone is:

$$S_{\beta o} = K_\phi^2 \frac{A^2}{2} J_1^2(\beta). \tag{3.44}$$

According to Expressions (3.44) and (3.36), the output signal noise spectrum density ratio of the tone can be solved, namely,

$$\left(\frac{S_{\beta o}}{\Phi}\right)_o = \frac{S_{\beta o}}{\Phi_0} = A^2 \frac{J_1^2(\beta)}{\Phi_n}. \tag{3.45}$$

According to Expressions (3.45) and (3.42), one can obtain:

$$\frac{\left(S_{\beta \mathrm{o}}/\Phi\right)_{\mathrm{o}}}{\left(S_{\beta \mathrm{i}}/\Phi\right)_{\mathrm{i}}}=2, \tag{3.46}$$

where $S_{\beta \mathrm{o}}$ is power spectrum density of the output signal; $S_{\beta \mathrm{i}}$ is single-side power spectrum density of the input signal. This expression implies the physical meaning that: for the product demodulation, the noise power will double (because the noise is uncorrelated) and signal power increases by four times (because the signal is correlated) with the effect of frequency folding.

According to Expression (3.44), one can obtain:

$$\left(\frac{S_{\beta o}}{\Phi}\right)_{\mathrm{o}}=\frac{A^2}{2}\frac{2J_1^2(\beta)}{\Phi_n}=M\frac{A^2}{2}\frac{1}{\Phi_n}=M\left(\frac{C_i}{\Phi_n}\right)=M\left(\frac{P_i}{\Phi_n}\right), \tag{3.47}$$

where $C_{\mathrm{i}}=\frac{A^2}{2}$ is carrier power; $M=2J_1{}^2(\beta)$ is modulation loss. The physical meaning of modulation loss is to determine how many times (S/φ) value of output modulated signal (tone) is lower than the total S/φ value of input carrier in this modulation mode.

Usually, TT&C channel does not have a limiter. Reasons are summarized as follows.

(1) For low $(C/N)_{\mathrm{i}}$, the change to $(C/N)_{\mathrm{i}}$ will cause the change of demodulation output tone amplitude and then the change of tone amplitude will cause the change of tone phase due to the effect of amplitude/phase conversion. This will lead to the ranging error.
(2) Output SNR would worsen by 1.05 dB due to amplitude limiting.
(3) As a nonlinear device, the limiter requires an amplitude/group delay conversion, thus causing the ranging error.

The above analysis is made for the ideal product demodulator. However, the phase discriminator is often used for practical analog circuit, which is a non-ideal product demodulator. Therefore, the analysis of phase discriminator must be made.

3.1.2.2 Demodulation for Phase Discriminator with Non-ideal Sinusoidal Characteristic

When an ideal multiplier is used, output signal is $B_{\mathrm{o}}(t)=\frac{K_{\Phi}A}{2}\sin\left[\Phi_s(t)\right]$. When the product demodulator is implemented by the phase discriminator, its phase detection characteristic often deviates from the sinusoidal feature of the above mentioned expression (including the impact of nonlinear before the phase discriminator, which will cause its product to multiply with the ideal sinusoidal characteristic

$K(u) \cdot \frac{1}{2} \sin[\Phi_s(t)]$ deviating from sinusoidal characteristic). One distorted sine function can be expanded by Fourier series to be:

$$B_o(t) = \frac{K_\varphi A}{2} \sum_{y=1}^{\infty} A_y \sin[\Phi_s(t)] = \sum_{y=1}^{\infty} B_{0y}(t). \tag{3.48}$$

For single frequency signal PM, $\Phi_s(t) = m \sin \omega_m t$, and given that $K_\varphi A$ is a constant. For the convenience of analysis, given $K_\varphi A = 1$, one can obtain from the above expression:

$$\begin{aligned} B_{01}(t) &= \frac{A_1}{2} \sin[m \sin \omega_m t] = \frac{A_1}{2} \sum_{n=1}^{\infty} J_{2n-1}(m) \sin[(2n-1)\omega_m t] \\ &\vdots \\ B_{0x}(t) &= \frac{A_x}{2} \sin[X_m \sin \omega_m t] = \frac{A_x}{2} \sum_{n=1}^{\infty} J_{2n-1}(X_m) \sin[(2n-1)\omega_m t] \\ B_o(t) &= \sum_{x=1}^{\infty} B_{0y}(t) = \sum_{y=1}^{\infty} \left[\frac{A_y}{2}\right] \sum_{n=1}^{\infty} [J_{2n-1}(y_m) \sin[(2n-1)\omega_m t]]. \end{aligned} \tag{3.49}$$

It can be seen that when the phase discrimination characteristics deviates from sinusoidal characteristic, output signals of the phase discriminator only contains odd harmonics, but their amplitude changes.

3.1.3 Two-Way Carrier Acquisition in C&T – "Frequency Sweep to Acquisition" and "Following Sweep Slope Determination" [1]

Two-way carrier acquisition refers to the transponder on the flight vehicle that acquires uplink carrier signals transmitted by ground station and the ground station acquires downlink carrier signals forwarded by the transponder. "Two-Way" refers to "uplink and downlink". Two-way carrier acquisition is also called "TWCA" for short, which is a very special and important operation in the C&T system. Orbit-measurement and information transmission can be performed only after the "two-way carrier acquisition" is completed. In a unified carrier system, the ground transmitter typically uses frequency sweep to complete two-way carrier frequency acquisition. When the ground transmitter sweeps to the frequency band within acquisition zone of transponder PLL, the transponder will acquire and sweep following the rule of frequency sweep of ground transmitter. When the transponder sweeps to the frequency band within the acquisition zone of main carrier loop of the ground receiver, the main carrier loop of the ground receiver will acquire and sweep following it. Then uplink and downlink carriers are acquired and locked and the following sweep is achieved. This following sweep can be used to determine whether uplink and downlink carriers are acquired and locked (called "Following Sweep Determination"). Then the transmitter stops frequency sweep and returns to zero. After the transmitter returns to its center frequency, the two-way carrier acquisition process is completed. At this same, velocity measurement can be carried out and the measured data are useful.

The following characteristics of TWCA can be used as the basis for TWCA determination [2]:

(1) "Double lock" Determination

The most significant indication of the "two-way carrier acquisition" is completion of transponder PLL lock and ground receiver carrier ring lock. If both are "locked", "two-way carrier acquisition" is achieved. For this method, the advantage is very direct, and disadvantage is the lock indication of transponder needs to go through satellite telemetering → downlink TT&C receiver → TT&C telemetry terminal → monitoring subsystem for transmission which involves many subsystems and leads to longer TWCA time and lower reliability.

(2) "Following Sweep" Determination

For uplink acquisition, the solution that the ground station transmits uplink carrier for frequency sweep is usually used in order to simplify satellite-borne equipment. Since uplink and downlink acquisitions are completed, the VCO in the receiver will sweep frequency following the transmitter. This can be taken as the basis for TWCA determination.

(3) "Loopback" Determination

After the TWCA has completed, uplink and downlink will form a loop. In this case, at the output of the downlink demodulator, there shall be output corresponding to the uplink modulation signals. So the loopback comparison between "Modulation in" and "Modulation out" can be used for TWCA determination. For example, when the high tone of 100 kHz is modulated to the uplink, the downlink demodulator shall provide 100 kHz output (the lock indication of high tone PLL is also available). Its disadvantage is that modulation signals should be added to the uplink and the downlink involves the demodulator and 100 kHz filter. The advantage is a simple determination.

Based on the above analysis, the typical TWCA solution will use:

(1) Auto TWCA: Use "Following Sweep" decision and "Double Lock" dual-decision.
(2) Manual TWCA: In order to ensure reliability of two-way carrier acquisition, a triple decision is used for manual TWCA, namely, "following sweep indicator" decision, "double lock indicator" decision, or "following sweep electric indicator" decision.
(3) TWCA keeping: it uses "Loopback" decision (because uplink modulation signals have been available) or "double lock indication".

Currently, a better solution is the digital, software-based "following sweep slope" decision method. Its principle is described as the following:

Since the TWCA is completed, uplink and downlink frequencies will conform to following relationship:

$$\dot{f}_d = m\dot{f}_u, \tag{3.50}$$

where m is coherent retransmission ratio, a known quantity; $\dot{f}_u$ is rate of uplink frequency sweep, a known quantity. When the digital sweep solution is used, f_u has very high accuracy and stability, and can be viewed as a known constant. *fd* is change rate of downlink frequency.

When the ground receiver uses a second-order loop, VCO will perform the frequency sweep without static error for linear sweep frequency, that is:

$$\dot{f}_{VCO} = \dot{f}_d = m\dot{f}_u. \tag{3.51}$$

Right side of Expression (3.51) is a known constant. So when $\dot{f}_{DCO}$ meets Expression (3.51), it determined the following sweep has been achieved. However, in practice, there is the interference $n(t)$ in the $\dot{f}_{DCO}$, namely,

$$\dot{f}_{DCO} = m\dot{f}_u + n(t). \tag{3.52}$$

The $n(t)$ contains the following components: (1) the interference caused by signal phase noise, expressed as $n_T(t)$; (2) the interference caused by receiver thermal noise, expressed as $n_R(t)$; (3) digital quantized noise, expressed as $n_D(t)$; (4) effect of target Doppler frequency change rate, expressed as $n_d(t)$.

The *S/N* ratio in TWCA determination depends on the first item (signal) and the second item (interference) at the right side of Expression (3.52). *S* is the slope of frequency sweep signal; *N* is the frequency fluctuation noise. The higher *S/N* is, the higher the acquisition probability and the lower false alarm probability. *S/N* is related to frequency sweep slope $\dot{f}_u$ and $\dot{f}_{DCO}$ algorithm. The faster the frequency sweep rate, the higher *S/N*.

(1) Following Sweep Slope Algorithm

The following sweep slope algorithm is shown in Fig. 3.4 for linear sweep signal.

In Fig. 3.4, $f(t)$ is triangular wave frequency sweep signal. The sweep rate of $f(t)$ is $\dot{f}$. Since the sweep oscillator has been digitalized, very high linear accuracy and stability for that can be achieved. $C(t)$ is sampling signal with sampling time $T_1 = T_2 = T_3 \ldots = T$, which is continuous sampling. The measurement method is:

1) Measure the average frequency $\bar{f}_1, \bar{f}_2, \bar{f}_3, \ldots \bar{f}_{n-1}, \bar{f}_n$ in each sampling time T;
2) For linear sweep, the frequency sweep slope is:

$$\dot{f} = a = \frac{\bar{f}_2 - \bar{f}_1}{T} = \frac{\bar{f}_3 - \bar{f}_2}{T} = \cdots = \frac{\bar{f}_n - \bar{f}_{n-1}}{T}. \tag{3.53}$$

$$\left(\bar{f}_n - \bar{f}_{n-1}\right)^2 = a^2T^2 \tag{3.54}$$

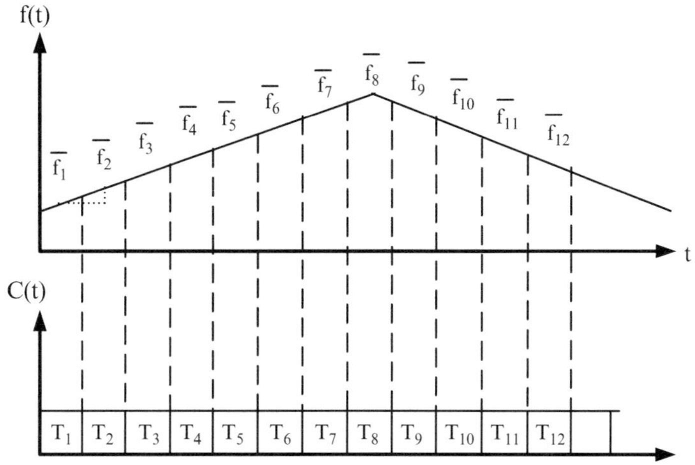

Fig. 3.4 Linear frequency sweep and Allan variance calculation

For linear frequency sweep, $\left\langle \left(\bar{f}_n - \bar{f}_{n-1}\right)^2 \right\rangle = \left(\bar{f}_n - \bar{f}_{n-1}\right)^2 = a^2T^2$, where $\langle \ldots \rangle$ is the average value. Because Allan variance is:

$$\sigma_{ys}^2(T) = \frac{1}{2}\left\langle \left(\frac{\bar{f}_n}{f_0} - \frac{\bar{f}_{n-1}}{f_0}\right)^2 \right\rangle = \frac{1}{2f_0^2}\left\langle \left(\bar{f}_n - \bar{f}_{n-1}\right)^2 \right\rangle. \tag{3.55}$$

Therefore $\left\langle \left(\bar{f}_n - \bar{f}_{n-1}\right)^2 \right\rangle = 2f_0^2\sigma_y^2(T)$. To substitute Expression (3.54), one can obtain:

$$a = \frac{\sqrt{2}}{T} f_0 \sigma_{ys}(T). \tag{3.56}$$

Expression (3.56) gives a meaningful result: the sweep slope of the linear frequency sweep signal equals the product of Allan variance and $\frac{\sqrt{2}}{T} f_0$. Therefore, once the Allan variance $\sigma_{ys}(T)$ is measured, the sweep rate a can be obtained (T and f_0 are known). Advantages of this algorithm are:

1) The effect of Doppler frequency shift is eliminated. When the target moves at a constant speed, given Doppler frequency shift f_d, according to Expression (3.55),

$$\sigma_{ys}^2(T) = \frac{1}{2f_0^2}\left\langle \left[\bar{f}_n(-f_d) - \left(\bar{f}_{n-1} - f_d\right)\right]^2 \right\rangle = \frac{1}{2f_0^2}\left\langle \left(\bar{f}_n - \bar{f}_{n-1}\right)^2 \right\rangle.$$

   It can be seen that the value of Allan variance is unchanged. For slow frequency components, most effects will be offset due to subtraction of adjacent sampling values.

2) Among all frequency variances, Allan variance has the shortest measurement time. So the above calculation can be finished within a short time, thus significantly shortening the two-way carrier acquisition time.

3) The effect of signal phase noise $n_T(t)$. This method associates the short-term stability performance of the signal with (S/N) of the frequency sweep determination to facilitate the engineering design and measurement. The relation derivation process is as follows: prior to initiation of the "two-way carrier acquisition" process, the carrier PLL of the ground receiver locks the beacon signal transmitted by the spacecraft. At this time, its short-term stability is $\sigma_{ynT}(T)$, which is one of the main reasons for false alarm and is used to develop the determination threshold. After the two-way carrier acquisition process has been initiated, the ground station transmits the frequency sweep signals to make transponder and the ground receiver following sweep. Given the following sweep signal $\sigma_{ys}^2(T)$ then

$$\frac{S}{N}=\frac{\sigma^2{}_{ys}(T)}{\sigma^2{}_{ynT}(T)}=\frac{a^2T^2}{2f_0^2\sigma_{ynT}^2(T)}. \tag{3.57}$$

4) With respect to receiver noise $n_R(t)$, the frequency fluctuation noise realizes $H(\omega)=[\sin^2(\omega T/2)]/(\omega T/2)$ filtering and thus increases S/N of following sweep determination signals. The characteristics of $H(\omega)$ are shown in Fig. 3.5.

The phase noise spectrum density caused by the receiver noise (given additive white Gaussian noise) is $S_\Phi(\omega)=N_0/P_s$, where P_s/N_0 is the signal noise power spectrum density ratio of the received signal. The Allan variance of the frequency fluctuation noise is:

$$\begin{aligned}\sigma_{ynR}^2(T)&=\frac{1}{2\pi\omega_0^2}\int_{-\omega_L}^{+\omega_L}S_\varphi(\omega)\omega^2|H(\omega)|^2\mathrm{d}\omega=\frac{N_0}{\pi\omega_0^2p_s}\int_0^{\omega_L}\frac{\sin^4(\omega T/2)}{(\omega T/2)}\mathrm{d}\omega\\&=\frac{2N_0}{\pi\omega_0^2T^3p_s}\left[\frac{3\omega_LT}{4}-\frac{3}{4}\sin\omega_LT-\sin^3\frac{\omega_LT}{2}\cos\frac{\omega_LT}{2}\right].\end{aligned} \tag{3.58}$$

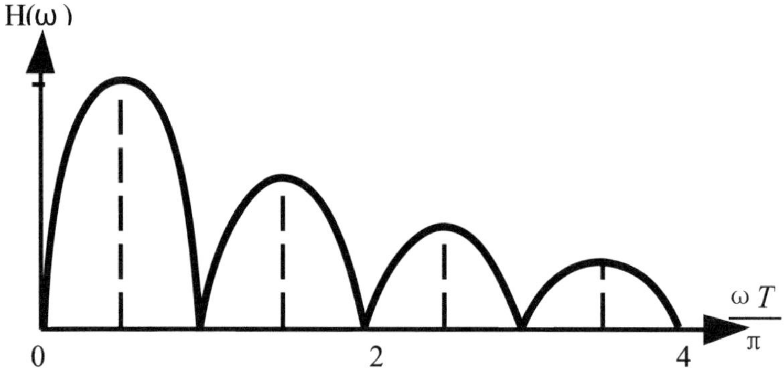

Fig. 3.5 Frequency fluctuation characteristics of Allan variance in frequency-domain

where ω_L is PLL bandwidth. Then, *S*/*N* of the following sweep determination signal is:

$$\frac{S}{N}=\frac{\sigma^2{}_{ys}(T)}{\sigma^2{}_{ynR}(T)}=\frac{a^2T^2}{2f_0^2\sigma_{ynR}^2(T)}$$
$$=a^2\pi^3T^5P_S/N_0\left[\frac{3\omega_LT}{4}-\frac{3}{4}\sin\omega_LT-\sin^3\frac{\omega_LT}{2}\cos\frac{\omega_LT}{2}\right]. \tag{3.59}$$

Total noise power is $(\sigma_{ynT}^2+\sigma_{ynR}^2)$. However, in actual operation, $\sigma_{ynR}\gg\sigma_{ynT}$, and digital quantized noise $n_D(t)$ can be ignored. Either computer calculation or graphical solution is suitable for Expression (3.59). In practical engineering, the said algorithm can be simplified to further simplify the solution and reduce calculation time. Since there are positive and negative sweep rate, $\langle|\bar{f}_n-\bar{f}_{n-1}|\rangle$ is taken in calculation. Considering $n(t)$, double thresholds H_0' and H_0'' must be used to reduce false alarm. As shown in Fig. 3.6, with the following sweep determined according to $H_0''>\langle|a|\rangle>H_0'$, the narrower "double-threshold" spacing is, the lower error acquisition probability is and the lower the corresponding acquisition probability. In order to ensure reliable two-way carrier acquisition, error acquisition probability shall be reduced as much as possible. The effect of $n(t)$ should be taken into account for double-threshold spacing. For the targets with higher acceleration, n_d effect will be mainly considered, namely,

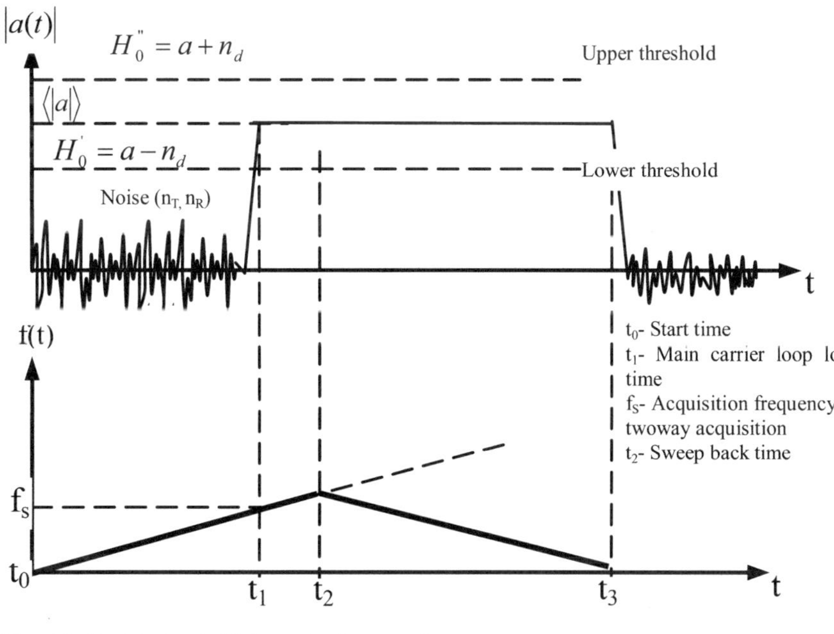

Fig. 3.6 Double-threshold determination waveform

$H''_0 \approx a + n_d$, $H'_0 \approx a - n_d$, where $a = m\dot{f}_u$ is known. For the targets with small acceleration, the effect of $n_R(t)$ will be mainly considered. Rectangular pulse in Fig. 3.6 is the "following sweep" determination signal. After returned to zero, another circuit will perform "two-way carrier acquisition holding". The above calculation and determination can be implemented using the software without hardware device, so it can contribute to economic benefits.

(2) Two-way carrier acquisition procedure

Without regard to time delay of space transmission, two-way carrier acquisition procedure is as shown in Fig. 3.7.

The meanings of symbols in Fig. 3.7 are as follows:

$t_{\text{angle guide}}$ – Point to target time after angle guide;

$t_{\text{beacon acquisition}}$ – the time when the ground carrier loop acquires transponder beacon after the ground antenna points to the target. FFT guide and PLL acquisition time is T_F;

$t_{\text{angle acquisition}}$ – the time of target acquisition & tracking by the angle tracking system. Angle acquisition time is T_A;

t_0 – start time, which corresponds to the center frequency f_0 of transmitting frequency of ground station;

t_1 – time acquired by carrier loop of ground station. T_f is acquisition time to frequency sweep.

$$T_f = t_1 - t_0 = \frac{(f_s - f_0)}{\dot{f}} + T_F, \tag{3.60}$$

where f_s – the center frequency of receiving frequency to transponder PLL, including frequency instability and inaccuracy of transponder VCXO as well as target Doppler frequency f_d. $\dot{f}$ is the sweep rate;

t_2 – finish time for following sweep determination. At this time, the frequency sweep is stopped. Following sweep determination time T_D is $T_D = t_2 - t_1$.

t_3 – start time for "zero return" sweep. Where, $T_S = t_3 - t_2$ is "pause" time for frequency sweep, which is used to prevent each PLL losing lock due to the step of frequency change rate when zero return sweep direction changes from positive to negative.

t_4 – "zero return" finish time. $T_W = t_4 - t_3$ is zero return time. In order to guarantee normal tracking of the PLL, original sweep rate (the direction may be opposite) shall be kept during T_W to return the zero position at a uniform speed. The frequency hop can't occur at the zero return point, otherwise it will result in instantaneous lock lost. It can be seen from Fig. 3.7 that $T_W = (t_2 - t_1) + (t_1 - t_0) = T_D + T_f$ and two-way carrier acquisition time T_{DD} is $(t_4 - t_0)$. $T_{DD} = 2T_f + 2T_D + T_S + T_{\text{delay}} + T_{\text{transmission}}$, where T_{delay} is the circuit delay time, $T_{\text{transmission}}$ is the wave propagation time.

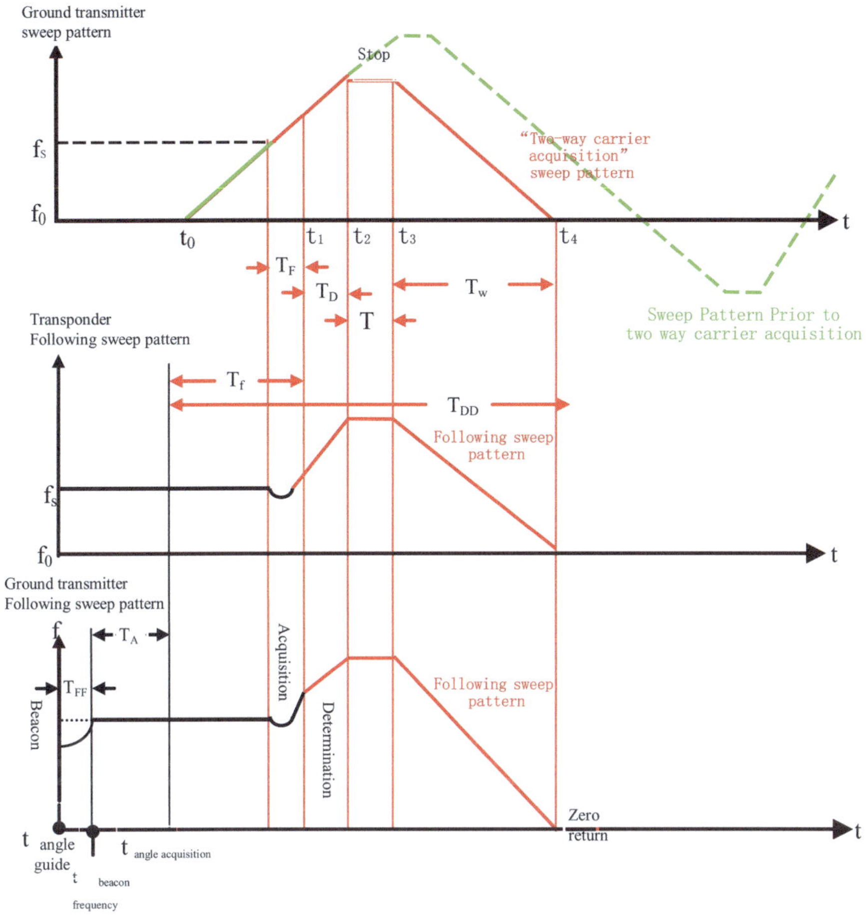

Fig. 3.7 "Two-way carrier acquisition" procedure chart

For above "two-way carrier acquisition" sweep pattern, the following measures can be used in order to facilitate two-way carrier acquisition.

1) "Direction selection to start frequency sweep": Before starting the sweep, select the sweep direction toward f_s value. If it moves in an opposite direction, additional $T/2$ sweep is required.
2) "Shortest path uniform-speed zero return": In order to shorten time, it should drive the zero return direction toward f_0 value direction. This is the shortest path for zero return at original speed.

DDS device can be used to provide frequency sweep, which is programmable and can be implemented by software:

① It should select the sweep direction upon preset Doppler frequency prior to starting frequency sweeping; ② For "zero return", the zero return direction will be selected according to whether the difference between the frequency at stop-sweep time and center frequency f_0. is positive or negative; ③ Pause 100 ms for "Stop-Sweep" and then perorm "zero return" frequency sweep; ④ Because the DDS is of very high short-term stability and long-term stability and can be used as a part of the frequency source, "zero return" can be implemented by directly making DDS frequency to the center frequency at a uniform speed. Moreover, the frequency sweep with DDS has very high accuracy in sweep linearity and sweep slope and can accurately control sweep start point, stop point, sweep rate, and scope in a programmable way. This can facilitate "calculation of following sweep slope" and fast two-way carrier acquisition. It is obvious that the frequency sweep with DDS and the digital carrier loop is the base for realizing this solution.

3.1.4 Combined Interference in the Unified Carrier C&T System – "Modulation/Demodulation Integration Characteristic Analysis Method" [3]

There are mainly two types of combined interference. One is multi-subcarrier combined interference caused by nonlinearity of multiple subcarriers phase modulation/phase demodulation, under which multiple subcarriers are a unified carrier with constant envelope and the nonlinearity of the carrier channel has small impacts. The other is multi-carrier combined interference caused by the nonlinearity of the carrier channel when multi-carrier signals operate simultaneously, which is likely to affect system performance. The former is a special topic in the unified carrier TT&C system; the latter is a general topic universally existing in the channels.

(1) Multi-subcarrier combined interference

The core of the unified carrier system is to use the subcarrier frequency division system to provide transmission of various types of information. Like other frequency division system schemes, one special challenge of this system is combined interference of multi-channel subcarrier signals. This is substantially a challenge involving the integration of functions by the USB system.

So far, most papers and literatures gave an analysis of combined interference for PM carrier $S(t)$ frequency spectrum according to the expansion of the Bessel function, as shown in Fig. 3.8.

Some studies think that the combined components in the modulated carrier $S(t)$ transmitted by the phase modulator would cause interference and propose a boundary condition for the signal design that combined components in $S(t)$ should be low enough to meet specific requirements. Even some take the view that the interference among these subcarriers is inevitable and the combined interference may be reduced only by controlling the number and modulation degree of the subcarriers.

In fact, from the perspective of system architecture, $S(t)$ is only an intermediate signal in the transmission link and the system finally requires demodulated output $B(t)$. That is to say, in the signal transmission process, the carriers are only used as carrying subcarriers to transmit signals to the receiving end. The requirements for

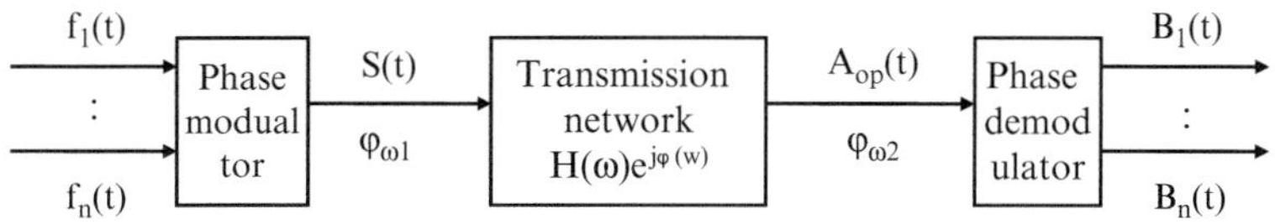

Fig. 3.8 Phase modulation/demodulation transmission model of USB channels

distortion and interference during transmission are finally embodied on $B(t)$, instead of carrier spectrum. Therefore, we should study the modulation/demodulation characteristics of the entire channel and the effect of transmission network $\dot{H}(\omega)$ on output signal $B(t)$. The analysis results show that combined components in video spectrum at the output end of the demodulator are much fewer than those in carrier spectrum. So the limit on USB multi-channel capacity and maximum modulation degree finally depends on non-overlapping of multi-channel modulated subcarrier spectrum in the baseband under the condition of appropriate design.

1) Conditions for nondistortion and no combined interference

In Fig. 3.8, PM characteristic can be expressed as $\varphi_{\omega 1} = K_1 f(t)$; phase demodulation characteristic can be expressed as:

$$\begin{aligned} B(t) &= K_2 \varphi_{\omega 2}, \quad \text{if } H(\omega) = 1, \\ B(t) &= K_1 K_2 f(t), \end{aligned} \tag{3.61}$$

where K_1 K_2 is modulation/demodulation integration characteristic. When K_1 K_2 is a constant, the distortion won't occur to $B(t)$ relative to $f(t)$. The transmission waveform relationship is shown in Fig. 3.9.

It can be seen from Fig. 3.9 that when $f(t)$ is a sine wave, so is $B(t)$. It is obvious that if $f(t)$ contains multiple sine waves, $B(t)$ will also contain corresponding sine waves. No combined interference will occur between $f(t)$ and $B(t)$.

In practical circuits, it is very hard to meet the condition $\dot{H}(\omega) = 1$. In general the condition to be met is $H(\omega) \approx$ a constant, $\phi(\omega) \approx K_3 \omega = \tau \omega$ (phase-frequency characteristic linearity, τ is group delay). Then input signal $S(t)$ and output signal $A_{op}(t)$ of the transmission network as shown in Fig. 3.8 are:

$$\begin{aligned} S(t) &= A \cos \left[\omega_0 t + \sum_{i=1}^{p} m_i \sin \omega_i t \right] \\ &= A \sum_{n_1=-\infty}^{+\infty} \cdots \sum_{np=-\infty}^{+\infty} \prod_{i=1}^{p} [J_{ni}(m_i)] \cos \left[\omega_0 t + \sum_{i=1}^{p} (\omega_i t) \right]. \end{aligned} \tag{3.62}$$

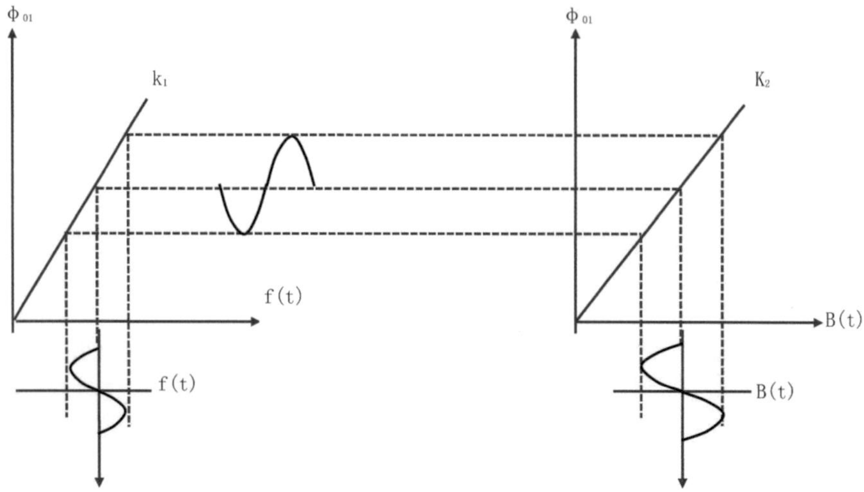

Fig. 3.9 Modulation → demodulation diagram

After network transmission under the condition of $H(\omega) = H$ constant and $\phi(\omega) = \tau\omega$

$$\begin{aligned} A_{\mathrm{op}}(t) &= AH \sum_{n_1=-\infty}^{+\infty} \cdots \sum_{n_1=-\infty}^{+\infty} \prod_{i=1}^{p} [J_{ni}(mi)] \cos\left[\omega_0(t+\tau) + \sum_{i=1}^{p} \omega_i(t+\tau)\right] \\ &= AH \cos\left[\omega_0(t+\tau) + \sum_{i=1}^{p} m_i \sin\omega_i(t+\tau)\right]. \end{aligned} \tag{3.63}$$

It can be seen that the subcarrier signal in $A_{\mathrm{op}}(t)$ does not produce distortion, but produces a time delay. The subcarrier signal output from $A_{\mathrm{op}}(t)$ which has performed linear phase demodulation will not produce distortion or create combined interference, but only produces a time delay τ.

Therefore, in the transmission channel, the conditions for the subcarrier signal not producing distortion and combined interference are:

- Modulation/demodulation characteristic is linear;
- Phase-frequency characteristic of the transmission network is linear;
- Amplitude-frequency characteristic of the transmission network is flat.

2) Effect of modulation-demodulation nonlinearity

Nonlinear characteristic of the phase modulator can be expressed as power series:

$$\varphi = \varphi_0 + b_1 x + b_2 x^2 + b_3 x^3 + \cdots$$

Nonlinear characteristic of the phase demodulator can be expressed as:

$$y = C_o + C_1(\varphi - \varphi_0) + C_2(\varphi - \varphi_0)^2 + C_3(\varphi - \varphi_0)^3 + \cdots \tag{3.64}$$

Combined the above two expressions, one can obtain modulation-demodulation integration characteristic:

$$y = C_o + b_1 C_1 x + \left(b_2 C_1 + b_1{}^2 C_2\right) x^2 + \left(b_3 C_1 + 2b_1 b_2 C_2 + b_1{}^3 C_3\right) x^3 + \cdots$$

Assume

$$C_o = a_o, \quad b_1 c_1 = a_1, \quad \left(b_2 C_1 + b_1{}^2 C_2\right) = a_2, \quad \left(b_3 C_1 + 2b_1 b_2 C_2 + b_1{}^3 C_3\right) = a_3,$$

Then

$$y = a_o + a_1 x + a_2 x^2 + a_3 x^3 + \cdots,$$

where a_1, a_2 and a_3 contain nonlinear coefficient of modulation characteristic and demodulation characteristic. Anyone of these nonlinearities will lead to integration characteristic nonlinearity, but they may provide mutual compensation. When $b_1 C_1 =$ a constant, $(b_2\ C_1 + b_1{}^2\ C_2) = 0$, $(b_3 C_1 + 2b_1\ b_2\ C_2 + b_1{}^3\ C_3) = 0$, the modulation-demodulation integration characteristic doesn't have nonlinear distortion. If nonlinearity exists, $K_1\ K_2 \neq$ a constant, but is a function of voltage U.

If the quadratic term of nonlinear distortion is only taken into account, $B(t)$ can be expressed as:

$$B(t) = K_1 K_2 f(t) = a_o + a_1 f(t) + a_2 f^2(t). \tag{3.65}$$

It is obvious that if a sine wave $f(t) = A\cos\Omega t$ is input, $B(t)$ will not be a sine wave, but will produce a second harmonic item. It is well known that the second harmonic distortion coefficient is:

$$k_2 = \left(\frac{a_2}{2a_1}\right) A,$$

where k_2 is the ratio of second harmonic item to fundamental harmonic item. Due to the nonlinear coefficient of the characteristic curve,

$$\eta = \frac{a_2 u^2}{a_1 u} = \frac{a_2}{a_1 A}.$$

When we assume that nonlinearity is half-and-half of ± deviations, $\eta = \pm (a_2/2a_1A)$, then $k_2 = \eta$.

It is obvious that if K_1K_2 is nonlinear, multi-subcarrier input will produce combined interference. The quadratic term of nonlinear distortion is [4]:

$$a_2u^2 = \sum_{i=1}^{N}\sum_{j=1}^{N} a_2\left(\frac{A_ia_i}{2}\right)\cos\left[(\omega_i+\omega_j)t+\theta_i+\theta_j\right]$$
$$+\sum_{i=1}^{N}\sum_{j=1}^{N} a_2\left(\frac{A_ia_j}{2}\right)\cos\left[(\omega_i+\omega_j)t+\theta_i-\theta_j\right]. \tag{3.66}$$

where $(\omega_i+\omega_j)$ and $(\omega_i-\omega_j)$ combined components are called second-order combined components, the amplitude of that is $a_2(A_iA_j/2)$, and (a_2/a_1) $(A_iA_j/2)$ is second-order inter-modulation carrier inter-modulation ratio.

Similarly, n-order term of nonlinear characteristic will produce n-order combined interference. It need not be repeated here. However, with the order increase, combined frequency component will decrease. So it is not the main part of the study. The abovementioned analysis method and conclusion are not only applicable to nonlinear integration modulation-demodulation characteristic, but also to nonlinear modulation characteristic and nonlinear demodulation characteristic.

3) Effect of product demodulator and amplitude/phase frequency characteristic

A product demodulator is often used in the TT&C. With the sinusoidal characteristic, the product demodulator is a nonlinear phase demodulator. Its phase modulation-demodulation model is as shown in Fig. 3.8.

In case of a subcarrier:

$$f(t) = A_{m_1}\sin\omega_{m_1}t$$
$$S(t) = A\cos\left[\omega_0 t + m_1\sin\omega_{m_1}t\right].$$

If $\dot{H}(\omega) = 1$ it is the transmission without distortion. Here,

$$A_{\text{op}}(t) = S(t) = A\cos\left[\omega_0 t + m_1\sin\omega_{m_1}t\right].$$

The product demodulation will implement $-\sin\omega_0 t \times A_{\text{op}}(t)$. The multiplying result is:

$$-A\sin\omega_0 t\cos\left[\omega_0 t + m_1\sin\omega_{m_1}t\right].$$

The difference frequency term is low-frequency output signal $B(t)$:

$$B(t) = \frac{A}{2}\sin\left(m_1\sin\omega_{m_1}t\right) = A\sum_{n=1}^{\infty}J_{2n-1}(m_1)\sin(2n-1)\omega_{m_1}t.$$

The above expression shows the nonlinear effect of sinusoidal characteristic on $m_1 \sin \omega_{m_1} t$. In the output $B(t)$, there is only odd-order term, but no even-order term. This spectrum is different from carrier $S(t)$ spectrum. $S(t)$ contains all the terms of $n = 1 \sim \infty$. Therefore, it can't use the carrier frequency spectrum to assess the combined interference.

In case of two subcarriers,

$$f(t) = A_{m_1} \sin \omega_{m_1} t + A_{m_2} \sin \omega_{m_2} t \tag{3.67}$$

$$S(t) = A \sum_{n=-\infty}^{+\infty} \sum_{k=-\infty}^{+\infty} J_n(m_1) J_k(m_2) \cdot \cos(\omega_0 + n\omega_{m_1} + k\omega_{m_2})t. \tag{3.68}$$

According to the property of Bessel function:

$$J_{-n}(m) = (-1)^n J_n(m), \quad J_{-k}(m) = (-1)^k J_k(m).$$

It can be obtained from Expression (3.68) that $S(t)$ spectrum has the following characteristics: ① even-order amplitudes symmetric to ω_0 are in the same direction; ② the sidebands generated by the combination of even order and even order and symmetric to ω_0 are in the same direction; ③ the sidebands generated by the combination of odd order and odd order and symmetric to ω_0 are in the same direction; ④ odd order symmetric to ω_0 is in reverse direction; ⑤ the sidebands generated by the combination of even order and odd order and symmetric to ω_0 are in the reverse direction. The physical meaning of product demodulation is to mix two quadrature phase coherent carrier signals and make two symmetrical sidebands shift 90° in the reverse direction respectively and then superpose, so that reverse-phase sidebands will add by means of in-phase and in-phase sidebands will subtract by means of reverse phase. Therefore, in theory, ①, ② and ③ will be neutralized to zero and will not cause interference, and ④ and ⑤ will be added to provide output (including fundamental wave and other interference caused by sinusoidal demodulation characteristics). However, it is not always applicable in practical situations, so the additional effect of transmission network and product demodulator must be taken into account. Because the second-order interference is of maximum component, we can discuss it firstly and by analogy for the rest. There are two types of second-order interference:

① Second-order high-order interference, namely, $n = \pm 2$ or $k = \pm 2$. In this case, it is symmetrical same-direction sideband and amplitude is $J_2(m) = J_{-2}(m)$.
② Second-order intermodulation interference, namely, $n = \pm 1$ or $k = \pm 1$ combined component. In this case, it is symmetrical same-direction sideband and amplitude is $J_{-1}(m_2) J_1(m_1) = J_{-1}(m_1) J_1(m_2)$. For the said interference, Ω_n is used uniformly to represent the interference frequency and An is used to represent spectral line amplitude, so PM wave can be expressed as:

$$\begin{aligned} S(t) = & J_0(m_1) J_0(m_2) \cos \omega_0 t + J_1(m_1) \cos(\omega_0 + \Omega_1)t \\ & - J_1(m_1) \cos(\omega_0 - \Omega_1)t - A_n \cos(\omega_0 + \Omega_h)t - A_n \cos(\omega_0 - \Omega_n). \end{aligned} \tag{3.69}$$

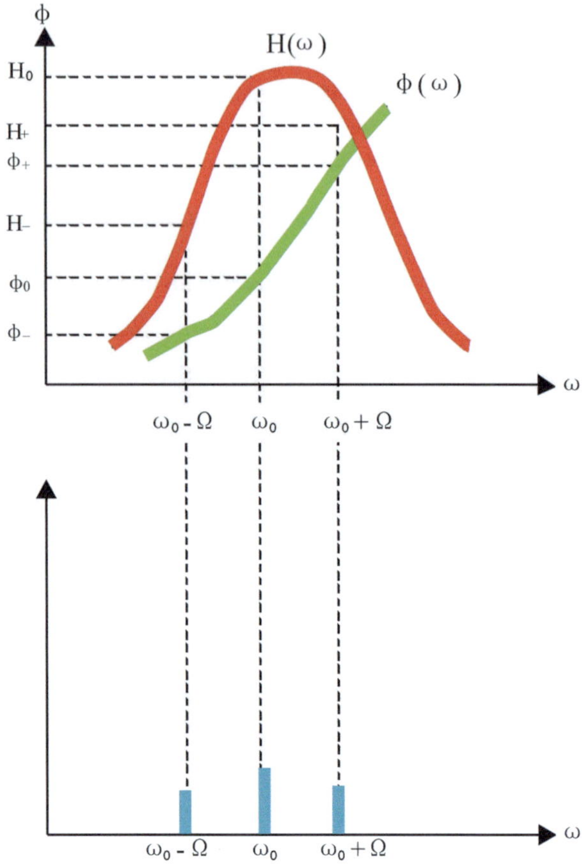

Fig. 3.10 Amplitude-phase frequency characteristic and PM wave spectrum

Through the amplitude-phase frequency response as shown in Fig. 3.10, the output signals are:

$$\begin{aligned} A_{op}(t) = & H_0 J_0(m_2) J_0(m_1)\cos(\omega_0 t + \phi_0) + H_{+1} J_1(m_1)\cos\left[(\omega_0 + \Omega_1)t + \phi_{+1}\right] \\ & - H_{-1} J_1(m_1)\cos\left[(\omega_0 - \Omega_1)t + \phi_{-1}\right] + H_{+n} A_n \cos\left[(\omega_0 + \Omega_n)t + \phi_{+n}\right] \\ & - H_{-n} A_n \cos\left[(\omega_0 - \Omega_n)t + \phi_{-n}\right]. \end{aligned} \tag{3.70}$$

In demodulation, reference signal $C(t)$ is reference signal coherent to and in quadrature with the residual carrier $H_0 J_0(m_1) J_0(m_2)\cos(\omega_0 t + \phi_0)$ and locked by PLL. If the non-orthogonal phase difference $\Delta\Phi$ occurs, then:

$$C(t) = \cos\left[\omega_0 t + \Phi_0 + \frac{\pi}{2} + \Delta\Phi\right], \tag{3.71}$$

where $\pi/2$ is orthogonal requirement; $\Delta\Phi$ is non-orthogonal phase difference. Product demodulation will perform the product operation of $A_{\text{op}}(t) \times C(t)$

trigonometric function. If only Ω_1 and Ω_n components are considered, the $B_0(t)$ can be expressed as:

$$\begin{aligned} B_o(t) = & \left(\frac{1}{2}\right) J_1(m) H_{+1} \sin\left[\Omega_1 t + (\Phi_{+1} - \Phi_0) - \Delta\Phi\right] \\ & + \left(\frac{1}{2}\right) J_1(m) H_{-1} \sin\left[\Omega_1 t + (\Phi_0 - \Phi_{-1}) + \Delta\Phi\right] \\ & + \left(\frac{1}{2}\right) A_n H_{+n} \sin\left[\Omega_n t + (\Phi_{+n} - \Phi_0) - \Delta\Phi\right] \\ & - \left(\frac{1}{2}\right) A_n H_{-n} \sin\left[\Omega_n t + (\Phi_0 - \Phi_{-n}) + \Delta\Phi\right]. \end{aligned} \tag{3.72}$$

Expression (3.72) shows that the demodulation produces two components to Ω_1 frequency, i.e., the one has $[(\Phi_{+1} - \Phi_0) - \Delta\Phi]$ phase shift, and the other has $[(\Phi_0 - \Phi_{-1}) + \Delta\Phi]$ phase shift. Similarly, for the two components of Ω_n frequency, they have $[(\Phi_{+n} - \Phi_0) - \Delta\Phi]$ and $[(\Phi_0 - \Phi_{-n}) + \Delta\Phi]$ phase shift, respectively. The output Ω_1 and Ω_n components are the vector sum of their two components, respectively, as shown in Fig. 3.11 [5, 6]. One can obtain:

$$B_n = \left(\frac{A_n}{2}\right)\left\{H_{+n}^2 + H_{-n}^2 - 2H_{+n}H_{-n}\cos\left[(\Phi_0 - \Phi_{-n}) - (\Phi_{+n} - \Phi_0) + 2\Delta\Phi\right]\right\}^{\frac{1}{2}} \tag{3.73}$$

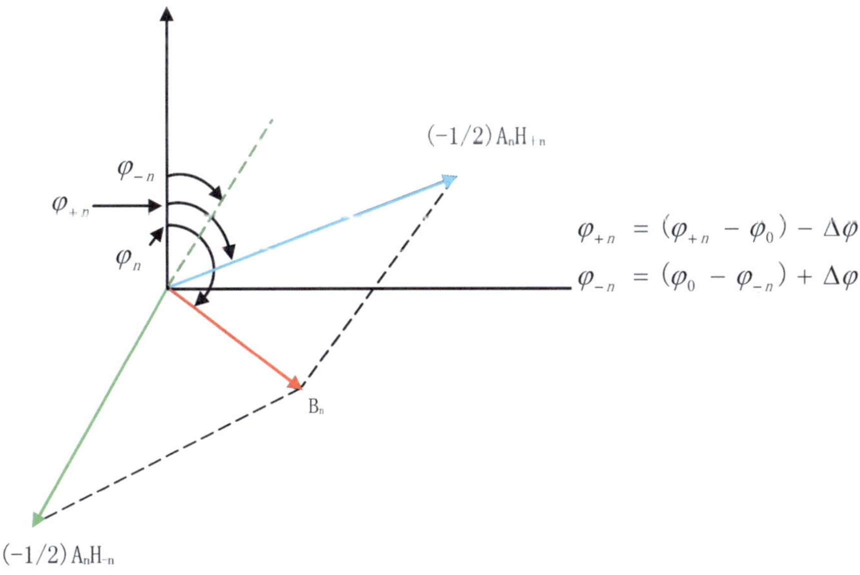

Fig. 3.11 Product demodulation vector diagram

$$\Phi_n = tg^{-1}\frac{H_{-n}\sin[(\Phi_0-\Phi_{-n})+\Delta\Phi+\pi]+H_{+n}\sin[(\Phi_{+n}-\Phi_0)-\Delta\Phi]}{H_{-n}\sin[(\Phi_0-\Phi_{-n})+\Delta\Phi+\pi]+H_{+n}\cos[(\Phi_{+n}-\Phi_0)-\Delta\Phi]}. \tag{3.74}$$

It can be seen from Expression (3.73) that only when amplitude-frequency characteristic is of even symmetry (i.e., $H+n=H-n$), phase-frequency characteristic is of odd symmetry (i.e., $(\Phi_{+n}-\Phi_0)=(\Phi_o-\Phi-n)$), and product demodulation is in quadrature (i.e., $\Delta\Phi=0$), the second-order interference is zero, and otherwise there will be some output. So the above mentioned performance shall be controlled to reduce combined interference. This "symmetric" analysis method is a special case under the above conditions for "nondistortion and no combined interference". The reason is that: for USB, the subcarriers are discrete and independent, instead of distributed in the whole baseband.

(2) Effect of nonlinear channel in case of single angle-modulation carrier

A general expression of angle-modulation wave (including FM and PM) can be written as:

$$f_i(t) = A_c\cos[\omega_0 t+\Phi(t)],$$

where $\Phi(t)$ can be phase angle modulation of any mode (e.g., BPSK, QPSK, OQPSK, MSK, and FM), and A_C is of constant envelope.

In order to analyze nonlinear effect, the expression of $f_0(t)$ in time-domain can be obtained by substituting $f_i(t)$ into a nonlinear expression expressed in power series:

$$\begin{aligned} f_0(t) &= a_1 f_i(t)+a_2 f_i^2(t)+a_3 f_i^3(t) \\ &= a_1A_C\cos[\omega_0 t+\Phi(t)]+a_2A_C^2\cos^2[\omega_0 t+\Phi(t)]+a_3A_C^3\cos^3[\omega_0 t+\Phi(t)]. \end{aligned} \tag{3.75}$$

With the formulas of trigonometric function: $\cos^3[\omega_0 t+\Gamma(t)] = \frac{1}{2}\{1+\cos 2[\omega_0 t+\Phi(t)]\}$ and $\cos^3[\omega_0 t+\Phi(t)] = \frac{1}{2}\{\cos 3[\omega_0 t+\Phi(t)]+3\cos[\omega_0 t+\Phi(t)]\}$, the Expression (3.75) can be expanded as:

$$\begin{aligned} f_0(t) &= \frac{1}{2}a_2A_c^2+\left(a_1A_C+\frac{4}{3}a_3A_C^3\right)\cos[\omega_0 t+\Phi(t)] \\ &\quad+\frac{1}{2}a_3A_C^2\cos[2\omega_0 t+2\Phi(t)]+\frac{1}{4}a_3A_C^3\cos[3\omega_0 t+3\Phi(t)]. \end{aligned} \tag{3.76}$$

It can be seen that $f_0(t)$ contains three angle-modulation waves with the center of ω_0, $2\omega_0$, and $3\omega_0$ respectively. ω_0 components can be filtered by a filter. Then:

$$f_0(t) = \left(a_1A_C+\frac{4}{3}a_3A_C^3\right)\cos[\omega_0 t+\Phi(t)]. \tag{3.77}$$

According to analysis results of Expression (3.77), $\Phi(t)$ has no distortion in time domain, and only carrier amplitude of ω_0 is compressed (a_3 is a minus sign) but is still constant-envelope. This is one important characteristic of phase angle modulation which can operate in nonlinear amplitude state. The physical meaning of that is very clear, that is, the information is carried on the phase angle instead of amplitude. However, because the information is carried on the phase angle, phase nonlinearity has a very great impact. If the angle-modulation carrier is filtered out the portion of signal spectrums by the band-limiting filter, then it becomes the signals which are modulated in angle and amplitude. In this case, the AM/PM conversion and AM/$\tau(\omega)$ conversion will cause the modulated signals to distort.

The above analysis shows that the unified carrier system can operate in nonlinear transmission state, because the carrier is constant-envelope. This is the significant advantage for it.

(3) Multi-carrier combined interference

The USB system often operates simultaneously with other carrier signals. In this case, the nonlinear amplitude will produce the new frequency components and the interference caused by amplitude/phase conversion. When the multiple signals are transmitted, the high-order components and combined frequency components will be generated. These new frequency components will cause frequency interference. Combined interference is the problem which involves overall system. The solution for that includes two aspects: one is to design signal rationally to reduce harmful interference and the other is to improve the performance of equipment linearity and the anti-interference ability against combined interference.

If the combined frequency component is to become an effective interference, two conditions must be met:

1) The combined frequency component falls within the useful signal bandwidth.
2) The amplitude of combined interference component is high enough to make signal/interference lower than the threshold value.

Therefore, combined interference can be reduced by frequency selection and amplitude suppression. This is a routine problem in channel design and there have been many simulation methods.

3.2 Digital Signal Transmission Technology in C&T

3.2.1 Overview

Telemetry, command, and remote sensing constitute the main contents for information transmission by flight vehicles, and the telemetry and command itself are important parts of TT&C. Although the telemetry, command and remote sensing operations transmit digital signals, they have different transmission features. The remote sensing operation mainly transmits high speed digital transmission

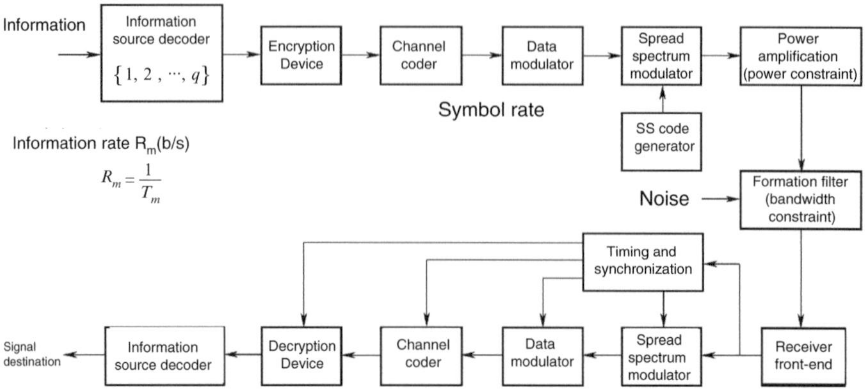

Fig. 3.12 Block diagram of typical digital information transmission system

signals, most of which have limited power and require high power efficiency. The information source coding currently is a research focus. The command transmission is low-speed data transmission, so it needs to improve the safety, reliability, confidentiality, error control, encryption, and anti-interception. The telemetry operation transmits the intermediate-speed digital transmission signals and needs to overcome the challenge of multi-channel communication. When the transmission channels for telemetry are too much to limit bandwidth, some technical challenges such as bandwidth efficiency and constant envelop modulation must be dealt with.

Typical digital information transmission system is as shown in Fig. 3.12.

Different from ground communication, space information transmission has many unique features:

(1) High-dynamic. The target to be connected is a high-velocity flight vehicle. The maximum velocity of satellites is about 7 km/s and the deep space explorer has a higher velocity. So received signals have higher Doppler frequency and Doppler frequency change rate.
(2) Strict restriction of size, weight, and power consumption and poor operating environment to the devices on the flight vehicle. The payload (or known as terminal station in information transmission system) is required to be lower in volume, weight, and power.
(3) Long distance and large time delay. The distance to the geosynchronous satellite is about 4×10^4 km, about 4×10^5 km to the moon, and farther to deep space explorer. This will lead to higher path loss and higher performance requirement of EIRP and G/T.
(4) Higher data transmission rate. More information acquiring by payload on the flight vehicle are higher data transmission rate requirements.
(5) Power constraint system. Due to long distance, high data transmission rate, and restriction of EIRP performance on the flight vehicle, the power of signals received by ground station is lower. This is a bottleneck for space information transmission.

(6) The frequency band is generally unrestricted. Generally, the flight vehicle is designed for special purposes. Its payload mostly operates in single carrier. Moreover, it has higher operating frequency and wider available band. Therefore, it generally is not band limited, nor shaping filtered. The carrier is constant envelope, so the power amplifier can operate in a saturated mode, thus reducing volume, weight, and power consumption.
(7) Close relationship between information transmission and TT&C. For this "point-to-point" information transmission, since one transmission terminal is on the flight vehicle, it is required to guarantee normal performance of it such as EIRP and G/T through TT&C. Moreover, in order to acquire and track the targets, the orbit data of vehicle obtained by TT&C should be used frequently.

The above mentioned special features lead many TT&C and information transmission technologies to the frontier of the information transmission era, for example, high-speed data transmission technology up to 1 Gb/s, cryogenic low-noise receiving technology approximating absolute zero, error correction coding/decoding technology close to the Shannon limit, super higher power transmitter technology, giant antenna technology, millimeter wave & laser communication technology, and high-accuracy ranging, velocity-measurement, and location technology.

3.2.2 Optimum Transmission Response of Digital Signal Transmission [7]

The optimum transmission response is discussed here under linear system conditions. In actual channel, the system will be linear in the case of smaller signals and gradually become nonlinear with increasing of signal input. For the nonlinear system, we typically discuss it based on the existing linear system and then analyze the additional deterioration caused by nonlinearity.

It is generally known that in terms of frequency domain, when the amplitude frequency feature $H(\omega)$ is constant (namely, $H(\omega) = \text{constant}$) and phase frequency feature is linear (namely, $\varphi(\omega) = \tau_0\omega$, τ_0 is constant) for the frequency band in question, the waveform distortion won't occur for the signals going through this frequency band. Otherwise, the waveform distortion will occur and thus cause intersymbol interference and SNR descending. Therefore, we analyze the impact of linear distortion upon bit error rate that is equivalent to the impact of the system amplitude/phase frequency features on bit error rate. Phase-frequency feature $\varphi(\omega)$ can also be expressed by group time delay, namely, $t(\omega) = \mathrm{d}\varphi(\omega)/\mathrm{d}\omega$. Since group time delay intuitively expresses time delay of group signal (refer to the signals containing a group of spectrum), and digital signals contain a lot of spectrum components, therefore, in digital communication, group time delay is frequently used to describe phase-frequency feature. If the time delay is the same (namely, $t(\omega) = t_0$, t_0 being constant) and the amplitude remains unchanged for the spectrum component in digital signal passing the system, the output waveform generated from the sum of spectrum components output by the system is not distorted and only time delay t_o is generated.

In ideal linear digital transmission system, the bit error rate is dependent upon E_b/N_0 for a given modulation/demodulation mode. The ideal system in question refers to match filtering and linear system without inter-symbol interference. However, the actual system may not achieve "ideal" status due to various deterioration factors influencing error code. Therefore, it requires higher E_b/N_0 to achieve the same bit error rate, and the increased value is considered as deterioration amount of $E_b/N_{0.}$ Since E_b/N_0 increment equals to (C/N) increment, so it is usually described by SNR deterioration. We define $L_e = [(C/N)_{\text{actual}} - (C/N)_{\text{theory}}]$ as equivalent C/N deterioration (dB in unit). In actual system, the impact of various factors on bit error rate can be evaluated by equivalent C/N deterioration, and total impacts of various factors is the sum of dB of each $L_{e.}$

Modulation/demodulation mode shall be determined in combination with the following factors: (1) bit rate, allowable channel bandwidth, and corresponding spectrum availability; (2) adaptability to transmission distortion (such as nonlinearity); (3) complexity, reliability, and cost of equipment; (4) anti-interference and anti-multipath capability.

3.2.2.1 Optimum Receiving of Binary Digital Signal

Optimum receiving refers to the best method to receive and decide "0" and "1" against the background of AWGN, for which Nudist's first criteria is usually used. Such criteria is implemented as follows: the receiver demodulate "0" and "1" signals which are subject to match filtering. The sample is performed at "0" and "1" transition point and then the sample results will be compared with a certain threshold value. It is "1" if higher than such threshold, or it is "0". Therefore, the filtering feature (namely, amplitude/phase transmission frequency response) and threshold value must be carefully selected. The standard evaluating selection is called "criteria". The optimum criterion is established based on the aspect of receiving quality evaluation. Different aspect of quality evaluation will result in different optimum criteria, such as "maximum output SNR criteria", "minimum RMS error criteria", "maximum posteriori probability", and "minimum error probability criteria", etc.

As described in reference [8]: to achieve "maximum SNR criterion", it is required to design a filter with transmission feature $H(f)$ meeting the following conjugacy equality relation, namely,

$$H(f) = \left[S(f)e^{j\omega t_0}\right]^* = S^*(f)e^{j\omega T_S}. \tag{3.78}$$

Expression (3.78) shows that, for the filter with the maximum output SNR at sampling point, its transfer function is conjugated with signal spectrum. Physical meanings of conjugacy: output noise power spectrum of the filter is coordinated with power spectrum form of input signal. From the perspective of time domain, impulse response of the filter is consistent with input signal waveform, so it is called matching filter. Its impulse response is:

$$h(t) = F^{-1}[H(f)] = S(T_S - t). \tag{3.79}$$

Expression (3.79) shows that impulse response compatible with match filtering design requirements shall be equal to waveform function of transmission receiving signal $S(t)$ that is folded and delayed T_s.

These references also point out: "optimum receiver", "minimum error probability receiver", "maximum SNR receiver", "matching filter" of binary signals contains the same meaning. The system design should meet the following conditions:

(1) The amplitude-frequency transmission features of the system are conjugated with signal spectrum.
(2) Phase-frequency feature of the system is linear.
(3) The threshold decision value is half of code pulse value, and the sampling point is the momentum at which code cell is ended.

3.2.2.2 Optimum Receiving of "Non-bandlimited" Linear System

Definition of "non-bandlimited": the bandwidth of channel is unlimited (or little limitation) in relation to signal spectrum. For the perspective of reception, it means that all spectrum of the signal won't be limited by bandwidth of channel (band limitation) and the signal can be passed through in whole or in substantial parts. Its equivalent transmission mode is as shown in Fig. 3.13.

In Fig. 3.13, $H_T(f)$ represents transmitting filter, $H_R(f)$ represents receiving filter; modulation/demodulation is "non-bandlimited" and linear. $S_i(t)$ and $S_o(t)$ respectively are input and output baseband signals; matching filters achieves the optimum receiving. $S_i(t)$ is binary digital signal of "0" and "1", and its code element waveform is shown in Figs. 3.3, 3.4, 3.5, 3.8, 3.9, 3.10, 3.11, 3.12, 3.13, and 3.14:

The pulse code is ideal rectangular signal. When appearance probability of "0" and "1" equals to 1/2, Fourier transformation of $S_i(t)$ is as follows:

$$S(f) = AT_S \frac{\sin(\pi f T_S)}{\pi f T_S} e^{-j\pi f T_S}. \tag{3.80}$$

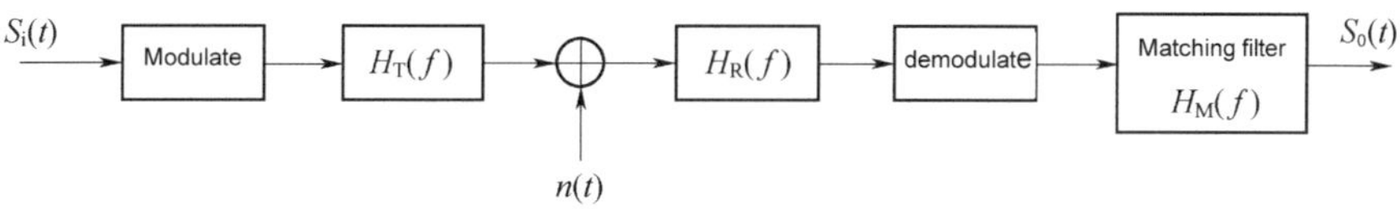

Fig. 3.13 Transmission model of "non-bandlimited" linear system

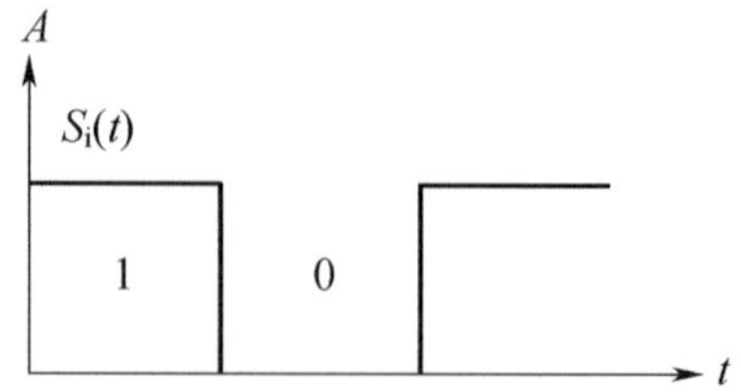

Fig. 3.14 Rectangular pulse code signal

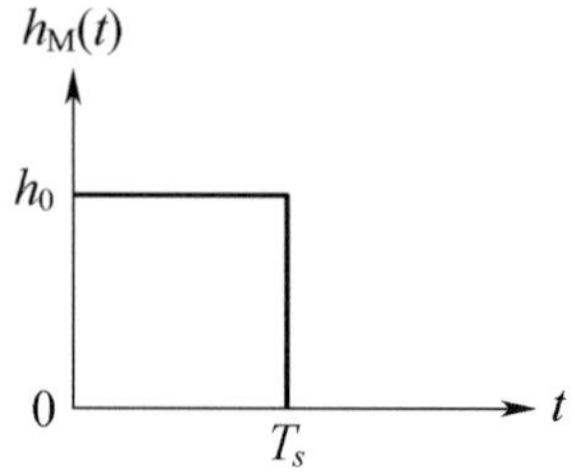

Fig. 3.15 Impulse response "integral-reset" filter

According to the foresaid optimum receiving criteria, the transfer function of its matching filter is determined upon expression (3.78), i.e.,

$$S^*(f) = AT_S \frac{\sin(\pi f T_S)}{\pi f T_S} e^{j\pi f T_S}$$

$$\begin{aligned} H(f) &= S^*(f)e^{-j2\pi f T_S} = \mathrm{A}T_S \frac{\sin(\pi f T_S)}{\pi f T_S} e^{j\pi f T_S} e^{-j2\pi f T_S} \\ &= 2AT_S \frac{\sin(\pi f T_S)}{\pi f T_S} e^{-j\pi f T_S}. \end{aligned} \tag{3.81}$$

An impulse response of "integral-reset" filter is as shown in Fig. 3.15.

Its frequency response $H_M(f)$ is Fourier transformation of $h_m(f)$, namely,

$$H_M(f) = h_0 T_S \frac{\sin(\pi f T_S)}{\pi f T_S} e^{-j\pi f T_S}. \tag{3.82}$$

It is shown that:

$$H(f) = \frac{2A}{h_0} H_M(f) \tag{3.83}$$

The two filters are the same in frequency response and the difference between them are only a constant. Therefore, "integral-reset" filter is matching filter of non-bandlimited binary non-reset rectangular pulse code signal. Due to the non-bandlimited feature, there is no inter-symbol interference under Nudist's first criteria.

The principle diagram of "integral-reset" filter is shown in Fig. 3.16.

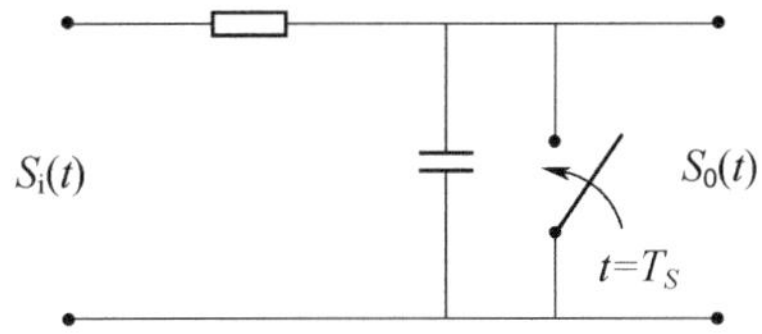

Fig. 3.16 Principle diagram of "integral-reset" filter applicable to rectangular pulse code matching filter

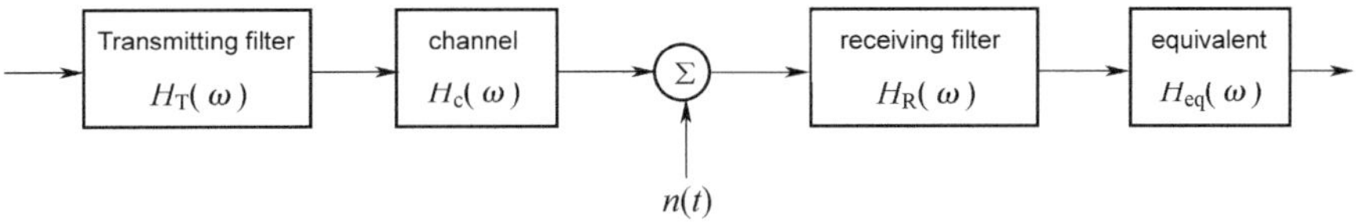

Fig. 3.17 Baseband transmission system model

In Fig. 3.16, the switch operates for reset and release. It closes at T_s and disconnects quickly.

In digital circuit, integral can be achieved through "accumulation" and release is achieved at T_S by code clock reset.

It is noted that the foresaid matching filter is only applicable to non-bandlimited and non-reset rectangular pulse code sequence. For other band-limitation conditions, "integral-reset" filter cannot provide optimum receiving of match filtering.

3.2.2.3 Optimum Transmission Response of "Bandlimited" Linear System

In Sect. 3.2.2.2, it is assumed that the channel bandwidth is unlimited. In fact, band-pass filter typically is set in the channel and used to limit signal bandwidth. The band-pass filter can provide the following functions:

(1) Increase number of channel for multi-path communication by improving spectrum utilization within the frequency band. For example, the formation filter with the raised cosine roll-off feature is used to reduce signal spectrum while the Nyquist's first criteria without inter-symbol interference will be met.
(2) Limit out-of-band power and harmonic at transmitting end to prevent harmonic from interfering.
(3) Filter noise on the repeater to increase SNR.

In Categories I and II of space communication, the items (1) and (2) above are not complied with seriously as in other communication modes.

Baseband transmission mode of digital transmission system is shown in Fig. 3.17.

Optimum transmission of "bandlimited" linear channel must meet the following three requirements:

(1) Transmitting spectrum must be limited within certain bandwidth range. To improve spectrum utilization and reduce interference between adjacent

channels, the framework for transmission spectrum of carrier keying signal can be set to limit the power spectrum within this framework. Definition of transmission spectrum framework actually brings about requirements for bandwidth and out-of-band attenuation of $H_T(\omega)$, which can be implemented through "formation network".

(2) Receiving/transmitting synthesis frequency response shall meet the conjugate match filtering conditions as discussed above, namely,

$$H_T(\omega) = BH_R(\omega)\mathrm{e}^{-j\omega t_0}. \tag{3.84}$$

(3) Receiving/transmitting synthesis response must satisfy Nyquist condition. According to baseband transmission principle, the receiving/transmitting synthesis response must meet the following condition in order to remove inter-symbol interference at sampling point.

$$H_T(\omega)H_R(\omega) = CN(\omega)\mathrm{e}^{-j\omega t_0}, \tag{3.85}$$

where C is constant, t_0 is delay time; $N(\omega)$ is the response meeting Nyquist's first criteria.

According to paragraphs (2) and (3) above, $H_T(\omega)$ and $H_R(\omega)$ must meet square root Nyquist response, namely,

$$\begin{aligned} H_T(\omega) &= C_1\sqrt{N(\omega)\mathrm{e}^{-j\omega t_1}} \\ H_R(\omega) &= C_2\sqrt{N(\omega)\mathrm{e}^{-j\omega t_2}}, \end{aligned} \tag{3.86}$$

where C_1 and C_2 are constant; t_1 and t_2 are delay time.

According to without inter-symbol interference criteria for sampling value decision, i.e., Nyquist's first criteria, the baseband transmission feature usually employs:

Amplitude-frequency feature:

$$N(\omega) = \begin{cases} T_S & 0 \le |\omega| \le \dfrac{(1-\alpha)\pi}{T_S} \\ \dfrac{T_S}{2}\left[1 + \sin\dfrac{T_S}{2\alpha}\left(\dfrac{\pi}{T_S} - \omega\right)\right] & \dfrac{(1-\alpha)\pi}{T_S} \le |\omega| \le \dfrac{(1+\alpha)\pi}{T_S} \\ 0 & |\omega| \ge \dfrac{(1+\alpha)\pi}{T_S} \end{cases}. \tag{3.87}$$

where $N(\omega) = H_T(\omega) \cdot H_C(\omega) \cdot H_{\mathrm{eq}}(\omega) \cdot H_R(\omega)$ is total amplitude-frequency feature; T_s is width of code element; α is roll-off efficient.

Expression (3.87) is considered as raised cosine roll-off filter, and the waveform output by such filter is called raised cosine roll-off wave waveform, with amplitude-frequency response as shown in Fig. 3.18.

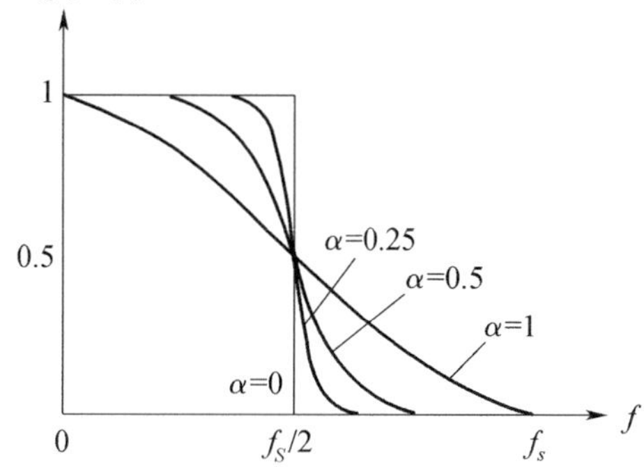

Fig. 3.18 Nyquist response of raised cosine roll-off

In Fig. 3.18, α is roll-off efficient and can be taken between 0 and 1. The lower α is, the narrower the bandwidth of $N(f)$. $f_N = f_s/2$ is called Nyquist bandwidth. Whatever α is, the amplitude-frequency feature at f_N is 6 dB smaller than that at f_0. When $\alpha = 0, f_s b/s$ symbol rate can be transmitted within $f_s/2$ Nyquist bandwidth. Therefore, the first condition meeting the optimum transmission response actually is equivalent to select appropriate roll-off coefficient to make the transmitting spectrum meet the framework requirement. The transmitting frequency spectrum with certain response requirement to formation can use both frequency domain and time domain methods.

According to expression (3.86), there is square root relation between $N(\omega)$, $H_r(\omega)$ and $H_t(\omega)$, so H_r and $H_t(\omega)$ are also called root cosine roll-off filter.

3.2.3 Digital Modulation/Demodulation Technology

3.2.3.1 Requirements of Vehicle C&T and Information Transmission System for Modulator/Demodulator

The basic function of modulator is to shift the frequency spectrum, i.e., shift the baseband signal to the channel of corresponding frequency band for transmission. Such function can be achieved by two steps: firstly, modulate the baseband signal containing information onto appropriate carrier; secondly, shift that to RF band applicable to channel transmission through up-conversion. These two steps can be performed at the same time. The information transmission has the basic requirements to modulation/demodulation for frequency band utilization and power utilization. The power efficiency and bandwidth efficiency for modulator design are two very important performance specifications used to compare different modulation modes.

(1) High power efficiency (also called power availability): refers to SNR (E_b/N_0) required by transmitting per bit information under certain bit error rate. It is also

characterized by bit error rate function of E_b/N_0 as independent variable, namely, when E_b/N_0 is given the smaller bit error rate, the higher the power efficiency of the modulation mode becomes. The smaller the deterioration of power amplification nonlinearity to bit error rate, the higher the power efficiency. For information transmission, the flight vehicle is required to transmit power as high as possible. However, because the space on the flight vehicle is very limited, the transmitter on the vehicle is required to minimize the volume, weight, and power consumption and provide high efficiency. Thus, such system on the vehicle typically is power limitation system, whose power amplifier shall operate at full saturation and signals approximate constant envelope status in order to achieve high conversion efficiency. For the transmitter operating on nonlinearity status, if the signal output by the modulator has fluctuated envelope, then the bit error rate may increase due to (AM-PM) and (AM-AM), and the frequency spectrum may be spread and thus violate the requirements of telecommunication authorities (such as ITU) for signal spectrum. Therefore, it is essential to perform constant envelope modulation.

(2) High bandwidth efficiency, also known as bandwidth utilization: refers to the ratio of data rate to bandwidth ((b/s)/Hz), namely, it means the bit number which is allowed to be transmitted within 1 Hz bandwidth under given bit error rate. The information transmission system that has limited available frequency band is a limited bandwidth system. To provide more channels (such as telemetry) or higher bit rate (such as remote sensing) within limited frequency band, the frequency band utilization shall be improved to increase bandwidth efficiency, namely, the signals output by modulator are required for high information rate ((b/s)/Hz) as possible within unit frequency band. This means that the main lobe of power spectrum of modulated wave shall occupy signal energy as much as possible, and the lobe shall be narrow and have rapid roll-off characteristics, as shown in Fig. 3.19. Additionally, out-of-band attenuation shall be strong and sidelobe shall be smaller, which will cause smaller interference to other paths, as shown in Fig. 3.20.

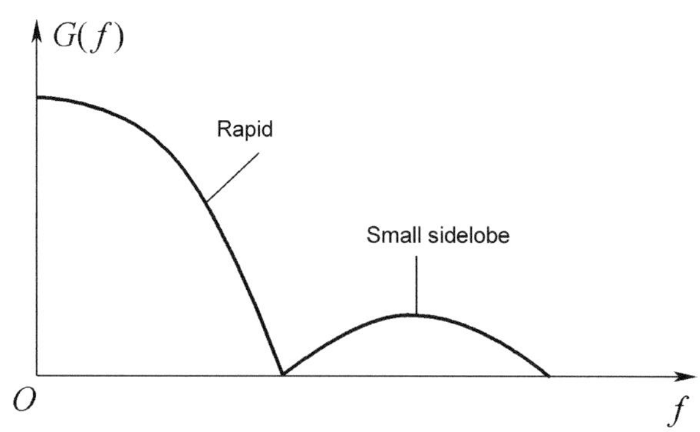

Fig. 3.19 Power spectrum of modulation signals

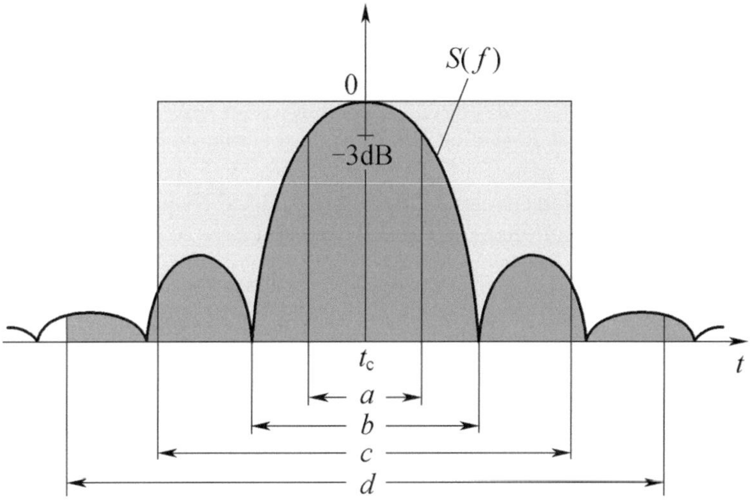

Fig. 3.20 Typical spectrum bandwidth measurement methods. (*a*) half-power bandwidth; (*b*) null-to-null value bandwidth; (*c*) equivalent noise bandwidth; (*d*) partial power capacity bandwidth

For bandwidth definition, several methods are provided currently.

1) Half-power bandwidth: signal power spectrum is located within ±3 dB bandwidth with the center of peak.
2) Null-to-null bandwidth: the width of main frequency spectrum lobe
3) Bandwidth of equivalent noise $B_{eq,}$ it is expressed as:

$$B_{eq} = \frac{\int_{-\infty}^{\infty} |S(f)|^2 df}{\max_f |S(f)|^2}. \tag{3.88}$$

where $S(f)$ is power spectrum density (PSD).

4) Partial power capacity bandwidth: the bandwidth containing designated percentage of signal power.

In addition to the foresaid two main performance specification requirements, the modulator and demodulator are required to be simple and practicable, small influence of undesired factors, low demodulation threshold, small volume, and low price, etc.

For modulator design, there are modulators with high power efficiency (such as constant envelope modulator) and modulators with high bandwidth efficiency (such as Nyquist bandlimited modulator). However, power efficient usually conflicts with bandwidth efficiency, so sometimes we can only make a compromise between them. In this case, a quasi-constant envelope modulator is produced. Based on this concept, we will discuss the digital modulation/demodulation technologies that are frequently used in flight vehicle C&T and information transmission.

3.2.3.2 Modulation Mode with the Highest Bandwidth Efficiency-Nyquist Bandlimited Modulation

Block diagram of Nyquist bandlimited modulation is as shown in Fig. 3.21.

In Fig. 3.21, Nyquist formation filtering plays a role of bandwidth limitation to improve bandwidth efficiency. Such modulator has the highest bandwidth efficiency and modulation envelope fluctuation [9]. Ideal Nyquist bandlimited filter is a rectangular baseband filter and can be expressed as:

$$\dot{S}(\omega) = \begin{cases} 0 & |\omega| > \pi T \\ S_0 T = \text{Constant} & |\omega| \leq \pi/T \end{cases}, \tag{3.89}$$

where T is width of code element; and S_0 is gain constant.

This is the transfer function of ideal low-pass filter with phase feature constantly being 0, and its impulse response shall be sinx/x waveform, as shown in Fig. 3.22. Such waveform will not be 0 for $t=0$ and be 0 at other sampling times ($t=KT$, $K \neq 0$), which means that there is no inter-symbol interference when such waveform is used as receiving waveform. It is shown from Fig. 3.22 and expression (3.89) that when the width of distortionless transmission code element is a sequence of T, the required minimum bandwidth shall be $1/(2T)$ (angle frequency is π/T). Generally, $1/(2T)$ is named as Nyquist bandwidth, and T as Nyquist time interval. If the transmission rate of sample is $1/T$, the required frequency width is $1/(2T)$. The frequency band utilization will be $2(b/s)$/Hz, when impulse response of ideal low-pass filter is used as receiving waveform and the sample sequence is binary signal. If the sample sequence is n-element signal, the frequency band utilization

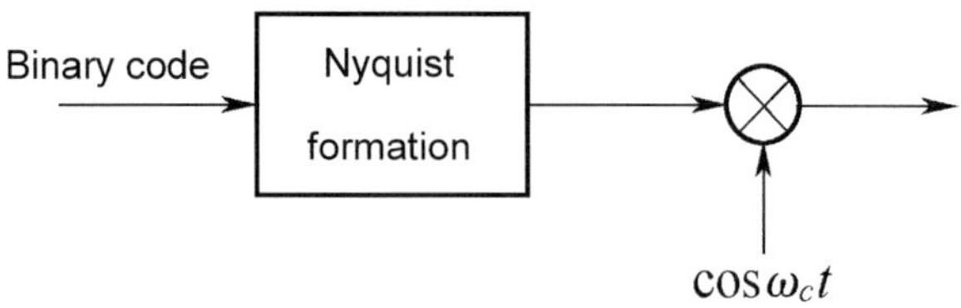

Fig. 3.21 Nyquist bandlimited modulation

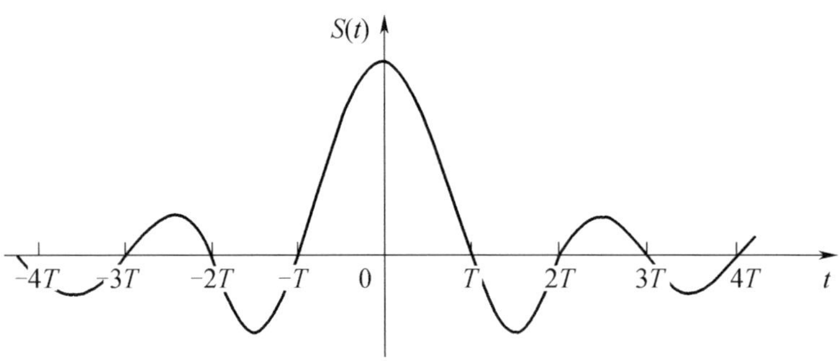

Fig. 3.22 Ideal low-pass impulse response

Fig. 3.23 Raised cosine roll-off signals with different roll-off coefficient

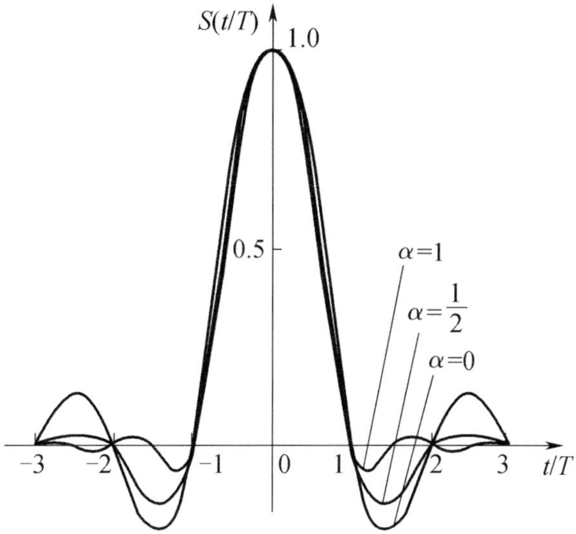

shall be $(2\log_2 n)(b/s)$Hz. It is the highest frequency band utilization achieved under distortionless sample.

The distortionless transfer conditions of sample value expressed by expression (3.89) and Fig. 3.22 only has physical meaning and cannot actually be achieved in, because it requires the transfer function to have unlimited cliffy transition zone. In practice, the waveform without crosstalk with the center of π/T and transition zone of odd symmetry raised cosine shape is widely applied and generally called raised cosine roll-off signals (note: here "roll-off" refers to spectrum transition features but not waveform shape). The waveform is as shown in Fig. 3.23 in case of different roll-off coefficients [9]; where, $\alpha = 0$ is corresponding to Nyquist band-limitation with the narrowest bandwidth.

Figure 3.23 shows normalized waveform. It can be seen from Fig. 3.23 that the cross interference of raised cosine roll-off signal is always 0 before and after sample point. Therefore, the distortionless transmission conditions of sample are satisfied. The lower the roll-off coefficient is, the higher the waveform fluctuates, but the transmission band will reduce; conversely, the higher α is, the lower the waveform fluctuates, but frequency band increases. There are two extreme cases: when $\alpha = 0$, namely, in case of Nyquist band-limitation with the narrowest band, there is no crosstalk waveform; when $\alpha = 1$, the occupied frequency band is the widest, being two times of such band width when $\alpha = 0$, therefore; the band utilization decreases to 50 %, i.e., $1(b/s)$/Hz (in case of binary sample sequence).

Considering that receiving waveform will be sampled again in regeneration decision to obtain distortionless sample, but it is impossible to implement ideal instantaneous sampling, namely, it is impossible to eliminate all error at the sampling time and sampling pulse width will not be 0. Therefore, in order to reduce influence caused by sampling timing pulse error, the roll-off coefficient α cannot be too small and usually ≥ 0.2.

3.2.3.3 Modulation Mode with High Power Efficiency

Constant envelop modulation is often used to obtain high power efficiency. The constant envelope modulation methods include: frequency shift keying (including BFSK, MFSK), phase shift keying PSK (including BPSK, DPSK, QPSK, OQPSK, and DQPSK), continuous phase modulation (including MSK, GMSK, TFM, etc.). Their spectrum bandwidth varies from each other. In case the band is not limited, the envelopes of foregoing various angle modulation signals are constant, and their envelop fluctuation are different after being bandlimited. In the section below, the common constant envelope modulation mode frequently used in C&T communication will be introduced.

(1) BPSK (binary phase shift keying)

BPSK modulation: code "0" and "1" are represented respectively with initial phase of carrier 0° and 180°, and it is modulation of suppression carrier.

BPSK demodulator mainly includes: squared loop and Costas loop demodulator.

(2) DPSK (binary differential phase shift keying)

DPSK modulation: code "0" and "1" are expressed by relative change of carrier phase. In common, the phase is unchanged when code is "0", and the phase has a 180° change when code is "1". It is also modulation of suppression carrier.

Since DPSK is to make comparison before and after the code, there is no phase fuzzy. However, when noise before and after the code overlaps, increment in noise will increase bit error rate. Therefore, bit error rate of DPSK will be 1 times higher than BPSK.

(3) QPSK (quadrature phase shift keying)

M-ary phase shift keying means that carrier phase has M kinds of values. When $M = 4$, it is QPSK, and the phase value is 0, $\pi/2$, π and $3\pi/4$, respectively corresponding to 11, 01, 00, and 10 after the input serial code stream undergoes serial-parallel conversion. Since two-channel parallel sequences with a half rate is formed after serial-parallel conversion, the occupied bandwidth is halved, therefore, the bandwidth utilization of QPSK is two times that of BPSK, but power utilization will not change.

Similarly, MPSK totally has M phase values quantized per (2π/M). Compared with BPSK, the standard gain of its bandwidth efficiency is $\log_2$M but the power efficiency is unchanged. The rate reduces to $\log_2 M$ times.

Compared with BPSK, the shortage of QPSK is to solve four kinds of fuzzy. Likewise, MPSK is also required to solve M-times fuzzy.

Another kind of QPSK modification is UQPSK (unbalanced quadrature phase shift keying modulation), which modulate different symbol rates in I and Q channels. Meanwhile, different power shall be distributed when I and Q are integrated. Such modulation method is frequently used in synthesis of different code rate information.

(4) OQPSK (offset quadrature phase shift keying)

OQPSK modulation is formed through offsetting the data code of I and Q channels in QPSK at a width T_b of input data bit.

Since the change status of two data stream does not appear simultaneously, polarity hop of one-channel data stream appears among the interval of sign bit of another channel data stream; thereby eliminating the possibility of $S(t)$ phase 180° hop. The phase hop is only limited within ±90° and the integral points may change every T_b seconds.

(5) CPM (Continuous phase modulation)

CPM means that the phase of the modulated wave changes continuously, with high bandwidth efficiency. Currently, the main modes are as follows:

1) Minimum shift keying modulation (MSK): Such mode changes the rectangular modulation waveform of OQPSK to a half-sine pulse for modulating carrier, thus forming MSK modulation. The mathematical expression of such modulation wave is the same as that of frequency shift keying (FSK) signals with modulation index being 0.5. The minimum modulation index that common FSK may achieve is 1, which may be smaller if the aforesaid modulation method is adopted, therefore, it is called the minimum frequency shift keying. The expression of power spectrum density is:

$$S_{\mathrm{MSK}}(f) = \frac{16P_{\mathrm{C}}T_{\mathrm{b}}}{\pi^2}\left[\frac{\cos 2\pi(f - f_{\mathrm{C}})T_{\mathrm{b}}}{1 - [(f - f_{\mathrm{C}})T_{\mathrm{b}}]^2}\right]^2, \tag{3.90}$$

where T_b refers to width of code; f_C is carrier frequency; P_C means carrier power.

Compared with OQPSK, the main lobe of its spectrum is wider and the side lobe is lower.

2) Gauss MSK (GMSK) modulation: filter the OQPSK-modulated pulse with Gauss-type low-pass filter and then modulate the carrier, thereby forming GMSK modulation. Frequency response of Gauss low-pass filter is:

$$G(f) = \frac{1}{2T_{\mathrm{b}}}\exp\left[-\frac{f^2\ln 2}{2B_{\mathrm{b}}^2}\right], \tag{3.91}$$

where $B_bG(f)$ is 3 dB bandwidth of $G(f)$.

Main lobe of GMSK spectrum is narrower than that of MASK, but its side lobe is very low.

3) Multi-modulation index continuous phase modulation system (CPM modulation system): CPM system is in material FM and alternatively uses each code with several phase modulation index in order, therefore, the phase modulation indexes are multiple, thereby realizing continuous phase modulation. Its frequency spectrum is more concentrated and bandwidth efficiency is 1.5 times higher than that of FQPSK system. It is clear that such modulation system witnesses further improvement and has a prospective future.

(6) BFSK (binary frequency shift keying)

FSK modulation was widely applied in TT&C and information transmission system and currently is used in command system. Compared with PSK, it is characterized by simple receiver without complicated phase-lock loop. For it, only two band-pass filter and envelope detector are required to promptly recover original baseband signal, without phase fuzzy problems arising out of PSK system. Weak points of FSK system: power efficiency and bandwidth efficiency are poor. Under the same BER condition, E_b/N_0 is 3 dB higher than PSK, and the bandwidth occupied by FSK is 1 times PSK.

(7) Bit error rate calculation of digital modulation systems

Calculation formula of bit error rate P_e of several common digital modulation systems and occupied bandwidth is as shown in Table 3.2 [10].

(8) Comparison among several constant envelope modulation

Constant envelope modulation modes consist of QPSK, OQPSK, MSK, and GMSK. Frequency spectrum diagram of the four modulation modes is shown in 3.24, and the comparison information as listed in Table 3.3 [10].

After the four systems are band-limited, in the case of band-limitation $B_{IF} = 1.5/T_b$, the bit error performance caused by non-linear channel is shown in Fig. 3.25 [10].

Table 3.3, Figs. 3.24 and 3.25 show that within the band-limited non-linear system, the performance modulated by MSK and GMSK is superior to QPSK and OQPSK, namely, in the case of band-limitation, the constant envelope characteristics of the former is superior to the latter.

Table 3.2 P_e calculation formula of common digital modulation systems and occupied bandwidth

Modulation mode	P_e	Baseband bandwidth W	Modulated carrier bandwidth B_c
BPSK	$\frac{1}{2}\mathrm{erfc}\sqrt{\frac{E_b}{N_0}}$	$R_b = \frac{1}{T_b}$	$2R_b$
QPSK	$\frac{1}{2}\mathrm{erfc}\sqrt{\frac{E_b}{N_0}}$	$R_b = \frac{1}{2T_b}$	R_b
MSK	$\frac{1}{2}\mathrm{erfc}\sqrt{\frac{E_b}{N_0}}$	$R_b = \frac{1}{2T_b}$	$1.5R_b$
DPSK	$\frac{1}{2}\mathrm{e}^{-E_b/N_0}$	$R_b = \frac{1}{T_b}$	R_b
Coherent FSK	$\frac{1}{2}\mathrm{erfc}\sqrt{\frac{E_b}{N_0}}$	$R_b = \frac{1}{T_b}$	$4R_b$
Non-coherent FSK	$\frac{1}{2}e^{-E_b/2N_0}$	$R_b = \frac{1}{T_b}$	$4R_b = \frac{1}{T_b}$
Sampling detection of baseband signal integral	$\frac{1}{2}\mathrm{erfc}\sqrt{\frac{E_b}{N_0}}$	$R_b = \frac{1}{T_b}$	R_b

Note: P_e is bit error rate per each information bit, W is occupied bandwidth of baseband digital signal (main lobe bandwidth, H_Z), B_c is occupied bandwidth of side carrier/carrier modulated by baseband digital signal, R_b is data rate of information bit (b/s), R_s is symbol rate of channel (S/s), where, the numerator S represents symbolic number, denominator s represents second

Table 3.3 Comparison among QPSK, OQPSK, MSK, and GMSK

S/N	Item	QPSK	OQPSK	MSK	GMSK*
1.	Power within $B_{IF} = 1.5/T_b$	94 %	94 %	99.5 %	99.9 %
2.	Envelope variation of bandlimited signal	100 %	30 %	0	0
3.	BER performance loss of ideal QPSK relevant to linear channel	0 dB	0 dB	0 dB	0.1 dB
4.	BER performance loss of ideal QPSK relevant to non-linear channel	1.1 dB	1.0 dB	0.3 dB	0.5 dB
5.	Sidelobe level generated after bandpass amplitude comparison	−15 dB	−25 dB	−30 dB	−35 dB
6.	Whether RF filtering is required before linear power amplification is conducted?	Yes	Yes	Yes	No
7.	Sensitivity of BER performance upon timing error	Middle	Middle	Low	Low
8.	Hardware complexity	100	110	120	120

Note: * represents IF or RF filtering is not required for GMSK, only baseband filtering is needed; corresponding to baseband bandwidth, the performance is $B_b = 0.5/T_b$; the other three systems requires RF filtering

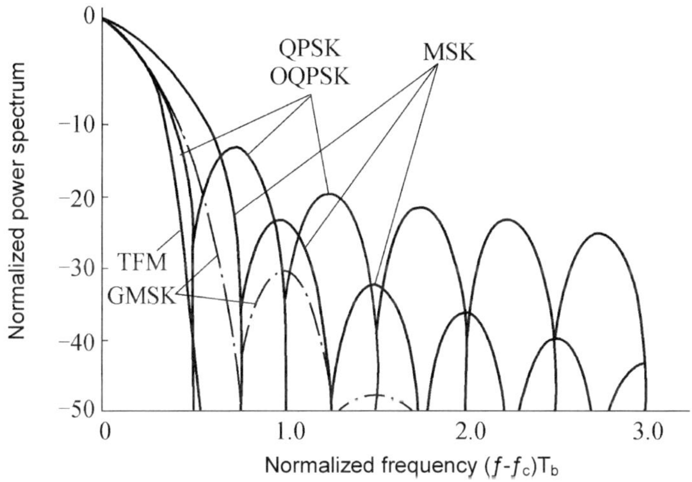

Fig. 3.24 Power spectrum density distribution of the four modulated system signals

3.2.3.4 Quasi-constant Envelope Modulation

It is a compromise between constant envelope and non-constant envelope.

Due to the aforesaid negative influence caused by band-limitation and nonlinearity, it is necessary to seek other modulation systems which shall ensure the envelope amplitude is constant after the carrier passes through band-limitation, which is called quasi-constant envelope modulation. Owing to small envelope fluctuation and AM, therefore, the influence of AM/PM and AM/$\tau(\omega)$ upon BER is very small.

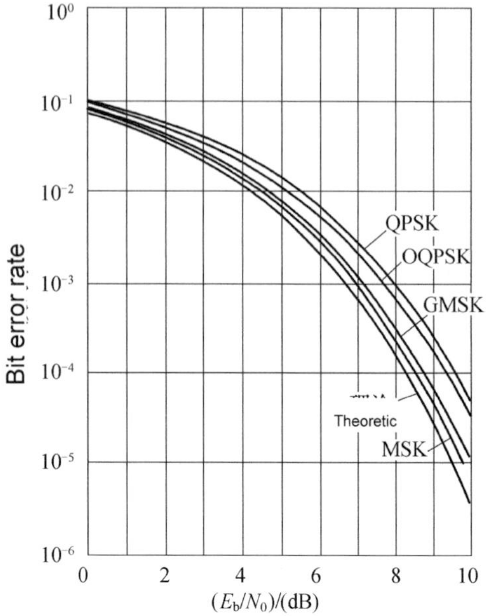

Fig. 3.25 Simulated curve of BER performance under non-linear channel in case of $B_{\mathrm{IF}} = 1.5/\mathrm{Ts}$

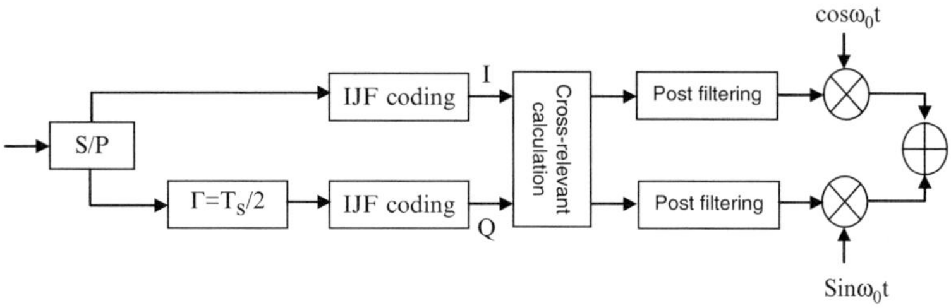

Fig. 3.26 Principle block diagram of FQPSK-B modulation

There are many quasi-constant envelope modulation modes, in which, the quadrature phase shift keying (FQPSK) for which relevant patent is issued to Feher as the prospective future. The principle block diagram is shown in Fig. 3.26.

In Fig. 3.26, Ts means symbol duration; Q-branch delayed $Ts/2$ implements OQPSK modulation, and I and Q branches are coded to raised-cosine waveform through IJF to achieve the purpose of no intra-pulse interference. If such waveform is directly used to conduct I and Q modulation, its envelope will accept 3 dB fluctuation. To achieve constant envelope, cross-relevant calculation is made for I and Q channels. Such cross-relevant make the carrier amplitude constantly approximate. According to the above, FQPSK can realize band-limited filtering without inter-symbol interference and can make modulated carrier envelope constant. Thereafter, a FQPSK-B system which adds post filter to cross calculation unit is developed. Its constant envelope feature is not so sound but the frequency spectrum is concentrated. Currently, FQPSK-B has been listed into IRIG-106-2000 telemetry standards and is evaluated as follows: "the standard transmission method of digital signals is FQPSK-B when the more efficient bandwidth utilization is required. FQPSK-B approximates constant envelope and is applicable to nonlinearity amplification channel, with very small spectrum expansion and mild detection performance degradation." According to this, in multi-channel signal transmission with band-limited, FQPSK-B is a better system.

Reference [9] analyzes the power spectrum density distribution different modulation made and average bit error rate characteristics of FQPSK. The results shows that FOPSK-B is best from frequency spectrum measurement, and the others are successively: FQPSK-B, GMSK ($BT_b = 0.5$), FOPSK, MSK, OQPSK. As an example, if "OQPSK + 0.5 roll-off filtering" is improved to be FQPSK-B, for -30 dB inter-channel interference, the spectrum utilization increases by three times, correspondingly, the bit rate can increase three times or the band-limited bandwidth can be narrowed by three times; but when FOSK is adopted to decrease bandwidth, it would cause loss to the error rate. For example, when $E_b/N_0 = 9.5$ dB, the error rate of OQPSK is 10^{-5}, and the error rate of FOPSK network code is 10^{-4}, OQPSK receptor of FOPSK is 10^{-3}.

3.2.3.5 Comparison of Various Modulation Modes [11, 12]

CCSDS developed a research through which frequently used high bandwidth efficiency modulation modes are compared. The modulation modes include: PCM/PM/NRZ, PCM/PM/biphase, QPSK, MSK, 8-PSK, BPSK/NRZ, BPSK/biphase, OQPSK, GMSK, and FQPSK-B. Research results shows that [9]:

① The bandwidth of FQPSK-B is the narrowest (the bandwidth efficiency is the highest) and only has small E_b/N_0 loss compared with BPSK/NRZ;
② The bandwidth efficiency of GMSK ranks as second place;
③ When high bandwidth efficiency is required, grid coding and 8-PSK is a good project.
④ When code rate is not very high, Turbo coding and BPSK/NRZ modulation combining with coding efficiency being 1/3 s is a modulation method with high power efficiency.

3.2.4 Channel Coding/Decoding Technology

3.2.4.1 Characteristics of Channel Coding/Decoding in Vehicle C&T and Information Transmission System

In TT&C and information transmission system, channel coding technology is widely applied to obtain coding gain, thereby reducing requirements for E_b/N_0.

Channel coding of vehicle CC&T has unique features. Adoption of channel coding technology is attributed to the following five aspects (same physical meaning but different requirements):

(1) Increase operating distance: preconditioned that the transmitting power and ground receiving BER of the flight vehicle is not changed, the operating distance is improved due to acquisition of coding gain. This is very important to deep space TT&C.
(2) Reduce transmitting power of flight vehicle: under the conditions of unchanged operating distance and bit error rate, transmitting power may be reduced, which is important to the vehicle TT&C and information transmission system with limited volume, weight, and power consumption.
(3) Increase communication capacity: with increase in transmission information channels, the transmission code rate increases accordingly. After coding gain of error-correction coding is obtained, the higher code rate can be transmitted under the same BER and transmitting power in power-limited system.
(4) Anti-interference: it has anti-interference function against continuous random error interference. RS coding has anti-interference function against burst interference like flame interference. Interleave coding has anti-interference function against the burst interference caused by multi-channel interference such as deep fading.
(5) Anti-interception and encryption: establish a code library with different coding modes, including plain code and secret code. In the case of application, these codes shall be frequently changed to obtain confidential performance.

Figure 3.27 shows several typical channel coding performance curves in CCSDS Standard [13]. According to such figure, in the case of 10^{-5} BER, RS(225, 233) + CONV(7,1/3) can obtain coding gain of (9.6 − 2.4)dB = 7.2 dB, with Turbo code being 9.7 dB, which shows that the coding gain stays at a high level and close to Shannon limit.

3.2.4.2 Basic Concept of "Shannon Limit" and Channel Coding

(1) Shannon's Capacity Theorem

Channel coding can obtain coding gain which is limited. Its maximum is limited by "Shannon's Capacity Theorem".

Shannon's second law (error-correction coding theorem) proves that when the information transmission rate R_m is lower than channel capacity C (also known as error-free capacity), a coding proposal can be used to implement error-free transmission. Channel capacity is a function of the channel features and, in additive white Gaussian noise, can be expressed as:

$$C = B\ \log_2\left(1 + \frac{P}{N_0 B}\right), \tag{3.92}$$

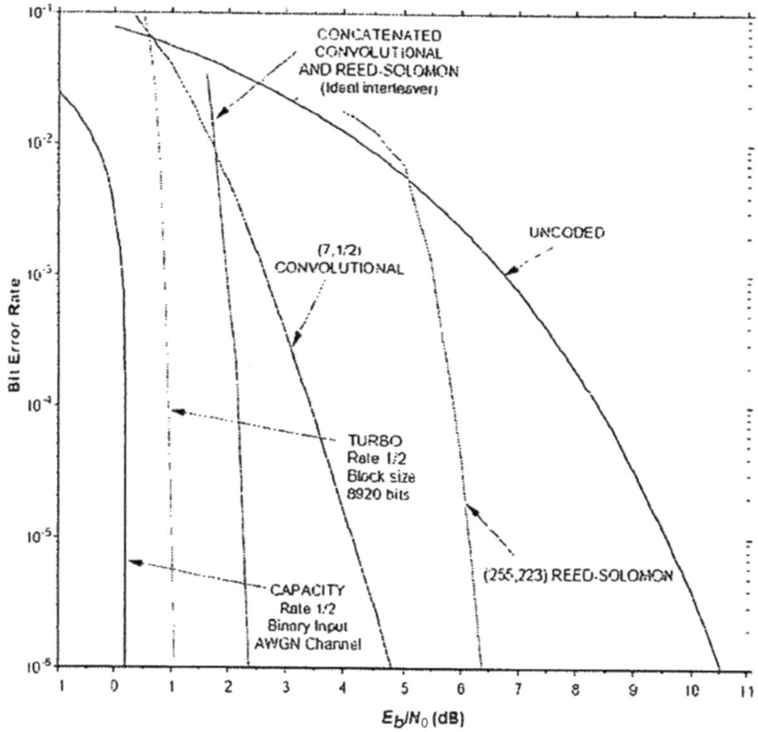

Fig. 3.27 Channel coding performance recommended by CCSDS Standard

where C is channel capacity (bit/s); B represents transmission bandwidth (Hz); P means power of receiving signals (W); N_0 means power spectrum density of single sideband noise (W/Hz).

After normalizing bandwidth B, the result is:

$$\frac{C}{B} = \log_2\left[1 + \frac{E_b}{N_0}\left(\frac{R_m}{B}\right)\right]. \tag{3.93}$$

Because of $P = E_b R_m$, (E_b denotes signal energy per bit at receiving end, R_m denotes transmission bit rate per second); therefore, according to Shannon's second theorem, let C in Expression (3.93) equal to R_m ($C = R_m$, meaning that transmission rate equals to channel capacity), we can obtain the relation curve between E_b/N_0 and R_m/B, as shown in Fig. 3.28. The curve represents whether arbitrarily small error probability can be realized when the system operates on a parameter point (E_b/N_0, R_m/B), the boundary is Shannon Capacity Limit in general meaning and represents the arbitrarily small error probability that can be realized through coding when the system operates on the left of or over the curve. Therefore, there is a minimum E_b/N_0 for any R_m/B, the arbitrarily small error probability can be realized only when it

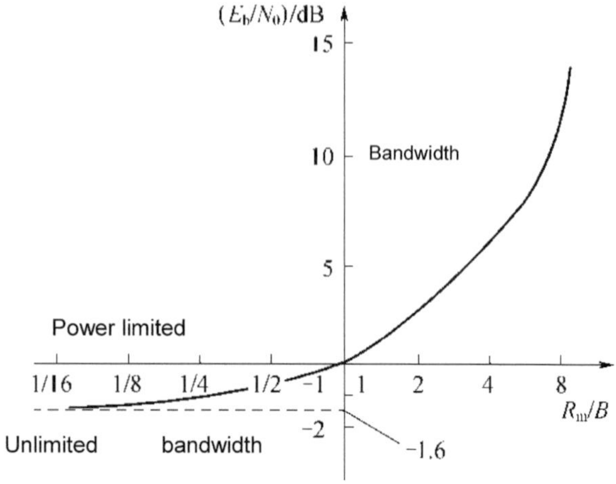

Fig. 3.28 "Shannon limit" curve

equals or is larger than the minimum E_b/N_0. It can be realized through error-correction coding, for example, the arbitrarily small error probability can be realized through adopting Turbo code or lengthen block length in block coding. Figure 3.28 shows that for fixed $R_{m,}$ when bandwidth B is very large, the required E_b/N_0 approximates to the minimum value -1.6 dB, also known as Shannon limit; namely, the minimum limit realizing 0 bit error rate is -1.6 dB. If the minimum E_b/N_0 cannot be satisfied, it is impossible to realize error-free transmission whatever non-bandlimited channel or forward error correction is employed.

Approaching Shannon Capacity Limit is a goal pursued by error-correction researchers for a long term and currently approximated by Turbo code and LDPC code. However, appropriate matching of information source and channel shall be a precondition for such realization; furthermore, some costs may be spent in terms of bandwidth, time delay, and complexity.

(2) Basic Concept of Channel Coding

Add redundancy code into information code to be transmitted according to proper rules to obtain reliability of information code in transmission. The code to be transmitted by information source is called information code (corresponding code rate is known as information rate), the added redundancy code is redundancy check code, the code which is generated after being encoded is called symbol code (corresponding code rate means symbol rate). The aim of channel coding is to obtain the highest reliability in the expense of adding the least redundancy code, thus obtaining coding gain. Definition of coding gain: the difference between SNR (E_b/N_0) required by realization of the same bit error rate in absence/presence of coding. For $E_b/N_{0,}$ E_b means energy per information code before channel coding is conducted not the energy of the channel symbol code, therefore, bandwidth increment and E_b/N_0 loss cannot be caused due to increase in symbol rate after channel coding.

According to the rule adding redundancy code, they can be divided into linearity and nonlinearity and are respectively called linear code and non-linear code. According to functions implemented by check point, they fall into error detection code with function of finding error and error correction code with function of automatic error correction.

In TT&C and information transmission, convolution code, convolution and RS concatenated code, interleave coding, Turbo code and LDPC code are mainly used and introduced below. Most of these encoders and decoders are available in markets; therefore, their basic principles and circuits are not introduced but their application will be emphasized below.

3.2.4.3 RS Encoding

RS code is also called Reed-Solomon code, and is a kind of special non-binary BCH code.

(1) Basic parameters of RS code

1) Input information can be divided into a group of k multiplied by m bit(k•m bit), and each group has k symbols, and each symbol is composed of m bit not 1 bit in BCH code.
2) Code length: $n = (2^m - 1)$ symbols or $m(2^m - 1)$ bit, namely, a symbol has binary m bit.

Information section: k symbols; or $k \cdot m$ bits;
Check section: $n - k = 2t$ *symbols; or* $m(n - k) = 2mt$ bit.
Minimum distance: $d = 2t + 1$ symbols, or $md = (2t + 1)$ bit.

(2) Error-correction capability of RS code

1) RS code is capable of simultaneously correcting random and burst errors, and the burst error-correction capability is higher.
2) The error pattern that can be corrected by RS code are as follows:

Single burst with total length $b_1 = (\mathrm{t} - 1)m + 1$ bit;
Two bursts with total length $b_2 = (\mathrm{t} - 3)m + 3$ bit;
...
i bursts with total length $b_i = (t - 2i + 1)m + 2i - 1$ bit.

The coding gain obtained by RS code is as shown in Fig. 3.29. Such figure shows the coding bit error probability curve of RS code with m respectively being 4, 5, 6, 7, and 8 and coding rate R approximating 3/4, it is a function of SNR E_b/N_0. Figure 3.29 shows that in the case of BER $P_b = 10^{-5}$, the coding gain of RS code varies between 1.5 dB (when $m = 4$) and 3.4 dB ($m = 8$, corresponding to code length = 28 − 1 = 255 symbols).

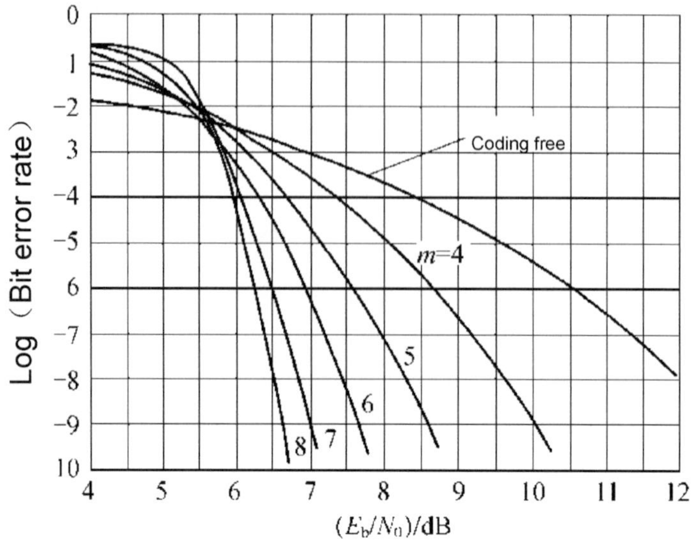

Fig. 3.29 BER of RS code in case of $R = 3/4$

For RS(255,223) code, its code length = 255, the code length of information section $K = 223$, code rate $R = k/n = 223/255 = 0.875$. The performance of RS (255,233) code recommended in CCSDS is as shown in Fig. 3.27.

3.2.4.4 Error Detection Code

Cyclic code is suitable for error detection. It has strong error-detection capability and easy operation and is commonly used in command information transmission.

(1) Error-detection function of cyclic redundancy check (CRC) code

1) Be able to find error with burst length less than $n - k + 1$;
2) Be able to find most errors with burst length equal to $n - k + 1$, including $2^{-(n-k-1)}$ error not detected;
3) Be able to find most errors with burst length equal to $n - k + 1$, including $2^{-(n-k)}$ error not detected;
4) All errors with code distance from permissible code block less than or equal to $d_{\min} - 1$;
5) All odd-numbered errors.

(2) Common CRC code

Commonly CRC codes which have been listed as international standard include the following four codes:

1) CRC-12: generated polynomial is $g(x) = 1 + x + x^2 + x^3 + x^{11} + x^{12}$
2) CRC-16: generated polynomial is $g(x) = 1 + x^2 + x^{15} + x^{16}$

3) CRC-CCITT: generated polynomial is $g(x)=1+x^5+x^{12}+x^{16}$
4) CRC-32: generated polynomial is $g(x)=1+x+x^2+x^4+x^5+x^7+x^8+x^{10}+x^{11}x^{12}+x^{16}+x^{22}+x^{23}+x^{26}+x^{32}$

where CRC-12 is used when the character size is 6 bit, and the other three are used when the character size is 8 bit.

3.2.4.5 BCH Coding

BCH code is mainly used in TT&C and information transmission system, such as BCH code (63, 56) recommended in CCSDS command. BCH code is a kind of important cyclic code and can correct several independent random errors in a block, being binary linear cyclic code. It is named with the initial letter of three discoverers' names and is widely used, featuring strong error-correction capability and simply structured.

To generate polynomial of BCH code:

$$g(x) = \text{LCM}[m_1(x), m_3(x), \ldots, m_{2t-1}(x)],$$

where t represents the number of corrected errors; $m_i(t)$ is irreducible polynomial, LCM means the least common multiple.

The code generated thereby is called BCH code. The minimum distance of BCH codes is $d \geq d_0 = 2t+1$, where d_0 refers to the designed distance. It can correct t independent random errors. BCH code can be divided into two categories:

(1) Factor of code length $n=2^m-1$, called primitive BCH code or narrow-sense BCH code.
(2) Factor of code length $n=2^m-1$, called non-primitive BCH code or broad-sense BCH code.

3.2.4.6 Convolution Code

(1) Convolution code

Convolution code was put forward by Elisa in 1955. The coding method can be expressed by convolution calculation (hence its name).

Different from linear block code and cyclic code, convolution code is a kind of non-block code. Convolution code is memory coding and has a memory system for any designated time period, n output of its encoder is not only relevant to k input within such period, but also in connection with the first m input stored in the encoder. Convolution code usually is expressed as (n,k,m) code or also as (n,k,m) or (N,R), with limitation length (memory length) being $N=m+1$. m is the sections registered in encoder, with code rate being $R=k/n$.

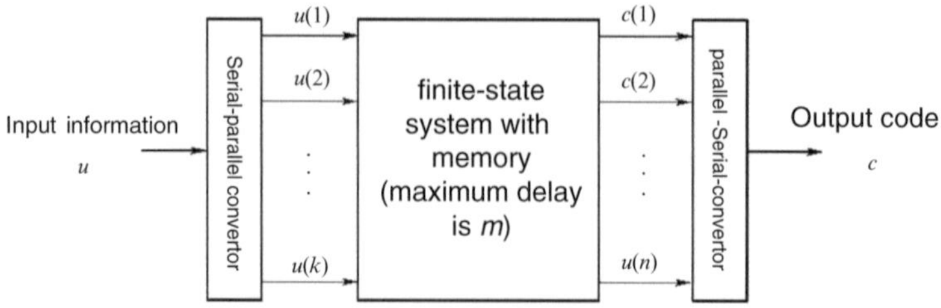

Fig. 3.30 Principle diagram of convolution code decoder

Typical structure of convolution code is a limited-status memory system consisting of k input terminals, n output terminals and shifting register with m sections. It is commonly called time sequence network. As shown in Fig. 3.30.

(2) Viterbi decoding

Decoding of convolution code falls into algebraic and probability decoding. As probability decoding, Viterbi decoding is a common decoding method and was introduced by Viterbi in 1967. Forney pointed out that it is materially the maximum likelihood for decoding.

(3) Soft decision decoder

To improve decoding performance, the decision is modified to be a soft decision. Soft decision refers to multi-level decision, and hard decision means dual-level decision. In case of strong noise, useful information may be lost in deciding. After multi-level soft decision is employed, the performance may improve by 1.5–2 dB. In practice, considering the complexity of multi-level realization, 4–8 levels is usually applied. Coding gain obtained by convolution coding is as shown in Fig. 3.31. The performance of convolution code recommended in CCSDS is shown in Fig. 3.27.

3.2.4.7 Concatenated Code

Concatenated code is a special effective method constructing long code with short code. The long code so constructed is composed of two short codes (i.e., inner code and outer code), without complicated coding circuit. Inner code may be designed to be of common complexity and to obtain appropriate bit error rate (typical value ranges between 10^{-3} and 10^{-4}). Outer code shall be designed to correct most of errors left after inner code error correction and thus obtain approximated error code free performance. In principal, block code or convolution code can be adopted for inner code or outer code. At present, the commonly used and optimum performance mix is as follows: inner code is convolution code with short constraint length adopting soft decision Viterbi decoding; outer code is RS code with powerful error correction capability, and can correct intra-byte and inter-byte inner codes but fails to correct all random and burst errors.

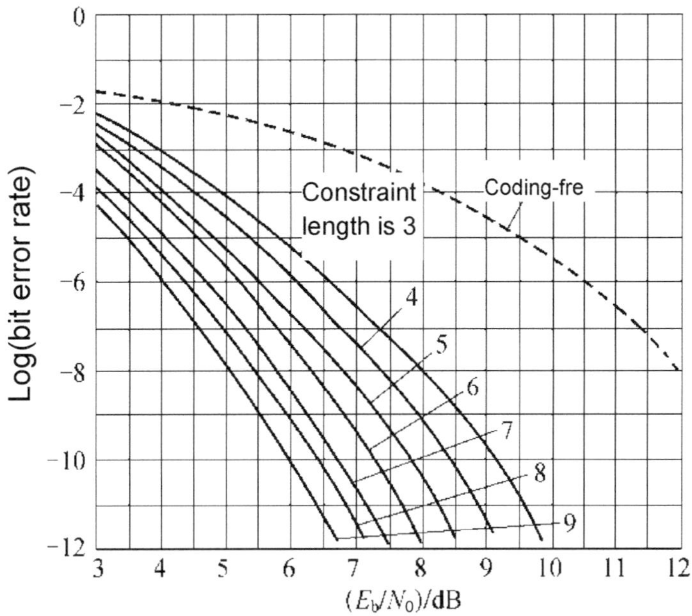

Fig. 3.31 Bit error rate of convolution code using software decision channel with 1/2 code rate

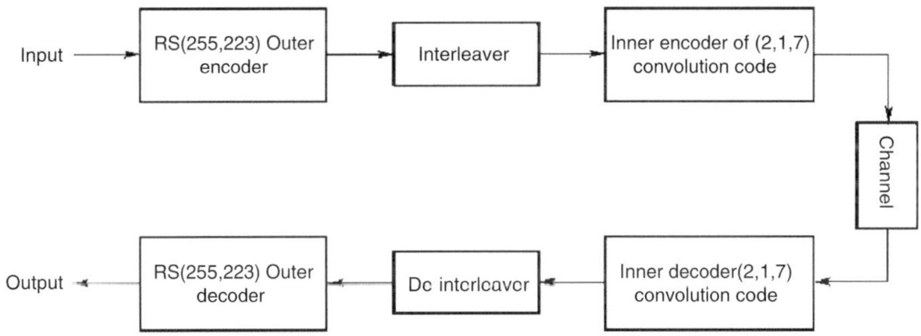

Fig. 3.32 Block diagram of typical concatenated code

In 1984, NASA adopted concatenated code with (2,1,7) convolution code as inner codes and (255,223) RS code as outer codes, and added an interleaver between inner codes and outer codes, with interleaving depth about 2–8 outer code blocks. Its performance can enable bit error rate: $P_b \leq 10^{-6}$ in case of $E_b/N_0 = 2.53$ dB. In 1987, CCSDS telemetry series standards were established based on such code. The block diagram of typical standard concatenated code system is shown in Fig. 3.32. Performance of such concatenated code is as shown in Fig. 3.27.

3.2.4.8 Interleave Coding

The coding methods introduced above are designed for random independent errors. In case of burst error, a series of errors will be caused. RS code has strong error correction capability for burst errors, but it is disabled for long burst errors more than tens of bits.

Design idea of interleave code is different from that of error-correction code. All error-correction codes are designed to adapt to error interference, namely, corresponding error-correction code is used for relevant interference. In case of random independent error interference, BCH code and convolution code can be adopted; in case of burst error interference, fire code, RS code, and other block/convolution codes specially correcting burst errors can be used. Interleave coding is designed to change channel interference rather than adapting to the channel interference. Through interleaving and de-interleaving, it modifies a memory burst error channel into a random independent error channel without memory and then makes error correction with error-correction code correcting random independent errors.

Block diagram of interleave code realization is shown in Fig. 3.33.

For example, for the code with block length being $L = M \times N$, $M - n$ rows of matrix can be constructed. After entry of code sequence, the interleave matrix memory is written in by row and read by column. After being sent to burst channel and entering de-interleave matrix memory, it is written in by row and read by line. Conversion between lines and rows can convert the burst channel into an equivalent independent channel. Such block periodic inter-leaver includes the following features:

(1) In code sequence, any burst errors with length being $L \leq M$ will, after being interleaved, become single independent errors separated at least by $N - 1$ bit.
(2) In code sequence, any burst errors with length being $L > M$ will, after being de-interleaved, convert long burst error into short burst errors with length being $L_1 = [L/M]$.
(3) Under the conditions that interleaving and de-interleaving conversion is completed and channel time delay not counted, the code sequence will generate $2MN$ symbols of time delay, half of which will be respectively located at receiving terminal and transmitting terminals.
(4) Under special circumstance, k random independent error sequences with cycle being M will, after being interleaved and de-interleaved, generate burst error with length as L.

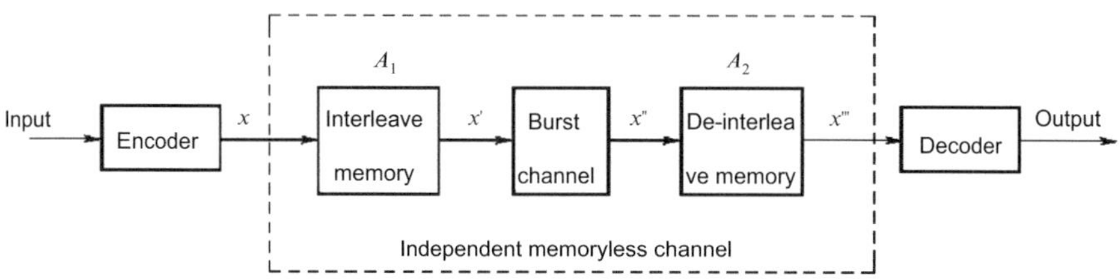

Fig. 3.33 Block diagram of block interleaver realization

According to the foregoing features of the block interleaver, it is an effective method for overcoming deep fading bit error and has been widely used. Especially, in case of low-elevation operation and much multi-path interference, it can effectively reduce error codes. However, it will bring about additional 2*MN* symbols of time delay, which will impose negative influence upon real-time communication, and improve equipment complexity for realization.

3.2.4.9 Turbo Code

The feed iterative structure of parallel concatenated convolution code used by the Turbo code decoder is similar to the principle of turbine motor, so it is called Turbo code. Soft-in/soft-out (SISO) iterative algorithm is adopted as decoding algorithm. Turbo code has sound performance approximating to Shannon capacity limit.

(1) Strong points of Turbo code

1) The interleaver at transmitting terminal plays a role of random code weight distribution, namely, random coding.
2) The interleaver at receiving terminal, together with corresponding multiple iterative decoding, allows for random decoding and can convert the burst error into random independent error for the attenuation channel with burst errors.
3) Long codes can be formed with short codes through concatenated coding and decoding which is random due to randomness of interleaving.
4) Parallel concatenated structure and the optimum multiple iterative soft-in/soft-out algorithm greatly improve the decoding performance.

(2) Weak points of Turbo code

1) Decoding equipment is very sophisticated. The improvement algorithm to balance decoding performance and complexity shall be sought.
2) Decoding time delay is large, which imposes small influence in deep space TT&C. However, such delay shall be considered duly in low or high-orbit satellite TT&C applications.
3) Floor effect is generated in case of low bit error rate, which is mainly attributed to small free distance of Turbo code.

The performance of Turbo code recommended by CCSDS is shown in Fig. 3.34, the used interleaving length is 16,384 bit. In such figure, the representative Turbo code in CCSDS is compared with the currently used concatenated code.

Different interleaving length is adopted for Turbo code, the error-correction effect is different accordingly. Additionally, when decoding is made, different iteration times will generate different error-correction effect. The more iteration times there are, the better error-correction effect. However, large time delay will be introduced and code length increased.

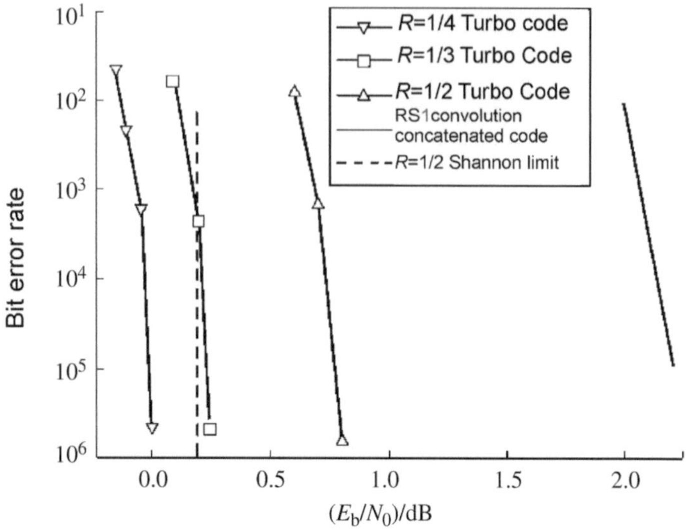

Fig. 3.34 Bit error features of Turbo code recommended by CCSDS

3.2.4.10 Low Density Parity Check Code (LDPC Code)

Compared with Turbo code, LDPC decoder can be implemented in parallel, thereby realizing high-speed decoding. Additionally, LDPC code employs more flexible design mode. Furthermore, LDPC code, as a code with the strongest error-correction capability in channel coding, also has the performance approaching Shannon capacity limit. Its decoder has a simple structure and can obtain extremely high throughput capacity at the expense of a little resource. It has been recommended by CCSDS for engineering application. Its strong points include:

(1) It is close to the theoretical limit and can obtain more coding gain.
(2) The generation matrix of such code word is cyclic on a block basis, enjoying low decoding complexity.
(3) Provide serialization features, namely, one encoder and one decoder can implement encoding and decoding with different code length and code rate.

Recommendation by CCSDS is as shown in Table 3.4, the recommended code rates include 1/2, 2/3, and 4/5, and the information bit length is 1,024 bit, 4,096 bit, and 16,384 bit. In the recommended code word, the fixed information bit length allow the code word user to focus on the generated data frame length not code rate.

Figures 3.35, 3.36, and 3.37 [14] show the performance of LDPC code listed in Table 3.4. According to these figures, the performance difference among 1/2, 2/3, and 4/5 LDPC code is about 1 dB when information bit length is 1,024 bit and bit error rate 10^{-7}; the performance difference among 1,024 bit, 4,096 bit, and 16,384 bit codons is about 0.6 dB when code rate is 1/2 and bit error rate 10^{-7}.

Table 3.4 Code length and code rate of LDPC code recommended by CCSDS

Information bit length/bit	Code length/bit		
	1/2 code rate	2/3 code rate	4/5 code rate
1,024	2,048	1,536	1,280
4,096	8,192	6,144	5,120
16,384	32,768	24,576	20,480

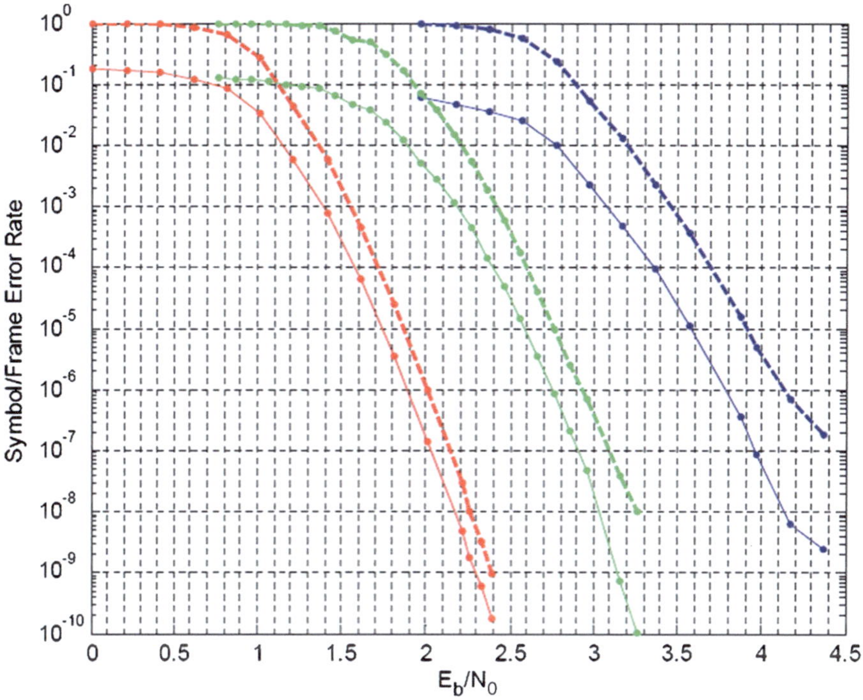

Fig. 3.35 Performance of length $k = 1{,}024$ LDPC codes: rate 1/2, 2/3, 4/5 (*left* to *right*)

3.2.4.11 Unification of Error-Correction Coding and Modulation

Power and bandwidth are important specifications weighting efficiency of communication systems. In modern digital communication system design and practice, the mainstream purpose is to improve reliability of digital communication systems that are preconditioned and whose transmission power is not increased and bandwidth not occupied. Under the condition of limited power, small power is desired to achieve system performance. For this, error-correction coding may be employed. However, the modulation rate of the modulator will increase, thus enlarging the transmission bandwidth. Under the condition of bandwidth constraint, high-order modulation mode may be used to improve frequency utilization. However, if it is

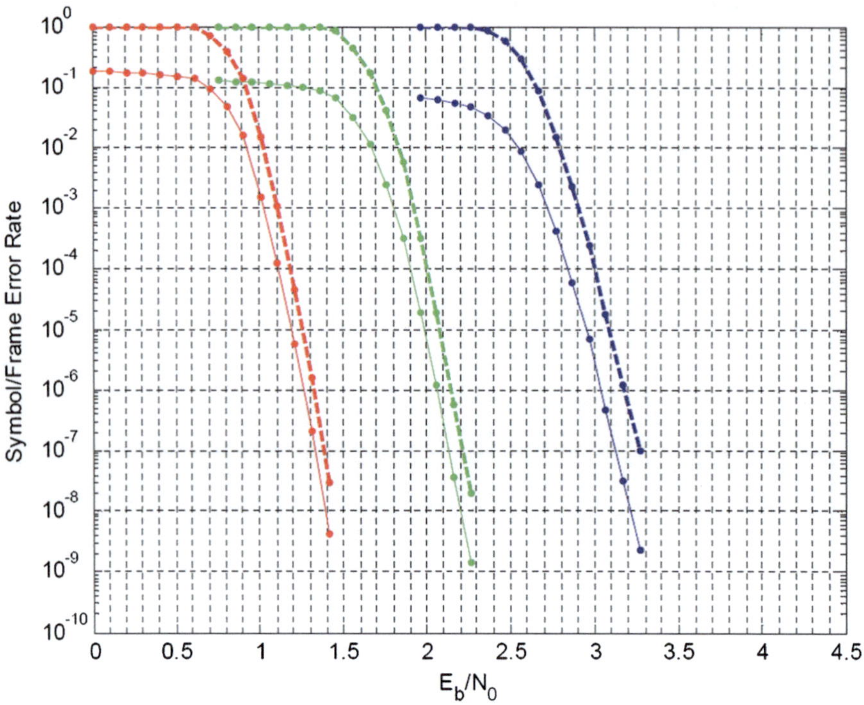

Fig. 3.36 Performance of length $k = 4{,}096$ LDPC codes: rate 1/2, 2/3, 4/5 (*left* to *right*)

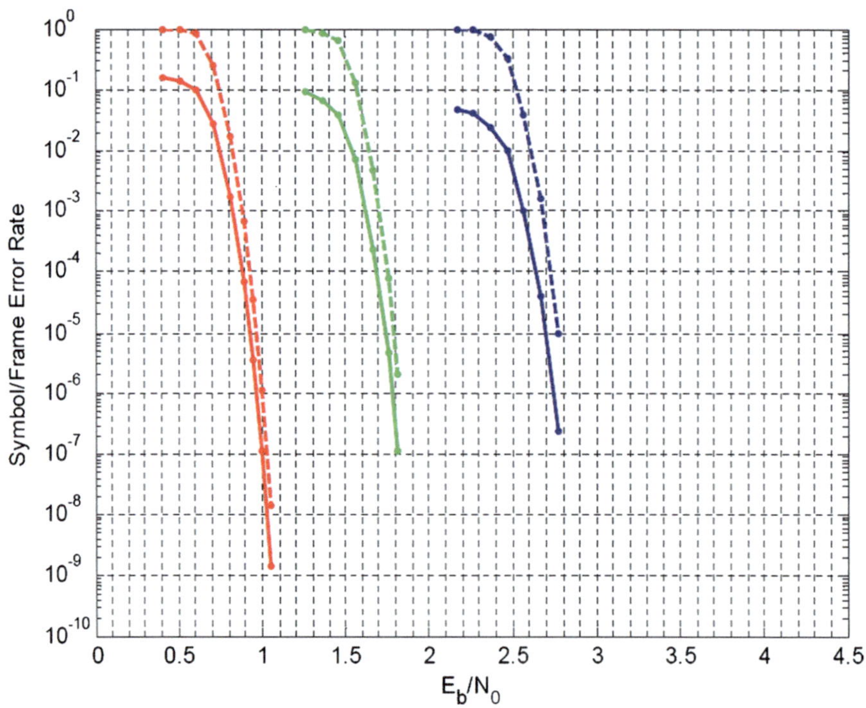

Fig. 3.37 Performance of length $k = 16{,}384$ LDPC codes: rate 1/2, 2/3 M 4/5 (*left* to *right*)

required to ensure the same bit error performance, signal power must be improved. In conventional digital communication system, the error-correction coding and modulation is designed separately, therefore, the requirements for power efficiency and bandwidth efficiency cannot be met.

Trellis coded modulation (TCM) provide solutions to solve the conflict between power efficiency and bandwidth efficiency. It maps each code segment after convolution coding as one signal of a group with 2^{n+1}($n+1$ means code length) modulation signal points for signal demodulation at receiving terminal, and then it is transformed to be code sequence which will then be sent to decoder via reverse mapping. Tests approve that TCM can obtain 3.6 dB coding gain under the conditions that bandwidth is not increased and information transmission rate not changed. Since both modulated signals and convolution code can be deemed as Trellis code, it is called Trellis coded modulation. TCM can provide redundancy required by encoding through expanding modulation signal, without increasing signal frequency bandwidth. Therefore, for the bandlimited communication system, integrated coded modulation scheme can be used to improve power efficiency and frequency band utilization. The system may employ convolution code or block code. Convolution code is used in TCM, which is attributed to the fact that convolution code is more easily realized than block code and the performance improvement so acquired is similar. Different from a conventional receiving system which implements demodulation and then decoding successively, TCM implements demodulation and decoding at the same time. Optimum decoding (decoding) of TCM is based on the grid chart of signals; namely, correct path will be found in the grid chart after the channel output terminal receives receiving signal sequence. Correct path is found through Viterbi algorithm. The actual quasi-optimum intercept Viterbi algorithm is frequently used in practice. Since its inception, TCM has been widely researched and applied in theoretical and practical fields.

3.2.5 *Impacts of Noise on Data Transmission BER – Amplitude Noise Equivalent Method*

3.2.5.1 Relationships Between AWGN and PSK Signal BER

In data transmission system, the LNA noise, antenna noise, and feeder noise all are additive white Gaussian noise (AWGN).

In the case of BPSK, the BER can be expressed as below if the system adopts matched filter without inter-symbol interference (ISI):

$$P_E = \frac{1}{2}\mathrm{erfc}\left(\sqrt{\frac{E_\mathrm{b}}{N_0}}\right) = \theta\left(\sqrt{\frac{2E_\mathrm{b}}{N_0}}\right), \tag{3.94}$$

where P_E is the BER, $erfc = 1 - erf(u)$ is the complementary error function, θ is Gaussian integral, and $E_\mathrm{b} = A^2 T_S$ is the code pulse energy.

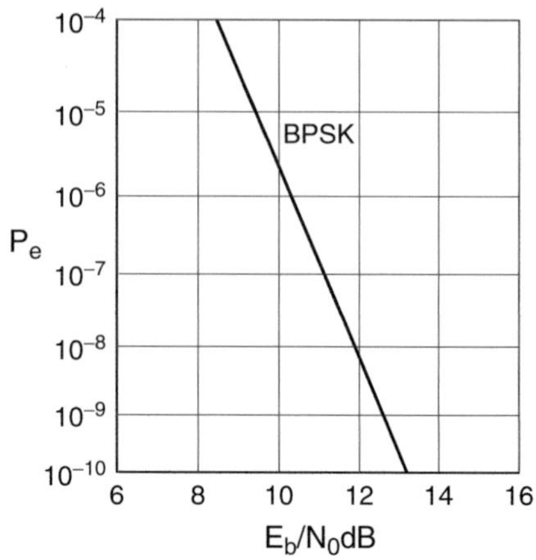

Fig. 3.38 Relationship between E_b/N_0 and BER

Table 3.5 Calculated values of BER

P_E	10^{-4}	10^{-5}	10^{-6}	10^{-7}
E_b/N_0	8.43	9.61	10.54	11.3

$P_E - E_b/N_0$ curve is shown in Fig. 3.38

Table 3.5 lists the calculated values of the BER:where,

$$\frac{E_b}{N_0} = \frac{C}{(N_0 b_r)}. \tag{3.95}$$

C/N_0 is the ratio of carrier power and noise power spectral density, and b_r is the code rate.

3.2.5.2 Relationships of Signal Source Phase Noise and BER – "Amplitude – Noise Equivalent Method" [15]

The signal source phase noise herein refers to the phase noises provided with all frequency sources in the transmitting/receiving channel, which include the master oscillator, receiving/transmitting local oscillator, atomic clock, and reference sources at all frequencies. The phase noises produced by these sources may add the phase modulation noise interference to the modulated signal before modulation, thereby affecting the BER.

The phase noise spectral power density caused by the amplitude WGN aforesaid is a plain white phase noise, with the phase noise power spectral density φ_{no} of its single sideband to be:

$$\varphi_{no} = \frac{N_0}{C} \left(\text{unit : rad}^2/\text{Hz, dB : dBc/Hz}\right)$$

$$\sigma_\varphi^2 = 2\varphi_{no} B_n = 2\left(\frac{N_0}{C}\right) B_n \ \left(\text{unit : rad}^2\right), \tag{3.96}$$

where σ_φ^2 is the phase jitter RMS value caused by the amplitude noise, and B_n is the baseband equivalent noise bandwidth of the output signal of the receiving system. Normally, if BPSK works at the optimum bandwidth, i.e., $B_n \approx 0.5b_r$, the phase noise power spectral density curve is as shown in Fig. 3.39.

Substituting Expression (3.96) into Expression (3.95), then the equivalent E_b/N_0 caused by white phase noise power spectral density φ_{n0} under BPSK is obtained, that is:

$$\frac{E_b}{N_0} = 1(\varphi_{no} b_r). \tag{3.97}$$

The essence of PSK error introduced by amplitude noise is that the amplitude noise triggers PSK phase jitter (with RMS value of σ_φ^2) which disturbs the amplitude of the digital demodulated signal, then the error is generated. Therefore, the impact of phase noise against BER is equivalent to that of the same phase noise against amplitude noise E_b/N_0, hence the E_b/N_0 shown in Fig. 3.38 may be replaced with $1/(\varphi_{no} b_r)$ in Expression (3.97), to obtain the relationships of white phase noise and BER, as shown in Fig. 3.40. Observed from the diagram, the phase noise has greater impact on the BER under high speed data transmission (b_r, greater value) and high carrier frequency (Φ_{n0}, greater value):

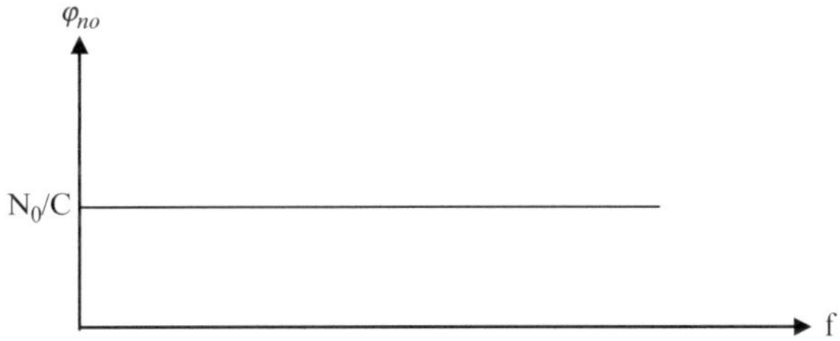

Fig. 3.39 White phase noise

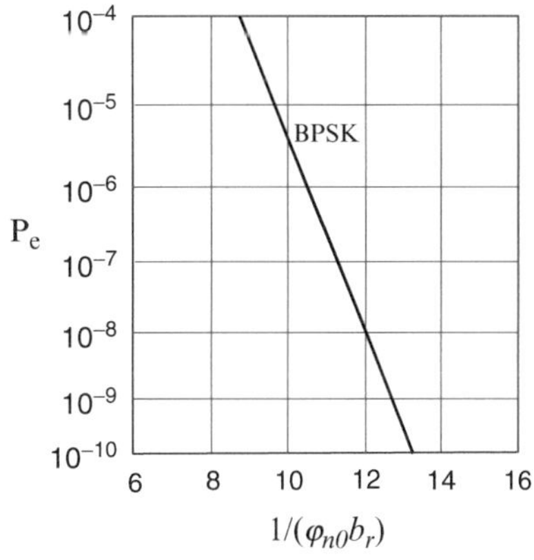

Fig. 3.40 Relationship between white phase noise and BER

If the phase noise is not white noise, an equivalent single-side white noise spectral density Φ_{no} with approximately constant value can be used to substitute $\varphi_{no}(f)$ and the BER can still be obtained through Fig. 3.40. Thus, the requirement of BER on phase noise can be converted to the requirement on Φ_{no}. The condition for Φ_{no} to substitute $\varphi_{no}(f)$ is:

$$\Phi_{no}B_n = \int_0^{+B_n} \varphi_{no}(f)\mathrm{d}f, \tag{3.98}$$

where Φ_{no} is the single-side noise spectrum density.

Solution of Φ_{no} with Expression (3.98) contains three methods below:

(1) If the measured value φ_{n0}(f) is known, then the integral is solved by computer.
(2) If the phase noise power law spectrum curve in Fig. 3.41 is known, the Expression (3.98) is solved with graphing method, that is, let the area at both sides of this expression be equivalent to substitute the integral of, $\varphi_{n0}(f)$ as a result, Φ_{no} is obtained if A = B, and then Φ_{no} is used to substitute φ_{n0} in Fig. 3.40 and approximate the BER P_e caused by phase noise. The testing result is properly consistent with that obtained by the calculation method above. As shown in Fig. 3.41, if b_r is high, the main component of φ_{n0}(f) is the white phase noise spectral density Φ_p and $\Phi_{no} \approx \Phi_p$.

It is worth pointing out that the assumption of constant $\varphi_{n0}(f)$ at high f end in Fig. 3.41 is just a theoretical power-law spectrum model (normally phase noise indexes are proposed based on this model). In fact, the phase noise of the actual signal source is not always a constant value at high f end (due to the influence of frequency responses of phased-locked loop and other circuits) and may be present to have a cutoff region as shown with dotted line in Fig. 3.41. This will make the equivalent Φ_{n0} be decreased to $\varphi_{n0}(b_r'/b_r)$ (b'_r is the bandwidth of cutoff region),

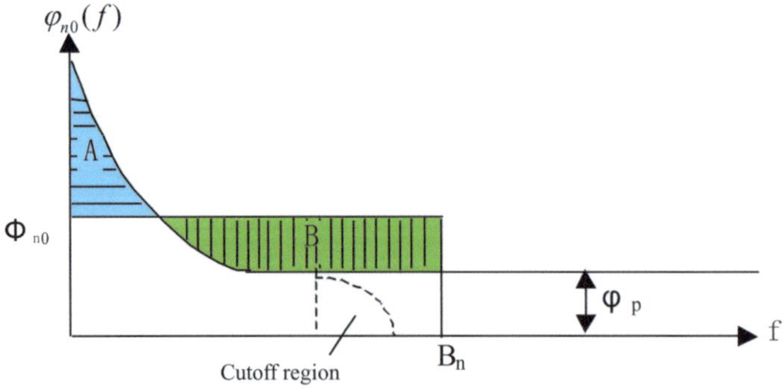

Fig. 3.41 Equivalent area method for white phase noise

while correspondingly the equivalent $E_b/N_0 = 1/\varphi_{n0} b_r^{'}$ and consequently the Φ_{n0} that fulfills the requirement of BER becomes:

$$\Phi_{no} \approx \frac{N_0}{E_b} \frac{1}{b_r^{'}}. \tag{3.99}$$

More importantly the curve in Fig. 3.41 shall be measured by spectrum analyzer, so that Φ_{n0} is calculated with that curve, especially $\varphi_{n0}(f)$ near b_r, which may obtain a different value from the specification and results in different Φ_{n0} solved by the equivalent area method, therefore, such impact on BER correspondingly is different. Currently, the phase noise with high frequency offset is not yet able to be measured with spectrum analyzer, so it is difficult to verify the results above with testing method.

(3) If the condition of phase noise spectral density in Expression (2.32) is known, it is substituted into Expression (3.98) for solving integral:

$$\Phi_{no} = \frac{1}{b_r}\left[h_{+2}\left(b_r - f_f\right) + h_{+1}\ln\frac{f_5}{f_4} + h_o\left(\frac{1}{f_3} - \frac{1}{f_4}\right) + \frac{h_{-1}}{2}\left(\frac{1}{f_2^2} - \frac{1}{f_3^2}\right) + \frac{h_{-2}}{3}\left(\frac{1}{f_1^3} - \frac{1}{f_2^3}\right)\right]. \tag{3.100}$$

Expression (3.100) is used to solve the characteristics of power law spectrum shown in Figs. 3.42, 3.43 and 3.44, and the allowable Max. phase noise envelope under all data transmission rates if $P_E = 10^{-7}$:

Learned from the analyses above, if high-end f has greater phase noise which worsens the BER, then the signal source adding filter can be used to reduce the effects on phase noise at high-end f.

The deterioration value L_e of equivalent carrier-to-noise ratio (corresponding to the variation of E_b/N_0) caused by phase noise is analyzed as follows:

Using Expression (3.96) to calculate the phase jitter RMS σ_φ^2 caused by amplitude thermal noise:

$$\sigma_\varphi^2 = 2\frac{N_0}{C} B_n. \tag{3.101}$$

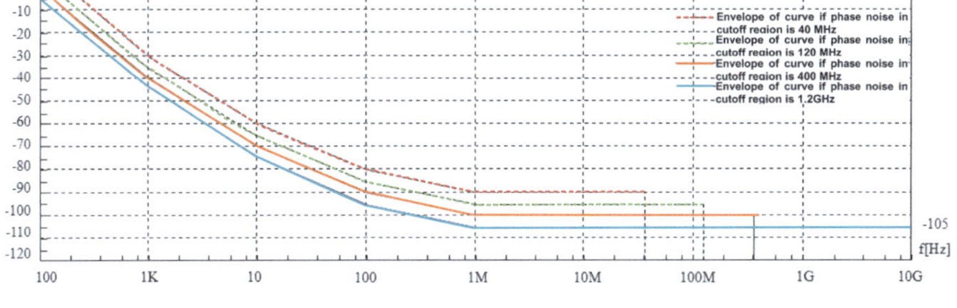

Fig. 3.42 Maximum allowable phase noise envelopes at a code rate of 1.2 Gbps with $P_E = 10^{-7}$

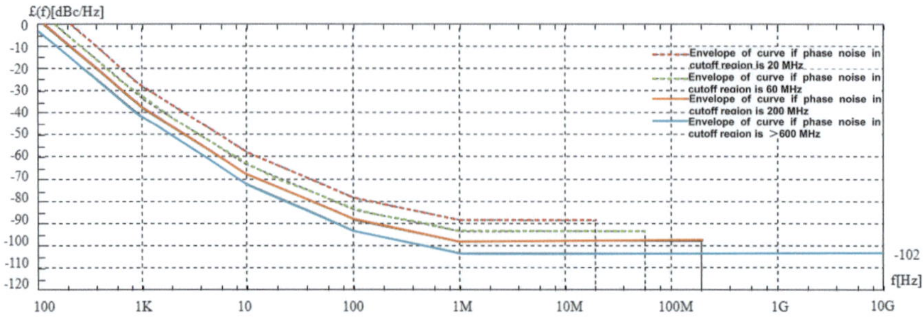

Fig. 3.43 Maximum allowable phase noise envelopes at a code rate of 600 Mbps with $P_E = 10^{-7}$

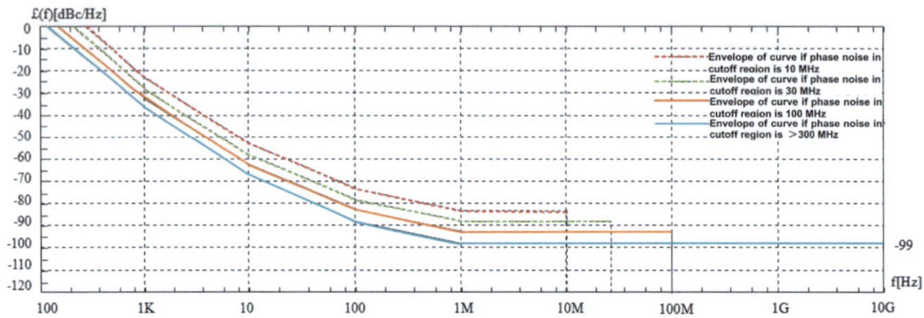

Fig. 3.44 Maximum allowable phase noise envelopes at a code rate of 300 Mbps with $P_E = 10^{-7}$

The jitter RMS σ_φ^2 caused by phase noise is:

$$\sigma_\varphi^2 = 2\Phi_{\text{no}} B_n. \tag{3.102}$$

Taking the sum of $(\sigma_\varphi^2 + \sigma_\Phi^2)$, then the phase jitter increment caused by addition of phase noise is equivalent to the deterioration value L_e of carrier-to-noise ratio:

$$L_e = \frac{\left[\sigma_\varphi^2 + \sigma_\Phi^2\right]}{\sigma_\varphi^2} = 1 + \frac{\sigma_\Phi^2}{\sigma_\varphi^2}.$$

Simultaneous calculating Expression (3.95), (3.99), (3.101), and (3.102) to obtain:

$$L_e = 1 + \frac{E_{\text{b}}}{N_0} b_{\text{r}} \Phi_{\text{no}} \tag{3.103}$$

If $B_n = b_{\text{r}}, \quad L_e = 1 + \frac{E_{\text{b}}}{N_0} b_{\text{r}}' \phi_{\text{no}}$.

The equivalent carrier-to-noise ratio deterioration value caused by equivalent white phase noise Φ_{no} is calculated with Expression (3.103) if the BER (corresponding to E_{b}/N_0) and code rate b_{r} are specified.

The Expression (3.103) is also used to solve the permitted phase jitter RMS σ_ϕ caused by phase noise:

$$\sigma_\Phi = \sqrt{2\Phi_{\rm no}B_n} = \sqrt{\frac{2(L_e-1)}{(E_{\rm b}/N_0)}\frac{B_n}{b_r b_r^{'}}}(\text{rad})$$

It is expressed with degree as:

$$\sigma_\Phi = 57.3\sqrt{\frac{2(L_e-1)}{(E_{\rm b}/N_0)}\frac{B_n}{b_r b_r^{'}}}(\circ). \tag{3.104}$$

3.2.6 Impacts of Linear Distortion on Data Transmission BER

The linear system may be described with amplitude/phase frequency characteristics. The code type will be distorted in the case of improper design of such characteristics and inter-symbol interference is produced, resulting in an increase of BER. Data transmission system includes the following two types.

(1) Non-band-limited system: For an ideal non-band-limited system, the phase frequency characteristic and amplitude frequency characteristic are constant, but in fact, there always has various "weak band-limited" filters (normally bandwidth is adopted to 1–1.5 of channel code rate) to filter all interferences and noises. Most space data transmission systems use this system.
(2) Nyquist shaping filter that is designed strictly by Nyquist criterion.

3.2.6.1 Relationships of Phase Frequency Characteristic and BER

The curves of BRE vs E_b/N_0 described in Sect. 3.2.5 are obtained under the condition of a linear phase frequency characteristic (that is, a flat group delay characteristic), but the group delay characteristic of an actual channel is often bent rather than flat. This will cause different components in the signal frequency spectrum to have different time delays, making their arrival times inconsistent through the channel. Consequently such generated waveform distortion will result in ISI due to the waveform tailing to the next code element (bits diffusion), which brings serious impacts on BER, especially in high-speed data transmission application.

Series of powers may be used to approach a bended group delay characteristic, that is:

$$\tau(\omega) = a_{\rm o} + a_1(\omega-\omega_o) + a_2(\omega-\omega_o)^3 + a_3(\omega-\omega_o)^3\cdots$$

Normally it takes the first order term and the quadratic term to approximate.

The first order term is:

$$D_1(f) = a_1(\omega - \omega_o).$$

The quadratic term is:

$$D_2(f) = a_1(\omega - \omega_o)^2.$$

When the carrier is conducted PSK modulation with one rectangle phase sequence, then the signal is expresses as:

$$S_i(t) = \sum_{n=1}^{\infty} \frac{b_n}{2} \cos(\omega_o + n\omega_s)t + \sum_{n=1}^{\infty} \frac{b_n}{2} \cos(\omega_o - n\omega_s)t,$$

where the rectangle phase sequence is $b_n = \frac{\sin(n\pi/2)}{n\pi/2}$. $\omega_o = 2\pi f_0$ is the carrier frequency, $\omega_s = 2\pi/T_s$. T_s is the width of code element.

When $S_i(t)$ is obtained by $\tau(f)$ of the group delay characteristic, each frequency spectrum component $(\omega_0 + \omega_n)$ will generate corresponding phase shift $\varphi + n$, then the output of channel is:

$$\begin{aligned} S_o(t) &= \sum_{n=1}^{\infty} \frac{b_n}{2} \cos[(\omega_o + n\omega_s)t + \varphi_n] + \sum_{n=1}^{\infty} \frac{b_n}{2} \cos[(\omega_o - n\omega_s)t + \varphi_{-n}] \\ &= \sqrt{x^2 + y^2 \cos\left(\omega_o t + \text{arc } tan \frac{y}{x}\right)}, \end{aligned} \tag{3.105}$$

where,

$$\begin{cases} x(t) = \sum_{n=1}^{\infty} \frac{b_n}{2} \cos\left(n\omega_s t + \varphi_{+n}\right) + \sum_{n=1}^{\infty} \frac{b_n}{2} \cos\left(n\omega_s t - \varphi_{-n}\right) \\ y(t) = \sum_{n=1}^{\infty} \frac{b_n}{2} \sin\left(n\omega_s t + \varphi_{+n}\right) + \sum_{n=1}^{\infty} \frac{b_n}{2} \sin\left(n\omega_s t - \varphi_{-n}\right) \end{cases},$$

where $x(t)$ and $y(t)$ are the in-phase component and the quadrature component respectively.

Observed from Expression (3.105), the code error is incurred by distortion of output rectangle pulse sequence.

In actual application, it is difficult to analyze the impact of group delay variable on BER under all modulation modes; on the contrary, it is much easier using the simulation method. The simulation below is conducted to analyze the BER deterioration degree caused by all group delay characteristics (linear, parabola, wave motion, etc.) under BPSK and QPSK. The linear group delay and parabolic delay normally relate to the bandpass filter installed in the equipment. The sinusoidal delay is usually caused by mismatched system impedance and multipath reflection.

In simulation, the symbol rate bandwidth is equal to symbol rate. If there is no group delay, then E_b/N_0 value is obtained while BER is 10^{-5}. Then, the group delay is changed to determine the added value of E_b/N_0 in case of the same BER, where the difference is the loss of E_b/N_0. It takes the product of symbol rate (b_r) and group delay (τ_D) as x-coordinate, and the loss L_D of E_b/N_0 as y-coordinate to indicate the degradation of BER, where group delay value is the biggest variable under full symbol rate bandwidths.

Reference [16] gives the fitted mathematical expression by simulation result curve, which can be available for reference. For the various group delay characteristics, the approximate relation of group delay value τ_D and L_D can be expressed by the following expressions:

(1) Linear group delay:

① For BPSK: $L_D = 0.1208(b_r\,\tau_D)^2 + 0.3537\,(b_r\,\tau_D) - 0.0315$. In the above expression, L_D is the loss of E_b/N_0, b_r is the symbol rate, τ_D is the biggest variation of the group delay in bandwidths of $1/b_r$, $(b_r\,\tau_D)$ is called normalization group delay.

② For QPSK: $L_D = 1.0322(b_r\,\tau_D)^2 + 1.1846(b_r\,\tau_D) - 0.1649$.

(2) Parabola group delay:

① For BPSK: $L_D = 0.2708\,(b_r\,\tau_D)^2 + 0.0284(b_r\,\tau_D) + 0.0016$.

② For QPSK: $L_D = 0.0227(b_r\,\tau_D)^2 + 0.4471(b_r\,\tau_D) - 0.0449$.

(3) Sine wave group delay:

① For one cycle fluctuation, BPSK: $L_D = 5.5227\,(b_r\,\tau_D)^2 + 1.2265(b_r\,\tau_D) - 0.0349$.

② For one cycle fluctuation, QPSK: $L_D = 12.2102\,(b_r\,\tau_D)^2 - 1.1859(b_r\,\tau_D) + 0.088$.

③ For two cycle fluctuation, BPSK: $L_D = 1.7333\,(b_r\,\tau_D)^2 + 0.3071(b_r\,\tau_D) - 0.0071$.

④ For two cycle fluctuation, QPSK:

$$L_D = 2.6227(b_r\tau_D)^2 - 0.1182(b_r\tau_D) + 0.0167. \tag{3.106}$$

The test results are available for reference since they are close to the expressions above.

3.2.6.2 Relationships of Amplitude Frequency Characteristic and BER

(1) Non-band-limited system

As mentioned above, both distortion of waveform and inter-symbol interference will appear if $H(\omega) \neq$ constant, resulting in increased BER. In this case, power series can be used to approximate the amplitude characteristics, that is:

$$H(f) = b_o + b_1 f + b_2 f_2 + b_2 f_2 + \cdots,$$

where f is the frequency deviation from f_0.

In engineering design, the condition can be satisfied by taking only the linear term and the quadratic term.

1) Linear amplitude distortion term:

$$H(f) = b_\mathrm{o} + b_1 f.$$

For convenient illustration, it can be normalized by b_0 to obtain $A(f) = 1 + \frac{b_1}{b_\mathrm{o}} f$.

Let $A_1 = b_1 f_p / 2b_\mathrm{o}$ and f_p be code rate, then

$$A(f) = 1 + 2A_1\left(\frac{f}{f_p}\right). \tag{3.107}$$

The C/N loss curve produced by the linear amplitude distortion is given in Ref. [8], as shown in Fig. 3.45.

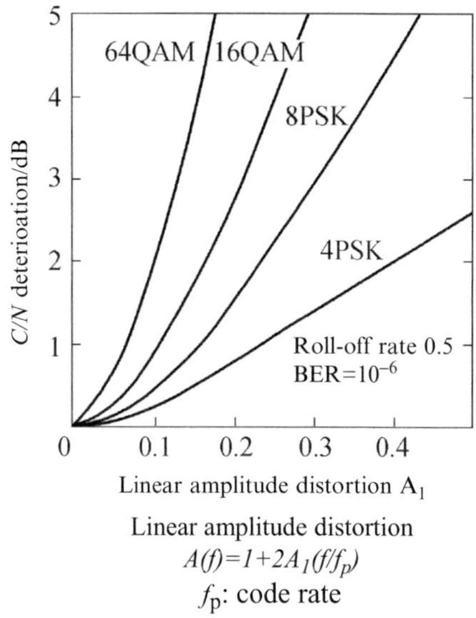

Fig. 3.45 Equivalent C/N deterioration amount caused by the linear amplitude distortion

2) Quadratic amplitude distortion term:

$$H(f) = b_o - b_o f^2,$$

normalized by b_0 to obtain $A(f) = 1 - \frac{b_2}{b_o} f^2$.

Let $\frac{b_2}{b_0} = \frac{4A_2}{f_p^2}$, that is $A_2 = \frac{b_2 f_p^2}{4b_0}$, then

$$A(f) = 1 - 4A_2 \left(\frac{f}{f_p} \right)^2. \tag{3.108}$$

The C/N loss curve produced by the quadratic amplitude distortion is given in Ref. [8], as shown in Fig. 3.46.

If the linear amplitude distortion term b_1/b_0 for normalization is given, $A_1 = \frac{b_1 f_p}{2b_o}$ can be used to solve A_1 and obtain the carrier-to-noise ratio loss L_{A1} caused by the linear amplitude distortion from Fig. 3.51.

In a similar way, if b_2/b_0 is given, $A_2 = b_2 f_p^2 / 4b_o$ is used to solve A_2 and obtain the carrier-to-noise ratio loss L_{A2} caused by the quadratic amplitude distortion from Fig. 3.52.

If a_1, a_2 and b_0, b_1, b_2 are not given by the group delay characteristics curve and amplitude frequency characteristics curve, the method described in Sect. 2.4.2 is able to solve each distortion coefficient above.

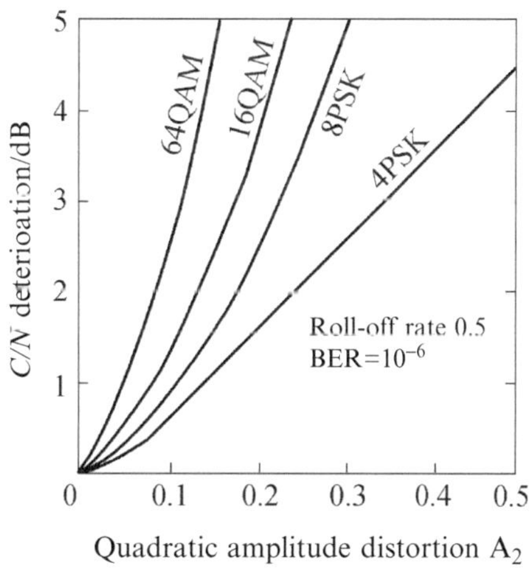

Fig. 3.46 Equivalent C/N deterioration amount caused by the quadratic amplitude distortion

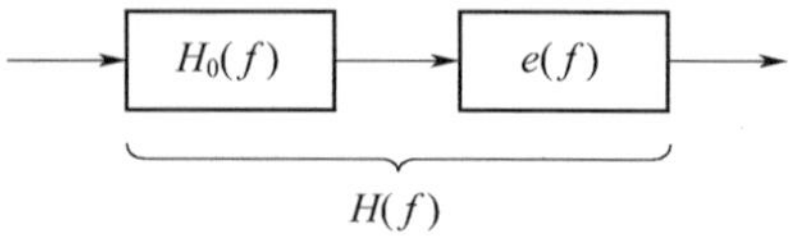

Fig. 3.47 Equivalence model of actual transmission function with errors

(2) Nyquist band-limited system

According to the Nyquist's first criterion and requirement of matched filtering, the filter on the *T*/*R* terminal shall satisfy the requirement of root cosine roll-off filter. Normally, the filter is impossible to meet all these requirements, there are always some errors, therefore, one transmission function $H(f)$ with errors may be combined with an ideal transmission function $H_0(f)$ and error transmission function $e(f)$ for equivalence, as shown in Fig. 3.47.

$$H(f) = H_o(f) \times e(f), \tag{3.109}$$

where $H_0(f)$ is the ideal characteristic, with a theoretic value of BER; $e(f)$ is the error characteristic deviated from the ideal characteristic, which will have an additional impacts on the BER, that is the deterioration of theoretic value.

Therefore, $e(f)$ shall be solved first, which is a curve not a straight line, and C/N deterioration amount then is solved from Fig. 3.45 or Fig. 3.46 according to the method of $H(f) \neq$ constant.

There is another design method for the situation where the ideal characteristic of a network and allowable frequency errors are given. In this case a certain form of filter may be used to approximate such ideal characteristic within the specified error range of frequency domain [17].

What's the real boundary of frequency domain errors? Since the error of frequency domain leads to the inter-symbol interference that causes BER, the Smith error expression is used to determine their relationship. After derivation, this expression is expressed as:

$$\frac{\sigma_i}{y_o} \leq 2M, \tag{3.110}$$

where y_0 is the sampling value for input signal y_i under inter-symbol interference, Φ_i is the RMS value of inter-symbol interference at the sampling point, and M is the module value of the allowable Max. frequency domain error.

M is determined by amplitude frequency error ε and phase frequency error κ. Under even distribution, $\varepsilon = \kappa$, $M^2 = 2\varepsilon^2 = 2\kappa$, derived from that, the relational expression of amplitude frequency characteristic $\varepsilon(f)$ and phase frequency characteristic error $\Delta\varphi(f)$ shall be obtained:

$$\varepsilon(f) \leq \frac{M}{\sqrt{2}} \tag{3.111}$$

$$\Delta\varphi(f) \leq 2\text{arc } sin \frac{M/\sqrt{2}}{2\sqrt{|C_0(f)|\times[|C_0(f)| + \varepsilon(f)]}}, \tag{3.112}$$

where $C_0(f)$ is the required transmission function for the ideal filter.

The example below describes how to apply the Smith error expression for integration with frequency domain method. For a baseband shaping network of PSK system, it uses the RMS value of inter-symbol interference that is not greater than 10 %; and according to the requirement of transmitting spectrum frame, if roll-off factor $\alpha = 0.25$, and the input signal of transmitting terminal is non-zero rectangle code pulse sequence, based on the previous analysis, the baseband shaping filter at *T*/*R* terminal shall have Nyquist's root-raised cosine roll-off characteristic, in order to eliminate inter-symbol interference and achieve *T*/*R* conjugate matching.

As mentioned above, the amplitude error frame and phase frequency error frame of the baseband shaping filter when $\alpha = 0.25$, $\Phi_i/y_0 \leq 10$ % are given in Ref. [8], as shown in Figs. 3.48 and 3.49. In Fig. 3.48, the transmitting filter and the receiving filter have different amplitude frequency error frames, the reason is that the derivation assumes the input code sequence is the impulse sequence and with even frequency spectrum, but in fact, the code is non-zero square wave and its frequency spectrum is sin*c* function, therefore, for satisfying the derivation condition above, one amplitude equalizer (mesh equalizer) shall be added on the transmitting terminal for weighing the inverse sin*c* based on the response of amplitude frequency transmitted by the filter. Figure 3.49 shows the error frame of phase frequency characteristics of the whole channel from transmitting to receiving section. With these frames, the network integration may be implemented with frequency domain, thus to make the amplitude frequency or the phase frequency is in the given frame. Obviously, different outcomes will be produced with different

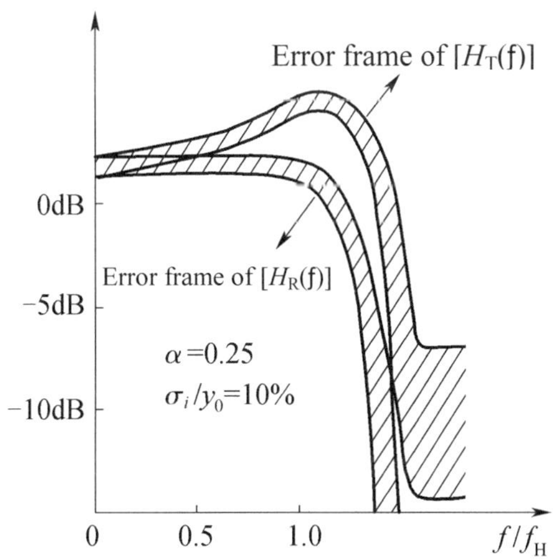

Fig. 3.48 Error frames of amplitude frequency characteristics

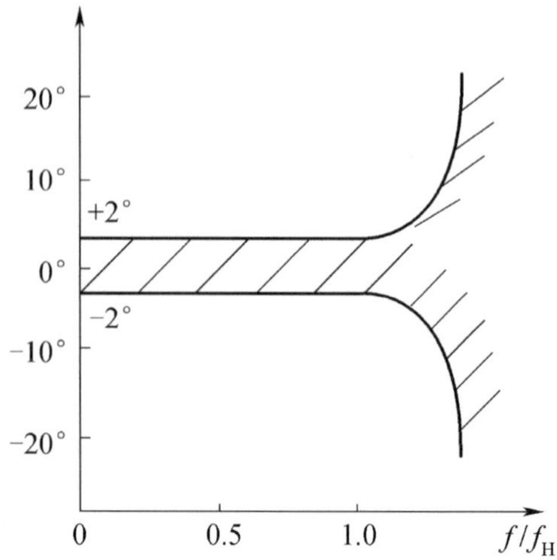

Fig. 3.49 Error frames of phase frequency characteristics

assumptions; nevertheless, this concept is valuable to use. Error frame is just an expression way for frequency response.

3.2.6.3 Relationships of Coherent Carrier Phase Error and BER [18]

When the phase error $\Delta\varphi$ is found with extraneous signal in carrier recovery, the output of coherent demodulation will be degraded.

(1) For BPSK signal

If static phase error is $\Delta\varphi$, then the output amplitude of demodulation shall be $u_{d,} = (A\cos\Delta\varphi)/2_D$, where A is carrier amplitude, and for BPSK the BER is:

$$P_{eB} = \frac{1}{2}\text{erfc}\left(\sqrt{\frac{E_b}{N_0}}\cos^2\Delta\varphi\right). \tag{3.113}$$

The incurred carrier-to-noise ratio is:

$$L_{\Delta\varphi} = 20\lg\frac{1}{\cos\Delta\varphi}. \tag{3.114}$$

If $L_{\Delta\varphi} \leq 1$ dB is required, $\Delta\varphi \leq 26°$ should be satisfied.

(2) For QPSK signal

QPSK is the quadrature sum of two-way BPSK signal. If $\Delta\varphi = 0$, the output of coherent demodulation in each way is normalized to the amplitude of $\cos(\pi/4)$; and when there is phase error $\Delta\varphi$, the output signal power for coherent demodulation in channel I and channel Q are $\cos^2((\pi/4)+\Delta\varphi)$ and $\cos^2((\pi/4)-\Delta\varphi)$ respectively, then the BER of QPSK signal in case of static phase error $\Delta\varphi$ should be:

$$P_{eB}=\frac{1}{4}\mathrm{erfc}\left[\sqrt{\frac{E_b}{N_0}}\cos\left(\frac{\pi}{4}+\Delta\varphi\right)\right]+\frac{1}{4}\mathrm{erfc}\left[\sqrt{\frac{E_b}{N_0}}\cos\left(\frac{\pi}{4}-\Delta\varphi\right)\right]. \tag{3.115}$$

The incurred carrier-to-noise ratio is:

$$L_{\Delta\varphi}=20\lg\frac{1}{|\cos\Delta\varphi|-|\sin\Delta\varphi|}. \tag{3.116}$$

If $\Delta\varphi<6°$, the calculated result from Expression (3.116) is $L_{\Delta\varphi}<1$ dB. Learned from that, QPSK has greater BER than BPSK, because $\Delta\varphi$ also causes cross interference for channel I and the channel Q. The degree of sensitivity against $\Delta\varphi$ ranks from QPSK, OQPSK, MSK to BPSK.

If lower SNR of the carrier recovery ring leads to stronger jitter of phase, the BER will be increased as well. Refer to Sect. 3.2.5.2.

3.2.6.4 Impacts of Phase Error of Phase Modulator on BER

Since signal loss and cross interference will be produced if the modulated value of the phase modulator deviates from the theoretic value $\Delta\varphi$, thus to bring out loss of carrier-to-noise ratio. The relationships of phase error and loss of carrier-to-noise is shown in Fig. 3.50 [18]. In addition, the parasitic amplitude modulation of the modulator shall be controlled to prevent the effect for BER from conversion of AM-PM.

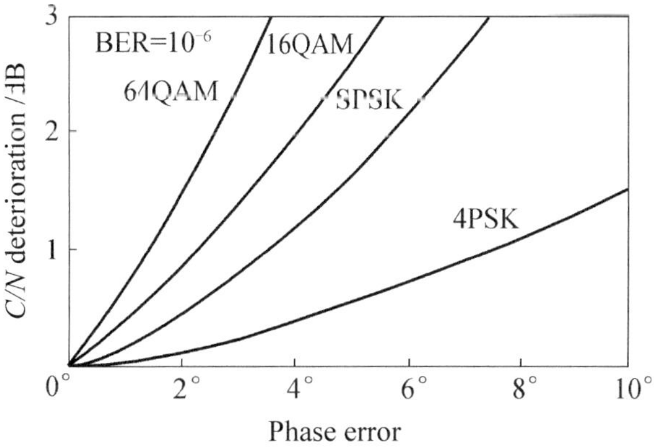

Fig. 3.50 Relationships of carrier phase error and carrier-to-noise ratio

Fig. 3.51 Equivalent *C/N* deterioration amount caused by timing phase error

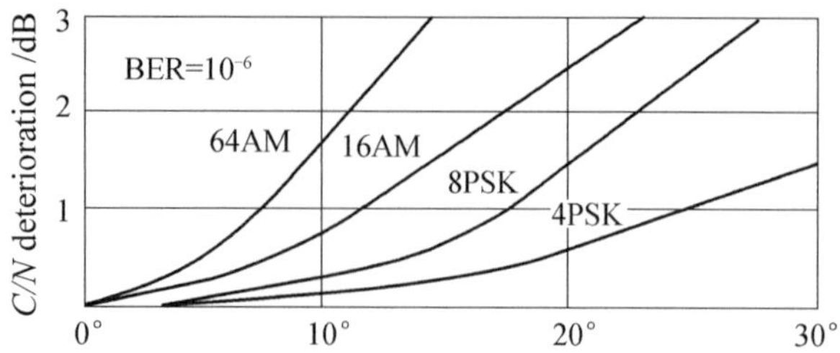

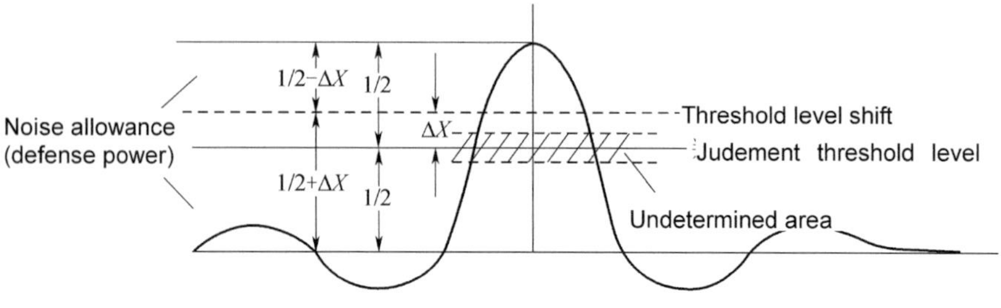

Fig. 3.52 Impacts of judgment level

3.2.6.5 Impacts of Deviation of Bit Timing Pulse on BER

Timing pulse is generated for bit synchronization. Its deviation or jitter will lead the sampling values of code pulses to deviate from their maximum values (this can appear in the contrary way, i.e., the code pulses jitter instead of sampling pulses), causing the BER to be increased and the equivalent *C/N* deteriorated. The amount of such deterioration is shown in Fig. 3.51 [19].

In Fig. 3.51, the X-coordinate is the phase deviation of timing pulse, and y-coordinate is the equivalent *C/N* deterioration amount △.

3.2.6.6 Relationships of Judgment Threshold Level Changes and BER

The judgment threshold normally takes the medium value. When the threshold changes (or imprecise), then the noise allowance of code 1 or 0 will be reduced, as shown in Fig. 3.52 [19].

If the threshold level has changed, the noise allowance of code "1" is reduced to (1/2+ΔX), but that of code "0" is increased to (1/2+ΔX), so that the increase of BER caused by the former is greater than the reduction of BER caused for the latter. Overall, the change leads to an increase of BER. Moreover, the level area that is hardly judged, i.e., undetermined area, shall be considered in high-speed data transmission. If the signal level is located in such area, it is equivalent that the signal amplitude is reduced, and so BER is increased by both factors. Refer to Fig. 3.53 for the equivalent *C/N* deterioration amount produced [19].

In Fig. 3.53, the *x*-coordinate refers to the percentage of threshold changes.

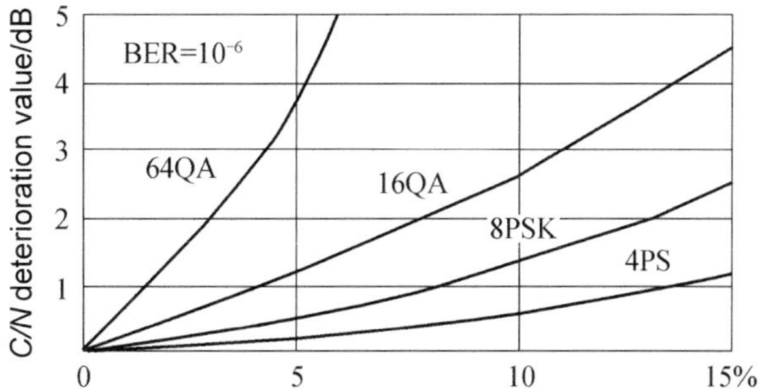

Fig. 3.53 *C/N* deterioration amount caused by judgment threshold changes

3.2.6.7 Impacts of Standing Wave and Multipath Reflection on BER

Essentially, the standing wave, multipath reflection and group delay fluctuation of the filter all are caused by the reflected signals. If any reflected signal is found in the transmission system, then the amplitude frequency characteristic and the phase frequency characteristic will be affected. Its analysis is described as follows:

For two-line model, the master signal is set as:

$$U_D(t) = E_0 \cos \omega t. \tag{3.117}$$

In the case of reflected wave, the reflected wave is set as:

$$U_r(t) = rE_0 \cos(\omega t - \omega\tau_0) = rE_0 \cos \omega t + rE_0 \sin \omega t(\sin \omega\tau_0), \tag{3.118}$$

where γ is the modulus of reflection coefficient, and τ_0 is the time delay of the reflecting wave relative to the master signal.

The composite signal of the master signal (direct wave) with echo wave is

$$\begin{aligned} U(t) &= U_D(t) + U_r(t) = E_0 \cos \omega t(1 + r\cos \omega\tau_0) + rE_0 \sin \omega t \sin \omega\tau_0 \\ &= E_0 H_r(\omega) \cos[\omega t + \varphi_r(\omega)], \end{aligned} \tag{3.119}$$

where,

$$H_r(\omega) = \sqrt{(1 + r\cos\omega\tau_0)^2 + (r\sin\omega\tau_0)^2} = \sqrt{1 + 2r\cos\omega\tau_0 + r^2} \tag{3.120}$$

$$\varphi_r(\omega) = -\arctan\frac{r\sin\omega\tau_0}{1 + r\cos\omega\tau_0} \tag{3.121}$$

$$\tau(\omega) = \frac{\mathrm{d}\varphi(\omega)}{\mathrm{d}\omega} = \frac{\tau r(r + \cos\omega\tau_0)}{1 + r^2 + 2r\cos\omega\tau_0}. \tag{3.122}$$

Its maximum group delay is found at $\omega\tau_0 = (2n+1)\pi$, and the minimum value is at $n\pi$.

From Expressions (3.120) and (3.121) we can see that when the main signal wave $U_D(t) = E_o\cos\omega t$ passes through a transmission channel with amplitude

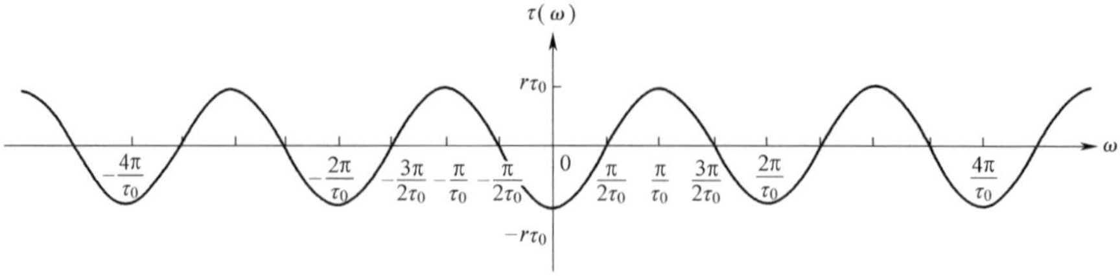

Fig. 3.54 Group delay fluctuation caused by reflection

frequency characteristic of $H_r(\omega)$ and phase frequency characteristic of $\varphi_r(\omega)$, due to superposition of echo waves, they will bring an impact on the BER in the way described above. Therefore, the analysis method aforesaid may be cited here.

The reflection loss for each part normally will be specified, so during transmission process, if $r \ll 1$, we can obtain that from Expressions (3.120) and (3.129):

$$H_r(\omega) \approx \sqrt{1 + r\cos\omega\tau_0} \tag{3.123}$$

$$\varphi_r(\omega) \approx -r\sin\omega\tau_0 \tag{3.124}$$

$$\tau(\omega) = \frac{\mathrm{d}\varphi(\omega)}{\mathrm{d}\omega} = -r\tau_0\cos\omega\tau_0. \tag{3.125}$$

Learned from that, the standing wave will affect the fluctuated amplitude frequency characteristic, phase frequency characteristic as well as group delay characteristic. Refer to Fig. 3.54 for group delay characteristic.

Multipath effect is also a kind of reflection. When the reflection coefficient is small, it can be estimated similarly. For aerospace information transmission system, since the target is movable, so τ_0 is not constant, as a result, the fluctuation characteristic is varied. When the fluctuation characteristic is deteriorated to the extent that the BER is increased as well, on the basis that the BER is increased with increase of $\tau(\omega)/T_c$, consequently, the multipath effect will greatly affect the BER at high code-speed. In addition, the multipath effect will significantly worsen the detectability of cross polarization, thereby causing serious interference between the two quadrature polarized channels.

3.2.7 *Impacts of Non-linear Distortion on Data Transmission BER*

3.2.7.1 Characteristics of Channel Nonlinearity

1. Definitions of nonlinearity indexes

The nonlinearity indexes mainly are described as follows: gain 1 dB compression point power, the third-order intermodulation coefficient, amplitude-phase conversion (AM-PM), amplitude-group delay conversion [AM/$\tau(\omega)$], amplitude modulation - amplitude modulation conversion and the third-order interception point, among which are expressed against the same nonlinearity with different methods and are interchangeable. In data transmission system, the first four terms are highly relevant, the first two of which describe the nonlinearity of amplitude while the latter two are related to the nonlinearity of phase (under nonlinearity status, the change of output phase is varied with the change of input amplitude); for transmission of phase-modulated signal, the latter two are more important.

2. Mathematic expressions of nonlinearity indexes [15]

Use an infinitely-many-term power series to approximate the non-linear curve, that is:

$$f_0(t) = a_0 + \sum_{n=1}^{\infty} a_n f_i^n(t), \quad n \text{ is positive integral} \tag{3.126}$$

where $f_0(t)$ and $f_i(t)$ respectively are the output signal and input signal of the channel, and a_n is the coefficient of each order, which is a complex number normally.

It assumes that the input signal is the sum of two sine signals at constant amplitude, $f_i(t) = (A/\sqrt{2})\cos\omega_1 t + (A/\sqrt{2})\cos\omega_2 t$, and the input power retains the power that is at amplitude A of the single carrier signal: $P_{\text{in}} = (1/2)A^2$, then substitute the two signals into Expression (3.126) to develop many new frequency components (the product of nonlinearity cross modulation). For slight nonlinearity channel, taking account for filtering effect of the filter, only the fundamental wave term and the third-order cross modulation term are usually found in the output signal, that is:

$$\begin{aligned} f_0(t) - \frac{A}{2}\left(a_1 + \frac{9}{8}a_3A_2\right)(\cos\omega_1 t + \cos\omega_2 t) + \frac{3a_3A^3}{8\sqrt{2}}\left[\cos(2\omega_1 - \omega_2)t\right. \\ \left. + \cos(2\omega_2 - \omega_1)t + \cos(2\omega_1 + \omega_2)t + \cos(2\omega_2 + \omega_1)t\right] \end{aligned} \tag{3.127}$$

According to Expression (3.127) and the definition of the third-order modulation coefficient, we can obtain:

$$I_{\text{mr3}} = \left(\frac{3a_3A^3}{8\sqrt{2}}\right) \Big/ \left[\frac{S}{\sqrt{2}}\left(a_1 + \frac{8}{9}a_3A_2\right)\right] + \frac{3a_3A^3}{8a_1 + 9a_3A_2}. \tag{3.128}$$

Since α_1 and α_2 has opposite phase, then I_{mr3} shall be a negative value. If the input signal $f_i(t)$ is a SF sine signal, $f_1(t) = A\cos w_0 t$, and substitute it into Expression (3.126) and develop the expression to obtain:

$$f_0(t) = \left(a_1 A + \frac{3}{4} a_3 A^3\right) \cos \omega_0 t + \frac{a_2 A^2}{2} \cos 2\omega_0 t + \frac{1}{4} a_3 A^3 \cos 3\omega_0 t. \quad (3.129)$$

Learned from Expression (3.129), if the channel's operating bandwidth is less than one octave, only the fundamental wave appears in the output signal. Consequently, the nonlinearity characteristic just affects the change of fundamental wave gain, that is $(a_1 + 3/4 a_3 A^2)$, and $\left(a_1 + \frac{3}{4} a_3 A^3\right)$ is always less than the linear gain α_1; and the difference of them is varied with the increase of input power $A^2/2$, which is the gain compression effect of non-linear channel. The compression characteristic defines the ratio of the amplitude post compression and prior to compression, it is expressed as:

$$q = \frac{\left(a_1 A + \frac{3}{4} a_3 A^3\right)}{a_1 A} = 1 + \frac{3}{4} \frac{a_3}{a_2} A^2. \quad (3.130)$$

Because a_3 is opposite to a_1 (opposite phase), then $q < 1$, it is related to the input signal A, when $q = -1$ dB, it means that the gain is compressed with 1 dB, correspondingly, $A^2/2$ is the 1 dB input power of the compression point.

If a_1 and a_3 in Expression (3.129) are considered as complex numbers, take a_1 for reference and let its phase be 0°, and α_3 corresponds to φ_3, then the output fundamental wave shall be:

$$f_0(t) = \left[a_1 A + \frac{3}{4} |a_3| A^3 e^{j\varphi 3}\right] \cos \omega_0 t = g(A) \cos\left[\omega_0 t + \theta(A)\right], \quad (3.131)$$

where,

$$g(A) = \left[\left(a_1 A + \frac{3}{4} A^3 |a_3| \cos \varphi_3\right)^2 + \left(\frac{3}{4} A^3 |a_3| \sin \varphi_3\right)^2\right]^{\frac{1}{2}} \quad (3.132)$$

$$\theta(A) = -\arctan\left[\frac{\frac{3}{4} |a_3| A^3 \sin \varphi_3}{a_1 A + \frac{3}{4} |a_3| A^3 \cos \varphi_3}\right], \quad (3.133)$$

where $\theta(A)$ refers to amplitude-phase conversion (AM-PM), and $g(A)$ refers to AM-AM conversion.

The physical meaning of that is: when the carrier amplitude A changes, under nonlinearity effect, it will lead to the variation of carrier phase $\theta(A)$ with amplitude compression of $g(A)$, as well as phase angular distortion, but if it is a constant envelope signal (A retains the same), then $\theta(A)$ is also a constant value and no phase angular distortion is produced. In addition, if there is no φ_3, let $\theta(A) = 0$, similarly, no phase angular distortion is found. That is to say, when the envelope is not constant, only amplitude nonlinearity will not cause distortion of angle, but there is angle distortion in the case of phase nonlinearity, which will affect the BER and the change of frequency spectrum. Nevertheless, the amplitude nonlinearity always accompanies phase nonlinearity.

Create simultaneous expression of Expression (3.128) and Expression (3.130), there is:

$$I_{mr3} = \frac{q-1}{3q-1} \tag{3.134}$$

$$q = \frac{I_{mr3}-1}{3I_{mr3}-1}. \tag{3.135}$$

Expression (3.135) shows the relationships of gain compression and the third-order cross modulation. Since 1 dB compression point $q = 0.891$ (201 g $0.891 = 1$ dB) when inputting a single signal, it substitutes into Expression (3.135) to obtain the third-order cross modulation at 1 dB compression point:

$$I_{\mathrm{mr3}}(1\ \ dB) = -19.35\ \ \mathrm{dB}. \tag{3.136}$$

Because fitting all kinds of nonlinearity through power series only obtains the approximated value, the actual result will be different from the measured value given by the expression above.

3.2.7.2 Impacts of AM-PM on Bit Error Rate

(1) In case of BPSK

Under AM-PM conversion, when the non-linear channel inputs an phase angle modulation wave $f_i(t) = A\cos[\omega_0 t + \varphi(t)]$, then $f_{\mathrm{i}}(t)$ is substituted into Expression (3.131) and obtain:

$$f_0(t) = g(A)\cos\left[\omega_0 t + \varphi(t) + \theta(A)\right]. \tag{3.137}$$

Learned from Expression (3.137), the change of carrier amplitude A will produce $\theta(A)$ in carrier phase, which is an interference against $\varphi(t)$. If $\theta(A)$ is DC or slowly changed interference, it can be offset by phase-lock coherent demodulation; if $\theta(A)$ is the interference that rapidly changes and cannot be eliminated, then the spectrum spread and the bit error rate will be increased and deviated from its theoretic value.

Under ideal phase demodulation, the demodulated output signal just relates to the carrier phase, not to amplitude, but if the modulator is not good enough it produces a relationship of the demodulation output and carrier amplitude, then the interference will appear in $g(A)$, moreover, the bit error rate will be increased while C/N is decreased due to compression of g(A).

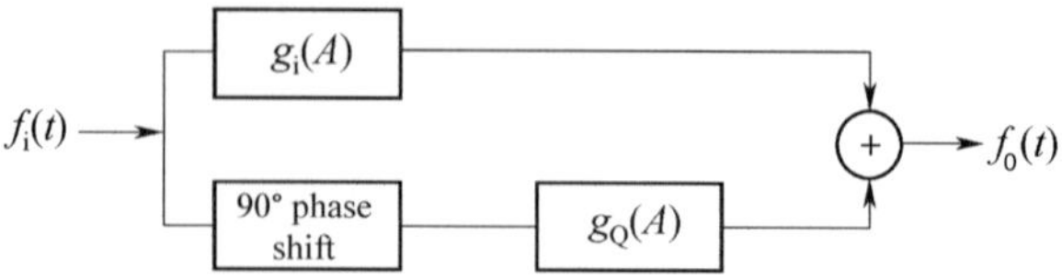

Fig. 3.55 Non-linear distortion model of in-phase/quadrature channel

(2) In the case of QPSK

AM-PM conversion still affects the quadrature modulated signal. It is analyzed that $\varphi(t)$ may not be considered during the research of quadrature interference, and the Expression (3.137) is developed with trigonometric function to obtain:

$$\begin{aligned} f_0(t) &= g(A)\cos\theta(A)\cos\omega_o t - g(A)\sin\theta(A)\sin\omega_o t \\ &= g(A)\cos\theta(A)\cos\omega_o t - g(A)\sin\theta(A)\cos\left(\omega_o t + \frac{\pi}{2}\right) \\ &= g_i(A)\cos\omega_o t - g_\theta(A)\sin\omega_o t. \end{aligned} \tag{3.138}$$

Expression (3.138) is represented with distortion model, as shown in Fig. 3.55. In Fig. 3.55, $g_i(A) = g(A)\cos\theta(A)$,$g_Q(A) = g(A)\sin\theta(A)$.

Learned from Fig. 3.55, when AM-PM conversion $\theta(A)$ occurs in the channel, the quadrature channel will generate the quadrature component $g_\theta(A)$, if $g(A)$ and $\theta(A)$ are variables and its spectrum is overlapped with the information spectrum, then the bit error rate will be increased.

If $\theta(A) = 0$, correspondingly, the interference of the quadrature channel $g_Q(A) = 0$, and $g_i(A) = g(A)$ for in-phase channel. Similarly, the interferences to channel Q and channel I are obtained from that, where the interference/signal ratio of the in-phase channel is $\frac{\Delta}{A} = \frac{g_Q(A)}{g_I(A)} = \frac{\sin\theta_A}{\cos\theta_A}$. Consequently, the greater the ratio, the more severe the spectrum spread, and the more deterioration the bit error rate of the two channels. As mentioned above, for data transmission system using quadrature modulation, the AM-PM conversion will cause to quadrature interference and lead to increase of the bit error rate, but all these factors are decided by the change of signal amplitude, therefore, the phase-modulation wave with constant envelope is needed.

3.2.7.3 Impacts of Band-Limited Non-constant Envelope on Bit Error Rate

In fact, the band-pass filter normally is installed in the channel to limit the signal bandwidth, which has the main functions as follows:

(1) Increase channels of multi-path communication by improving spectrum availability. For example, the shaping filter using raised cosine roll-off feature is used to narrow signal spectrum and, at the same time, meet the Nyquist's first criteria for no inter-symbol interference.

(2) Limit out-of-band power and harmonic at the transmitting parts to prevent interference of signals.
(3) Filter noise on a repeater to increase SNR.

Paragraph (1) and (2) above in category I and category II space communication are not restricted like others.

As mentioned above, in the ideal phase shift keying modes like BPSK, QPSK, and OQPSK, the envelope of the phase-modulated wave is constant, but after band-pass filter, part of spectrum has been filtered out, which leads to distortion of signal, so that the envelope of the phase-modulated wave will be varied as well. Taking BPSK as an example, during phase shift keying generating $0° \rightarrow 180°$ phase step point, its corresponding frequency spectrum is infinitely wide since the frequency is the differential of the phase. If it passes through a band-pass filter, the waveform will get distorted since not all the spectrum is passed. The resulted envelope drop and fluctuation are shown in Fig. 3.56 [20].

Observed from Fig. 3.56c, the envelope waveform resumes to the waveform before band-limiting after hard amplitude limiting (without phase nonlinearity) and correspondingly the spectrum is also close to that before band-limiting. Therefore, the functions mentioned in paragraph (1) and (2) above are meaningless. It is pointed out in Ref. [10] that the fall of envelopes in BPSK and QPSK will reach to 100 % under band-limiting (due to $0° \rightarrow 180°$ phase step), and that in OQPSK is up to 30 % (due to $0° \rightarrow 90°$ phase step).

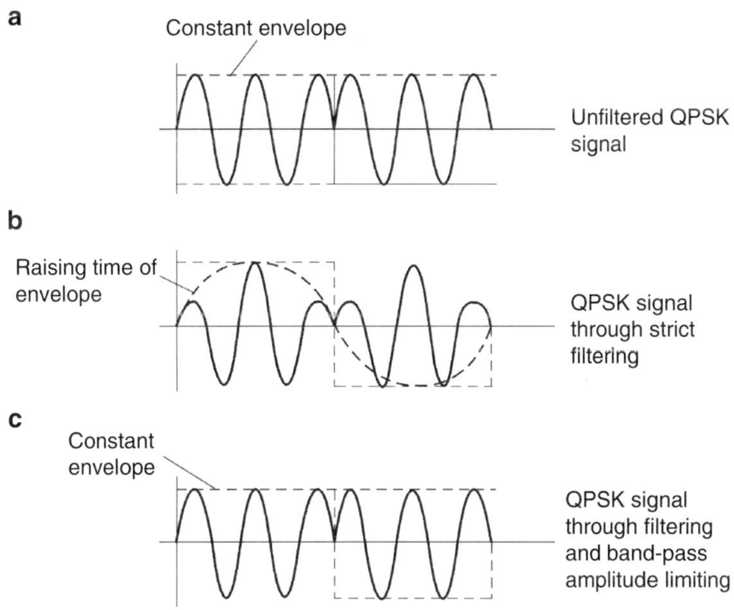

Fig. 3.56 Impact of band-limiting on envelope

Fig. 3.57 Impacts of different band-limiting conditions on bit error rate under nonlinearity effect

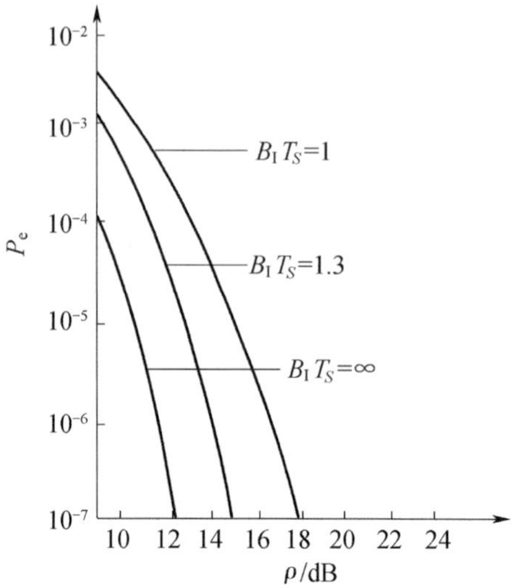

If the phase-modulated wave with fluctuated envelope output after band-limiting is undergoing a nonlinear effect (AM/PM, AM/$\tau(t)$), the phase of carrier will be changed, which causes the phase modulated signal to be distorted, results in spectrum spreading and inter-symbol interference, and increases the bit error rate as well. Reference [18] reveals a study of impacts of QPSK signal on bit error rate, that is, how the QPSK signal affects the bit error rate through power amplification of traveling-wave tube by bring fluctuation in envelope via different band-limiting conditions. The power amplifier of TWT operates in the non-linear status that saturated output level rolls back 1 dB. The band-limited filter is the four-section Chebyshev filter, where B_I is 3 dB bandwidth and T_s is the symbol width. In QPSK system, the relationships of it and total bit rate f_b are: $T_S = 2/f_b$, $B_I T_S = 2B_I/f_S$.

Learned from Fig. 3.57, under unlimited bandwidth ($B_I T_s = \infty$), the SNR corresponding to the bit error rate $P_e = 10^{-7}$ is close to the theoretic value because of constant envelope. If the bandwidth is narrower, i.e., $B_I T_s = 1$, since the nonlinearity of TWT causes greater fluctuation of envelope that leads to greater bit error rate, therefore, the SNR ρ has to raise to 17.5 dB to retain the same bit error rate of 10^{-7}, that is, the equivalent SNR deterioration value is 5.5 dB, which is considerable.

As mentioned above, because the band-limiting condition changes the carrier amplitude of phase-modulated signal, then the change of amplitude generates the inter-symbol interference through AM-PM nonlinearity effect so as to increase the bit error rate, which constitutes the physical meaning of the impacts of nonlinearity effect on bit error rate. If the amplitude is a constant envelope, then the bit error rate won't be changed, as a result, the modulation of constant envelope is a very important technique.

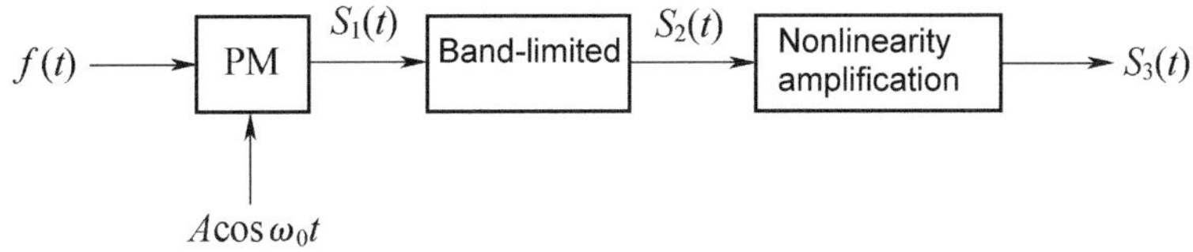

Fig. 3.58 Spectrum spreading analysis model

3.2.7.4 Frequency Spectrum Spreading Effect Caused by Channel Nonlinearity

The frequency spreading effect means that the spectrum of a band-limited digital-modulation signal is spread through a non-linear channel, which can be represented with the spectrum spreading of a single frequency phase-modulated signal, its analysis model is shown in Fig. 3.58.

In Fig. 3.58, $f(t) = A_m \sin \Omega t$ is the single frequency phase-modulated signal, $A\cos\omega_0 t$ is the carrier signal, the phase modulated wave after phase modulation $S_1(t)$ is the wideband signal containing $\omega_0 \pm n\Omega$. For limiting its bandwidth, the band-pass filter shall be set to only allow $\omega_0 \pm \Omega$ passing through while filtering out the signal beyond the bandwidth $\omega_0 \pm \Omega$, where such signal is output $S_3(t)$ after linear amplification.

$S_1(t)$ can be expressed with Bessel function as:

$$S_1(t) = A\cos(\omega_0 t + m\sin\Omega t) = A\sum_{n=-\infty}^{+\infty} J_n(m)\cos(\omega_0 + m\Omega)t\cdots, \quad (3.139)$$

where m is the phase-modulation index. As shown in Expression (3.139), $S_1(t)$ is a signal with wide spectrum, which can be output from $\omega_0 \pm \Omega$ band-pass filtering:

$$S_2(t) = AJ_0(m)\cos\omega_0 t + J_1(m)\cos(\omega_0 t + \Omega)t - J_1(m)\cos(\omega_0 t - \Omega)t. \quad (3.140)$$

At that time, $S_2(t)$ is a narrow-band signal that becomes a phase-modulated and amplitude-modulated signal, that is, there is the fluctuation of envelope. After passing through a nonlinear channel, $S_2(t)$ can be developed with a power series to obtain $S_3(t)$ (the third order approximation), that is:

$$\begin{aligned} S_3(t) &= a_1 s_2(t) + a_2 s_2^2(t) + a_3 s_2^3(t) \\ &= a_1 J_0(m)\cos\omega_0 t + J_1(m)\cos(\omega_0 + \Omega)t - J_1(m)\cos(\omega_0 - \Omega)t \\ &\quad + a_2[J_0(m)\cos\omega_0 t + J_1(m)\cos(\omega_0 + \Omega)t - J_1(m)\cos(\omega_0 - \Omega)t]^2 \\ &\quad + a_3[J_0(m)\cos\omega_0 t + J_1(m)\cos(\omega_0 + \Omega)t - J_1(m)\cos(\omega_0 - \Omega)t]^3. \end{aligned} \quad (3.141)$$

Then it is developed with $\cos^2 A = 1/2(1 + \cos 2A)$, $\cos^3 A = 1/4(\cos 3A - 3\cos A)$ to obtain many combined frequency components, in which components near the

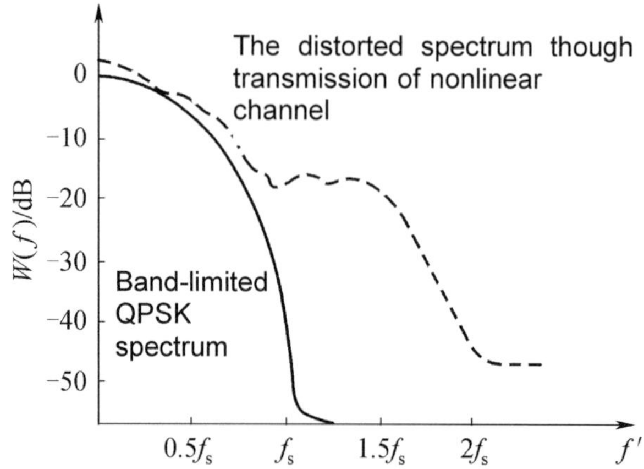

Fig. 3.59 Frequency spectrum spreading caused by channel nonlinearity

ω_0 fundamental wave include ω_0,$(\omega_0+\Omega)$, $(\omega_0-\Omega)$, $(\omega_0+2\Omega)$, $(\omega_0-2\Omega)$, $(\omega_0+3\Omega)$, and $(\omega_0$-$3\Omega)$, that means, the frequency spectrum has been spread.

It is a fact that more spectrum components produced in digital PM signal can be derived from the concept of single frequency signal. The digital-modulation signal will generate plenty of frequency components that cause the spectrum to spread after band-limiting and nonlinearity effect, and such spread will interfere with the neighboring channel. Therefore, the band-limiting is quite a critical issue for multiplex digital microwave communication system, but rarely affects the digital microwave communication for a single channel.

For a band-limited QPSK signal that is given in Ref. [18], the computer simulation results of the equivalent baseband spectrum after TWT nonlinear amplification is shown in Fig. 3.59.

As shown in Fig. 3.59, after nonlinear amplification, the side-lobe components suppressed by band-limited filtering get raised up again. The degree of side-lobe raising-up varies with that of nonlinearity.

3.2.7.5 Impacts of Nonlinearity on Frequency-Division Multiplex (FDM) Performance

The following three issues shall be necessarily taken into considerations under FDM operating conditions:

- Intermodulation noise: The nonlinearity effect will generate the new frequency components when inputting multiple carriers, if these components fall into the signal band, they will cause interfering noise. When generating the new frequency components include two factors like nonlinearity of amplitude (AM-AM) and nonlinearity of phase (AM-PM).
- Spectrum spreading.
- Compression of output power.

To maximize the output efficiency of the power amplifier on a satellite, it shall be operated close to the saturated points, nevertheless, the serious intermodulation noise will appear when operating on multiple carriers.

(1) Intermodulation caused by amplitude nonlinearity

The amplitude nonlinearity of power amplifier can be represented with power series. In a satellite transponder, since the bandwidth of carrier frequency is much greater than frequency band of the transponder and the even terms of the power series do not contain the fundamental wave components of a signal, the frequency components fallen within the bandwidth of the transponder will not contain the components combined with the even terms of power series and thus the nonlinearity of the power amplifier can be simplified as:

$$V_0 = a_1 V_i + a_3 V_i^3 + a_5 V_i^5 + a_7 V_i^7 + \cdots \tag{3.142}$$

Since the value of a_7 or above is very small (taking TWTA of № IS-IV Satellite as example, at the saturation point, $a_1 = 1.65$, $a_3 = -0.887$, $a_5 = 0.167$), it can be further approximated to:

$$V_0 = a_1 V_i + a_3 V_i^3 + a_5 V_i^5. \tag{3.143}$$

After FDM of channel M, multiple carriers are approximated to replace the input signals, that is:

$$V_i = \sum_{i=1}^{k} A_k \cos\left[\omega_i t + \varphi_i(t)\right], \quad k \geq m. \tag{3.144}$$

Substitute Expression (3.144) into Expression (3.143) to obtain the third-order and the fifth-order intermodulation component when input Kth sine wave, where the number of intermodulation products are listed in Table 3.6.

The symbols A, B, C, D… in Table 3.6 respectively refer to each input frequency, for example, $A = f_i$ and $A \neq B$, T_3 and T_5 refer to the main mode of the third-order and the fifth-order intermodulation product respectively.

Table 3.6 Number of intermodulation products at each order under k sinusoidal inputs

Intermodulation mode	Order sequence (m)	Number of frequencies in intermodulation (L)	Total number of intermodulation products under this mode
2A − B	3	2	$K(K-1)$
→A + B − C	3	3	$T_3 = K(K-1)(K-2)/2$
3A − B − C	5	3	$K(K-1)(K-2)/2$
A + 2B − 2C	5	3	$K(K-1)(K-2)$
→A + B + C − E	5	5	$T_5 = K(K-1)(K-2)(K-3)(K-4)/12$

These intermodulation products are the interfering noises, wherein T_3 and T_5 are main components. Reference [19] gives the approximate value of the intermodulation noise power P_m at the m order.

1) Hard limiter.

$$P_m/P \approx 1/m^2, \tag{3.145}$$

where P_m is the total power of intermodulation noise, P is the limiting power of hard limiter, and m is the intermodulation order.

2) Soft limiter. For *I*/*O* nonlinearity, the soft limiting characteristics are assumed as follows:

$$V_0 = (2P)^{1/2}\mathrm{erf}'|V_i|(\mathrm{sgn}V_i), \tag{3.146}$$

where $\mathrm{erf}'\left(\frac{X}{\sigma}\right) = \left(\frac{1}{2\pi}\right)^{\frac{1}{2}} \int_0^X \exp\left(-\frac{t^2}{2\sigma^2}\right)$ is a type of error function and under this condition then obtain

$$\frac{P_m}{P} \approx \frac{1}{m^2(1+y^2)^m},$$

where σ is the assigned parameter of the error function, and y refers to the scale of nonlinearity. When the maximum limiting output power is P, then $P/(1+y^2)$ means the output power at the operating point (that is the power returned back from the maximum limiting output power); if $y = 1$, the power at the operating point is $P/2$, that is the output power is returned back for 3 dB; if $y = 0$, that is, the hard limiting amplifier approximates the maximum saturated output. Table 3.7 lists the ratio of intermodulation noise and total power respectively under hard limiting and power returning of 3 dB conditions (power amplifier with error function).

Table 3.7 Relative powers of hard-limiting amplifier and error function amplifier with 3 dB output power return

Intermodulation order	Hard limiter		Error function limiter with 3 dB power return
	P_m/P	Approximation value $P_m/P \approx l/m^2$	$P'_m/P = P_m/P \bullet 2^m$
3	0.0980	0.1111	0.0123
5	0.0367	0.04	0.00115
7	0.0191	0.0204	0.000149
9	0.0117	0.0123	0.000023
11	0.0079	0.00826	–
13	0.00568	0.00592	–
15	0.00429	0.00444	–
Total	0.215	$(\pi^2/8 - 1) = 0.2336$	0.0136

For comparison, the approximation value of hard limiter is also given in this table, that is, $P_m/P \approx l/m^2$, and the number of input signals $K \gg 1$. In Table 3.7, P_m and P'_m respectively are the intermodulation noise power under the two conditions. Although the analyzed results above are approximate, it is clear that returning of power reduces the intermodulation noises learned from the table.

(2) Intermodulation noise caused by AM-PM conversion

New frequency components will be generated through AM-PM conversion, this is because the envelope is fluctuated while multiple carriers are input, thus use the following to create an additional phase shift in every carrier through AM-PM conversion, according to $\omega = \mathrm{d}\varphi/\mathrm{d}t$, so the new components are generated.

When the input $V_i(t)$ represents the sum of multiple sine waves, it is expressed as:

$$V_i(t) = \sum_{i=1}^{n} A_i \cos \omega_i t = \sum_{i=1}^{n} A_i \cos (\omega_0 t + \Omega_i t), \tag{3.147}$$

where ω_0 is the sine wave at the lowest frequency and $\Omega_i = \omega_i - \omega_0$.

In aggregation of multiple carriers, the envelope $A(t)$ of $V_i(t)$ and the phase $\varphi_0(t)$ are varied, and the Expression (3.147) is expressed as:

$$V_i(t) = A(t) \cos [\omega_0 t + \varphi_0(t)] \tag{3.148}$$

where,

$$A^2(t) = \left(\sum_{i=1}^{n} A_i \cos \Omega_i t \right)^2 + \left(\sum_{i=1}^{n} A_i \sin \Omega_i t \right)^2$$
$$\varphi_0(t) = \arctan \left(- \frac{\sum_{i=1}^{n} A_i \sin \Omega_i t}{\sum_{i=1}^{n} A_i \cos \Omega_i t} \right).$$

Put such $V_i(t)$ into a nonlinear system with AM-PM conversion, and then each output carrier will add a phase shift $\theta_A(t)$ caused by AM-PM conversion because of amplification variation of $A(t)$, that is:

$$V_0(t) = \sum_{i=1}^{n} A_i \cos [\omega_0 t + \theta_A(t)], \tag{3.149}$$

where $\theta_A(t) = \theta(A) \cdot A(t)$ and $\theta(A)$ is the amplitude conversion coefficient with unit of (°)/dB.

For the Kth carrier, there is:

$$V_{ok}(t) = A_k \cos [\omega_k t + \theta_A(t)] \approx A_i \cos \omega_k t - \theta_A(t) \sin \omega_k t, \tag{3.150}$$

where the second term is the intermodulation noise on ω_k carrier through AM-PM conversion.

The relationships of the intermodulation interfering noise P_3 and the signal P_s generated by the constant sine wave with three amplitudes that all are A is given in Ref. [18], under the circumstance of AM-PM conversion coefficient $\theta(A) = (k/2)$ $A^2(t)$:

$$\frac{P_S}{P_3} = \left(\frac{1}{2kA^2}\right)^2. \tag{3.151}$$

Since AM-PM conversion also exists in the linear section of the amplitude characteristic but AM-AM observes small, therefore, for the linear section (small signal), the intermodulation noise is mainly caused by AM-PM conversion; for the nonlinear section, the intermodulation noise is caused by the amplitude nonlinearity, and total noise is the sum of both. In Ref. [18], it is proved both in theoretical and experimental investigation that the total intermodulation noises caused by AM-AM and AM-PM may be calculated by multiplying the intermodulation noise generated through AM-AM and any coefficient that is greater than 1, and such coefficient only relates to the operating points of power amplifier.

3.2.7.6 Measures for Reducing Nonlinearity Impacts

(1) Using power amplifier with good linearity

For example of replacing solid-state power amplifier with TWT power amplifier. With the improvement of solid power amplifier tube and microwave integration technologies, the linearity, size weight, service life, and reliability of the solid power amplifier become much better than the traveling-wave tube. The linearity of solid-state power amplifier is compared with TWT (Fig. 3.60) in Ref. [19], where C/N deterioration value is the deterioration value obtained under band-limiting condition.

As shown in Fig. 3.60, the solid line depicts the solid-state power amplifier while the dotted line is the TWT power amplifier. Although this diagram has been published for years, it is still used for reference. For example, some specifications were given by a manufacturer as follows: the third order intermodulation of TWT is −24 dBc (power returning: 7 dB), and AM-PM is 2.4°/dB (power returning 6 dB); the third order intermodulation of solid-state power amplifier is −33 dBc (power returning 7 dB), and AM-PM is 1.7°/dB (saturated power); as mentioned above, the linearity of TWT power amplifier is poorer than that of the solid power amplifier, and AM-PM is greater due to the slow-wave line effect of TWT, in addition, TWT has another defect that its power voltage is higher and it is provided with a complicated power supply system.

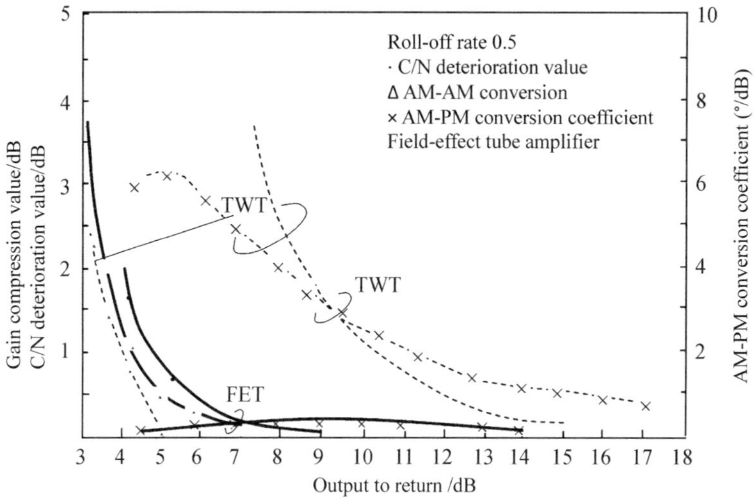

Fig. 3.60 Comparison of solid power amplifier with traveling-wave tube

(2) Methods for nonlinearity compensation

1) IF pre-distortion method. Inserting a network with an amplitude and phase characteristics just opposite to the power amplifier before it for compensation may make the overall characteristics remain linear.

   The pre-distortion network may use the soft-amplitude-limiting linearizer (LZ-LRZ) that is consisted of PIN or Scottky barrier tube or field-effect tube. The microwave integration technology is adopted to minimize the volume of LZ-LRZ.

2) Adaptive baseband distortion. The pre-distortion method cannot correct the variation of nonlinear characteristics caused by change of environmental temperature, aging of elements and components, change of DC bias, etc., but the adaptive baseband distortion can solve this problem. It can be implemented in digital baseband: firstly use the directional coupler at the output end of the power amplifier to sample the output signal, such sampled signal is demodulated and A/D converted and then sent to the linearizer in the baseband, where it is compared with the digitalized IF excitation voltage to figure out the gain from the IF excitation voltage to power amplifier; finally the gain deviation (if any) from the predetermined linear gain is used to adjust the gain of the pre-distortion network to maintain the overall linearity.

3) Feed-forward. Its difference from the adaptive baseband compensation is that it uses open-loop control, that is, after comparison of input signal with output signal to figure out the value ΔU of the output signal deviation from linear amplification, a ΔU value sent from the other in-phase sub-channel shall be applied to combine with output signal to generate the linear output.

4) Negative feedback. It uses the nonlinear distortion signal output from power amplifier to negatively feed back to the input end to offset part of linearity of the power amplifier itself. The cost of this method is that the output power is lower than that without feedback.

(3) No "band-limiting" or "weak band-limiting" prior to nonlinear power amplifier

One of the band-limiting functions is used for multiplex communication and preventing inter-channel interference. Since most operations in remote sensing & receiving, deep space communication and aerospace TT&C relate to a single channel or single address, this method herein is applied. There is another method where the band-limited filter is added after power amplifier, which is worth discussing under the circumstances of single frequency and small power.

(4) Constant envelope modulation

The constant envelope modulation is used to maintain constant of the envelope after band limitation so as to eliminate the impacts of nonlinearity. Currently, the mature modulation methods like MSK, GMSK, FQPSK-B, CPM, etc. can be adopted.

(5) Using *I*/*O* power compensation to maintain linearity of power amplifier

The power back-off method is not a good way due to poor power efficiency adding of volume, weight, and power consumption as well as the price, especially in restrictions of these factors for spacecraft application.

(6) Using energy diffusion signal

The inter-modulation noise will be generated when using multiple input carriers (for example, FDM). For the carrier with single frequency, its inter-modulation interfering signals are seriously discrete; for wide-spectrum signal, it is less impacted by the inter-modulation interference, therefore, the spreading modulation method can be applied to minimize the inter-modulation interference.

3.3 Information Transmission Techniques for Telemetry, Command and Remote Sensing

3.3.1 Information Transmission Techniques for Telemetry

3.3.1.1 Telemetry Overview

Telemetry is a measuring technique through which parameters (information) to be tested on the flight vehicle are measured and the measuring results are transmitted to the ground station for recording, displaying, and processing. The ground TT&C station receives and processes telemetry signals downloaded from the telemetry transmitter on the fight vehicle and restores telemetry data. These telemetry data reflect not only operating conditions of each system of the flight vehicle (systems of attitude control, dynamics, and energy), which is called engineering telemetry, but also the operating conditions of payloads, which is called mission payload telemetry. Engineering telemetry are generally the same with business telemetry in terms of telemetry equipment. The only difference lies in the amount of data to be processed by the terminal data processing computer. Typically, more parameters need to be processed in engineering telemetry.

Telemetry is the only means by which the flight vehicle, after take-off, can timely report its internal operating conditions to the TT&C station as well as the only method for the ground station to decide whether intervention with the subsystems on the vehicle is necessary.

Telemetry is featured with various information sources. For different measured objects, the frequency response in each channel can share a difference of 3–4 orders of magnitude. For example, frequency responses to temperature, pressure, supply voltage, and current in the instrument capsule are only 0.01–0.1 Hz, while those to structure vibration can reach several thousands of Hz. Considering such large amount of parameters to be processed, it is obviously impossible for each parameter to occupy a carrier for transmission. Therefore, the first technical problem to be solved in telemetry is how to combine hundreds of signals with such huge frequency response difference to realize multi-channel transmission, which is a key technique in telemetry. The development of telemetry multi-channel transmission system has experienced the following four stages.

(1) Frequency division multiplexing (FDM) system: FDM technique was adopted when aerospace telemetry started in the early 1950s, which uses multiple subcarriers to conduct FDM transmission. This system is similar to the aforesaid USB system, which is an analogue transmission type.
(2) Time division multiplexing (TDM) system: Started in the mid-1950s, the system is based on the sampling theorem raised by Claude Elwood Shannon. It uses the time division mode (also called commutator) to make samples of multiple-channel telemetry information to form discrete information, and to arrange each channel of information into different time intervals for transmission. By using multi-level frame segmenting technique, this system solves the difficulty of time division transmission of signals over thousands of channels in certain bandwidth. However, as an analogue system, it still has disadvantages of error accumulation and inconvenient processing.
(3) Digital Pulse-code Modulation (PCM) system: In the early 1960s, the PCM telemetry system was formed as a result of integrating PAM, analogue to digital conversion (A/D), digital to analogue conversion (D/A), and digital synchronization techniques. The digitalized telemetry system has many advantages. Telemetry information can be stored for the convenience of calculation and automatic data processing; error correction coding can be realized to digital coded signals; redundant information can be removed and source coding can be done to realize compression and information encryption; data reproduction is viable to avoid error accumulation and other disadvantages as a result of multi-forwarding.

(4) Packet telemetry system: From the mid-1980s to 1990s, the previously inefficient frame structuring mode of tandem connection of multi-level time division sampling switches is improved to the mode of packet switching in communication, namely, each information source is sampled, digitalized, and made into data for storage according to the requirements for frequency response and measuring. Then multiple such data are made into a uniform data stream by uniform frame structuring and packaging, and transmitted with the same uniform rate. This method increases the sampling rate for each information source, and decreases data redundancy, on the other hand, it also abandons the past transmission method where various time division sampling switches, subcarriers, and carriers are forced to be adopted just due to different frequency responses and data rates.

In analogue telemetry era, frequently used modulation systems are amplitude modulation (AM), frequency modulation (FM), etc. After digitalization of telemetry baseband signals, digital modulation and demodulation modes like amplitude shift keying (ASK), frequency shift keying (FSK), etc. are all adopted. At present, phase shift keying (PSK) and its variant modulation and demodulation modes tend to be used. Development tendency of the telemetry system at present is that PCM, packet telemetry and modulation and demodulation systems with high "bandwidth efficiency" are adopted.

How to produce digital telemetry data in TT&C system? At first, telemetry information is made into data, and then modulated onto the downloading carrier where data for downloading transmission are arranged to a certain format. Apart from telemetry parameters, information to express the format also needs to be transmitted, such as frame synchronous code, frame number, check bit, etc. At ground receiving terminal, the receiver will demodulate data and complete frame synchronization and code conversion, extract transmitted telemetry parameters (called de-formatting), add time mark, and then send them to the telemetry data processing computer. In practice, a very important part of telemetry system is the inversion of telemetry data, which inverts telemetry data streams into the actual physical parameters. In engineering, this part of work is generally performed in TT&C center, hence not included in TT&C station. However, for timely learning of situations, individual data are selected in TT&C station for inversion and displaying, especially those in relation to satellite tracking. In a high-level integrated management TT&C network, TT&C station transmission is transparent, and inversion of telemetry data on the station can be skipped; however, on business TT&C station, for the convenience of payload management, processing, displaying, and recording of telemetry data are all conducted on the station. Information about telemetry to be introduced herein mainly is based on engineering TT&C.

3.3.1.2 Basic Model of Telemetry Information Transmission System

There are various types of telemetry systems which fall into two categories according to their channels: wireless and wired telemetry. Virtually all telemetry systems adopt the wireless one. The functional block diagram of a typical wireless telemetry system is shown in Fig. 3.61.

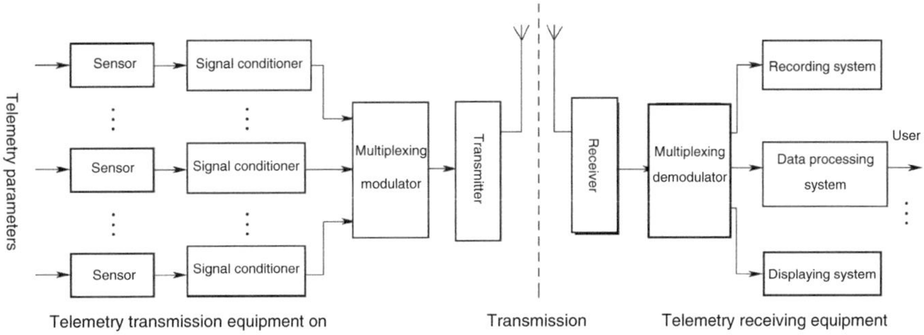

Fig. 3.61 Functional block diagram of wireless telemetry system

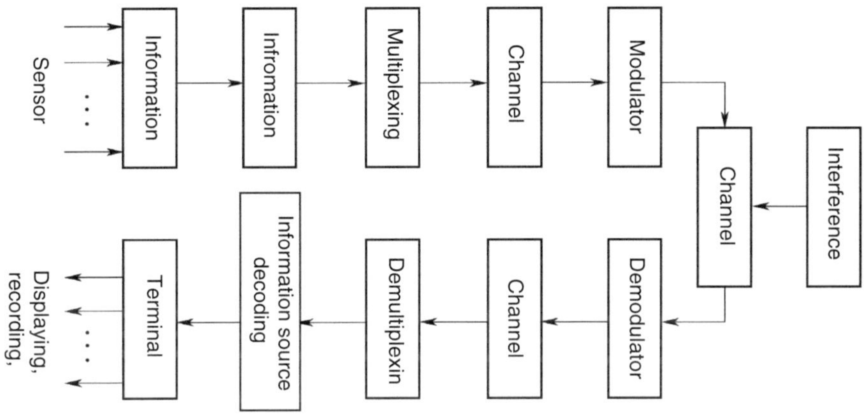

Fig. 3.62 Transmission model of telemetry information

In the figure above, the transmission terminal on the flight vehicle uses multiple sensors to turn into standardized signals suitable for transmission the multiple parameters to be measured, then after proper adjustment, it uses the multiplexing device to combine multi-channel telemetry signals in a certain multiplexing mode for modulation of the RF carrier. Then they are transmitted to the ground receiving station through the transmitter and antenna. At the ground station, the telemetry carrier signals, after received by the antenna, are sent into the telemetry receiver for carrier demodulation. Then each channel of original signals are restored through the demultiplexer, and recorded and displayed by the terminal equipment. After processed by the data processing system, final variables, data or diagrams are formed according to the user's requirement, and sent to the user. This is the whole process of telemetry.

The information transmission model as shown in Fig. 3.62 can summarize each kind of telemetry system.

Specific technical issues in each box in the figure above have been introduced in Sect. 3.2, this section only describes their characteristics in telemetry application.

(1) Information source

An information source is the source of transmitted telemetry information, where the measured parameters are turned into electrical signals in sensors. While at the receiving terminal, the telemetry signals are recorded, displayed, and processed by terminal equipment.

(2) Source coding and decoding

The main task of information source coding is to realize digitalization of analogue signals and improve effectiveness of digital signals. For example, if certain accuracy is required, the task is to express signals with the least symbol number and to improve information transmission efficiency by compressing frequency band. Information source decoding at the receiving terminal is the inverse transformation of information source coding.

(3) Multi-channel and demultiplexer

As mentioned above, only one channel is required for transmission of multi-channel signals for the reason of higher transmission efficiency under the context of multiple telemetry parameters. Multi-channel equipment is used to integrate each channel of signals, and demultiplex equipment at receiving terminal is used to disassemble integrated signals into each channel of signals.

The basic principle of signal transmission of radio telemetry systems is, in reality, the same as that of the radio multi-channel communication. Multiplexing modes applied extensively are FDM, TDM, and code division multiplexing (CDM), which are still main methods in telemetry systems.

1) FDM

FDM modulates various signals to be transmitted onto different subcarriers. When frequency intervals between each channel of modulated subcarriers are large enough, frequency spectrum of these subcarriers will not overlap again. After transmission on public channels, they can easily be separated by a set of band-pass filters. Thus the purpose of multi-channel signals sharing the same one channel is reached.

FDM uses second modulation where the modulation mode for subcarriers and carriers can be any one of amplitude modulation (AM), frequency modulation (FM), and phase modulation (PM). Different combinations of two modulations can form telemetry systems of multiple types, such as FM/FM, PSK/PM, etc. PSK/PM means: the first modulation is to subcarriers, the modulation system of which is PSK; the second one is to carriers with PM modulation system.

2) TDM

TDM is a multi-channel communication system which uses one channel, in chronological order, to transmit discrete information from multiple different channels. Its operating principle is that: The "time division switch" at transmission terminal samples each channel of continuous signals and arranges them together in chronological order to form multi-channel signals (called time division multiplexing). When the switch rotates (or operates) a cycle, it samples each channel of signals in turn once, which means a frame of signals has been transmitted. The "time division switch" at receiving terminal then separates the arranged multi-channel signals (called time division demultiplexing). This requires both the transmitting and receiving switches operate synchronously. Only by this can each sampling signal be sent into corresponding channels. The method to realize synchronization is that: Transmit a frame synchronization signal at transmission terminal, after it has been received at receiving terminal, control rotation period of the terminal's time division switch to be the same with that of the synchronization signal (namely frame synchronization). As the sequence of sampling signals is the same with that of the transmission terminal after signal synchronization of this frame, therefore, multi-channel signals can be separated. For convenience of identification of this frame synchronization signal, it shall be made distinguishing from the transmitted signals, for example, it can be attached to with a special mark, like pulse amplitude difference or a specific coded signal.

3) CDM

CDM uses different code patterns (or waveforms) to differentiate multiple (channels of) telemetry information. Code division multiple access (CDMA) telemetry is mainly applied in multi-target telemetry. For example, if all sequences of each telemetry data acquired from multiple targets, after being processed by direct sequence spread spectrum (DS) by using pseudorandom sequence (PN), can carry out PSK on the same carrier frequency, then the ground station only need one receiver to simultaneously receive telemetry signals transmitted by these targets. Major advantages of this kind of telemetry: first, it costs less than FDMA and TDMA systems, because there needs to be a receiver for each target using such two systems; second, it occupies less bandwidth and saves more radio frequency resources under the same technical requirements; third, it resists stronger interference and is harder to be intercepted, which proves better confidentiality.

(4) Channel coding and decoding

Channel coding can bring coding gain for telemetry channels as well as certain confidentiality and anti-interception and anti-jamming effects. For example, techniques like R-S error correction coding can decrease bit error rate as a result of burst interference caused by flame and stage separation in telemetry systems like missile and spacecraft. Its theoretical value of coding gain is between 1.5 and 3.4 dB (corresponding $m=4\sim8$). In the case of a severe burst interference, the coding gain value gained in a test is about 2 dB.

Channel coding techniques have been widely adopted in fight vehicle telemetry. Among them, convolutional coding/soft decision Viterbi decoding, Reed-Solomon code (R-S code), concatenated code of convolutional code and RS code, Turbo code, and LDPC code have been recommended by the CCSDS as the standards of channel coding. According to relevant regulations, the above-mentioned R-S code can be used independently for adaption of characteristics of rocket telemetry channels. Their detailed introduction is referred to in Sect. 3.2. Telemetry is just one of their applications.

(5) Modulator and demodulator

FM and PM are frequently used in telemetry.

1) FM: FM means the deviation of the instantaneous frequency of the modulated carrier from the center frequency varies linearly with the instantaneous value of the modulated signal. FM enjoys extremely extensive application in telemetry, featuring relatively stronger anti-jamming capability.
2) PM: FM means the deviation of the modulated carrier's phase from the reference phase (or initial phase) varies linearly with the instantaneous value of the modulated signal. The PM demodulator generally uses the coherent detector the threshold of which is very low. Therefore, PM is frequently adopted in deep space exploration.
3) Pulse modulation: Pulse modulation means the amplitude or width or phase of the modulated pulse sequence varies linearly with the instantaneous value of the modulated signal.

Apart from the above-mentioned carrier modulation modes, there is a processing process during telemetry information transmission which samples and quantifies continuous information, the name of which is pulse modulation. In this case, the modulated signal is a pulse sequence. When the amplitude, width, and repetition frequency of the pulse varies linearly along with the modulated signal, they are separately called pulse amplitude modulation (PAM), pulse duration modulation (PDM) and pulse position modulation (PPM). All three of these belong to analogue modulation and the corresponding telemetry system is analogue telemetry system. When pulse amplitude (namely, sampled point value of the modulated signal) of PAM signal is expressed by a set of binary codes, this is called pulse coding modulation (PCM).

The foresaid modulation techniques have been introduced in many books; therefore, they are not repeatedly described in this book. The following special problems exist in telemetry systems of flight vehicle telemetry.

1) Constant envelope modulation: Under certain output power requirement, the last stage power amplifier generally operates in amplitude limiting and non-linear status to reduce volume, weight, and consumption of the flight vehicle telemetry transmitter and to improve output power and efficiency of the same kind of power amplifier tube. The modulated carrier is required to be free from PM (namely, constant envelope) and to take constant envelope modulation because during non-linear operation, conversion from amplitude modulation to phase modulation (AM-PM) will cause inter-symbol interference and higher bit error rate for digital transmission.
2) Bandwidth limitation: As telemetry is required to transmit increasingly more channels and bit rate is getting larger and larger, transmission bandwidth, accordingly, shall be wider and wider. However, available frequency band is limited, and the system is bandwidth-limited. Therefore, adoption of high

"bandwidth efficiency" modulation mode can help compress signal spectrum width and improve high frequency band efficiency. In this modulation mode, bandwidth of the modulated signal is required to be narrow (frequency spectrum main lobe shall be narrow), out-of-band interference is required to be small (frequency spectrum sidelobe shall be small), and its envelope shall be constant or quasi-constant. These requirements make it possible for last stage power amplifier to operate in high power and efficient non-linear status. These modulation systems have been specifically introduced in Sect. 3.2.

Considering the above factors, constant envelope modulation systems like QPSK, OQPSK, MSK, and GMSK are recommended for telemetry, which are also standard spacecraft telemetry modulation systems suggested by the CCSDS. Besides, FQPSK modulation featuring narrow band and quasi-constant envelope has developed progressively-. FQPSK – B has become a part of IRIG - 106 -2000 telemetry standard.

(6) Telemetry terminal equipment

Telemetry terminal equipment is at the output end of telemetry receiving and demodulation equipment, with main functions as follows: ① recording of measured data; ② real-time monitoring; ③ data processing; ④ alarming: alarm signals are given when measuring parameters exceed pre-determined limit.

Category of telemetry terminal equipment is various. Common terminal equipment can be classified into the following types.

1) Telemetry data recording:

Telemetry data recording equipment can be a magnetic tape unit or computer hard disk and optical disk, etc. Advantages of this equipment: Data rate of information recording is high; magnetic tape can be used repeatedly, and its exchanging is convenient, more convenient and reliable in the case of cassette tape.

Telemetry data recording has two forms: analogue recording before inspection and digital recording after inspection. Usually both methods are combined to use, each as the other's supplement, to make sure of higher reliability. Analogue recording before inspection is to directly record on the analogue recorder the high frequency signals, the frequency of which has been converted but which has not been demodulated by receiver carriers. The advantage of this recording: recorded signals include original information which has not been demodulated and can be played back to improve data acquisition reliability. Digital recording after inspection records data that have been demodulated and whose frame and code have been synchronized. Recording medium can be hard disk, optical disk, magnetic disk, etc. For the convenience of recording medium exchange, cassette units are still common at present. Besides there is quick-look recording equipment which during spacecraft flight real-time processes and displays curves and values of telemetry data for real-time monitoring. It can demonstrate variation trends of each parameter, featuring simplicity, intuition, convenient interpretation, etc. Common quick-look recorders have two types: (1). electrostatic printer-plotter; (2). Thermo-sensitive recorder.

2) Telemetry data processing and displaying equipment

Telemetry equipment are usually capable of immediate processing for rapid provision of telemetry data processing results and equipped with necessary display workstation to process and display parameters necessary to be monitored. Commercially available business computers are often taken as processing and displaying equipment to reduce costs and improve openness.

3.3.1.3 Deep Space Telemetry Technique

(1) Characteristics of deep space telemetry

Data for deep space transmission includes that from command, telemetry, and detection. As transmission range is remote, generally, only a relatively low data transmission rate can be achieved. Therefore, detection data transmission in deep space TT&C is usually brought into telemetry.

As a target in deep space detection is extremely far from the earth, signals returning back to earth are intensely weak and time delay of received signals are quite large, therefore, deep space telemetry has featured the following characteristics.

1) Deep space data transmission is a power-limited system. Remote operating range results into extraordinarily weak received signals. The principal contradiction, here, is that limited power leads to low transmission bit rate and narrow frequency band; Therefore, it is a "power-limited" system rather than a "bandwidth-limited" system. Thus the design starting point of data transmission communication at present is how to improve "power availability", while "frequency band availability" is an issue to be considered for future development.
2) Receiving of low carrier-noise ratio: As signals arriving at antenna are very weak, the received carrier-to-noise (C/N) ratio is accordingly quite low. How to effectively utilize such valuable C/N ratio is the utmost important issue in deep space data transmission and receiving.
3) Error correction coding with high coding gain: Adoption of signal error correction coding can acquire coding gain, which is an important measure in low C/N receiving. However, as high coding gain is obtained, this makes operating C/N ratio meeting certain bit error rate become lower (corresponding to E_b/N_0), demodulation threshold problems like square loss are now become quite prominent.
4) Lower bit rate of data transmission: As range of deep space detection gets increasingly further, realizable bit transmission rate becomes gradually lower. This is the reason why receiving of extremely low bit rate is a special issue in deep space data transmission.
5) Residual carrier and suppressed carrier modulation modes: At present, the above-mentioned two modulation/demodulation modes both are frequently applied to deep space telemetry for the purpose of achieving better power efficiency respectively in low and medium/high bit rate and fitting different ranging systems.

6) Large Doppler frequency: Doppler frequency is larger than that of earth orbit spacecraft, because deep space detector may have very large velocity relative to the earth and can operate on higher frequency *Ka*/*X*- band.
7) Long acquisition time: Acquisition time is very long, even to about 1000 s, as Doppler frequency offset is large, *C*/*N* ratio is low and bandwidth of phase-locked loop is narrow.
8) Non-real-time transmission: The ground station can record firstly and then process transmitted data by using storage and forwarding method on deep space transponder, as time delay of transmission is very long (e.g., over 22 min from Mars to the earth) (data transmission under such condition is not real-time) and the time for observation is comparatively long.
9) Compression of information source data: Data compression is an important measure in deep space data transmission for the purpose of improving operating range by decreasing transmission rate.
10) High power transmitter: For the purpose of maximizing the receiver *C*/*N* value, the ground station shall use extra high-power transmitter and deep space transponder with large transmitting power but small volume, weight, and power consumption shall be equipped on deep space detector.
11) Adoption of "duplex" and "non-duplex" channels: During "duplex" operating, insertion loss of the duplexer will decrease system *G*/*T* value, which impacts huge influence on deep space station where noise is extremely low. Therefore, some deep space stations are supplemented with "non-duplex" channels. Namely, the duplexer is turned off to improve system *G*/*T* value in operating mode where only detection data is received.

(2) Signal waveform of deep space telemetry

1) Modulation mode

Residual carrier and suppressed carrier modulation modes are two main modulation modes adopted in deep space telemetry.

- Residual carrier modulation mode: The most frequently used modulation mode in deep space telemetry is that square wave subcarriers are used to carry out BPSK modulation on residual carriers. Apart from historical reasons, this practice also has the following advantages: ① residual carriers modulation scheme can share the same downlink power in performing multifunction, like two-way ranging and △ difference one-way ranging (△DOR); ② unmodulated spacecraft carrier is very useful to radio science study; ③ before cycle slipping appears, relatively low carrier loop SNR is allowed in residual carrier tracking, it will not be damaged by relatively more half-cycle slipping, which is exactly a shortcoming of suppressed carrier tracking; ④ subcarriers are capable of making data side band far from residual carriers in frequency domain, therefore, carrier tracking can be conducted without interference. Though subcarrier adoption takes more frequency spectrum than direct carrier modulation, it occupies less bandwidth in low data rate telemetry.

Sine wave subcarriers are used in class A (near earth) tasks; high order harmonic of sine wave subcarrier descends quickly, making occupied bandwidth less than that of square wave subcarrier which has the same frequency. The disadvantage of sine wave subcarrier is that the receiver can only restore fundamental wave energy while data energy transmitted in high order harmonics is lost.

Contrary to sine subcarrier, square wave subcarrier is able to restore all data energy in receiver bandwidth, which makes it achieve extensive application in deep space TT&C.

The mode of direct carrier modulation is comparatively better scheme in medium and high speed data telemetry. Occupied frequency spectrum bandwidth in this modulation mode is less than half of that of a similar square wave subcarrier system.

- Suppressed carrier modulation mode

  This mode includes:

① Suppressed carrier BPSK: Its performance under high data rate condition is almost as good as that of the residual subcarrier BPSK and it performs even better under medium data rate condition. Its bandwidth occupation is the same with that of the residual carrier BPSK which has no subcarriers. The fact that it cannot be used at the same time with ranging or ΔDOR is the shortcoming of suppressed carrier BPSK.

② Bandwidth efficiency of QPSK and OQPSK surpasses that of BPSK. For designated binary symbol rate, bandwidth occupation of QPSK or OQPSK carrier is only half of that of the BPSK modulation carrier (without subcarrier). However, QPSK and OQPSK share the disadvantage with suppressed carrier BPSK: They cannot be used at the same time with ranging or ΔDOR. One possible solution is to adopt UQPSK, uploading data in its one channel and transmitting ranging codes at the other channel.

Basic telemetry performance of QPSK and OQPSK is the same with that of the suppressed carrier BPSK under high data rate condition. Under condition of adoption of data signals of shaped filter, OQPSK has some advantages over QPSK in frequency spectrum width, popularizing it in satellite communication; but for data signals of unshaped filter, both of them are the same in performance and frequency spectrum occupation.

- Contrast between residual carrier and suppressed carrier: The contrast of these two kinds of carriers is mainly associated with bit rate. When frame error rate (FER) and bit rate are designated, the smaller P_T/N_0 requirement will be the criterion of better telemetry performance. Contrast conclusion: Usually residual carrier performs better in telemetry when the bit rate is extremely low, in particular, when bit rate is low and carrier loop bandwidth is comparatively larger (in the case that phase noise is large or Doppler dynamic is not compensated). Under medium bit rate condition, telemetry performance of suppressed carrier is far better than that of residual carrier. Under high bit rate condition, telemetry performance of suppressed carrier is 0.1 dB better. Of the two modulation modes, which is to be chosen by the ground station is sometimes not based only on telemetry performance. For example, residual carriers are required sometimes in radio science tests, or in some other applications, it is all-important to minimize capture time. Under these conditions, criterion to select residual carrier or suppressed carrier is based on faster capture speed.

The reason why residual carrier under low bit rate condition performs better than suppressed carrier is that residual carrier loop has no half cycle slipping. However, suppressed carrier loop will slip half cycle, this is why higher carrier loop SNR is needed to prevent such slipping which will cause error codes (though residual carrier loop will slip a full cycle, the chance for this is smaller than half cycle slipping, hence smaller error code).

Modulation index shall be designed to be optimal or at least near to obtain best residual carrier performance.

2) Coding mode:

Choice of coding mode is independent of modulation mode, which shall be considered based on four conditions which are coding gain, bandwidth, delay time, and error floor. In general, coding gain increases along with bandwidth and delay time. A deep space station has the following coding modes:

- No coding

Data free from coding require the least bandwidth and delay time. It mainly applies to the conditions where link margin is enough and bandwidth is restricted, leading to the circumstance that coding cannot be used.

- RS code

RS codes (255,223) used in deep space networks have good error correcting capability for burst errors. It can correct up to 16 symbol errors in each 255 symbols. It is corresponding to $16/255 = 6.2 \times 10^{-2}$ bit error rate (BER) at the decoder input end, and if input BER is lower than this value, it means wrong decoding. In this case, the corresponding E_b/N_0 is about 3 dB.

Its contribution to bandwidth expansion is not large. When E_b/N_0 is larger than 3 dB, its error correcting effect becomes small along with decreasing number of error symbols. This is the reason why it has only limited application under high E_b/N_0 condition.

- Convolutional code with short constraint length and 1/2 code rate

Convolutional code with short constraint length ($k = 7$) and 1/2 code rate is a kind of low delay time code which requires two times the bandwidth of uncoded data, but coding gain will be produced no matter what SNR is input. The real-time performance of its low delay time is more important than the method of using high delay time code to acquire high coding gain, which makes it a good choice in low rate emergency communication.

- Concatenated code of RS code and 1/2 code rate convolutional code

When 1/2 code rate convolutional coding is carried out on spacecraft RS coded data before transmission, the composed code bandwidth expansion is a slightly large, but under various SNR conditions, its performance is much better than that of any one of these two coding. This is because when it is close to threshold value, the convolutional decoder produces burst error due to losing lock, but the subsequent RS decoder can effectively correct such error. Error pulse produced when convolutional decoder is near threshold domain maybe go beyond error correcting capability of RS decoder. In this case "interleaving" technique can be used to relieve this problem by exchanging more delay time for better performance.

- Convolutional code with long constraint length and high code rate

Each deep space station can process long constraint length convolutional code with up to 1/6 code rate on 1 link each time. These codes are improved much more than codes with $k=1$ and $r=1/2$, but the cost is that bandwidth expands to 6 from 3. Like short constraint length codes, they can be cascaded with RS coding, to obtain better performance. Limited support from deep space ground station is the reason that they are not preferable for application.

- Turbo code and LDPC coding

Turbo and LDPC codes boast performance that is close to Shannon limit. The bandwidth expansion of Turbo code is from (slightly larger than) 2 to (slightly larger than) 6. Suggested code block sizes are 1,784, 3,568, 7,136 and 8,920 characters. Small block is for low data rate, but when data rate becomes larger and coding overhead gets smaller, big block is used. Their shortcomings: Processing throughput that has to be completed in decoding is large, and when frame bit error rate is about 10^{-6}, error floor will happen. Decoding complexity causes longer delay time, limiting frames that can be processed in a certain time interval and the code rate determined by processing time of each frame. LDPC code is mainly applicable to conditions with higher code rate. Its structure is simple and can be expressed by graphics, bringing many advantages to decoding. It has been recommended by CCSDS for application in deep space TT&C communication.

3.3.2 Command Information Transmission Technology

3.3.2.1 Overview of Command

Command is an abbreviation of remote control and also called Command Telecontrol. It refers to the process to send control commands through a transmission channel of a control station to a controlled object at a distance (i.e., remote) so as to make it generate a predetermined action. It can be classified into analog command system and digital coded command system in accordance with the different forms of command signals. The analog command system, with signals in the form of continuous wave, is suitable for control of continuously changed states. For example, the control of air jet duration of an engine's nozzle by pulse width belongs to analog command. The coded command system uses discrete code bits or symbols to represent the content and address of commands and is mainly adopted by command systems of current flight vehicles.

According to different applications, command is divided into missile (rocket) safety command and flight vehicle command. The task of safety command is to protect the safety of launching site and flight area, and to control and destroy the failure missile or rocket in flight. The features of safety command are of high safety and reliability, strong real-time, small number of commands, short time of task execution, and strong capability of anti-flame. Considering its high demand for safe reliability, special independent safety command systems with wide beam antenna are adopted for air and ground designs. The main task of flight vehicle command is to control flight vehicle's orbit, attitude, engine ignition, separation of all levels, astronaut escape, payload, etc. and to inject data into flight vehicle, such as the required flight orbit parameters, dynamical parameters, work procedures, computer programs, clock correction, and delay switching commands, etc. It is featured in a big number of commands, long time of task execution, and far working distance. Spacecraft's command is usually integrated with the TT&C and communication system with unified carrier. For orbital spacecraft there is less remote control and command is rather simple (such as command on orbital satellite); but for combat weapon the requirement of remote control on their maneuvering is pretty high (such as UAV control, missile weapon command), so it plays an important role and has been a key part of command and control system in military information system.

Command has three features compared with telemetry: Firstly, command signal is discontinuously transmitted and is issued only when needed to take measures on spacecraft's orbit, attitude, and working states of its internal subsystems; Secondly, each command signal is transmitted in one complete command, different from traditional telemetry where signal is transmitted in unit of each sampling point; Thirdly, reliability of transmitting command signal is much higher than that for telemetry since any malfunction may cause disastrous effect.

For the expression method of baseband signal of command, initially, the command is represented by different combinations of n oscillators in different frequency. When there is a large number of commands, the hardware for such command coding method will become very complicated and difficult to add auxiliary information other than commands. Different combinations of "0", "1" binary sequence are used to represent command after digital technology matured in the 1960s. It is similar to telemetry in both modulation and demodulation; but it is different from telemetry in transmission mode besides discontinuous transmission and has a bit error rate lower than telemetry in 2 orders of magnitude. As the frame length of command is short, in order to ensure no mistake appears, sometimes special channel encoding mode or ARQ is used where the commands received by spacecraft are returned to ground station through telemetering channel or special feed-back channel and then executed after proving to be correct by comparison.

CCSDS command standard was formulated in the 1980s, which expands the function of command composition, employs the frame format structure similar to telemetering signal source package, and improves the detection and error correction capability.

Some command systems do not have the capability of tracking spacecraft and need to use angular position data from radar or other equipment to drive and align antenna to spacecraft for operation.

3.3.2.2 Composition and Work Process

The basic composition of a command system is as shown in Fig. 3.63.

Operational contents of a command system can be summarized into three points: command's formation, transmission, and execution. Comparing Fig. 3.61 with Fig. 3.63, it can be seen that, viewing from information transmission, command system is basically the same as telemetry system, with the differences being: discontinuous transmission, block coding low code rate, short command frame, high real-time, high reliability requirement, and multiple protection measures to be taken.

(1) Formation of command and data

Command has two operational modes: command control and data injection. According to command plan, real-time flight state and work state information of controlled object provided by tracking and telemetry system, computer of

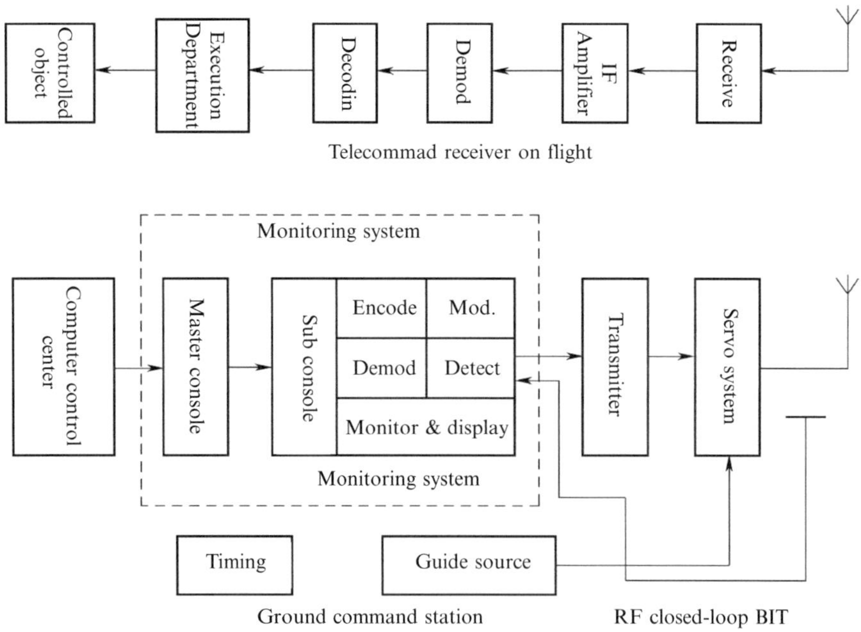

Fig. 3.63 Composition block diagram of command system

command and control center automatically calculates and generates command program, information, and data to be injected (formed under the participation of control personnel), which are then sent to console of command system. Command information shall be converted into digital command signal by convertor before being sent to encoder. There are two encoding modes in the encoder, i.e., static coding and dynamic coding. For the static coding, it prepares command number for the preselected command code according to stipulated data format and stores it in the encoder, then outputs the command through lookup table according to controlled target and command number. It is simple and reliable, but needs big storage capacity and is difficult for change. For the dynamic coding, it respectively stores each part of command code and combines each part of command code when transmitting the command. Its advantages and disadvantages are contrary to those of static coding. When there is long command, big number, limited memory or some part of command code shall be often changed, static coding shall be used. The generated command code shall be sent to encryption unit for encryption, code pattern conversion and channel coding, and then to modulator for modulation; while the injected data can be directly sent to modulator after code pattern conversion.

The control command generated by encoder cannot be transmitted until there is no error after verification judgment. Verification method includes video closed-loop BIT (self-loop), RF ringlet BIT, big loop verification (through equipment located on flight vehicle closed-loop). In case of RF closed-loop BIT, coupling output of high power amplifier forms wired closed-loop to test the correctness of control command and uplink transmission. If it is not correct, close control gate and switch off transmitter to stop issuing wrong command. In case of big loop verification, it is necessary to demodulate command code or inject data on the flight vehicle, which is transmitted back to ground station through downlink telemetry channel for bit-by-bit comparison of *R/T* command codes to form uplink/downlink big loop verification. Console of command system, in addition to issuing the operation and control commands (also can be operated and controlled by measurement and control center), conducts integrated control and monitoring on its own equipment.

(2) Transmission of control command and data information

Transmission system is a radio digital information transmission system, whose functions include command information coding, channel coding for carrier signal modulation, up-conversion, power amplification and RF signal radiation from ground antenna to flight vehicle, then received by receiving antenna of flight vehicle, the signal passes through command receiver, demodulator, channel decoding and is output to detection equipment, which includes decoder, command judgment device and control gate. Decoder conducts decoding identification on output of demodulator, detects command and data, sends to command judgment device for anti-interference judgment and confidential judgment. It pre-stores ground transmitting command into command judgment device and compares with actual received command, opens the control gate after no error judgment, to send correct command signal to execution device or send correct data to computer on the flight vehicle.

(3) Execution of control command

There are two cases for execution of control command: One case is without stringent accuracy requirement on execution time; another case is having strict requirement on execution time, such as command control of spin stable GEO satellite, which is called synchronous control since there is a strict synchronization requirement on control signal with satellite spin speed.

3.3.2.3 Command System

Command can be classified into the following systems according to its signal mode, error control, equipment scheme, etc.

(1) Modulation system

A triple modulation system is in common use for aircraft command system, i.e., pulse code modulation (PCM), secondary modulation and N carrier modulation.

1) Pulse code modulation (PCM). All command information is firstly under pulse code modulation (PCM), which is processed in binary code.
2) Secondary modulation. Secondary modulation is conducted on command commands and data codes before carrier modulation to improve confidentiality and anti-interference of command. There are three types of secondary modulation: ①. subcarrier modulation; ②. direct sequence spread spectrum modulation (DS-SS); ③. high alphabet modulation.
3) Carrier modulation. The carrier modulation is often applied to flight vehicle's command system, it includes FM and PM. Subcarrier and high alphabet secondary modulation signals can be combined with these kinds of carrier modulation, while DS-SS commands usually adopts carrier PM system.

(2) Multi-channel transmission system

The essence of multi-command transmission of command system is multi-channel communication. Code division multiplex system is used for command system of most flight vehicles.

(3) Equipment scheme

According to equipment comprehensive utilization, ground command system is classified into independent system and unified comprehensive system, with the latter using uplink integrated channel and integrated baseband of command and orbit measuring system.

(4) Error control system

Error may occur in command signal transmission due to interference and noise. Once a command error occurs, it may cause irreparable loss and so error possibility of command must be required to be very low. Generally, possibility of missing command shall be less than 10^{-6}, possibility of false command shall be less than 10^{-8}, while for safety command the requirements may be higher.

Error possibility of command can be reduced through two aspects: (1) decrease bit error rate; (2) take measures in encoding, receiving, and judgment rules and use corresponding error control technology to decrease possibility of error command. It is hard to meet the requirements by taking only first and the second method should be combined. In general, the first method can decrease bit error rate to 10^{-6}–10^{-5} and the second method can decrease error command possibility to arbitrarily small (theoretically) at the cost of increasing command time and equipment complexity.

(5) Station distribution scheme

For flight vehicle remote control, sometimes one station is enough but in most cases multiple stations collaborative work is needed. So ground command system can be classified into single station system and multiple station system.

Safety command is to control the active segment of flight target with a small airspace to be covered. So one ground command station is enough to meet requirements and single station system is often used. However, a multiple station system may be better for improving reliability and anti-flame attenuation, etc.

For vehicles flying at medium and low orbit, one ground command station can only provide limited orbit coverage, so a multiple station system should be used for its control.

3.3.2.4 Basic Model of Command Information Transmission System

Command information transmission is a key part of command system. The ordinary information transmission modes include analog transmission and digital transmission. Compared with analog transmission, digital transmission has stronger anti-interference, higher flexibility, easier encryption, higher reliability, etc., and so it is widely used for command system. The analog transmission mode is only used in certain situations where higher requirements for real-time is required and the analog quantities are more suitable for representing the control parameters. This book mainly introduces the digital transmission mode of command system.

The model of command information transmission system is shown in Fig. 3.64.

(1) Command generation

It is the information source of command information system. The command information includes command and injection data. Computer or operator uses command generation equipment to generate command signal according to comparison of pre-designed state with actual state data of commanded object. The output command signal is generally a group of binary coded digital signal, which is also called command code. Each group of command code is matched with a corresponding command number. Either command code or command symbol can be used for transmission from telemetry center to command station, but only command code can be used for transmitting command information to flight vehicle.

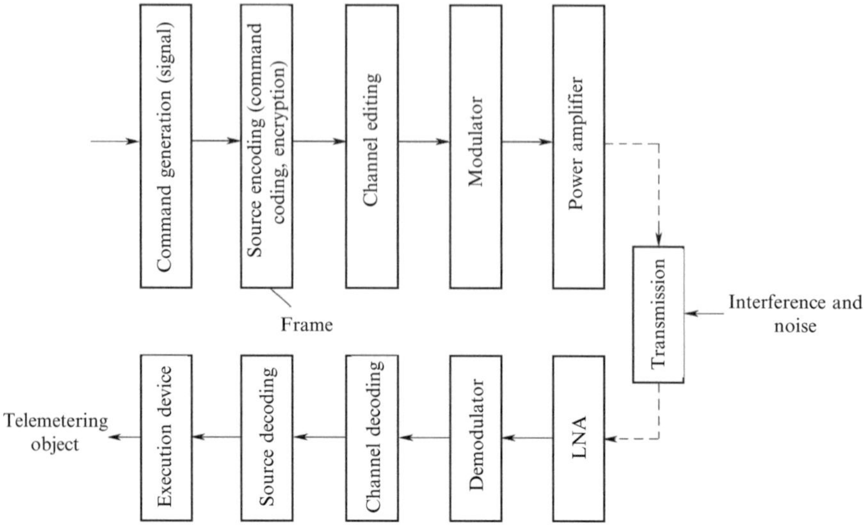

Fig. 3.64 Model of command information transmission system

Such command information source is typically intermittent and there is no need for continuous transmission. The time of each transmission is called command work period. The subsequent transmission system is digital information transmission system.

(2) Source encoding and decoding

Source encoding in command information transmission system has its special features, including functions of command encoding encryption and format formation, etc. Corresponding command decoder includes command decoding, decryption, de-formatting, etc.

1) Command encoding and decoding

Commands are compiled into code groups at transmission end to differentiate different commands and improve the reliability, anti-interference, and confidentiality of command signal. Combined coding is often used to increase the capacity of remote controlling command. Commonly used combined coding includes frequency combined coding and pulse combined coding. Pulse combined coding is mostly used for missile and spacecraft command. It edits command signal into code group and forms code command. Its advantages include high capacity, strong anti-interference, good confidentiality, and convenience for implementation with digital logical circuit and computer, etc.

At receiving end, the demodulated encoding command signal is converted to original command signal. The main components of command decoder are the decoding matrix composed of switches and staggered input and output lines. The number of input lines is the number of the binary bits of a command. For example, for a command composed of 8-bit binary number, the number of input lines of decoder is 8. The number of output lines is the number of commands. For example, 8-bit binary number can compose 256 commands and the number of decoder output line is 256. The corresponding command encoder is its inverse process. Command decoder has now been made into integrated module in a small volume and can be realized through computer. Encoding method includes static coding (pre-prepared) and dynamic coding (arithmetic coded according to preset coding rules before issuing).

2) Encryption and decryption

The above mentioned code information command is still poor in confidentiality. Command system plays a very important role for implementing flight vehicle control. With the development of space technology, command information transmission channel changes from the single channel transmission from domestic ground station to spacecraft to multi-channel transmission for foreign ground station and relay satellite retransmission. Therefore, spacecraft command safety becomes very important. With the development of electronic reconnaissance and countermeasures, the requirement on data protection technology becomes higher and higher. Encryption method is required to protect command information safety. Transmission of command information is performed through space link, which makes a command system with an open feature. In the case of command on flight vehicle, the enemy can intercept the transmitted command signal, analyze and steal command information content, forge command information, which forms a serious threat to our flight vehicle.

Data encryption and encryption authentication are commonly used for command information safety protection.

3) Formation of command and data format

The format of remote controlling command and data is usually composed of boot, address synchronization character, command code/data, and end character, as shown in Fig. 3.65.

The function of boot is to establish bit synchronization for demodulation and decoding process of command receiving equipment. Some receiving equipment does not need setup time for demodulation and pre-built bit synchronization is not required for decoding. For such equipment, leading boot is not needed.

The address synchronization character is used for either address differentiation or information synchronization. It only accomplishes information synchronization in single target control. The function of information synchronization is to indicate start bit of command code character or data in the binary data stream to the decoder to ensure correct decoding. It is also used as address code in multi-target command.

Boot	Address synchronization character	Command code/data	End character

Fig. 3.65 General format of remote controlling command data

The command code/data is the substantive control information to be transmitted through command channel.

The end character indicates that such frame information has been transmitted, can output decoding results and close decoder. End character cannot be used for command system with fixed length of command and data.

The time during which command channel is operated once is called command work period. The beginning of each command work period has a boot sequence, which makes the demodulator of carrier and sub-carrier of flight vehicle enter stable and bit synchronization state. If certain command frames or a sequence of command frame are intermittently transmitted in one command work period, idle sequences shall be used to fill the intervals to maintain synchronization.

(3) Channel encoding/decoding

The function of channel encoding in command information transmission system is the same as that of typical digital information transmission system, which is used to reduce channel bit error rate with certain role of encryption as well. As command has a high reliability requirement, channel encoding technology is not enough to reduce bit error rate and multiple protection measures shall be taken, but encoding/decoding scheme should be selected in accordance with command characteristics. As command has features of "spurt" and "intermittent", block code is mostly used when the time for transmitting command is short. Linear block code specified in CCSDS proposal is: information length $k = 32$, 40, 48 or 56, monitoring bit length 7, e.g., (63. 56) BCH block code is used; CRC code is often inserted into the serial injection data frame for error detection. Convolutional encoding/Viterbi decoding is used in case of long command or continuous transmission. RS coding can be used in case of serious error burst out. Such BCH coding, CRC coding, convolutional encoding, and RS encoding have already been introduced in the section of channel encoding/decoding of the book and so no more details need to be repeated here.

(4) Modulation/demodulation

Command information transmission system is a power-constrained system, featuring low code rate, little prominence in multiplex transmission and limited band but high requirements on reliability and anti-interference.

Due to the low code rate, command system often uses secondary modulation, such as PSK, FSK, ASK keying modes for primary subcarrier modulation and PM, FM, AM modes for secondary RF modulation. ASK – AM system asks for simple equipment but has poor anti-interference; FSK – FM system requires complex equipment but has better anti-interference; PSK – PM system requires more complex equipment and has best anti-interference. For modulation systems, there is corresponding stipulation in relevant standards, e.g., PSK – PM system is adopted for low-altitude spacecraft and FSK – PM/FM system is adopted for synchronous satellite. Under the same conditions of BER and code rate, the signal power efficiency and bandwidth use efficiency of FSK are not as good as PSK, but FSK does not need phase-locked loop in demodulation and so has a good anti-burst interference.

For secondary modulation system where command data is firstly modulated on the subcarrier, the subcarrier frequency f_{sc} shall be selected as integral times of R_b, with f_{sc}/R_b between 4 and 64. CCSDS stipulates f_{sc} shall be chosen between 1 and 8 kHz. The subcarrier frequency stability of command shall be better than: long-term stability $\pm 5 \times 10^{-5}$/24 h; short-term stability $\pm 1 \times 10^{-5}$/s; in the case of PSK, the phase of subcarrier makes 180° stepped with the level of input data changed between "0" and "1". For "1" bit, the front edge of bit corresponds to the positive zero crossing point of subcarrier PSK waveform. For "0" bit, the front edge of bit corresponds to the negative zero crossing point of subcarrier PSK waveform. During the engineering, the phase stepped tolerance is limited within 10° of subcarrier sinusoidal signal crossing zero point. Command information is transmitted intermittently. To avoid missing command information before subcarrier demodulation, 15 leading boot codes of "0" and "1" shall be transmitted before transmitting address and synchronization character of primary command frame, to ensure bit synchronization of demodulator. Sometimes leading boot code can be selected between 0 and 128 b and alternate codes of "0" and "1" can be used for leading boot code. In addition, a string of alternate codes of "0" and "1" can be sent to modulate subcarrier and maintain normal operation of bit synchronizer, so as to keep ground-air bit synchronization during the idle period when primary command frame has been transmitted and the next group of command frame is not yet transmitted.

High alphabet modulation and spread spectrum modulation are also used as modulation modes for command information, where spread spectrum modulation has advantages such as high confidentiality, anti-interference, anti-interception, etc. and has been widely used, for details of which please refer to Chap. 4.

(5) Linearity of power amplifier and low-noise amplifier

As command information transmission system is a power constrained system rather than a band limited system, band-limited filtering is not used in the transmission system, while PM, FM, and other angle modulation are the most used modulation modes. Signal's envelops are constant since they do not pass band-limited filters and there is little impact from channel nonlinearity. Besides, the transmission power is required to be as high as possible to increase its reliability, so the power amplifier often works in saturation state. It also works at burst state to increase its confidentiality. These are the embodiment of command particularity.

3.3.2.5 Command Error Control Techniques

In flight vehicle command system, the error command probability of command transmission P_{error} (or command frame receiving error probability P_E) is required to be less than 10^{-9}–10^{-8}, and the missing command probability $P_{missing}$ (or command frame rejection probability P_r) less than 10^{-7}–10^{-6}. Obviously, the performances of channel bit error rate of a general information transmission system cannot meet the requirements of such error command probability. Therefore, in order to meet the above required $P_{missing}$ and P_{error}, command error control measures must be adopted for the transmission channel. During selecting error control system, the requirement on false command probability (P_{false}) is to be met and the complexity of equipment and real-time of command are also to be considered.

There are several classification methods for flight vehicle error control. Generally, it is divided into two kinds: space-ground link feedback verification method and forward error correction method which includes retransmission accumulation decision system, forward error correction system, forward error detection system and error detection & retransmission system. Space-ground link feedback verification method uses the telemetry equipment in the flight vehicle to feed back the command information received by the telemetry receiver in the space to the ground station for comparison, check and retransmission in case of any error. Forward error prevention means the forward command code group transmitted to the flight vehicle by the ground station has some error prevention capability. This code group performs error prevention encoding on the ground and there is a corresponding decoder on the flight vehicle. In order to improve command reliability and simplify the command equipment on the flight vehicle, space-ground loop is generally adopted by the command system of geostationary satellite. Forward error prevention system is generally adopted by near-earth flight vehicle and missile safety control system, where retransmission accumulation decision system, forward error correction system, and error detection & retransmission system are often used. Next, we will focus on the above-mentioned systems.

(1) Feedback check command system for space-ground loop

This system is characterized by taking the contents to be controlled as "preparatory command" and the actions to be executed as "action command". The basic process is that the ground command station sends command preparatory command code to the flight vehicle, which will save but not execute it for the moment after reception and then return the original code of "preparatory command" to the ground through telemetry. The ground command station will compare it with the original one, and send the "cancel" command and repeat the above process in case any error is found. It will not send "action command" pulse until the link comparison is correct. After the flight vehicle receives the "action command" pulse, it will produce the prescribed actions according to the contents of stored "preparatory command" code. The advantage of this system is the ability to obtain a low error probability by using a simple coding method and to achieve synchronous control function by controlling the phase accuracy of "action command" since it is a pulse signal. The disadvantage of this system is the long time of feedback verification, so the control speed is rather slow. It is hard to be applied on the flight vehicles flying at medium and low earth orbit with a short controllable time, but for stationary satellite it is not a problem.

The characteristics of this system are:

1) Both command channel and feedback channel (generally completed by telemetry) are required.

2) Under the same error control performance, compared with forward error correction system, encoding and decoding equipment of this system are simple, especially the decoding equipment.
3) The theoretical expected value of the error control probability of the system is: $P_{im}(\text{error}) = P_{i0}(\text{error}) \frac{1-P_{i0}^{N+1}(\text{retransmission})}{1-P_{i0}(\text{retransmission})}$, and it is impossible to be better than P_{i0} (error), where P_{i0} (error), the error command probability of command code, depends on basic code design; and N is the number of retransmission.
4) Error detection depends on feedback verification.
5) Error correction is implemented through retransmission.
6) It is easy to control the time of command action and its duration as well as action repetition times. It is applicable to complete proportional control, synchronous control and on-off control with high time accuracy, which cannot be easily realized by forward error correction system.
7) Compared with forward error correction system, the time to be spent is much longer.

Due to the above characteristics of the space-ground loop, it is hard to be applied to those targets with a short time passing through the ground TT&C station. However, this error control system is, without exception, widely applied to synchronous orbit satellites with a long controllable time.

(2) Forward error correction command system

Forward error correction command system is characterized by the combining of "preparatory command" and "action command", i.e., the identifiable command will be immediately executed once it is received by the flight vehicle. Due to the error caused by distortion and interference during transmission, the command received by the flight vehicle may be the one transmitted by the ground station (the correct command) or may not be the one transmitted by the ground station (the error command). It may even be the one generated by interference when the ground station does not transmit any command (the false command). It may also be such situation where a command transmitted by the ground station is not received by the flight vehicle (the missing command). Thus, some error actions may be generated. In order to reduce probabilities of such errors, it is required to work hard on coding techniques, which may definitely make the coding and decoding more complicated, the command length longer, and the accurate control of executed action phase more difficult as well. However, although the command code is a bit longer, the transmission time, compared with the aforesaid feedback verification system, is much shorter. This advantage is very precious to the medium and low earth orbit flight vehicles with a short controllable time. Therefore, forward error correction system now is widely applied to the command systems for the medium and low earth orbit flight vehicles both abroad and at home.

For a channel error correction coding system, the calculation of its bit error rate has been described in Sect. 3.2. If the bit error rate without error correction coding is P_e at certain C/N_0 and it decreases to P_e' when error correction coding is used, the missing command probability for a command code with a length of n is

$$P_{\text{missing}} = nP_e'.$$

Missing command is quite harmful to flight vehicle command. For example, for UAV command, there are three kinds of commands to flight trajectory, i.e., climbing command (command code is simply set as 01), diving command (set as 10), and centering command (set as 00). Centering command refers to a situation of no command when the elevating rudder will return to the position parallel to the tailplane. Generally the output command will be coded as 00 once missing command occurs, resulting in a "centering" action. If a missing command suddenly appears (it may be caused by error code, PLL losing lock or signal fadeout interruption) when sending a series of climbing signals to UAV, the elevating rudder will suddenly return to the horizontal position from nose-up trim. This is equivalent to apply a downward diving force to the climbing flight vehicle to change the flight path. If the flight vehicle is landing at this moment, this diving force will make the flight vehicle hit the ground and cause damage. It happens when multi-path interference is most serious and the multi-path fading will easily cause missing command during landing. Therefore, it is very important to overcome missing command for UAV command. Compared with error command and false command, missing command can appear if there is one error bit in an n-bit command, while error command and false command will be generated only when each bit in the n bits is the same as the command code in command table, so the probabilities of the latter two are much lower. There are two measures in solving missing error: one is to reduce missing command probability and the other is to reduce its effect on command executor once missing command has occurred, such as taking command "memory" measures.

(3) Error detection & retransmission system

If one command is transmitted but by judgment it is not received or correctly received by the command receiver, the ground command system will retransmit it several times, and the command receiver will execute the command as long as it receives the command for only one time. This is error detection & retransmission system. It is a kind of error control system generally used in flight vehicle command, especially in medium and low earth orbit flight vehicle command. In order to ensure the execution of telecontrol command, another system called "continuous retransmission system" is evolved from this system, i.e., it will continue to retransmit without error detection judgment to ensure the execution of command. It is used in "safety control" detonation command.

(4) Comparison of these systems

For an engineering special case, error probabilities of the following error control modes are calculated as shown in Table 3.8.

From Table 3.8 it can be seen that:

1) Error detection code can only reduce error command probability, but cannot decrease the missing command and the error command probability and false command probability generated by itself.

Table 3.8 Table of error probability comparison of error control modes

Control mode \ Item		Missing command probability $P_{missing}$	Error command probability P_{error}	False command probability P_{false}
Without error control		3.2×10^{-3}	3.2×10^{-3}	7×10^{-8}
Error detection code		3.2×10^{-3}	3.6×10^{-12}	7.8×10^{-8}
Error correction code		4.96×10^{-6}	4.96×10^{-9}	2.3×10^{-6}
Retransmission accumulation	Three times of retransmission One time of decision	3.3×10^{-8}	1.08×10^{-11}	7×10^{-8}
	Three times of retransmission Two times of decision	3.07×10^{-5}	3.9×10^{-23}	4.9×10^{-15}
Link verification		3.2×10^{-7}	3.6×10^{-23}	0

2) Error correction code can obviously decrease the missing command and error command, but it will increase the false command probability.
3) Repeated retransmission and repeated accumulation decision mode can obviously decrease the missing, error, and false command probability.
4) Loop feedback verification command system can obviously decrease these three kinds of error probability and it can absolutely prevent its own false command probability and error command probability.

3.3.3 Remote Sensing Information Transmission Technique

3.3.3.1 Remote Sensing Overview

Remote sensing refers to using special instruments or equipment through remote sensing method (other than contact method) at long distance to receive and record the electromagnetic information transmitted or reflected by the target, and then detecting and identifying the target nature through transmitting, processing, analyzing, and interpreting the information. Remote sensing has a more generalized definition, in which physical detection, mechanical detection, and seismic wave detection are included, but remote sensing does not include the general signal reconnaissance. However, as for the signal forms of the information transmission most of them are digital signal, so the technical problem is the same.

The main differences among remote sensing, telemetry, and command are: remote sensing using non-contact detection and special instruments and equipment (also called remote sensor); while telemetry often uses contact sensor, which is the information source of telemetry information system, and command often uses contact executor, which is the information destination of command information system. Generally, remote sensing, command and telemetry are integrated in one TT&C system. Only in remote sensing system, the TT&C is generally served for business TT&C and dynamic orbit determination method is generally used to determine the orbit of an orbit flight vehicle. The information transmission systems of remote sensing, command and telemetry are similar in that they are all digital information transmission systems in contemporary digitized information age. This book aims at explaining their different and special issues based on this unified theory.

According to its sensing information, remote sensing can be classified into image remote sensing and non-image remote sensing. Image remote sensing is the main sensing information at present, and it can be classified into active remote sensing (e.g., radar) and passive remote sensing (e.g., optical photography and electronic and photoelectric scanning imaging). Results of non-image remote sensing are curve, measured data, etc. Remote sensing can also be classified into space remote sensing, aerial remote sensing, etc. according to its carrier.

Remote sensing has a wide range of application. For civil use, it can be applied to resource survey, ground mapping, disaster monitoring and reporting, crop estimation, forest fire monitoring, pest monitoring, production management, etc. It is very important to national economy. For military use, it can be widely used in military reconnaissance, camouflage identification, guided attack, etc.

For remote sensing imaging observation, the flight vehicle is required to fly in an orbit at an altitude of 200–1,000 km to improve the ground resolution, no matter what it is applied to, land, ocean, national defense or scientific research. Multispectral visible light remote sensor and infrared remote sensor are required to take pictures of earth's surface. Ground object imaging picture is a still image, whether it is square or rectangle picture. High resolution remote sensor data will be transmitted in a large-capacity and at a high speed, while telemetry and command data are generally transmitted at a low speed. This is the biggest difference between remote sensing data transmission and telemetry, command data transmission.

For an optical remote sensor, its data rate is:

$$S_b = \frac{WV_g n_b B_n}{R_g^2}, \tag{3.152}$$

where: S_b is the data rate of the original video signal of the remote sensor; W is swath width; V_g is sub-satellite point velocity; n_b is quantized bit number of each sample (pixel); B_n is the number of spectral channel; R_g is the resolution of ground pixel.

For a 1 m panchromatic/4 m 3-channel panchromatic/multispectral CCD camera at an altitude of 700 km, its original data rate can be up to 2 Gb/s under 8b quantized bit number and 30 km swath width. For a 0.1 m resolution panchromatic CCD camera, its original data rate will exceed 50 Gb/s under 8b quantized bit number and 10 km swath width.

For satellite-borne SAR, its data rate of original video signal is

$$S_b = 2Bn_b k_s = \frac{cn_b k_s}{\rho_r}, \tag{3.153}$$

where: p_r is range resolution; c is light velocity; n_b is quantized bit number of samples; k_s is oversampling coefficient, generally $k \approx 1.2$; B is the bandwidth of radar transmission waveform.

It can be calculated that its original data rate can be up to 3 Gb/s under single polarization, 1 m resolution, 8 b quantized bit. If the resolution is improved to 0.3 m, its original data rate can exceed 9 Gb/s.

Fine remote sensing is required to meet the requirements of fine detection, precise calibration, and accurate positioning. High resolution measurement is required for fine detection. Generally, compared with conventional detection, fine detection will generate more data information to be transmitted. It can be roughly deemed that the data transmission rate of data transmission system is directly proportional to the number of ground imaging unit, quantized bit number of video output, and spectral number (or band, polarization number).

At present, high resolution earth observation is a hot spot for remote sensing development, ultra high-speed data transmission (bit rate up to 2 Gb/s) is definitely the development trend and the Ka-band exploitation is its development direction.

Main factors for data transmission are considered: ① bit rate requirement for satellite-earth transmission; ② bit error rate requirement; ③ minimizing ratio of bit energy required to noise power spectrum density; ④ minimizing system bandwidth to be required; ⑤ realizing strong anti-interference performance; ⑥ reducing system complexity, calculation amount and cost.

3.3.3.2 Basic Model and Main Technical Issues of Remote Sensing Information Transmission

The model of remote sensing information transmission is shown in Fig. 3.66.

The contents of each block in Fig. 3.66, have been introduced in Sect. 3.2 in details. In this section, only the special issues concerning remote sensing information transmission will be introduced.

(1) Information source

At present, remote sensing information are mainly image information and generally they are digital signals coming from camera, synthetic aperture radar, cloud picture imager, optical sensor, infrared sensor, optical mechanical scanning imager, push-broom scanning imager, CCD camera, etc.

(2) Source coding and decoding

Information source coding is to convert the input information into the digital sequence required by the next loop. In order to reduce the bit rate of the digital

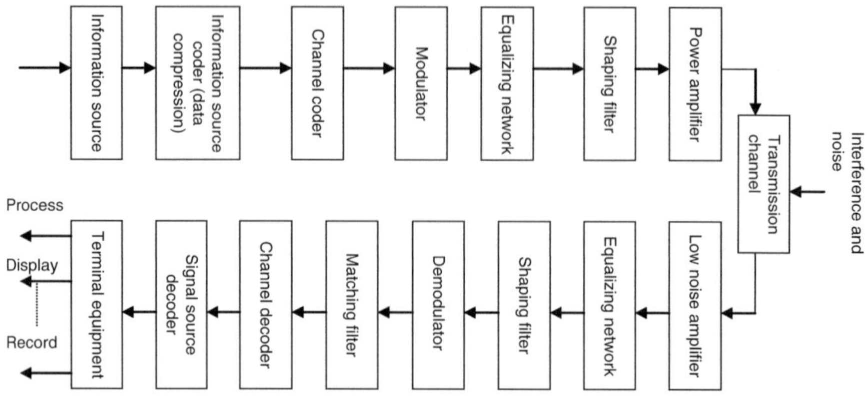

Fig. 3.66 Model of remote sensing information transmission system

sequence to reduce bandwidth, save power and storage capacity, one of the important tasks of information source coding is data compression. Through the processing by data compression coder, the correlation between symbols is removed to make data statistics independent. It decreases the redundancy in the information to make it have the largest information amount so as to realize the bit rate compression. While at the port for information source decoding, it is required to restore the original information correctly or with a certain loss tolerance.

For such remote sensing tasks as monitoring and reconnaissance, it is required to obtain high resolution 2D or 3D image. With the rapid development of digital image technology, image resolution is continuously improving, data size is growing and the data rate to be transmitted by image information transmission system is higher and higher. All these exert great pressure upon information transmission system. Therefore, it is required to compress images, obtain a high compression ratio, and compress images without any loss. If it is difficult to realize lossless compression due to large information amount, lossy compression with a little distortion is allowed. The basic conditions for image compression is there are redundancies among original image data, including the correlation between adjacent pixels in the image (spatial redundancy), the correlation among successive frames of image signals (time redundancy). According to the features of human's insensitive vision to rapidly changing edge which has frequency redundancy due to its high frequency component, poor distinguishing capability to colors and sensitivity to brightness, conduct image compression to make the quality after decompression have a good subjective assessment quality.

There are many methods for image compression, e.g., discrete cosine transformation (DCT), dynamic quantization, DPCM and run-length coding, Huffman coding, inter-frame prediction, motion compensation, wavelet transformation, etc. Readers can refer to relevant books. In engineering application, a series of coding standards for image compression have been developed.

(3) Channel coding/decoding techniques

Because the power in remote sensing information transmission system is limited and extremely low bit error rate is required for some special data services, it is required to select an error control method with strong error correction capability, and at the same time take the realization complexity of the selected error correction coding into consideration, especially the realizability of high-speed coding/ decoding at high bit rate.

For a satellite remote sensing high-speed data transmission system with a limited bandwidth, the method of combining multi-system modulation and low complexity coding can be used to realize the error control under limited power and band (e.g., TCM/8PSK, or TCM for short).

(4) Modulation/demodulation mode

As viewed from characteristics of the remote sensing information transmission system of flight vehicles, the modem must have high power utilization rate, strong anti-nonlinear capability, easy synchronization and networking performance, low power consumption, and small volume. For satellite remote sensing system, generally its power is limited but the frequency band is non-limited, so modulators with constant envelope are used. Modulation modes which can make the satellite power amplifier operate in saturation conditions at a high power and high efficiency mainly are QPSK, OQPSK, MSK, TCM adopting 8PSK, etc. As the case stands, most remote sensing satellites and data communication satellites use QPSK modulation/demodulation technology for data transmission.

Besides, TCM modulation system integrates the advantages of both frequency band utilization and power utilization and it is a proper modulation system for satellite remote sensing data transmission. This modulation technology combining modulation and channel coding is a developing direction of satellite communication modulation/demodulation technology.

With the further improvement of transmission bit rate, multi-system modulation also has been improved. MPSK is widely used due to its advantage of high frequency band utilization, while, MDPSK is more widely used than MPSK because it is free of phase ambiguity in demodulation. For remote sensing satellites launched into space, their data transmission systems are generally 4DPSK. Compared with QPSK, 8PSK is more advantageous in frequency band utilization, so it is full of application potentiality in ultra high-speed data transmission system. In summary, multi-phase system has the following features:

1) When channel symbol rate is the same, the bandwidth of a multi-phase system is the same as that of two-phase system, but the information rate of the multi-phase system is $\log_2 M$ times that of two-phase system.
2) Bit error rate of multi-phase system is higher than that of two-phase system, and it will increase with the increase of M.
3) Multi-phase system belongs to constant envelope modulation and can make full use of the transmitter power.

There are also some new methods:

1) QAM: quadrature carrier amplitude phase keying, is a double controlled digital modulation of both amplitude and phase. Four-level QAM (4QAM) uses quadrature carrier modulation technology to transmit amplitude shift keying (ASK) signal, which can improve the frequency band utilization by 1 time, up to 2(b/s)/Hz. If combined with the multi-system (16APSK/16QAM etc.), the frequency band utilization can be further improved. Its disadvantage is that the envelope is not constant.
2) Single-carrier frequency domain equalization: it still uses QPSK, OQPSK, etc., but it uses frequency domain equalization technology to reduce the influence of channel group delay on bit error rate.
3) MSK: continuous phase binary frequency shift keying with 0.5 shifting ratio. MSK modulated signal has a constant envelope, frequency shift of signal is strictly equal to $\pm 1/4T_b$ and the corresponding modulation index is 0.5. One symbol period of the signal should be integral multiple of 1/4 carrier cycle, and the phase of the signal should be linearly change accurate $\pm\pi/2$ in reference to carrier phase. At the moment of symbol conversion, signal phase is successive and signal waveform is free of hop. MSK signal has higher spectrum utilization and stronger anti-noise performance than that of 2PSK.
4) $\pi/4$DQPSK: relative phase shift modulation of limited phase hop, with the following features:

   - There are eight phases, namely 0, $\pi/4$, $2\pi/4$, $3\pi/4$, $4\pi/4$, $5\pi/4$, $6\pi/4$ and $7\pi/4$;
   - Adjacent phases are free of 180° hop, and envelope fluctuation is small, which is favorable to reduce out-of-band radiation.
   - Adjacent phase shift can only be $\pi/4$, $3\pi/4$, $5\pi/4$, $7\pi/4$, therefore, after 2b information is input, the phase of output signal will certainly hop between odd number group and even number group and cannot always be in odd number group or even number group.

$\pi/4$ DQPSK demodulation can adopt coherent demodulation, noncoherent demodulation and frequency discrimination demodulation. The latter two demodulations have simple hardware circuit and good anti-interference capability to random frequency modulation caused by fading channel.

The bit error rate formulas of common multi-system modulations are listed in Table 3.9.

E_s is the average signal energy of unit code element and E_b is the average signal energy of unit bit. For M system, there is:

$$E_b = E_S/\log_2 M,$$

when $M=4$, $E_b=Es/2$; when $M=8$, $E_b=Es/3$; when $M=16$, $E_b=Es/4$.

Table 3.9 E_b/N_0 comparison of digital modulation modes commonly used in ultra-high-speed transmission system with bit error rate of 10^{-5}

Digital modulation/demodulation mode	E_b/N_0(dB)	Calculation formula of symbol error rate P_S and bit error rate P_B
8PSK coherent demodulation	13.2	$P_S = 2Q\left[\sqrt{\frac{6E_b}{N_0}\sin^2\left(\frac{\pi}{8}\right)}\right]$ $P_B \approx \frac{P_S}{2}$
16QAM coherent demodulation	13.5	$P_S = 1.5Q\left[2\sqrt{\frac{E_b}{N_0}}\right]$ $P_B \approx \frac{P_S}{2}$

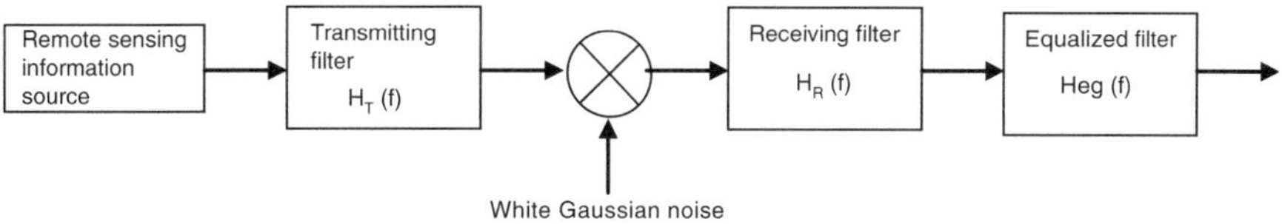

Fig. 3.67 Model of band-limited remote sensing information transmission system

(5) Shaping filter, matched filter and equalized network

At present, remote sensing information transmission includes non-band-limited transmission and band-limited transmission, corresponding to power limited and bandwidth limited respectively. The transmission channel model of band-limited transmission in which only band-limited filter is included is shown in Fig. 3.67.

Transmitting filter and receiving filter together constitute the shaping filter of raised cosine roll-off filtering. When $|H_R(f)| = |H_T(f)|$, they will complete matched filtering and be called root cosine roll-off filter. $H_R(f)$ and $H_T(f)$ make digital information transmission system complete matched filtering and Nyquist filtering. $H_{eg}(W)$ is equalized filter and its function is to equalize and calibrate the amplitude/phase characteristics introduced by other factors in the channel so as to make them have flat amplitude-frequency characteristics and group delay characteristics. Because the bit rate of remote sensing information is high and the bit error rate caused in transmission is directly proportional to data transmission rate multiplied by group delay fluctuating value, the equalization is very important to high-speed data transmission of remote sensing. It is very difficult to implement the wideband equalized network due to the wide bandwidth. This is a special problem in remote sensing information transmission. Adaptive equalization shall be adopted for obtaining good performance due to the poor operating condition on flight vehicle, great changes of channel frequency characteristics and signal amplitude and phase changes caused by multi-path fading (if any). Adaptive equalization is to

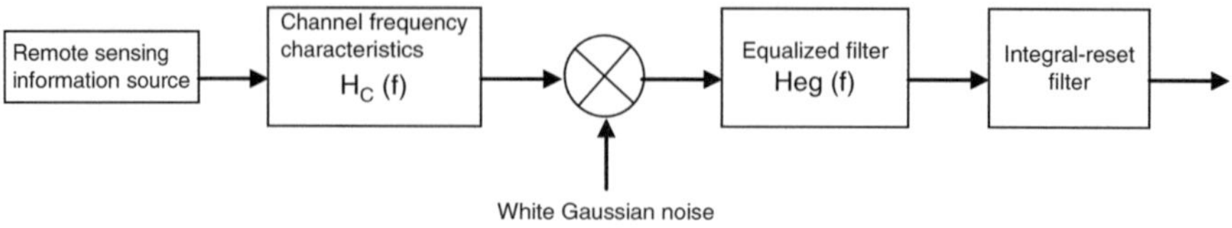

Fig. 3.68 Model of non-band-limited remote sensing information transmission system

automatically adjust parameters of the equalizer according to changes of channels and signals. This technology is more and more widely applied in remote sensing high-speed data transmission.

The transmission system model of remote sensing information transmission system operated in non-band-limited state is shown in Fig. 3.68.

Channel frequency characteristics include the amplitude/phase frequency characteristics introduced by other non-ideal factors. Equalized filter is used to equalize and calibrate $Hc(f)$ to make it have flat amplitude-frequency characteristics and group delay characteristics in information bandwidth, so non-bandwidth-limited code pulse signal doesn't generate inter-symbol interference. Integral-reset filter is used to minimize the bit error rate of matched filter under the function of white Gaussian noise.

(6) Operating band selection

At present, most domestic and foreign remote sensing satellite data transmission systems operate at X-band. However, according to the regulations on frequency utilization and distribution of International Telecommunication Union, the available bandwidth in X-band is less than 400 MHz.

The common modulation mode in X-band is DQPSK. For DQPSK, the minimum allowable value in 1 dB bandwidth is 0.8 times of channel data rate at present. For dual-channel, the guard interval between the two channels shall be reserved. There are two ways to improve frequency band utilization: ① dual-channel + 8PSK; ② dual-channel + polarization multiplexing + QPSK, however such improvement is limited. The only way to realize a higher data rate is to improve the operating frequency of data transmission system. In Ka-band, the available bandwidth is 3 GHz. 2.4 Gb/s or higher rate can be achieved by adopting DQPSK modulation, dual-channel, and polarization multiplexing. For parabolic antennas having the same aperture, when the operating band is improved to Ka-band, antenna gain will increase, but the aperture loss transmitted by Ka-band will increase too. The two just cancel out each other, so band increase can't strengthen the receiving signal. However, if directional parabolic antenna is also used on the satellite, the antenna gain will increase with the increase of frequency band, and the signal will be strengthened. When channel data rate is low, fixed earth-matched beam form is generally selected by satellite data transmission. Beam coverage can be up to 130° and the gain at the maximum direction is about 7 dB only. In ultra high-speed data transmission system, high gain antenna must be selected. There are two forms of common high gain antenna: electric scanning phased array antenna and mechanical scanning antenna. The former has no movement and no disturbance to satellite attitude, and the latter, relying on dual axial driving mechanism, has a larger pointing range than that of the phased array antenna.

If a transmission rate higher than that of MW band is planned to be realized, optical communication is the only choice.

(7) Quasi real-time transmission

When data rate is too high to conduct real-time transmission, the fast write-in and slow read-out method through a buffer memory can be used to decrease the transmission speed of data transmission. This operating mode of data transmission system shall match the application mode, i.e., substituting the intermittent strip imaging mode for the traditional continuous strip imaging mode to make a detailed survey of the region of interest.

In summary, the main technical issues of remote sensing information transmission are:

1) Satellite remote sensing data compression technology, including visible light, SAR, hyperspectral remote sensing data compression technology.
2) New technical studies on high bit rate coding and modem, including the studies on increasing operating bands (e.g., Ka-band) and modulation mode (e.g., QPSK/8PSK/TCM/160 AM etc.).
3) Intelligent processing technology of satellite remote sensing images, including image pre-processing, automatic target detection, automatic target classification, adaptive data processing, adaptive compression, automatic formation of data frame, satellite region of interest (ROI) detection, and compression technology.
4) Anti-interference technology for data transmission system.
5) Adaptive modulation and coding technology, LDPC coding technology.
6) Single-carrier frequency domain equalization and multi-carrier OFDM high-speed data transmission technology.
7) Pointing controllable high gain phased array antenna technology.

3.3.3.3 High-Speed Data Modulation/Demodulation

High bit rate data transmission of flight vehicle is mainly used to directly or by the relay satellite transmit the high resolution remote sensing data to the ground. It is the key technology in remote sensing information transmission, and its technical difficulties include two aspects. One is the realization of high-speed circuit for high-speed data transmission modem. The other is the transmission characteristics have great impact on bit error rate of data transmission. In high-speed data transmission, the impacts on amplitude/phase frequency characteristics, group delay equalization and phase noise are greater than that in low-speed data transmission and the technical difficulties are greater too. All these issues are included in high-speed demodulator and high-speed data transmission channel.

The principle circuit block diagram of high-speed modem is the same as that of low-speed modem. The difference is how to achieve high-speed in specific circuit. The measures generally used are:

(1) High-speed data frequency domain parallel processing. According to the current devices and technical level, it is of great difficulty in technology when processing speed is up to 1 Gb/s. A common method is using multi-channel parallel processing to reduce the processing speed of each channel.
(2) Fast algorithm and solution selection for digital signal processing.
(3) High-speed *A*/*D* sampling.
   A/*D* sampling is a bottleneck for the demodulator implementing all kinds of algorithm. The quality of *A*/*D* sampling, especially *A*/*D* sampling accuracy and *A*/*D* output SNR, directly affects the demodulation quality. Meanwhile, it is also important to make sure that there is no crosstalk between high-speed output data. Dual-channel 8b and ultra high-speed A/D device with extremely high sampling speed are generally used.
(4) Strict PCB design. Advanced high-speed PCB design software is used for PCB wiring and layout. The PCB design software is required to strictly design impedance matching, time delay, crosstalk, EMI, etc. based on strict simulation.
(5) Modular design, equipping each module with a shielding box and other measures can be used to improve the high frequency performance of the circuit and reduce the influence of distribution parameter and high frequency interference.

The difficulties in high-speed data transmission design are to solve the technical challenges under an environment where the received signals have a large Doppler shift (up to 2,400 kHz), such as the tracking demodulation receiver technique for non-steady-state phase error data transmission signals in high-speed satellites or other platforms, the parallel demodulation algorithm for high-speed modulation signals, and the high-speed error correction and decoding algorithm with continuously variable rate. There is great difficulty in the development of demodulator due to the high demodulation rate and high IF frequency. Firstly, low SNR stable operation problem must be solved, i.e., the locking threshold of the demodulator must be decreased to $E_b/N_0 = 5$ dB or below. Second, the carrier-to-noise ratio loss caused by the imperfection of the demodulator itself must be reduced. The fundamental way of improving the performance of the demodulator is to further decrease the amount of carrier-to-noise ratio deterioration introduced by each unit or component of the demodulator. The following measures are included: IF transmission channel, IF filter and baseband receiving filter of the demodulator shall have flat and good amplitude-frequency response and group delay characteristics; the input saturation level, input/output impedances and return loss should be reasonably designed; local reference source and frequency synthesizer should have low phase noise and stray; both the carrier and clock recovery circuit should use digital PLL to ensure low operating threshold and minimum RMS phase jitter error. Besides, a good isolation is required for high-speed data conversion outputs and interfaces to avoid introducing crosstalk and causing bit error.

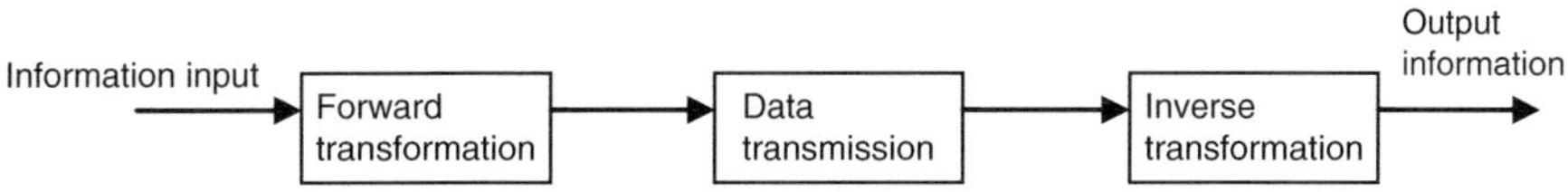

Fig. 3.69 Model of remote sensing system

3.3.3.4 Function and Composition of Remote Sensing Information Receiving System

The model of a remote sensing system is as shown in Fig. 3.75.

In Fig. 3.69, the input remote sensing information will be forward transformed into digital data, transmitted to the receiving station through data transmission system, and then inverse transformed into remote sensing information. Data transmission system includes the transmitting system on the platform and its important part ground receiving system.

(1) Main functions

The main functions of remote sensing satellite information receiving system are:

1) Tracking the X-band or Ka-band broadcast type remote sensing satellite or authorized transmission type remote sensing satellite, and receiving the remote sensing data transmitted by the receiving satellite at the same time.
2) Function of S-band telemetry.
3) S/Ka (or S/X) dual-band auto-tracking and auto-switching capability, X-band single dot frequency double rotation and multiple dot frequency single rotation auto-tracking capability; whole hemisphere tracking and receiving capability, ensuring no target loss in vertex tracking.
4) Receiving, generating and recording remote sensing image signal, recording original data, formatted data and quick-look image data of the whole transit period, generating data for browse and auxiliary data, capability of real-time recording, post playback, and processing of original data.
5) Capability of quick-look display of visible light image data.
6) Automatically receiving the tasks, plans, orbit information, etc. from departments, conducting conflict detection of multi-satellite receiving plan according to a variety of rules, automatically organizing the implementation of tasks and plans, automatically generating local tasks and plans.
7) Capability of timing and time keeping.
8) Capability of automatic monitoring and management, to put all monitorable equipment of the whole system under the centralized monitoring and management.
9) Function of system closed-loop detection, equipment condition monitoring, unit fault location, etc.
10) Having certain scalability and upgradeability.

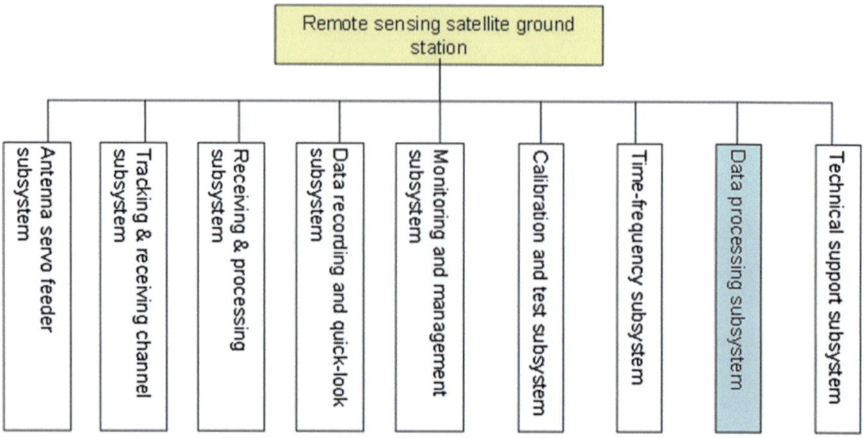

Fig. 3.70 System composition

(2) Composition of remote sensing information receiving system

As shown in Fig. 3.70, the system is generally composed of antenna servo feeder subsystem, tracking & receiving channel subsystem, receiving & processing subsystem, data recording and quick-look subsystem, monitoring and management subsystem, calibration and test subsystem, time-frequency subsystem, data processing subsystem and technical support subsystem. It is mainly used to receive, record and archive the downstream remote sensing data of low earth orbit remote sensing satellite and complete the quick-look display of optical remote sensing image data.

1) Antenna servo feeder subsystem is composed of antenna feeder subsystem, mechanical structure subsystem, antenna control subsystem, monitoring subsystem, and accessory equipment.
2) Tracking & receiving channel subsystem is composed of Ka-band data receiving channel, S-band telemetry/tracking channel, Ka-band tracking channel, Ka-band upstream test channel, and IF equalization matrix.
3) Receiving & processing subsystem is composed of high-speed data demodulator (including test modulator), low-speed demodulator, two sets of integrated baseband, tracking signal switch, data distribution unit, and decryption equipment. It mainly completes the receiving, demodulation, decoding, decryption, and formatting of the data for transmission, receiving of S-band telemetry signals, data receiving of Ka-band, and implementation of functions of angle tracking receivers of S-band and Ka-band
4) Data recording and quick-look subsystem is composed of data recording equipment, optical quick-look equipment and magnetic tape unit. It mainly completes the receiving, recording, achieve, quick-look display and playback of satellite downstream remote sensing data.

5) Monitoring and management subsystem is composed of task management equipment, system monitoring equipment, integrated information display equipment, database management equipment, etc. It mainly completes the management of system task implementation, real-time monitoring of the operating state of each subsystem, calculation and prediction of satellite orbit, and the management of service data.
6) Calibration and test subsystem is composed of test modulator, Ka-band test up-converter, beacon, etc. It mainly completes system calibration and test and provides the system with BIT and performance test means and methods to keep the system in a good operating state for a long time.
7) The main task of time-frequency subsystem is providing the system with standard time and standard frequency of high accuracy and high stability.
8) Data processing subsystem is composed of image product generating equipment, product storage and management equipment, etc. It mainly completes the production of optical image products of all levels (0–3 level), product storage and management, etc.
9) Technical support subsystem is composed of power supply equipment, environment protection equipment, lightning protection equipment, earthing device, etc. Its main task is providing supports to each system task.

References

1. Liu Jiaxing (2004) A new two-way carrier acquisition method for USB-follow-scan slope decision method. Acta Electron Sinica 32(3):499–501
2. Liu Jiaxing, Yang Hongjun (2007) A new two-way carrier acquisition method for USB. ITC Proceedings, Las Vegas, NV, USA
3. Liu Jiaxing (1997) Combined interference of PM system of multi-subcarrier modulation. Acta Electron Sinica 25(3):111–113
4. CCSDS Green Book (1989) Season 2. Adv Modul Tech (10)
5. Liu Jiaxing (1980) Tone ranging time delay in linear four terminal network and product demodulator. J CETC 2(3):20–24
6. Liu Jiaxing (2009) Group delay characteristics analysis method for range error and its applications in deep space TT&C systems. J Spacecr TT&C Technol 28(2):22–26
7. Ziemer RE et al (2005) Introduction to digital communication (trans: Yin Changchuan, Hao Jianjun). China Machine Press, Beijing
8. Feng Yumin (2003) Communication system principles. Tsinghua University Press, Beijing
9. Simon MK (2003) Bandwith-efficient digital with application to deep space communications (trans: Xia Yun, Sun Cheng). Tsinghua University Press, Beijing
10. Jiang Chang (2006) The introduction of space communication and tracking technology. Beijing University of Technology Press, Beijing
11. DSMS Telecommunications Link Design Handbook, JPL, 2007
12. Joseph, YH (1998) Deep space telecommunication system engineering [M]. Plenum press, New York
13. CCSDS 100.0-G-1
14. CCSDS 131.1-0-2
15. Liu Jiaxing (2005) Impacts of linear distortion and noise on data transmission bit error rate in TDRSS. Album of Telecommunication Engineering

16. Azizzadeh A, Mohammadi L (2008) Degradation of BER by Group Delay in digital phase modulation. In: 2008 I.E. fourth advanced international conference on telecommunication
17. Yang Zhixing (1988) Digital microwave relay channel and circuit. Post & Telecom Press, Beijing
18. Niu Jilie (1989) Principle and design of digital microwave relay communication system. Beijing University of Posts and Telecommunications Press, Beijing
19. Fu Haiyang (1998) SDH digital microwave transmission system. Post & Telecom Press, Beijing
20. Wang Yunfei, etc. (1991) Digital microwave communication. Post & Telecom Press, Beijing

Chapter 4
Spread Spectrum TT&C

4.1 General

Spread spectrum is an information transmission technology. The spread spectrum TT&C system is a system which realizes TT&C and information transmission using direct sequence (DS) spread spectrum, frequency hopping (FH) spread spectrum, time hopping (TH) spread spectrum or any combination of them. PN codes are the typical spread spectrum signals. For example, PN codes in a DS TT&C system are used for spread spectrum of telemetry & telecommand information and for integration of ranging signals and information. With respect to the orbit-measuring function, the system also uses PN codes for ranging and carriers for Doppler velocity-measuring, and then conducts angle tracking for the carrier signals after spread spectrum by PN codes.

The meaning of spread spectrum:

(1) The bandwidth of a signal after spectrum spreading must be larger (generally much larger) than that of the information. Moreover, the spread spectrum code and information code are mutually independent. Information codes are used to modulate the spread spectrum codes (e.g., modulo-2 addition) to realize spread spectrum.
(2) Relevant de-spreading process must be performed during receiving and demodulation, that is, using native codes to conduct correlation operation of transmitted spread spectrum codes. In general, native codes are obtained through the code loop.

Although some modulation systems can make the bandwidth of modulated waves larger than information bandwidth, they cannot be deemed as spread spectrum because they do not meet the above-mentioned conditions. For example, either FM or BPSK is not spread spectrum even though they can generate wideband carrier.

J. Liu, *Spacecraft TT&C and Information Transmission Theory and Technologies*,
Springer Aerospace Technology, DOI 10.1007/978-3-662-43865-7_4

Through the spread spectrum technique, the system can obtain higher performance. Such increment in performance is called the processing gain of the spread spectrum system, which approximates the ratio of spread bandwidth to original information bandwidth. It describes the differences of system performance with and without the spread spectrum system.

4.2 Features of Spread Spectrum TT&C

The core issue of spread spectrum TT&C is the introduction of PN code in spread spectrum (chaotic code may be introduced for further development). Due to the wideband (high code rate) and randomness of the PN code, the spread spectrum TT&C system is characterized by low power spectrum density, anti-interception, anti-interference, privacy, multiple access communication, ranging and covert TT&C. The model of PN code direct sequence spread spectrum TT&C system is shown in Fig. 4.1.

As shown in Fig. 4.1, the code loop is often used to give synchronous local PN codes and complete ranging; the carrier recovery loop is used to give velocity-measuring signals.

The block diagram of spread spectrum modulation channel is shown in Fig. 4.2. In Fig. 4.2:

(1) Data rate: uncoded data bits per second before channel coding and after packing & framing.
(2) Post-coding symbol rate: binary rate after forward error correction (FEC) coding.
(3) Channel symbol rate: binary rate after the distribution of channel symbols in multi-channel system, namely, the actual code rate transmitted in the channel. Shown in the figure is a QPSK modulation mode.
(4) Information rate: generally refers to the rate before an information source encoder in the information source.

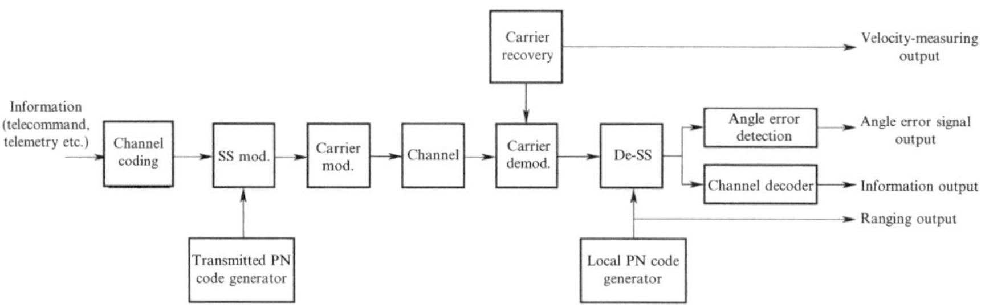

Fig. 4.1 PN code direct sequence spread spectrum TT&C system model

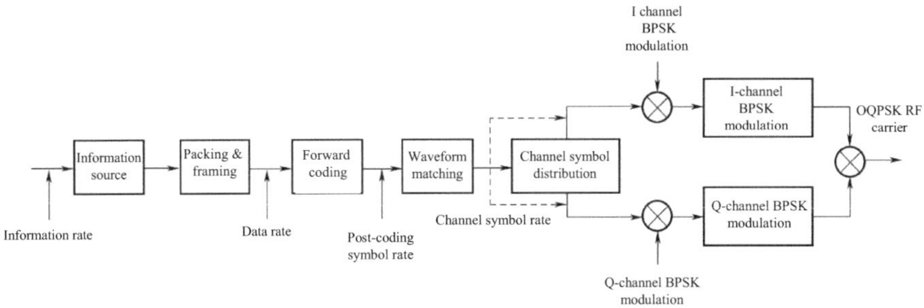

Fig. 4.2 Block diagram of spread spectrum modulation channel

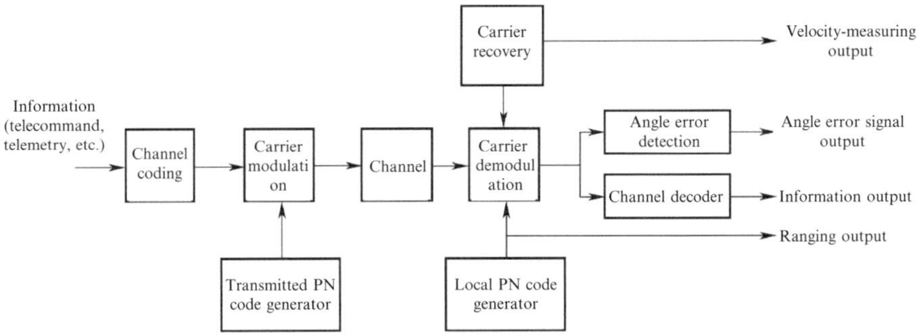

Fig. 4.3 Only PN code TT&C system model

Before the appearance of spread spectrum TT&C system, an only PN code ranging has been used in the TT&C system, whose model is shown in Fig. 4.3. For the sake of facilitating discussion, the said system is called only PN code TT&C system. Its input information is modulated to the carrier through multiple subcarriers or UQPSK. Through comparison of Fig. 4.2 with Fig. 4.3, it can be seen that the two systems are basically the same in the orbit-measuring part, except parts of information spectrum spreading and de-spreading. The introduction of PN code in spread spectrum brings characteristics of PN code for the spread spectrum signal, so the orbit-measuring parts of these two systems have similar input signals. Their bandwidths are determined by the PN code and their signals have the characteristics of PN codes such as randomness, multiple access capability, and low power spectrum density.

Similarities of PN code spread spectrum TT&C system and only PN code TT&C system include:

(1) Both systems have very wide bandwidth of modulated carrier, which are determined by PN code rate.
(2) Their transmitted signals have very low power spectrum density and can be used to realize covert TT&C, counter-reconnaissance, anti-interception, and for reducing the interference to other electronic equipment as well.

(3) They can use the coherence of their PN codes for counter-interference or interference suppression, except for Gaussian white noise.
(4) They can use various PN code patterns to achieve confidentiality.
(5) They can use various PN code patterns to realize multi-target TT&C.
(6) Their wide PN code spectrum has anti-multipath capability and can realize multipath diversity reception.
(7) They both use PN codes for ranging.
(8) They both use the carrier recovery loop to measure the velocity.
(9) While extracting angle-measuring error signals in a low *S/N*, they both use PN code coherent reception to decrease bandwidth and increase the *S/N*.
(10) They both have shortcomings such as high occupancy of bandwidth and complicated system implementation.

Differences between PN code spread spectrum TT&C system and only PN code TT&C system include:

(1) The PN code spread spectrum TT&C system is to conduct spread spectrum of telecommand and telemetry information to provide a series of advantages of spread spectrum communication, while the only PN code TT&C system does not conduct spectrum spreading of information.
(2) The PN code spread spectrum TT&C system integrates ranging signals and information and unites various types of information with a unified data flow to realize unified TT&C in time domain, while the only PN code TT&C system fails in such integration and unification.
(3) Due to square loss in receiving spread spectrum signals, receiver acquisition sensitivity, and orbit-measuring accuracy in a low *S/N* both are decreased. This is a very important issue in low *C/N* receiving.
(4) De-spreading may cause circuit complexity and de-spreading loss.

To sum up, spread spectrum TT&C provides advantages of spread spectrum for telecommand and telemetry and can realize the unification of ranging and telecommand and telemetry signals, but has square loss, circuit complexity, and other disadvantages in tracking & orbit-measuring, compared with the PN code TT&C.

Spread spectrum TT&C differs from spread spectrum communication. Their differences are:

(1) The spread spectrum TT&C system should use PN codes to conduct high-accuracy ranging. In order to meet ambiguity-resolving requirements, the PN code should have a longer length. However, this may lead to some technical issues such as difficulty in long-code acquisition in a high Doppler frequency and a low *S/N*. Phase drift error of the code loop should be decreased in order to reduce ranging system error. This requires the minimal distortion of the PN code and a higher amplitude/phase frequency characteristic of the system. In order to reduce random ranging error, the code clock frequency of the PN code should be as high as possible and the bandwidth of the code loop should be as narrow as possible.

(2) The spread spectrum TT&C system should use the carrier after de-spreading and carrier recovery to conduct a high-dynamic and high-accuracy Doppler velocity measurement. Then, velocity-measuring error caused by the short-term instability and effects of de-spreading and carrier recovery loop on the velocity-measuring accuracy should be considered.
(3) The spread spectrum TT&C system should use the carrier after de-spreading and carrier recovery to conduct a high-accuracy angle measurement. Input *S*/*N* is very low due to the use of spread spectrum signals. Angle tracking may be a key technique in such a low *S*/*N*. While the mono-pulse angle-measuring system is employed, the sum channel and difference channel should be de-spread. The *S*/*N* of the difference channel is lower and the technology is more difficult.
(4) Compared with the communication system, the TT&C system has a longer operating range, so its *S*/*N* is very low. Various losses must be reduced as much as possible, of which square loss is a major issue, in such case, in TT&C at a very long distance, the application of spread spectrum TT&C system may be limited (e.g., deep-space TT&C).

In summary, the difference between spread spectrum TT&C and spread spectrum communication is orbit-measuring, and the orbit-measuring involves a special key issue, namely, accuracy measurement.

4.3 Basic Methods for Spread Spectrum TT&C

SS, short for the spread spectrum, is a technique used to spread the bandwidth of the signal spectrum, whose main feature is that the bandwidth of the spread spectrum code sequence after spectrum spreading becomes much wider than that of information code element before spreading. It realizes anti-interference and encryption by introducing redundancy and increasing randomness in the deterministic signals and can transmit signals with a low power spectrum density to realize concealment. According to the Shannon Equation:

$$C = W\log_2\left(1 + \frac{S}{N}\right).$$

If the channel capacity *C* remains unchanged, increasing bandwidth *W* can make the signal to noise ratio (*S*/*N*) decrease. If the signal spectrum is spread, signal power per unit band will decrease, that is, the power spectrum density of the signal will decrease to even lower than that of the noise and thus make it difficult for the enemy to detect signals in the background noise. Therefore, spread spectrum technique is widely used in military TT&C.

Spread spectrum techniques include direct sequence spread spectrum (DSSS), frequency hopping spread spectrum (FHSS), and time hopping spread spectrum (THSS) [2].

The most typical method to generate spread spectrum signals is to conduct second modulation of the modulated carrier by a spread signal with a wide bandwidth. Spread spectrum modulation may be phase modulation, quick change of carrier frequency or any combination of them and other techniques. If the phase modulation is used for spectrum spreading, resulting signals are called DSSS signals; if quickly changing carrier frequency is used for spectrum spreading, resulting signals are called FHSS signals; if both DSSS and FHSS are used, resulting signals are called hybrid DS-FH spread spectrum signals. Spread spectrum signals selected for spectrum spreading should have some features in order to enable expected receivers to demodulate the transmitted signal conveniently and make unexpected receivers difficult to demodulate it as much as possible. These features also allow expected receivers to separate communication signals from interference signals. If the bandwidth of a spread spectrum signal is larger than data bandwidth, the transmission bandwidth for spread spectrum will be determined by such spread spectrum signal, almost independent of the data signal.

4.3.1 *Direct Sequence Spread Spectrum (DSSS)*

4.3.1.1 Binary Phase Shift Keying (BPSK) DSSS

BPSK modulation is the basis of other DSSS modulation modes, so a detailed introduction to this modulation mode is made here first. A constant-envelope modulated carrier with power P, angle frequency ω_0, and phase modulation $\theta_d(t)$ can be expressed as [2]:

$$S_d(t) = \sqrt{2P}\cos\left[\omega_0 t + \theta_d(t)\right]. \tag{4.1}$$

BPSK spectrum spreading is to be implemented by multiplying $S_d(t)$ by a $c(t)$, a function representing spread spectrum waveform. As shown in Fig. 4.4, the transmitted signal is:

$$S_t(t) = \sqrt{2P}c(t)\cos\left[\omega_0 t + \theta_d(t)\right]. \tag{4.2}$$

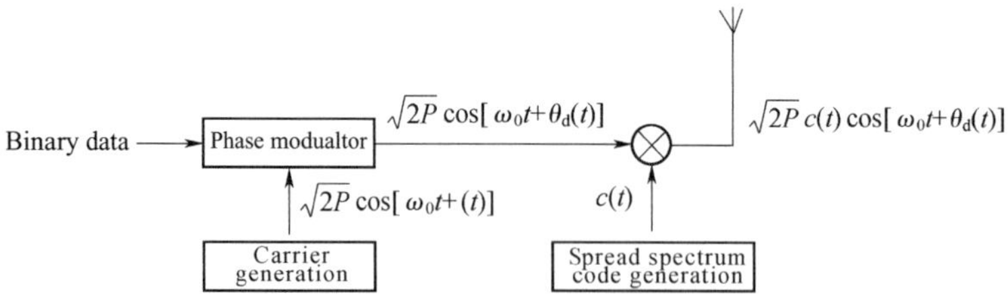

Fig. 4.4 BPSK DSSS transmitter with spectrum spread modulated carrier

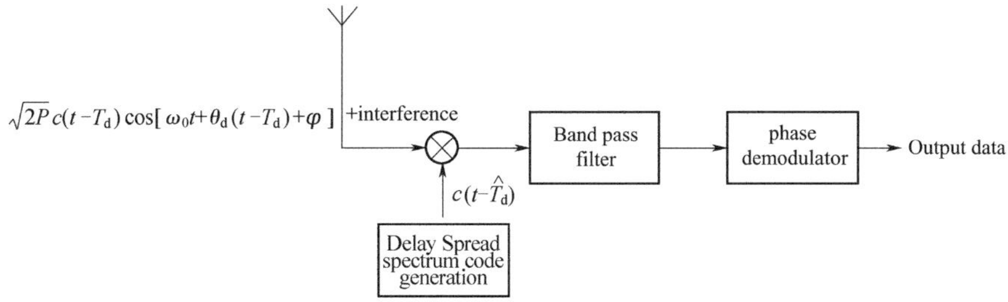

Fig. 4.5 BPSK DSSS receiver

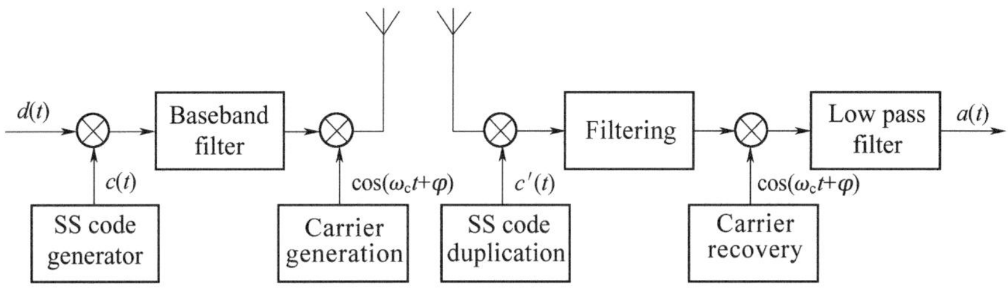

Fig. 4.6 Schematic diagram of DSSS with BPSK used for both spread spectrum and carrier modulation

After undistorted transmission for a transmission delay T_d, the signal $S_t(t)$ will be received together with some types of interference or Gaussian noise. Demodulation is completed through correlation or counter modulation of spread spectrum codes with a proper time delay, as shown in Fig. 4.5. This counter modulation or correlation of received signals by delayed spread spectrum waveform is called de-spreading which is the core issue of the spread spectrum system. Signal component output by de-spreading is:

$$\sqrt{2P}c(t-T_d)c(T-T_d)\cos[\omega_0 t+\theta_d(t-T_d)+\varphi], \tag{4.3}$$

where $\widehat{T}_d$ is the optimal estimation of transmission delay by the receiver.

If $T_d=\widehat{T}_d$, i.e., the spread spectrum codes at the receiving end synchronize with those at the transmitting end due to $c(t)=\pm 1$, the product $c(t-T_d)c\left(t-\widehat{T}_d\right)$ equals 1. Therefore, after correct synchronization, signal component from the de-spreader equals $S_d(t)$, except for random phase φ. $S_d(t)$ can be demodulated by a coherent phase demodulator.

The data modulation above is not necessarily BPSK. The forms of $\theta_d(t)$ are unlimited. BPSK modulation process of spread spectrum can also be implemented through modulo-2 addition and BPSK modulation to carrier by data and spread spectrum codes, as shown in Fig. 4.6.

As shown in the figure above, $d(t)$ refers to symbol sequence; $C'(t)$ refers to spread spectrum sequence. Spread spectrum is to realize modulo-2 addition in mathematics, equivalent to the multiplication in physics. Therefore, the spread signal is:

$$S_{\rm ss}(t) = c(t)d(t)\cos(\omega_c t + \varphi), \tag{4.4}$$

where the bandwidth of $S_{ss}(t)$ is $W_{\rm ss}$; the bandwidth of $d(t)\cos(\omega_c t+\varphi)$ is B; $W_{\rm ss} \gg B$ after being spread with $c(t)$.

If at the receiving end, spread spectrum code $c'(t)$ reproduced by the code loop is the same as the spread spectrum code $c(t)$ at the transmitting end, i.e., $c'(t) = c(t)$, the de-spread signal obtained from multiplication is:

$$S_1(t) = d(t)\cos(\omega_c t + \varphi), \tag{4.5}$$

where $d(t) = 1$, because (+1) and $(-1)^2$ both equal 1, and the effects of $c(t)$ is eliminated.

Original data signal can be recovered through demodulation. Received signals unmatched with the spread spectrum code will be filtered through cross-correlation operation. For example, a signal to be received which is accompanied by very strong narrowband interference will have broadband interference $i_{\rm w}$ after de-spreading (i.e., spread spectrum by $c(t)$), then after narrowband filtering, it will have only a small portion of residual interference $i_{\rm wr}$, as shown in Fig. 4.7.

Suppose the single-side power spectrum density of the interference after de-spread is N_0, the power of residual interference after filtering by a filter with bandwidth B is:

$$P(i_{\rm wr}) = N_0 B.$$

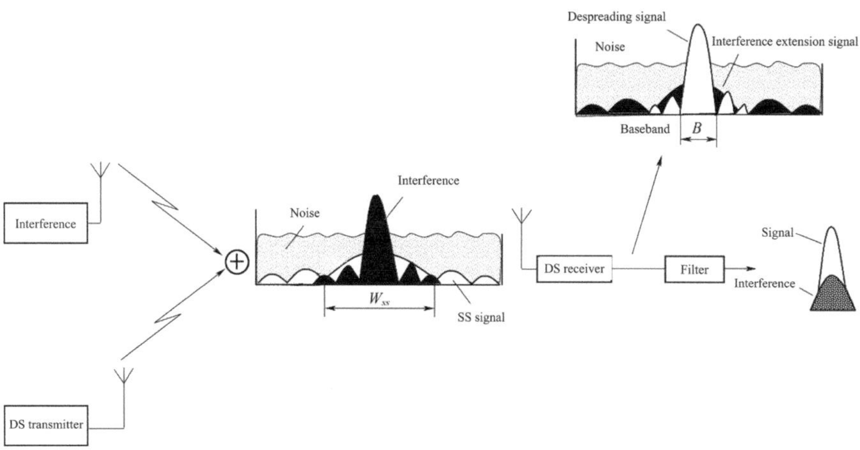

Fig. 4.7 Diagram of DSSS for resisting narrowband interference

The bandwidth of the spread spectrum signal is W_{ss}, therefore the whole power of unfiltered interference signal is:

$$P(i_{wr}) = N_0 W_{ss},$$

and

$$P(i_{wr}) = \frac{B}{W_{ss}} P(i_w) = \frac{1}{G} P(i_w), \tag{4.6}$$

where G refers to spread spectrum gain or processing gain, which is a very important parameter in the spread spectrum system.

Spread spectrum gain refers to the gain of anti-interference capability obtained by spreading the bandwidth of transmitted signals. Here, spread spectrum gain is:

$$G = \frac{W_{ss}}{B} = \frac{T_b}{T_c} = \frac{f_c}{f_b}. \tag{4.7}$$

Expression (4.7) indicates that the spread spectrum gain refers to the ratio of signal bandwidth W_{ss} after spectrum spreading to the bandwidth B before spreading, and that the gain, in time domain, represents the number of chips of PN code sequence contained per information bit, and in frequency, represents the ratio of PN code rate to information rate.

4.3.1.2 Quadrature Phase Shift Keying (QPSK) DSSS

One advantage of using QPSK is spectrum saving. However, in the spread spectrum system, an overriding consideration is not bandwidth efficiency but resistance to interference. The advantages of spread spectrum quadrature modulation are that modulated signals could neither be easily reconnoitered nor affected by some types of interference signals. Various techniques can be used to add spread spectrum modulation to the quadrature carrier.

(1) The information uses BPSK, while the spread spectrum code uses QPSK. When a system [2] as shown in Fig. 4.8 is used for QPSK spread spectrum modulation, a carrier modulation scheme with information BPSK/spread spectrum QPSK is provided. In Fig. 4.9, binary information data first carries out BPSK modulation to the carrier at frequency ω_0 ($\theta_d(t)$ may be random phase modulation). At any branch of the quadrature hybrid circuit, output power is a half of input power. Then spread spectrum code carries out QPSK modulation to the carrier. $C_1(t)$ and $C_2(t)$ are spread spectrum codes, which take a value of ± 1. Two codes are independent of each other. In design, their code rates are the same, synchronous and

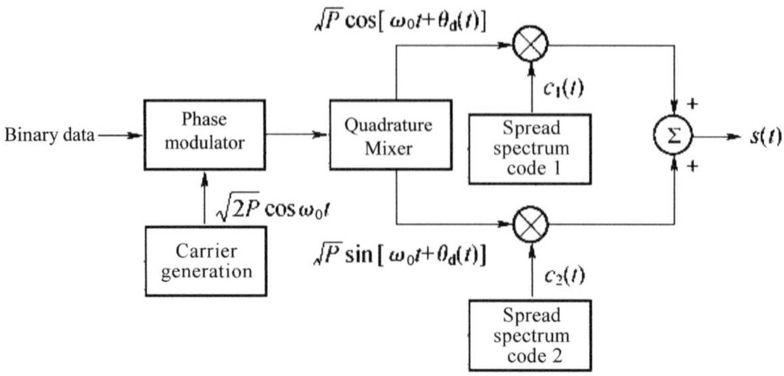

Fig. 4.8 QPSK spread spectrum modulator

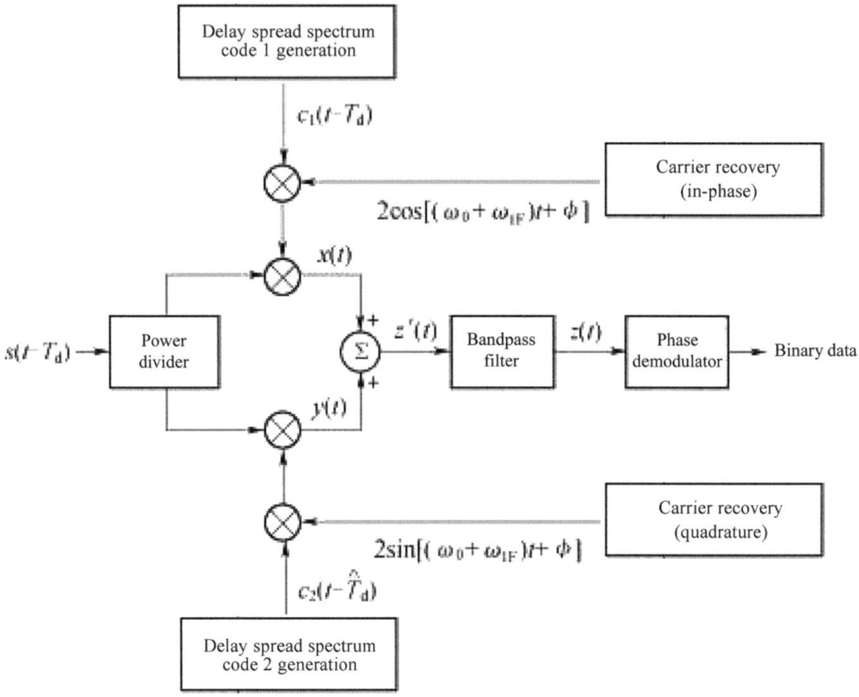

Fig. 4.9 QPSK spread spectrum receiver

coherent, but $\theta_d(t)$ and C(t) may be incoherent, then the output of spread spectrum QPSK modulator is:

$$\begin{aligned} s(t) &= \sqrt{P}c_1(t)\cos\left[\omega_0 t + \theta_d(t)\right] \\ &= \sqrt{P}c_2(t)\sin\left[\omega_0 t + \theta_d(t)\right] \approx a(t) + b(t). \end{aligned} \tag{4.8}$$

It can be seen from Expression (4.8) and Expression (4.2) that, except for different amplitudes and a potential phase shift, items included in the QPSK spread spectrum signal expression and the BPSK spread spectrum signal expression are the same, and two signals are quadrature, so the power spectrum of the sum signal is equal to the algebraic sum of two added BPSK powers. Its power spectrum form is the same as BPSK.

The receiver for transmitted signals in Expression (4.8) is shown in Fig. 4.9. If the phases of spread spectrum codes reproduced by the receiver code loop are synchronized, the output is:

$$z(t) = \sqrt{2P}\cos\left[\omega_{1F}t - \theta_d(t)\right]. \tag{4.9}$$

In the above derivations, the receiver is supposed to track well the phase of the incoming carrier. According to Expression (4.9), data modulation carrier has been fully recovered. That is to say, QPSK spread spectrum modulations added at the transmitting end have been fully removed by the de-spreading at the receiving end.

(2) Information and spread spectrum code both use QPSK, UQPSK, or OQPSK

Another exceptional case of QPSK spread spectrum modem is shown in Fig. 4.10. Information data and spread spectrum code both use QPSK, in which case, data modulation signal and spread spectrum modulation signal in in-phase channel and quadrature channel may either be different or the same. When these two channels have the same power, data rate, and spread spectrum code rate, the modulation is called balanced QPSK spread spectrum modulation, or dual-channel QPSK. Transmission waveform of dual-channel QPSK is:

$$s(t) = \sqrt{P}d_1(t)c_1(t)\cos{}_0t + \sqrt{P}d_2(t)c_2(t)\sin\omega_0 t. \tag{4.10}$$

Its total power is P. The receiver of this waveform is shown in Fig. 4.10b. A small but important change to the signal as shown in Expression (4.10) will obtain UQPSK spread spectrum. In-phase channel and quadrature channel of the signal have different power, so the signal is called unbalanced QPSK (UQPSK) spread spectrum signal. Then the transmitted signal becomes:

$$s(t) = 2\sqrt{P_1}d_1(t)c_1(t)\cos\omega_0 t + \sqrt{P_Q}d_2(t)c_2(t)\sin\omega_0 t. \tag{4.11}$$

In the QPSK spread spectrum modem, a delay of spread spectrum code at I and Q for half a code element can realize OQPSK spread spectrum.

QPSK spread spectrum modems discussed above are major ones actually used at present. There are many schemes through some changes, especially in implementation details. The performance may be better in some specific conditions. For example, MSK spread spectrum can be obtained by half sine weighting cosine weighting of chips at the upper and lower branches, respectively, as shown in Fig. 4.10.

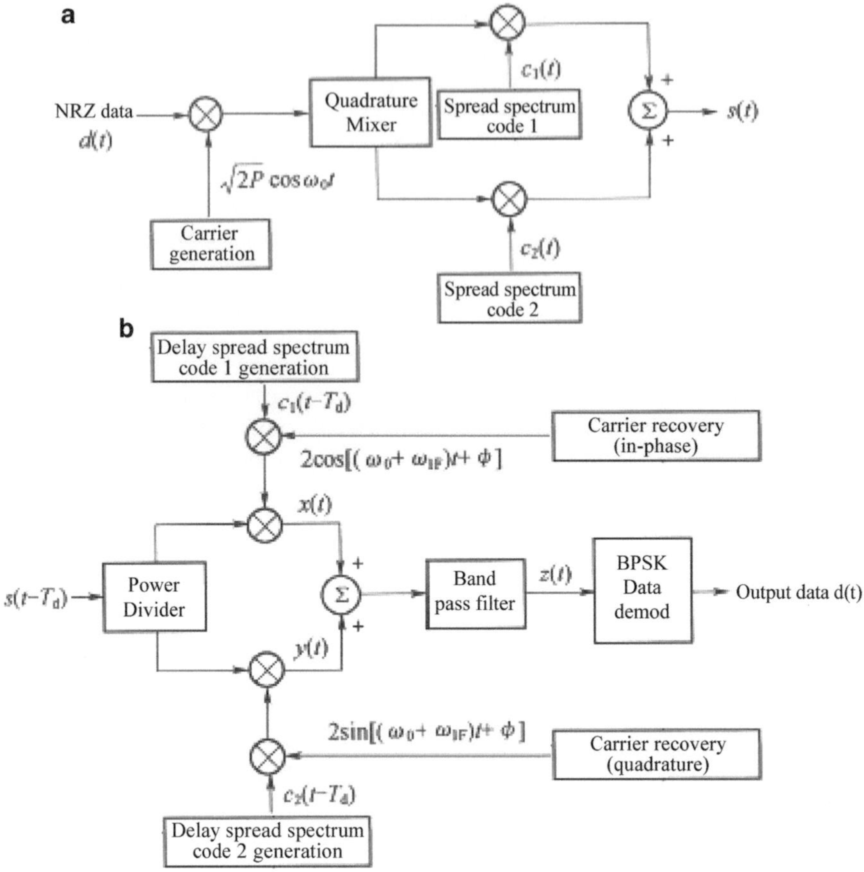

Fig. 4.10 Balanced QPSK DSSS Modem. (**a**) Transmitter; (**b**) receiver

4.3.2 *Frequency Hopping Spread Spectrum (FHSS)*

With respect to FHSS, the total band is divided into n non-overlapping sub-bands, of which one or more are selected by the spread spectrum code during signal transmission, in order to make the carrier randomly jump to another according to some rules. Figure 4.11 shows the basic principle of frequency hopping.

In one instant of transmitting a signal in a selected subband, the signal bandwidth B is determined. If total frequency hopping bandwidth and instantaneous signal bandwidth are W_{ss} and B, respectively, the processing gain of the frequency-hopping system is:

$$G = \frac{W_{ss}}{B}. \tag{4.12}$$

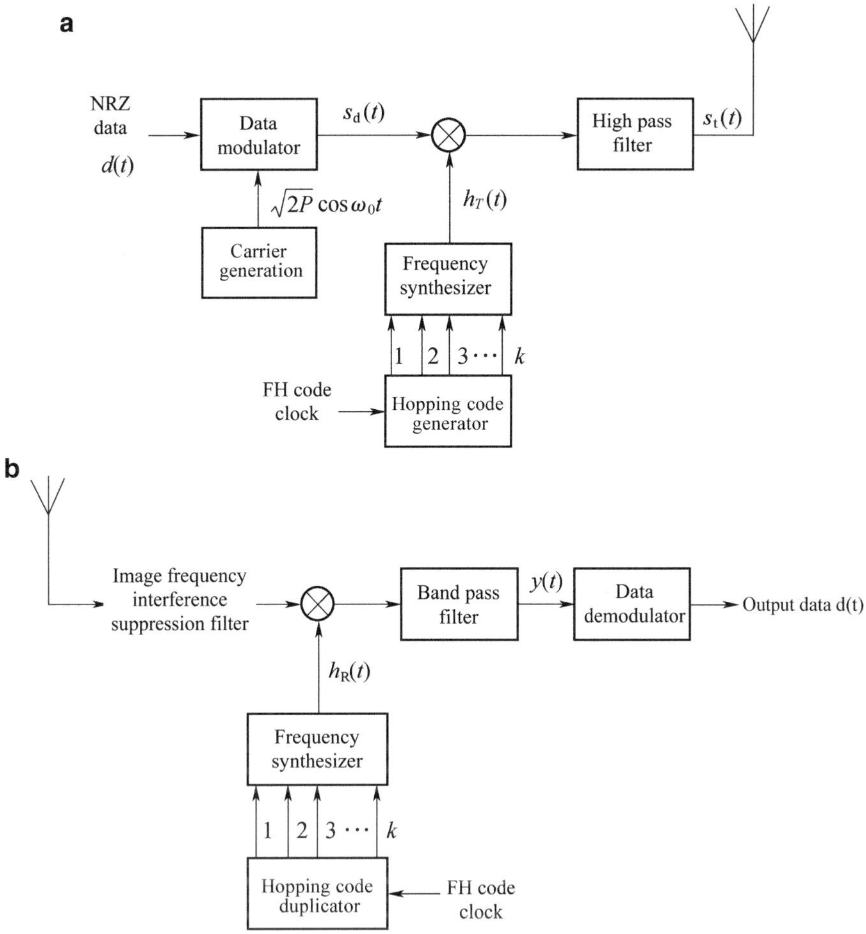

Fig. 4.11 FHSS modem. (**a**) Transmitter; (**b**) receiver

The frequency-hopping system also has strong anti-interference capability. Suppose all users hop at different carrier frequencies synchronously, if there are n subbands, the probability of interference in the subband of the target user is $1/n$. If there are i mutually interfering users, the probability of occurrence of at least one interference in the subband of the target user is:

$$P = 1 - \left(1 - \frac{1}{n}\right)^i = \frac{i}{n}.$$

In this sense, the bigger the number n of frequency hopping points, the higher the anti-interference capability.

According to the relationship between information bit rate and frequency hopping rate, fast frequency-hopping (FFH) and slow frequency-hopping (SFH) are included. If more than one frequency hopping occurs within one information bit width, the frequency hopping is called FFH; if one or more information bits are transmitted in a time interval of frequency hopping, the frequency hopping is called SFH. In practice, the frequency hopping at over 1,000 hops per second is sometimes called FFH.

Since it is very difficult to keep phase coherence of the carrier frequency due to its phase discontinuity in hopping, FHSS signals are often incoherently demodulated. The modulation modes include FSK and ASK. Coherent Doppler information could not be extracted due to incoherence, so it will be a challenge to carry out coherent carrier Doppler velocity measurement by using FHSS signals.

In frequency hopping, each carrier frequency is usually generated by a frequency synthesizer containing 2^k frequencies and the interval between frequencies in this frequency set approximates data modulation spectrum width, however, the said two conditions are not absolute. In such case, spread spectrum codes are used to control the hopping of the carrier frequency to form a frequency sequence. Because the frequency spectrum is spread due to the jumping of a transmitted signal from one frequency to another, this type of spread spectrum is called frequency hopping spread spectrum (FHSS). At the receiving end, frequency hopping will be removed by mixing (down-conversion) the received signal and the local oscillator signal synchronously hopping with the received signal. Block diagram of the transmitter and the receiver is shown in Fig. 4.11.

Since it is very difficult to establish a really coherent frequency synthesizer, most FHSS systems use incoherent or differentially coherent data modulation method. So the receiver does not need to conduct accurate recovery of the phase of the modulated carrier.

In SFH spread spectrum, a received signal is down-converted through a local oscillator (LO), and then the down-converter outputs a single-frequency sequence. In fact, there is an offset f_{IF} between LO frequency and the transmitting frequency, so the dehopping output is in the frequency f_{IF}. Then, the signal can be demodulated in the incoherent demodulation method.

In FFH spread spectrum, hopping frequencies in each symbol would change many times. The biggest benefit of using FFH is that frequency diversity gain can be obtained in each bit symbol, which is especially useful for an environment with some frequency bands under interference.

In the case of FFH, multiple frequencies are used to transmit one bit of information. Then if one frequency is under interference but most of the frequencies are not, a majority decision can also be used to reduce bit error rate and improve resistance to interference. The system can operate in different signal processing modes. One is a mode in which one decision is made upon receipt of a frequency hopping chip and data output is estimated based on the decisions of all K chips (the decision criterion may be a simple majority decision); the other is a mode in which each data output symbol is taken as a function of all received signals on K chips to calculate the likelihood probability and select the biggest value.

With regard to a given E_b/N_0, it is required to calculate that the receiver of each symbol transmission likelihood probability is the best receiver in a sense of minimum error probability. Both the performance and complexity of each possible processing mode are different. A processing mode which can best solve all involved issues must be selected. An FFH system is very useful under a signal fading environment or an environment with some frequency bands under interference, and can be conveniently used together with the error-correcting encoder.

4.3.3 Hybrid Spread Spectrum of DSSS and FHSS

The best method for FHSS is combined use of DSSS and FHSS to integrate their advantages in one system. This method is used in the military spread spectrum system. There are many methods to combine DSSS and FHSS. The following describes a simple hybrid system.

Figure 4.12 gives the principle block diagram of combined DS/FH spread spectrum modulation-demodulation in differential binary PSK data modulation. Due to the use of incoherent frequency hopping, data modulation must be incoherent or differentially coherent. Due to DPSK modulation, differential data coding should be completed before carrier modulation. DPSK-modulated carrier is first multiplied by DS spreading waveform $c(t)$ for direct sequence spread spectrum, and then carries out frequency hopping through FH upper-conversion.

A series of reverse conversion is completed at the receiving end, as shown in Fig. 4.12b. However, two types of spread spectrum control code words must be acquired and tracked, which can be mutually correlative.

Its processing gain is the product of processing gains of DSSS and FHSS.

4.3.4 Time Hopping Spread Spectrum

The time hopping spread spectrum system divides the time into many time slots of which one is selected in a pseudorandom way by the information symbol sequence for transmission. A symbol sequence with a length of T is compressed and transmitted within a time slot of τ, so the symbol width becomes narrowed. In frequency domain, that is to say the spectrum has been spread by T/τ times. Because it is worked in a burst way, the system is better at anti-reconnaissance and anti-interception. Its processing gain is related to the duty ratio of the burst (τ/T). The expression is:

$$G = \frac{T}{\tau}. \tag{4.13}$$

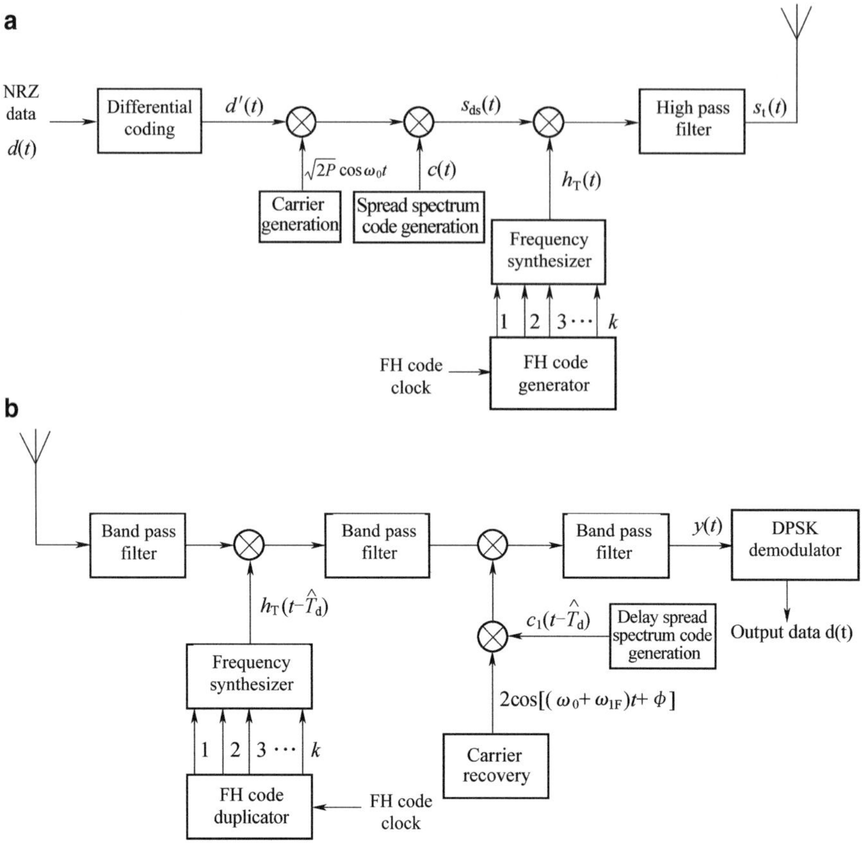

Fig. 4.12 Hybrid DSSS/FHSS modem. (**a**) Transmitter; (**b**) receiver

4.3.5 *Code Hopping*

The abovementioned DS, FH, and TH cannot stand away from spread spectrum codes. Therefore, in order to further improve the performance of the spread spectrum system, the design of spread spectrum codes will be a core issue. The introduction to "code hopping" here is to provide better confidentiality for the spread spectrum codes and thus to further improve "anti-interception", "anti-reconnaissance", and "anti-interference" capabilities of the spread spectrum system.

4.3.5.1 Concept of "Code Hopping"

Code hopping refers to a random hopping selection of a code group set ("code set") under the control of code hopping graphics to produce a new digital coding sequence which is hopping among various types of code patterns ("code hopping sequence"). Code patterns include chaotic code, pseudorandom code, Walsh code, Barker code, and so on. The chaotic code hopping is the most attractive technique, so the following introduction takes chaotic code hopping as an example.

Due to difficulty in synchronization of infinite chaotic codes, truncated-cycle chaotic codes are used. However, such use may damage the unpredictability of chaotic codes. The use of "chaotic code hopping" will significantly increase the number, type, and permutation patterns of "sequence", thereby greatly improving sequence security, privacy, and anti-interception to recover the loss of the said unpredictability.

Chaotic code hopping sequence can be composed of the following types of chaotic codes:

(1) All kinds of chaotic codes are generated by various iterative formulas, such as by the following mapping:
    Tent-Map, Chebyshev, Logistic-Map, modified Logistic-Map, SB, VB, IB, combined, embedded, cascade, and various PN codes.
(2) A great number of chaotic codes generated by different initial values and parameters of various iterative formulas.
(3) ($n!$) different chaotic code hopping sequences through full permutation of n chaotic codes under the control of code hopping graphics.
(4) Each chaotic code has a different code length to form a very long code sequence period which may be difficult to be intercepted. From the other point of view, when the code sequence period meets certain requirements (e.g., unambiguous distance), the width of code element may be decreased to obtain a higher spread spectrum gain and provide stronger anti-interference performance.

When the number of chaotic code patterns keeps increasing and the period of chaotic code hopping sequence increases, an infinite chaotic sequence will approach, making it difficult to predict in engineering application. Although in theory, only infinite chaotic sequence is unpredictable, in engineering, any failure in detecting and intercepting the chaotic code hopping sequence within a required detection time can be deemed undetectable in engineering application, unnecessarily using the infinite chaotic sequence. For example, the period of 20 different truncated chaotic codes after full permutation can reach 10^7 years which may be "infinite" relative to the life cycle of one decoder.

4.3.5.2 Generation of Code Hopping Sequence

One feasible method is to store n code patterns of codes in n storage units and take out n codes in the designed hopping sequence, under the control of code hopping graphics, to form a continuous code hopping sequence. Taking chaotic codes as an example, its block diagram is shown in Fig. 4.13.

The main principle of this scheme is to treat the storage unit as a chaotic code generator. The length of chaotic code hopping sequence generated by the storage unit shall be limited by the scale of the storage unit.

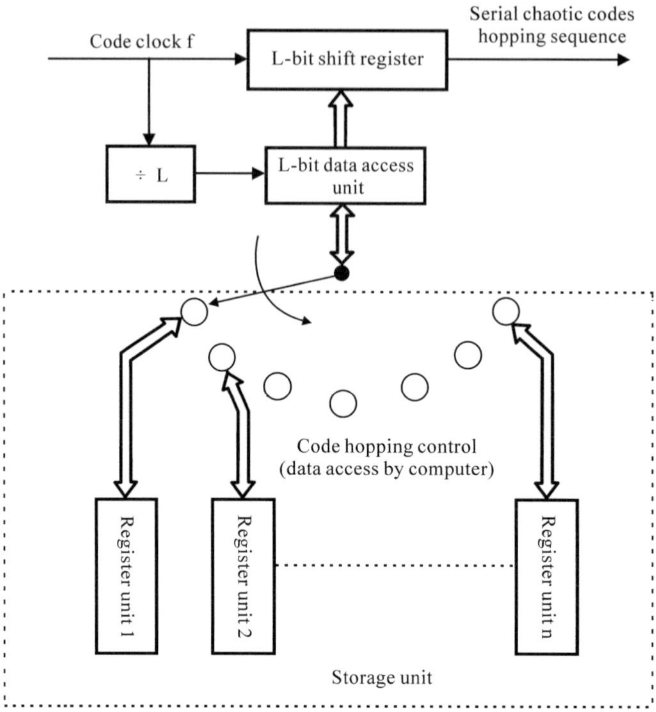

Fig. 4.13 Principle of chaotic code hopping sequence generation

Since chaotic code hopping can use shorter truncated chaotic codes, changes to chaotic codes brought about by the change of initial values and parameters in each operation with iterative formula could not be reflected in a short time period. Therefore, DSP may be used to generate chaotic frequency hopping sequences. Simplifying the circuit and unlimited by the scale of the storage unit, this scheme will generate infinite and nonperiodic chaotic code hopping sequences and provide better privacy and anti-interception capabilities.

4.3.5.3 Synchronization of Code Hopping Sequence

Basic principles for synchronization implementation:

(1) Use a periodic and non-hopping leader sequence to realize acquisition and carry out code hopping.
(2) Take a specific feature code immediately followed by the leader sequence as "hopping starting marker", which is known and can be solved by the correlation receiver.

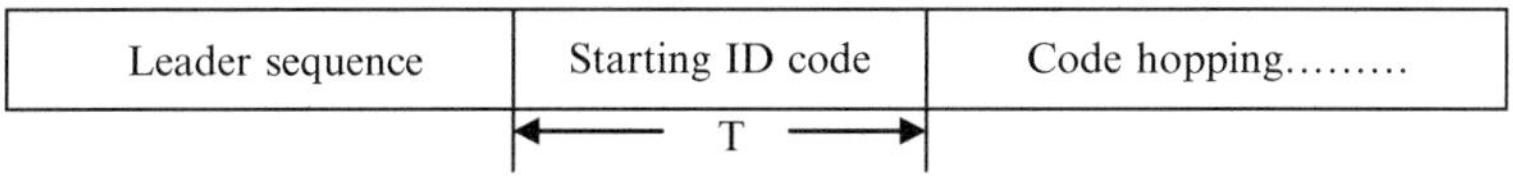

Fig. 4.14 Structure of a code hopping sequence

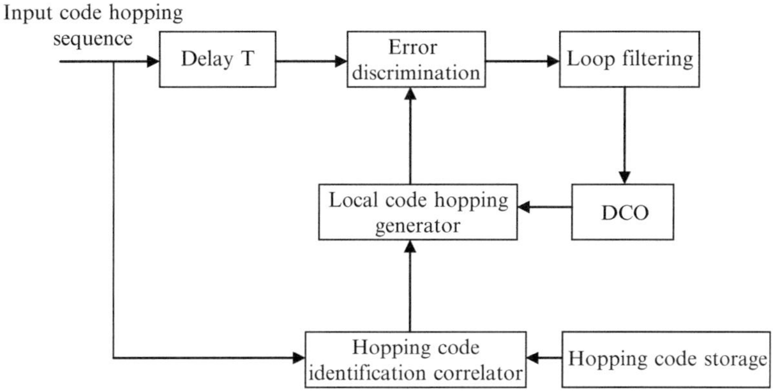

Fig. 4.15 Synchronization of chaotic code hopping sequence

(3) The transmitting end controls the start of code hopping. The receiving end starts code hopping at the receiving end once the "hopping starting marker" is received, with its code hopping rules being the same as that at the transmitting end so as to realize synchronization after code hopping.
(4) The code hopping generator at the receiving end is driven by clock frequency with Doppler at the receiving end.

Structure of a code hopping sequence is shown in Fig. 4.14.

Principle block diagram of synchronization of code hopping sequences is shown in Fig. 4.15.

Delay T is to give an identification time T to identify a hopping code. At the time end of starting hop recognition code, the receiving end starts local code hopping in order to synchronize received and transmitted code hopping sequences.

4.3.5.4 High-Order Full-Permutation Code

If based on the abovementioned n code hopping sequences, a full permutation is given to these n sequence codes, a new full-permutation code sequence can be obtained, with total number up to $[(n!)!]$. This sequence can be called second-order full-permutation code. Then by similar derivations, higher-order sequence code can be obtained to improve privacy and anti-interception performance.

The code hopping sequence produced in the said method can be used as the control code of "DSSS", "FHSS" or "THSS" to provide better anti-interception and anti-interference for the spread spectrum system.

4.4 Acquiring and Tracking of Spread Spectrum TT&C Signals

4.4.1 Acquiring and Tracking of DSSS Signals

For de-spreading of DSSS signals, synchronization must be established between the spread spectrum codes at transmitting end and those at the receiving end. In case of out-of-synchronization, de-spreading loss will be caused even if only one code chip exists. In this case, the signal energy arriving at the information code demodulator will be drastically reduced and PN code ranging error will be increased. Synchronization of PN codes between transmitting and receiving end is generally achieved at the receiving end through a code loop. The synchronization includes two steps: the first step is code acquisition, which is done by using the available pilot information to determine initial phase of PN code; the second step is code tracking, that is to maintain continuous synchronization tracking of PN code after initial phase of PN code is obtained. Acquiring and tracking of spread spectrum signals is a 2D signal duplication process (duplicating both code and carrier frequency), where the duplicated code is used for ranging and de-spreading and the duplicated carrier frequency is used for measuring velocity and demodulation.

Commonly used methods for acquiring PN code include:

(1) Serial search method: to search all phases and frequencies of the code until acquiring synchronization. Decision shall be made in respect to the signals output by de-spreading receiver to determine whether de-spreading happens for each time a reference phase/frequency is selected. If both selected phase and frequency are correct, de-spreading is achieved, the signal spectrum is narrowed and the narrowband filter will output signals, which shall be decided as correct acquiring. If the phase and frequency of the code are incorrect, the received signals cannot be de-spread and the output of the narrowband filter is very small; in this case, it is required to adjust the reference spread spectrum signal generator at receiving end to a new phase/frequency for the next decision. The simplified block diagram of this system is shown in Fig. 4.16.

According to Fig. 4.16, if the reference spread spectrum signal is almost aligned with the received code (namely, the difference is no more than $\pm$ 1/2 code chip), sufficient signal energy can pass through the filter after de-spreading, and the decision results will stop searching.

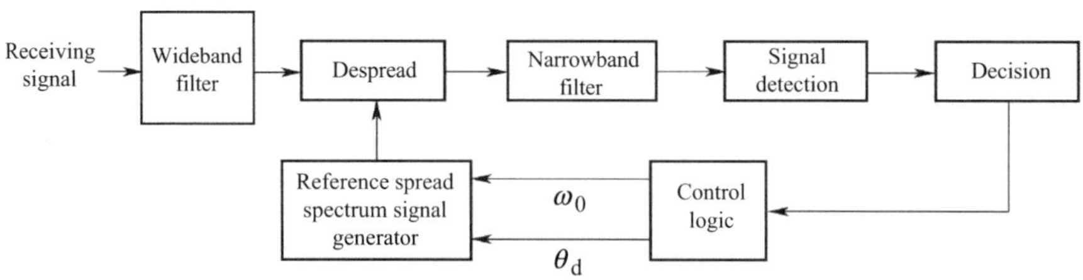

Fig. 4.16 The simplified block diagram of serial search

Reference [1] points out that when step serial search is used, if there is an initial uncertainty of C code chips, the average searching time is

$$T_{\mathrm{s}} = (C-1)T_{\mathrm{da}}\left(\frac{2-P_{\mathrm{d}}}{2P_{\mathrm{d}}}\right) + \frac{T_{\mathrm{i}}}{P_{\mathrm{d}}}, \tag{4.14}$$

where T_{i} is integral time of each unit; P_{d} is the detection probability when the correct unit is evaluated; $T_{\mathrm{da}} = T_{\mathrm{i}} + P_{\mathrm{fa}}T_{\mathrm{fa}}$ is the average residence time on incorrect phase unit; P_{fa} is false alarm probability (the probability of an incorrect unit being judged as correct); T_{fa} is the time necessary to refuse occurring incorrect unit.

The variance of searching time approximates to

$$\sigma_T^2 = T_{\mathrm{da}}^2 C^2\left(\frac{1}{12} - \frac{1}{P_{\mathrm{d}}} + \frac{1}{P_{\mathrm{d}}^2}\right), C \gg 1; \quad P_{\mathrm{d}} \approx 1; \quad P_{\mathrm{fa}} \ll 1. \tag{4.15}$$

One goal of spread spectrum system design is to minimize the average searching time of system synchronization. Expression (4.14) shows that the average synchronization time is a function of $P_{\mathrm{d}}, P_{\mathrm{fa}}, T_{\mathrm{i}}, T_{\mathrm{fa}}$ and C. The designer may appropriately control these variables including C. According to system requirements, the areas with uncertain phase (unit: s) is fixed, but these areas may still be divided into any quantities of units. The remaining four variables are not independent of each other, which brings some challenge to design. Large P_{d} and small P_{fa} reveal that T_{i} is relatively large. Therefore, there should be an optimum combination of $P_{\mathrm{d}}, P_{\mathrm{fa}}, T_{\mathrm{i}}$, T_{fa} which may minimize the average searching time. The minimum average searching time can be obtained when $P_{\mathrm{d}} = 1$. However, it is difficult to detect correct unit at the first scanning and so several times of uncertain scanning is required. Nevertheless, an appropriate P_{d} can make T_{i} far less than the value when P_{d} is large, thus reducing the average value of T_{s}. For analysis and design of acquiring spread spectrum system codes, you can see relative reference.

(2) Matched filter method: matched filter can output a pulse and corresponding sequence phase when receiving specific code symbol sequence. Once the pulse is received, the code generator at the receiving end will come to operate with the same initial phase as the received code stream, thus acquisition process is completed. Such solution requires a matched filter with very large time-bandwidth product. In the section below, a matched filter method (also called delay correlation matching acquisition method) [1], as shown in Fig. 4.17, is taken as an example for explaining their principle.

Assume the demodulated baseband signal is

$$\upsilon(t) = Ad(t-T_{\mathrm{d}})c(t-T_{\mathrm{d}}) + 2N(t)\cos(2\pi f_0 t + \varphi_0), \tag{4.16}$$

where $A_{\mathrm{c}}(t-T_{\mathrm{d}})$ is the received spread spectrum code sequence signal (assuming the data signal $d(t) = 1$ during observation period), which is sent to M-stage shift register for registering. Local reference spread spectrum sequence is saved into another M-stage register according to certain phase state. Then the data in

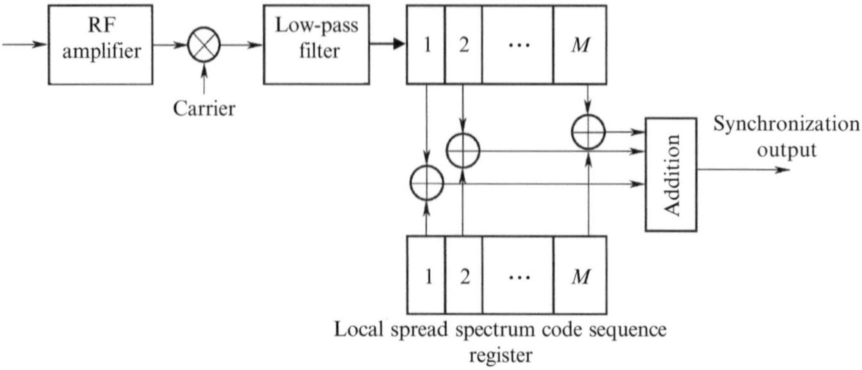

Fig. 4.17 Block diagram of baseband matched filtering method

corresponding bits of the two registers is used to calculate the correlation function value of the two sequences (conduct model-2 summation then add together). The low correlation value will be obtained if the phase states of the two code sequences are inconsistent.

At the moment, delayed shift shall be conducted to the received spread spectrum code sequence in M-stage shift register, thereby solving the correlation value under such phase state. In this way, delayed shift evaluation shall be made to the sequence phase and then correlation summation shall be conducted until the correlation summation value output is the maximum (in case of M units), the phase of receiving spread spectrum sequence is consistent with the phase state of local reference spread spectrum register, thereby implementing acquisition of spread spectrum code sequence. Its acquiring time is analyzed as follows.

It takes MT_s from receiving spread spectrum signal to loading M code elements of spread spectrum sequence into M-stage shift register, where T_s is the time during which each stage delay shift register loads a code element. A register saves a code element, so $T_s = T_c$. Therefore, the minimum synchronization acquisition time of the spread spectrum code sequence is the loading time of M-stage shift register; namely

$$T_{\mathrm{AC,min}} = MT_c. \tag{4.17}$$

When the width of code element is T_c and the number of correlators is N, the maximum synchronization acquisition time is

$$T_{\mathrm{AC,max}} = MT_c + (N-1)T_c. \tag{4.18}$$

The average synchronization acquisition time is

$$\overline{T}_{\mathrm{AC}} = MT_c + \frac{N-1}{2}T_c. \tag{4.19}$$

The average synchronization acquisition time obtained by such method is longer than that obtained by $2N$ correlators but the quantities of equipment decrease greatly. In addition, the average synchronization acquisition time obtained by such method is shorter than that obtained by phase search acquisition method but the quantities of equipment increase mildly.

The quantities of code phase and frequency to be searched during acquisition is directly proportional to uncertainty of transmission delay and relevant to the relative dynamic features of transmitting/receiving end. The transmission delay corresponds to the number of code chips, and the time width of code chips is reversely proportional to bit rate; therefore, the acquisition time is proportional to the clock frequency of spread spectrum code because the clock rate used by frequency hopping spread spectrum system is far fewer than transmission bandwidth. Therefore, the synchronization time of frequency hopping system is far fewer than the direct sequence spread spectrum with the same transmission bandwidth. This is the reason why some spread spectrum systems select frequency hopping. For frequency hopping spread spectrum, the initial synchronization technology is the same as the DSSS system.

The acquisition process can generally make the phase of receiving spread spectrum code accurate to $\pm 1/2 \sim \pm 1/4$ of the code chip. When the tracking loop of the code is closed, the tracking loop will eliminate the $\pm 1/2$ code chip's phase error.

The main difference between the phase-locked loop for carrier tracking and the phase-locked loop for code tracking ("code loop" in short) lies in the implementation of the error discriminators. For carrier tracking, the phase discriminator can be used as a simple multiplier. However, for code tracking loop, several multipliers, filters and envelope detectors are required to discriminate time difference, and thus it is also called a delay locked loop. Since the code loop is also a phase-locked loop, the way for analyzing spread spectrum code tracking system is usually by establishing different code tracking loop models similar to the conventional phase-locked model, and thus many existing phase-locking loop conclusions can be utilized.

Code tracking loop of spread spectrum system can be classified in several ways, such as video correlated loop, IF envelop correlated loop, coherent loop, and non-coherent loop. Coherent loop utilizes phase information of receiving carrier but non-coherent loop does not. Typical code tracking loop will utilize receiving signals and two spread spectrum waveforms (lead-lag) with different delays generated locally for correlation processing. Two independent branches are employed for two types of correlation calculation. Such calculation also can be conducted by time-sharing branch. The tracking loop utilizing two independent correlators is known as full time lead-lag tracking loop, and the tracking loop sharing a single tracking loop is called τ jittering lead-lag tracking loop. In the section below, two main loops are introduced.

(1) Video correlated loop: principle block diagram of full time lead-lag tracking loop achieved by video correlated loop in video is shown in Fig. 4.18 [2]. It conducts code loop tracking to carrier demodulated video signals. The principle diagram is the same as that of PN code tracking loop of pseudo-noise code ranging system.

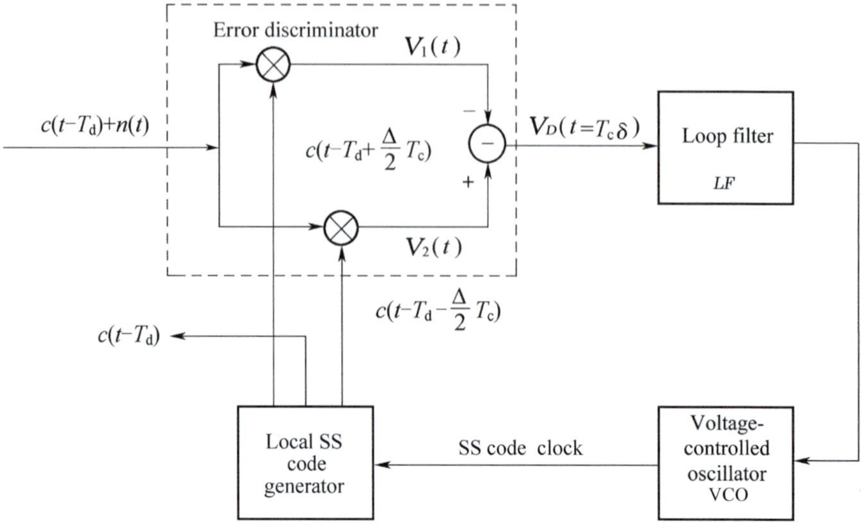

Fig. 4.18 Principle block diagram of video delay locked tracking loop

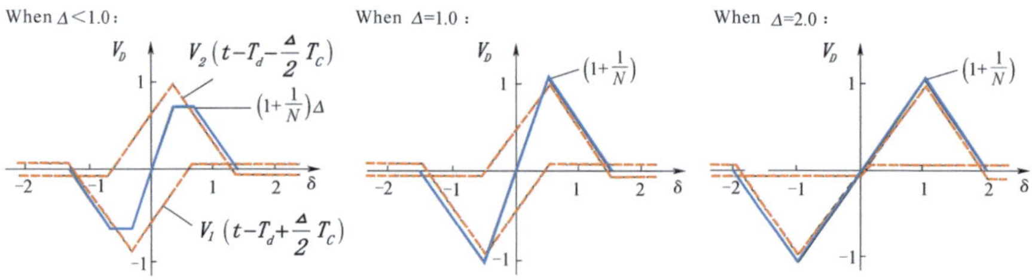

Fig. 4.19 Error characteristic relationship between DC output signal V_D of delay locked discriminator and time difference δ of received/transmitted codes at different Δ values

Figure 4.19 shows the relationship of DC output component of delay locked discriminator with the time difference δ between transmitted and received codes when delay difference τ between leading code and lagging code on two branches are taken as different values (represented by $\Delta = \tau/T_c$ when the code chip width T_c is taken as the unit of T_c). δ is the time difference with width of code element as the unit of T_c, namely, $\delta = t/T_c$, $V_\Delta(\delta)$ is the ratio of the difference between the two relative multipliers' outputs to the received code amplitude (namely, normalized). Two multipliers use leading codes and lagging codes respectively. In the case of baseband, the widest linear characteristics will be in $\Delta = 2T_c$. Output of time discriminator is voltage supplied to the voltage-controlled oscillator to facilitate lock of local spread spectrum code generator. Therefore, it is important to employ liner tracking range as wide as possible.

The difference between τ jittering lead-lag tracking loop and full time lead-lag tracking loop is that the former only uses a correlator and operates on time-sharing basis. Square wave switch signals are used to control the distribution correlator to operate on local lead codes or lag codes. For lagging codes, the output of the correlator should be NOT, and difference value between leading and lagging correlation shall be taken as the same as full-time loop. Frequency of square wave switch signal shall be low so that the transient effect occurring on conversion can be eliminated. Such conversion will cause 3 dB loss against difference signals, but the threshold of carrier loop is higher (generally 3 dB more than threshold of code loop), so the arising influence is weak. At present, various DLL discriminators are given by relevant documents and can be selected through demonstration when the equipment is developed. For example, "dot product power discriminator" utilizes three correlators, which optimizes performance and reduces operation. The advantage of τ jittering loop is that both leading signals and lagging signals use the same correlator. In this way, the loop tracking performance will not deteriorate due to unbalance between two correlated branches of leading/lagging reference codes. According to Fig. 4.19, when changes in amplitude, shape, and time delay of two branches are caused by various factors (including circuit parameters difference, temperature fluctuation), the null point and error slope output by the delay lock discriminator will change, thus generating ranging system error and causing de-spreading loss and degraded DLL performance. Therefore, ranging error and de-spreading loss can be reduced through employing τ jittering DLL.

Disadvantages of video correlation loop method: (1) it requires to demodulate video signals, bringing a challenge of coherent demodulation in case of low C/N. (2) Video signals include information signals, which will influence operation of delay lock loop. This problem can be solved by conducting delay lock tracking to IF signal (or zero IF signal) prior to demodulation. At this time, the local reference signal participating in correlated multiplying is a LO signal spread by the local spread spectrum code. The output signals after it multiplies the input signal will contain error components required by delay locking, which can be extracted through envelope detection (or square circuit) to implement delay lock tracking.

(2) Envelope correlated loop: The IF modulated signal after correlation processing contains delay error signal. In order to obtain delay error signal, it shall be processed by envelope detection to remove the carrier. So the loop shall include an envelope detector located at the output end of the multiplier in delay lock discriminator, making leading/lagging correlation processing only conducted to the envelope of the modulation signal. Therefore, such a loop is also called envelope correlation synchronization tracking loop.

Pre-detection integral shall be conducted before envelope detector. The reciprocal of the integral time shall not be less than the rate of information code. If the pre-detection integral time is extremely large, the envelope detector cannot detect amplitude of information code (or suffer loss). However, its amplitude will carry correlation value of locally duplicated code and input spread spectrum code. Loss of correlation value will influence performance of code loop. However, during the acquisition course before code synchronization is completed, the starting and ending time of pre-detection integral is not aligned with the leading and lagging edge of the information code, which will cause loss of correlation value. Therefore, the loss caused by such "alignment failure" shall be considered.

The rate of the information code limits increase in pre-detection integral time. Integral is added after envelope detection to ensure that the whole code loop achieves long integral time. There is a non-linear envelope detector between the integrator before/after envelope detection. When input SNR is low, it will deteriorate the output SNR. The relationship of input/output SNR meets square-law relation in case of low SNR, which is called square loss. The carrier tracking loop also suffers a similar loss. The method of reducing square loss is to improve SNR before code loop discriminator and carrier loop discriminator, namely, to increase pre-detection integral time. However, such integral time is limited by rate of information code. If there is no information code (namely, it is only PN code but not spread spectrum information code), the pre-detection integral time may be very long (its dynamic feature shall be considered during actual design), thereby decreasing square loss, which is considered as only PN code loop and PLL carrier loop. For this, an operating mode that can remove information codes can be designed, under such mode, the pre-detection integral time can be reduced, thus decreasing square loss and improving tracking threshold. In terms of acquisition tracking sensitivity, tracking threshold and orbit measuring accuracy, only PN code TT&C is better than spread spectrum TT&C, which is a disadvantage of spread spectrum TT&C compared with only PN code TT&C.

4.4.2 Acquiring and Tracking of Frequency Hopping Spread Spectrum Signals

It includes two parts: first, de-hopping of carrier frequency, namely, achieving synchronization of frequency synthesizers between receiving and transmitting ends; secondly, achieving synchronization of frequency hopping control codes.

The frequency hopping synchronization tracking loop of frequency hopping system is similar to the synchronization tracking loop of direct sequence system. The frequency hopping synchronization tracking loop is formed by replacing DSSS demodulator with frequency synthesizer in envelop correlation synchronization tracking loop. It is shown in Fig. 4.20 [1]. A frequency synthesizer and delay line are used to generate the local reference signal of leading and lagging correlation branches.

Due to discontinuity of carrier signal phase caused by frequency hopping, the code loop employs incoherent delay lock loop (also called envelop-correlated delay lock loop). The input signals are not baseband signals but direct sequence modulation signal not subject to de-spreading and demodulation; therefore, it is not required to generate coherent carrier before implementing synchronization tracking. Furthermore, the discriminator used in the loop is an energy detector which is not sensitive to the phase of information modulation and carrier; therefore, the phase discontinuity caused by frequency hopping will not bring about any influence.

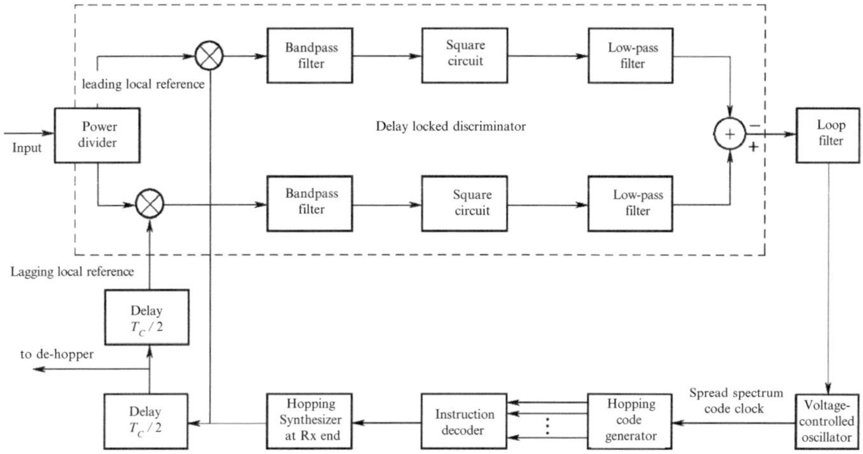

Fig. 4.20 Hopping code synchronization tracking loop of a frequency hopping system

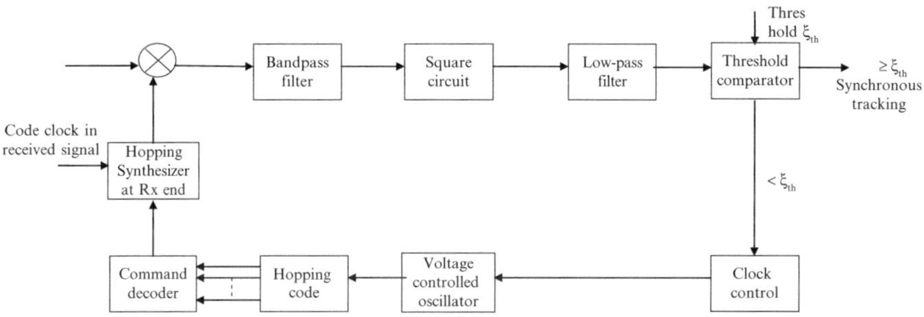

Fig. 4.21 Acquisition system of frequency hopping code

There are many acquisition methods for the frequency hopping system. Scc Fig. 4.21 for an acquisition method.

In Fig. 4.21, the receiving code clock is taken as the reference of frequency synthesizer. Since it has been carried with Doppler frequency shift, there is always an intermediate frequency difference between the frequency of the output signal of the frequency synthesizer and that of receiving signal and a maximum IF output after mixing as well.

If the phase of local frequency hopping pattern is not consistent with that of the received frequency hopping pattern (namely, the hopping times of frequency hopping are inconsistent), the signal after mixing will deviate from IF, such that the bigger or maximum IF output will not be generated and the output after squaring and integration will be very small, which controls the clock circuit to make the frequency hopping code sequence hopping by a phase of 1/2 code element, making the search continue till achieving consistence with the frequency hopping pattern of the received signals.

For flight vehicle TT&C and communication, there is a big Doppler frequency shift in the received carrier. To decrease the frequency de-hopping error arising, the code clock component in receiving signal will be used to conduct Doppler compensation to the frequency hopping frequency synthesizer at receiving end.

In frequency hopping TT&C system, the carrier phase discontinuity caused by frequency hopping will bring impact on the carrier Doppler velocity measuring and phase modulation signal, which is an unsolved technical problem.

4.4.3 Velocity Measuring of Frequency Hopping Spread Spectrum—"Two-Step Method" Frequency Hopping Velocity Measuring

The meanings of "two-step method":

Step 1: implement incoherent de-hopping. For frequency hopping technique, it is generally difficult to implement full-coherent de-hopping. The current mature method is conducting non-coherent frequency hopping synchronization. Although after achieving non-coherent carrier synchronization, the synchronized "frequency hopping instruction code", "code clock Doppler frequency" and ranging data can be obtained at the receiving end, the coherent carrier Doppler velocity measuring is still unavailable.

Step 2: implement "residual carrier velocity measuring". Employ "frequency hopping instruction code" and "code clock Doppler" obtained by the first step at the receiving end to control "frequency hopping synthesizer at the receiving end" to make it "quasi-synchronized" with the frequency of the received frequency hopping signal, thereby implementing non-coherent de-hopping. Since the remaining synchronization error is very small, the de-hopped IF signal is equivalent to a NFSK signal with noncontinuous phase and small modulation degree. However, there is a residual carrier in its spectrum, which can be used for velocity measuring by mature residual carrier loop.

After non-coherent de-hopping, the resulting residual carrier signal and its spectrum are as shown in Fig. 4.22.

Figure 4.22a shows the time-varying law of frequency hopping signal at transmitting end and its spectrum, where W_{ss} represents frequency hopping bandwidth, f_0 represents the central frequency in the frequency hopping bandwidth, used as reference frequency for carrier Doppler velocity measuring; T_c is the frequency hopping time interval, serving as the code element width of the frequency hopping control code output by the frequency hopping pattern generator. Figure 4.22b shows the time-varying law of IF signal after non-coherent de-hopping at receiving end and IF spectrum when the Doppler frequency shift is zero. Due to non-coherent de-hopping, frequency residual error exists, which has maximum value $W_{ss}/2$, far less than $W_{ss}/2$ before de-hopping. In the figure, $\omega_{ss} = W_{ss}/K$, where K is

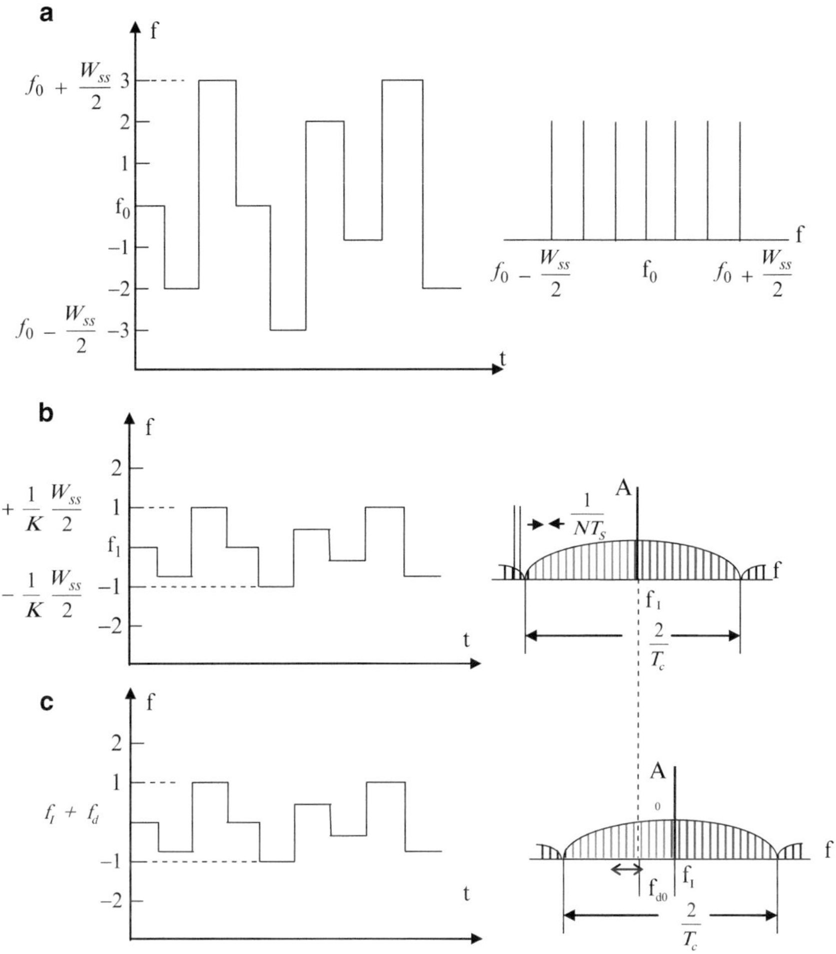

Fig. 4.22 Principle diagram of residual carrier formation. (**a**) Time-varying law of frequency hopping carrier at transmitting terminal and its carrier spectrum. (**b**) Time-varying law of IF signals after de-hopping at receiving terminal in case of $f_d = 0$ and its spectrum. (**c**) Time-varying law of IF signals after de-hopping at receiving terminal in case of f_{d0} and its spectrum

the de-hopping coefficient of frequency hopping. For different non-coherent synchronization schemes, $W_{ss}/2$ is different but always meets $\omega_{ss} \ll W_{ss}$. $W_{ss}/2$ is very small, the IF signal subject to de-hopping is a non-coherent FSK modulation signal (NFSK) with small modulation degree, with corresponding spectrum as shown in the right figure in Fig. 4.22b. Due to a residual carrier, its size is dependent upon the size of remaining frequency difference $W_{ss}/2$ after non-coherent de-hopping. The smaller $W_{ss}/2$ is, the bigger the residual carrier value. The following method can be used to analyze the size of residual carrier: based on Fourier transform, NFSK modulation signals can be divided into a sum of n harmonics

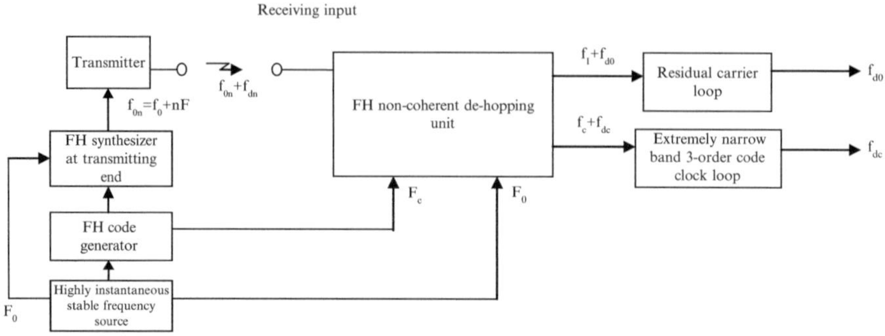

Fig. 4.23 Principle block diagram of "two-step method" frequency hopping velocity measuring

and one DC component. The effect of sub-harmonics is actually multi-frequency angle modulation to the carrier signals. According to the angle modulation theory described in Chap. 2, the residual carrier of multi-frequency angle modulation is $A_0=[J_0(\beta_1)\cdot J_0(\beta_2)*\ldots J_0(\beta_n)]$, where $J_0(\beta_1), J_0(\beta_2)\ldots J_0(\beta_n)$ are zeroth-order Bessel functions; $\beta_1,\beta_2\ldots\beta_n$ are angle modulation indicators of each sub-harmonics, which is in proportion to $\omega_{ss}/2$. The smaller $\omega_{ss}/2$ is, the bigger zeroth-order Bessel functions are, and the bigger the residual carrier A_0 is too. Figure 4.22c shows the situation under which Doppler frequency shift f_{do} exists. At this time, the residual carrier also shifts to (f_0+f_d). (f_0+f_d) can be measured through phase-locked loop tracking filter, and f_0 can be measured with the phase-locked loop at $f_{do}=0$; therefore, f_{do} can be obtained by figuring out the difference between the two.

The principle block diagram of frequency hopping velocity measuring by "two-step method" is shown in Fig. 4.23.

In Fig. 4.23, assuming that the central frequency of frequency hopping bandwidth at transmitting end is f_0, under the control of frequency hopping pattern with code clock of f_c, the carrier frequency is hopped by nF at both sides of f_0, where F is any one of the frequency intervals designed for frequency hopping, thus to implement FH spread spectrum. If the hopped frequency is f_{0n} and the reference frequency for velocity measuring is chosen as f_0, the design shall make the average frequency of all hopping points as close to f_0 as possible so as to reduce the system error. Namely, the better the hopping frequency balance is, the smaller the system error will be. After frequency hopping non-coherent synchronization is implemented by "step-1", under an ideal de-hopping, the output IF shall be a single carrier IF corresponding to f_0, which then can be used to carry out typical carrier Doppler velocity measuring. In fact, since it is a "quasi synchronization", the following issues exist:

(1) Synchronization error will generate NFSK modulated wave; but it has a small modulation degree, so there is a residual wave which can be used to for residual carrier Doppler velocity measuring.
(2) De-hopping loss: it will reduce the amplitude A_0 of residual carrier.

In this case, a narrowband filter can be used to filter out the residual carrier. The narrowband filter cannot only filter out the sideband spectrum of the residual carrier but also improve C/N_0, thereby improving velocity measuring accuracy. This residual carrier can be used for velocity measuring either with PLL or by typical "whole-circle measuring with fixed time" method.

Meanwhile, non-coherent synchronization also obtains code clock ($F_c + f_{dc}$), which can be used to conduct code clock Doppler velocity measuring, but its random error is rather big. The extreme narrowband code clock loop can be used to improve the SNR. In order to reduce the dynamic error introduced by the narrowband, third-order loop should be used. The measured value f_{dc} can be used for coarse velocity measuring or $\left(\frac{F_0}{F_c} f_{dc}\right)$, which can be used for Doppler compensation of frequency synthesizer frequency standard at receiving end, thereby further reducing the frequency synchronization error of frequency synthesizer at receiving end caused by Doppler effect.

The quasi-synchronous frequency error stated above is mainly caused by the synchronization error between "frequency hopping synthesizer at transmitting end" and "frequency hopping synthesizer at receiving end". To reduce the error, the two synthesizers shall:

1) Employ the same reference frequency source with good short-term frequency stability to reduce velocity measuring error and conduct Doppler compensation in the frequency synthesizer at receiving end.
2) Employ the same frequency synthesizer solution.
3) Have high frequency control accuracy and short control transition time.

After adopting this scheme, the random velocity measuring error is dependent upon C/N_0 and phase-locked loop of residual carrier. The velocity measuring system error may be deducted through calibration on ground station; namely, make the frequency of the synthesizer at transmitting end deviate by ΔF_0, and measure the frequency shift ΔF_{do} of the residual carrier at receiving end, then the system error ($\Delta F_{d0} - \Delta F_0$) is deducted.

4.5 Measuring Accuracy and Tracking Threshold for Direct Spread Spectrum TT&C

Measuring error and tracking threshold is caused by the same factor. The receiver will be out of lock when measuring error exceeds the specified limitation, thus forming tracking threshold. The code loop and carrier tracking loop is linear, especially near the threshold area. Therefore, the true tracking performance can be determined only when simulation is made under dynamic and SNR. However, the error formula is usually used for predication in engineering. There are many measuring error sources in each type of tracking loop. For predicating tracking threshold, it is only required to analyze the main error source [3].

4.5.1 Phase Error in Carrier Loop of Spread Spectrum Receiver

The main phase error source of the receiver PLL is phase jitter and dynamic error. According to the engineering design experience, the tracking threshold of PLL can be calculated according to the formula below:

$$3\sigma_{\mathrm{PLL}} = 3\sigma_{\mathrm{j}} + \theta_{\mathrm{e}} \leq 45°, \tag{4.20}$$

where σ_{j} is r.m.s value of phase jitter caused by all other sources other than dynamic error; θ_{e} is dynamic error of PLL tracking loop.

It can be seen from Expression (4.20) that the dynamic error is added to $3\sigma_{\mathrm{j}}$ phase jitter. Phase jitter is square root of the square sum of incoherent error sources which mainly include thermal noise and phase noise of the oscillator (also can be described by Allan variance). It is revealed from Expression (4.20) that 1σ PLL tracking threshold is 15° in terms of engineering. When the power ratio of carrier to noise C/N_0 decreases, the error output characteristic of PLL discriminator becomes flat and smooth when input phase error exceeds 45°. 1σ engineering threshold of PLL tracking loop is

$$\mathrm{PLL} = \sqrt{\sigma^2 t_{\mathrm{PLL}} + \theta^2{}_{\mathrm{A}}} + \frac{\theta_{\mathrm{e}}}{3} \leq 15°, \tag{4.21}$$

where $\sigma_{t\mathrm{PLL}}$ is $1\sigma_{\mathrm{PLL}}$ thermal noise phase error (°); θ_{A} is the phase jitter of oscillator caused by phase noise of the oscillator itself (°).

The other phase jitter sources may be transient or omitted, so thermal noise is usually taken as the only carrier tracking error source. Thermal noise phase error of PLL can be calculated according to the expression below:

$$\begin{aligned}\sigma_{\mathrm{iPLL}} &= \frac{360°}{2\pi}\sqrt{\frac{B_n}{C/N_0}\left(1+\frac{1}{2TC/N_0}\right)} = \frac{360°}{2\pi}\sqrt{B_n\varphi_n}(\circ)\\ \sigma_{\mathrm{iPLL}} &= \frac{360°}{2\pi}\sqrt{\frac{B_n}{C/N_0}\left(1+\frac{1}{2TC/N_0}\right)} = \frac{\lambda}{2\pi}\sqrt{B_n\varphi_n}(m),\end{aligned} \tag{4.22}$$

where B_n is noise bandwidth of carrier loop (Hz); C/N_0 is the ratio of carrier power to noise power spectral density (times); $\frac{N_0\left(1+\frac{1}{2TC/N_0}\right)}{C} = \varphi_n$ is phase noise power spectral density with consideration of square loss; T is pre-detection integral time (s); λ is wavelength of operating carrier (m).

Expressions (4.21) and (4.22) are independent of the loop filter order. The phase error in thermal noise is not directly independent upon the loop order but having indirect relations, which is attributed to the fact that different orders of loops have different dynamic characteristics (1-order loop is sensitive to velocity, 2-order loop sensitive to acceleration; 3-order loop sensitive to jerk). Generally speaking, increase in loop order will improve dynamic characteristics. Therefore, the bandwidth can be reduced by increasing loop orders. For the same minimum C/N_0, the phase error in thermal noise will decrease, and the dynamic characteristics will not get worse. When the error is expressed in degree, Expression (4.22) has nothing to do with carrier frequency. The phase error in thermal noise of carrier is tightly related to C/N_0, noise bandwidth and pre-detection integral time. In case of increase in C/N, the phase error in thermal noise will decrease. Additionally, decrease in noise bandwidth will reduce phase error in thermal noise. In Expression (4.22), $\mathrm{Ls} = \left[1 + \frac{1}{2TC/N_0}\right]$ is the foresaid square loss. In case of low SNR, squaring will cause SNR loss, which is the difference between "spread spectrum TT&C" and "non-spread spectrum TT&C". Apart from this, calculating other measuring errors is similar. Pre-detection integral time T is designed based on the bandwidth of information to be transmitted. $T \leq 1/R_b$; where R_b is information code rate. The higher the information code rate is, the smaller T becomes. The bigger the value of thermal noise phase error of code loop, the bigger the square loss Ls. If information codes are not transmitted; namely, there is no spread spectrum other than PN code, the T may be very big. Let $\mathrm{Ls} \to 0$ ($T \to \infty$ may be selected in theory). According to analysis above, for spread spectrum, T shall not be very big for transmitting information code and thus causing square loss. However, T of PN code TT&C system may be large, there is no square loss (or very small) in theory.

Carrier loop is not only used for velocity measuring but also for ranging. PLL thermal noise phase error is a part of carrier phase measuring values, so Expression (4.22) can also be expressed in unit of range. The velocity can be calculated from carrier Doppler or approximately obtained through the carrier phase changes between two range measuring values in a short time.

4.5.2 *Range Measurement Error in Spread Spectrum TT&C*

In spread spectrum TT&C system, range measurement error is mainly determined by code loop. The range measurement error in the code loop (i.e., Delay Locked Loop DLL) mainly includes thermal noise error and dynamic error. The tracking threshold of DLL is determined by the following expression.

$$3\sigma_{\mathrm{DLL}} = 3\sigma_{\mathrm{tDLL}} + R_e \leq d, \tag{4.23}$$

where σ_{DLL} is the root-mean-square error of all errors of the code loop (the unit is the number of chips); d is half of the constant slope range (sometimes shorted as linear range) of the delay-locked discriminator indicated by the number of chips; σ_{DLL} is the root-mean-square value of code tracking thermal noise phase error; R_e is the dynamic error of DLL tracking loop.

The root-mean-square value of thermal noise caused by incoherent DLL tracking jittering is

$$\sigma_{\mathrm{tDLL}} \approx \begin{cases} \sqrt{\dfrac{B_n}{2C/N_0}\Delta\left[1+\dfrac{2}{TC/N_0(2-\Delta)}\right]} & \left(\Delta \geq \dfrac{\pi R_c}{B_{fe}}\right) \\ \sqrt{\dfrac{B_n}{2C/N_0}\left[\dfrac{1}{B_{fe}T_c}+\dfrac{B_{fe}T_c}{\pi-1}\left(\Delta-\dfrac{1}{B_{fe}T_c}\right)^2\right]\times\left[1+\dfrac{2}{TC/N_0(2-\Delta)}\right]} & \left(\dfrac{R_c}{B_{fe}}<\Delta<\dfrac{\pi R_c}{B_{fe}}\right), \\ \sqrt{\dfrac{}{B_n 2C/N_0}\left(\dfrac{1}{B_{fe}T_c}\right)\left[1+\dfrac{1}{TC/N_0}\right]} & \left(\Delta \leq \dfrac{R_c}{B_{fe}}\right) \end{cases} \tag{4.24}$$

where the unit of the calculated result of the expression is the number of chips, $\Delta = T_D/T_c$, T_D is the interval between the lead code and the lag code, T_c is the width of the chip; B_n is the equivalent noise bandwidth of the code loop (Hz); C/N_0 is the power ratio of the carrier to the noise; T is the pre-detection integral time (s); and B_{fe} is front-end two-sided bandwidth (Hz).

When $\Delta = 1$, i.e., $T_D = T_c$, Expression (4.24) becomes:

$$\sigma_{\mathrm{tDLL}} = T_c\sqrt{\frac{B_n[1+2/T(C/N_0)]}{2C/N_0}}. \tag{4.25}$$

This is a special case frequently applied.

According to Expression (4.24), phase jitter of DLL is directly proportional to the square root of the noise bandwidth of the filter, but is independent of the order of the loop (decreasing B_n can reduce the jitter and allow for a lower C/N_0 threshold).

Besides, increasing the pre-detection integral time T will also allow for a lower C/N_0 threshold, but its effect is less than that of decreasing B_n. The part containing the pre-detection time T in the Expression is called squaring loss. Increasing pre-detection integral time will reduce the squaring loss. According to Expression (4.24) and Fig. 4.19, reducing the interval Δ between the correlators will increase the slope of the discriminator, so as to reduce DLL jitter, however the dynamic threshold is also reduced. Reducing the interval between the correlators will also help to decrease the impact of multi-path interference on the code tracking loop. According to Expression (4.24), when Δ is reduced to $\Delta \leq R_c/B_{fe}$, the further reduction of Δ has no meaning to reducing the thermal noise range measurement error.

If only PN code is used without spread spectrum, T_c can be increased to a very high value and Expression (4.24) will become:

$$\sigma_{\mathrm{tDLL}} = T_c\sqrt{\frac{B_n\Delta}{2C/N_0}}. \tag{4.26}$$

This is the difference between PN code only ranging and spread spectrum ranging.

4.5.3 Rate Measurement Error in Spread Spectrum TT&C

In spread spectrum TT&C, there is a squaring loss in PLL, so the phase noise power spectrum density N_ϕ of the carrier loop is determined by the following expression:

$$N_\phi = \varphi_n = \frac{N_0\left(1 + \frac{1}{2TP_s/N_0}\right)}{P_s}.$$

The relationship between N_ϕ and the rate measurement error was derived in Sect. 2.3.3.2 and has been given in Expression (2.101).

Using the same derivation method and substituting N_ϕ in the above expression we can obtain the expression of velocity measurement error introduced by thermal noise:

$$\sigma_{\mathrm{RN}} = \frac{c}{2\sqrt{2}\pi f_0 \tau}\sqrt{\frac{N_0 B_{\mathrm{L}}}{p_{\mathrm{s}}}\left(1 + \frac{1}{2TP_{\mathrm{s}}/N_0}\right)}. \tag{4.27}$$

Where, N_0 is the power spectrum density of thermal noise, P_{s} is the power of carrier signal, τ is the integral time, B_{L} is the bandwidth of PLL, f_0 is the carrier frequency, T_{D} is radio wave delayed time, T is pre-detection integral time.

4.6 "Double Spread Spectrum" and Its Application in TDRSS

4.6.1 Problems Raised

There is an S-band multiple access (SMA) scheme in TDRSS, where "SDMA + CDMA" multiple access scheme formed by the ground multi-beam is used. It transmits the signals of each antenna element in the satellite array antenna to the ground through the multiplexing satellite-ground link, and the array element signals are recovered under the condition of meeting the requirement of distortion. It is the key to the success of forming ground multi-beam. In American TDRSS, frequency division multiplexing (FDM) scheme is used to transmit the 30 array element signals of the array antenna on TDRS between the satellite and the ground. The disadvantages of this scheme are as follows:

(1) It limits the number of array elements so as to limit the gain increasing of satellite array antenna, which places high requirements on the G/T value and EIRP of the user satellite.

    In FDM scheme, the frequency division interval is 7.5 MHz, and 30 array elements will occupy a bandwidth of 30×7.5 MHz = 225 MHz. If the number of array element is increased to N, the occupied bandwidth will be $N \times 7.5$ MHz. However, the bandwidth assigned for SMA is limited, so N is limited.

(2) It places high requirements on the consistency of phase shifts and gains among channels in FDM, and will introduce an additional synthesis gain loss, which has become a technical key point of SMA.

(3) In FDM, the antenna array element signal output from "multiplexing" to "demultiplexing" shall be absolutely of the same frequency to ensure the in-phase addition after phase weighting. Therefore the 7.5 MHz interval frequency of comb frequency generator of 7.5 MHz interval on TDRS and the ground shall be coherent. This requirement is higher than that of FDM of general communication. It is also a technical difficulty.

(4) The multiplexing/demultiplexing scheme of FDM should have a complicated circuit, large volume and heavy weight, which is a major shortcoming to TDRS equipment.

The above-mentioned disadvantages are the key points for the improvement of TDRSS-SMA. On that account, the double spread spectrum of code division multiplexing is proposed.

4.6.2 "Code Division Multiplexing" and "Double Spread Spectrum"

Code division multiplexing (CDM) uses address code to distinguish the signal of each channel, and this address code can be PN code of good orthogonality, chaotic code or Walsh orthogonal code. The multiplexed signals share one frequency and the whole time slot, so only one channel is required. Its principle diagram is shown in Fig. 4.24.

In Fig. 4.24, the signals received by N array elements are spread spectrum signals containing M users. Their amplitudes are essentially the same. The PN code of each spread spectrum signal is used to distinguish the user satellite with a different address, so it is called the address code of "code division multiplex access" (CDMA) and its spread spectrum is called "CDMA spread spectrum" or

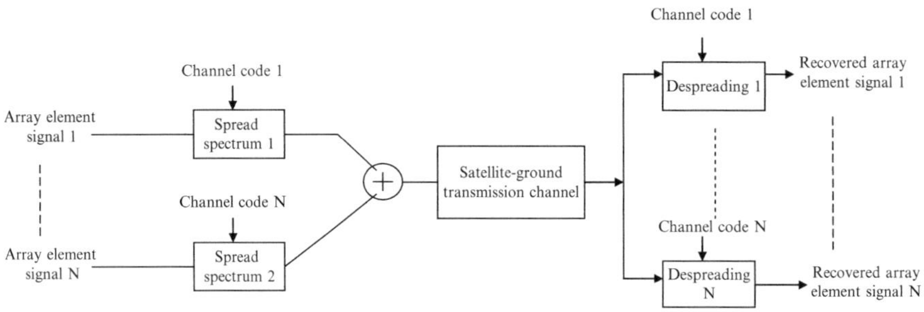

Fig. 4.24 Principle block diagram of code element multiplexing

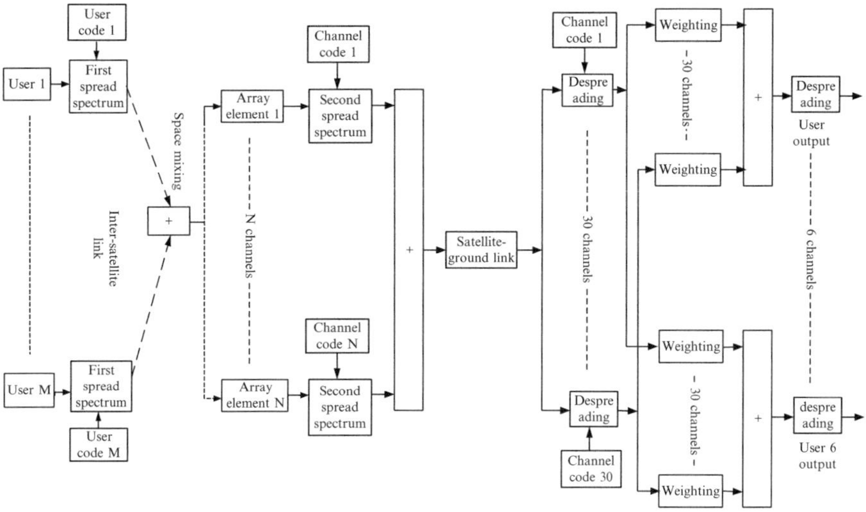

Fig. 4.25 System block diagram of "double spread spectrum"

"first spread spectrum". The code used for code division multiplexing in Fig. 4.24 is called the channel code of "code division multiplexing" and its spread spectrum is called "code division multiplexing spread spectrum" or "second spread spectrum". The former is used to distinguish the user satellites and the latter is used to distinguish the *N* array elements. Therefore, the signal transmitted in the satellite-ground transmission channel is a "double spread spectrum" signal. The system block diagram of this "double spread spectrum" is shown in Fig. 4.25.

In Fig. 4.25, the CDMA mode of the original TDRSS is still used for the multiple access of M user satellites. The user address code is used by each user satellite for the spread spectrum of data, called the first spread spectrum. The first spread spectrum signals transmitted by M user satellites are mixed in the air, and received by the array antenna on TDRS through inter-satellite transmission channel. The array antenna is composed of *N* array elements. The distance between the user satellite and TDRS is longer than the distance between the array elements, so the amplitudes of the mixed signals received by each array element are almost the same. CDM is used for the multi-channel merger of the output signal of *N* array elements, i.e., respectively using one channel address code (shorted as channel code) for the second spread spectrum of the output signal of *N* array element, multiplexing into one channel signal and then transmitting in one channel. The ground terminal station will divide the CDM signal into *N* channels after receiving it. Each channel uses its "channel code" to de-spread and recover the signal of each array elements on TDRS so as to form the adaptive multi-beam on the ground. Because there are M users, the output signal of each array element is divided into M channels. They will be used to form M beams and the output data of each user will be obtained after de-spreading.

According to the above description, CDM has the following advantages compared with FDM:

(1) Because the bandwidth occupied by CDM will not increase with the increase of the multiplexing number N, the bandwidth is no longer a limitation for increasing array elements number N, and the array antenna gain can be improved by increasing N.
(2) There is no phase/amplitude consistency problem for CDM as that in FDM multi-channel transmission since CDM uses only one channel for transmission. The synthesis gain loss caused by group delay, nonlinearity, multi-channel isolation, and interference is also free of this technical key point.
(3) N channels are turned into 1 channel, so the equipment is greatly simplified. Especially the volume, weight, power consumption, and technical difficulty of the satellite equipment are dramatically decreased.
(4) "Double spread spectrum" improves the anti-interference capability of the system.

However, "double spread spectrum" also has some new technical problems which are discussed as follows.

4.6.3 Main Technical Problems of "Double Spread Spectrum"

(1) Multiple access interference

The ratio of "S/N before interference" and "S/N after interference" is [4]:

$$L = \frac{\left\{\dfrac{N_0}{2E_b}\right\}^{-1}}{\left\{\dfrac{K-1}{3N} + \dfrac{N_0}{2E_b}\right\}^{-1}} = 1 + \left(\frac{K-1}{3N}\right)\left(\frac{2E_b}{N_0}\right), \tag{4.28}$$

where k is the number of the array elements, N is the processing gain of the second spread spectrum, and E_b/N_0 is the E_b/N_0 received by each array element.

It is clear that when the value N is high enough, the multiple access interference caused by "code division multiplexing" will be very small.

(2) About "near-far" effect

In the general CDMA communication, "near-far effect" is an important problem. With the magnitude of interference user A_2 getting increased, the amplitude of its brought multiple access interference to user A_1 becomes larger to increase the bit error rate. Besides, the sidelobe of the cross-correlation property of A_2 increases too. It will occur at a certain time difference (time delay) between A_2 and A_1. It may interfere with the code loop. Because of the different distance between the interference user and the expectation user, the ratio of A_2/A_1 may exceed 10. This is the

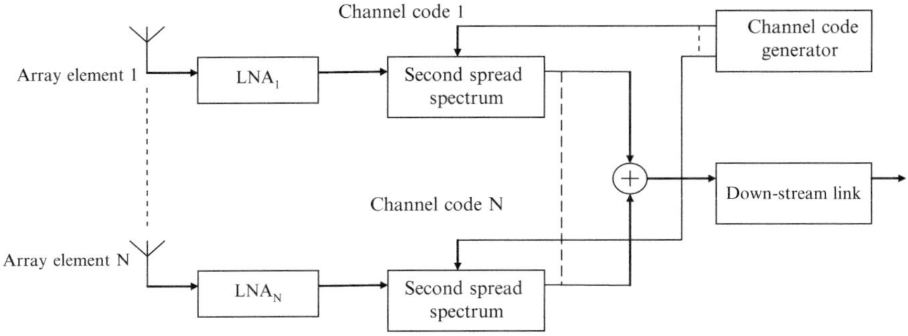

Fig. 4.26 Principle diagram of second spread spectrum

"near-far effect". Decreasing the correlation coefficient ρ_{12} may reduce the influence of the near-far effect. When $\rho_{12}=0$ (completely orthogonal signal), there is no influence of the near-far effect. In TDRSS-MA, long code length and gold code of small correlation is used to decrease ρ_{12}, so as to reduce the influence of the near-far effect.

After the design parameters are given, the multiple access interference of a specific engineering system can be calculated by computer simulation.

Because the amplitudes of the signals received by the array elements are almost the same, the "code division multiplexing" in this scheme is free of the near-far effect in the general multiple access communication.

(3) Second spread spectrum based on RF

In order to simplify the scheme, in this scheme, the second spread spectrum of TDRS is realized based on RF, as shown in Fig. 4.26.

In Fig. 4.26, the second spread spectrum is realized by using a "RF spread spectrum modulation unit". The front LNA is used to reduce the influence of the loss of rear spread spectrum modulation unit on the system noise temperature. The gain shall be as small as possible to reduce the influence caused by the phase shift and gain inconsistency among LNAs. "RF spread spectrum modulation unit" is realized by a binary phase-shift keyer composed of the microwave diode. The mathematical description of microwave spread spectrum is:

A modulating carrier of constant envelope with power of P, angular frequency of ω_0, and phase modulation of $\theta_d(t)$ can be expressed as:

$$S_d(t) = \sqrt{2P}\cos\left[\omega_o t + \theta_d(t)\right]. \tag{4.29}$$

BPSK spread spectrum can be completed by simply using $S_d(t)$ to multiply a function $c(t)$ representing the spread spectrum waveform. It can also be realized by microwave BPSK and the signal sent by it is:

$$S_t(t) = \sqrt{2P}c(t)\cos\left[\omega_o t + \theta_d(t)\right]. \tag{4.30}$$

Thus, the spread spectrum is realized.

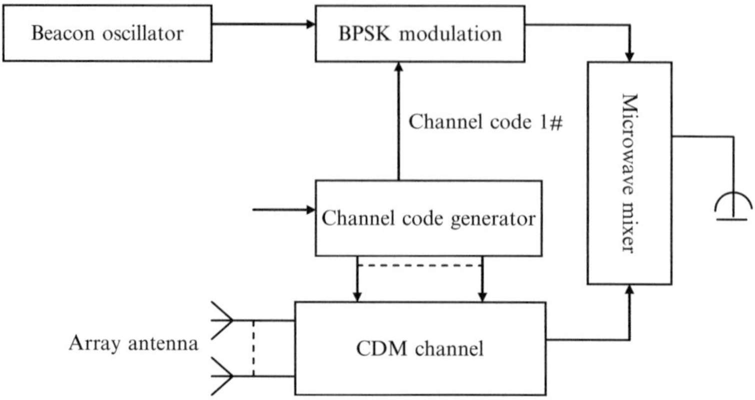

Fig. 4.27 Satellite scheme of beacon synchronization

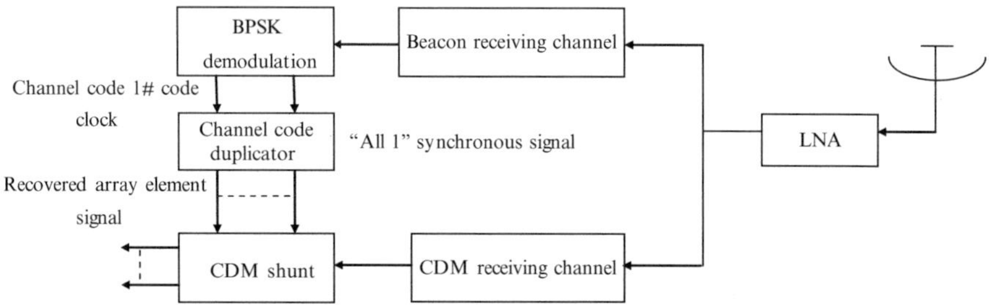

Fig. 4.28 Ground scheme of beacon synchronization

(4) Synchronization of "channel code"

1) Self-synchronization scheme: i.e., use the traditional method to capture and synchronize the "channel code". However it is hard to be realized because of the low C/N_0 of each array element signal. In addition, the equipment is complicated.
2) Beacon synchronization scheme: its block diagram is as shown in Figs. 4.27 and 4.28.

In Fig. 4.27, a beacon modulated with channel code 1# is used on TDRS and is transmitted to the ground at a proper EIRP in the bandwidth occupied by the original FDM to demodulate the channel code 1#. The whole synchronization includes two processes:

- Code clock synchronization: all the channel code generators on TDRS are driven by the same satellite code clock. Any channel code transmitted by the satellite beacon will be acquired and tracked by the code loop of the ground station for obtaining the code clock of channel code which will be used to drive the ground channel code generator.

- "All 1" channel code synchronization: "all 1" of all satellite channel code shall be synchronous in design. Once a channel code is demodulated by the ground station, all channel codes can be duplicated based on the known "all 1", code type and code clock. These channel codes, synchronous with the channel codes received array element signal, can be used for demodulation to separate the array element signal.

(5) Influence of the amplitude-phase characteristics and amplitude non-linearity of the transmission channel on the bit error rate of the channel code.

The influence of amplitude-phase characteristics will be greater with the increase of spread spectrum code rate. The non-linear amplitude-phase conversion and combination interference will affect the bit error rate due to the non-constant envelope of "CDMA" signal. Therefore, SMA shall operate in linear state, which is different from SSA and KSA which can operate in saturation state.

It is important to point out that in order to reduce the additional influence caused by double spread spectrum, the gain processing of double spread spectrum shall be improved as much as possible.

4.7 Chaotic Sequence and Chaotic Spread Spectrum TT&C [5]

4.7.1 Characteristics of Chaotic Sequence

The chaos theory proposed that the chaotic signal is not predictable and the unpredictability is the theoretical basis for its confidentiality, security, and anti-interception. The chaotic signal designed based on chaos theory has good security and anti-interception.

"Chaos" means if a behavior generated by a non-linear system has a sensitive dependence on a subtle change of the initial conditions, such behavior is called "chaotic" since it is unpredictable after a rather long period of time. For TT&C to speak, this non-linear system can be realized by a deterministic iterative formula. It produces chaotic signals, that is, by making minimal subtle changes of the initial value of the iterative formula, it can produce various random-like signals which are completely unrelated to each other. For example, the following iterative formula can be used.

$$x_{k+1} = 1 - 2x_k^2$$

By taking the initial values within $-1 < x_k < 1$ for iterative operation, it will produce the improved Logistic chaotic sequence. With the continuous proceeding of iteration, the sequence will be extended indefinitely. The frequency of iterative operation is code clock.

Chaotic sequence has the following characteristics:

(1) Complexity and diversity: Chaotic signal has a very complex internal structure and is sensitive to the initial conditions and parameters. The chaotic system can easily produce a variety of different chaotic signals. Thus, it is difficult to estimate the structure of the chaotic system or predict the chaotic signal in a long term. It is the high complexity and the difficulty in prediction of the chaotic signal that make it possess anti-interception and security.
(2) Random-like, aperiodicity, and wideband white spectrum characteristics: the chaotic signal is aperiodic, so its spectrum components are continuously distributed in the frequency band. Wideband signals can be used to resist the bad influence of the channel, such as frequency selective fading, anti-multipath interference, narrow band interference, etc.
(3) Good cross correlation: The chaotic signals produced by using different initial values and system parameters by different chaotic systems or the same chaotic system have the cross correlation function which disappears quickly. These signals can be deemed as the signals which are uncorrelated but satisfy the orthogonality in a certain sense.
(4) Extensive application prospect. These chaotic signals can be easily produced in a large number. A chaotic sequence can be produced only if one iterative formula and one initial value are given. Therefore, it has extensive application prospect in multi-target TT&C and has been successfully applied in CDMA system.
(5) Determinacy. The chaotic sequence is produced by the deterministic equation (in differential form or discrete form), and its production and receiving are controllable.
(6) Non-linearity and irreversibility. Chaos can only be produced in a non-linear system (non-linear mapping), and its mapping relationship is not one-to-one correspondence (or irreversible).
(7) Conditionality. In some non-linear systems, the occurrence of chaotic state relates not only to its mapping form but also to the parameters of the system equation. Only the parameter space under certain conditions corresponds to the chaotic state.

However, the above-mentioned continuous chaotic signals, i.e., the chaotic system expressed by the differential equation, are realized by the analog circuit. It is sensitive to the intrinsic parameters of the circuit and the error of chaotic regeneration algorithm, so it is difficult to be realized and not suitable for the actual application. The suitable one is the chaotic sequence. It is the discrete value sequence satisfying the chaotic rules. It may be the sampling value of the continuous chaotic signals, but more frequently, it is directly generated by the iterative process. Chaotic sequence has the pseudo randomness and initial value sensitivity similar to the continuous chaotic signal.

Compared with the analog chaotic TT&C, the chaotic sequence TT&C can conveniently use the mature technology of traditional pseudo-random code spread spectrum TT&C system. It has small realization difficulty and risk and the system is easily stable. Compared with the traditional pseudo-random code spread spectrum sequence, the chaotic sequence has not only the advantages of the pseudo-random code but also some special advantages such as good security, good anti-interception, and good anti-interference.

In actual engineering application, the chaotic sequence often has to be processed by the following ways:

(1) Binaryzation: convert it to a binary chaotic sequence from real value sequence. The method is to define a threshold function:

$$G_c(x) = \begin{cases} 0, & x < c \\ 1, & x \geq c \end{cases}.$$

Use this function to convert the analog chaotic sequence into a binary sequence $\{G_c(f(x_i))\}$.

(2) Truncation and periodization: truncate the infinite sequence according to period N to obtain a large number of spread spectrum sequences whose periods are N.

The traditional calculation shows that when the truncation period N of the chaotic sequence is small, the balance of the chaotic sequence is worse than that of Gold sequence. When the truncation period N becomes larger, for example $N \geq 255$, the balance of the chaotic sequence is improved. The distribution of cross-correlation of the chaotic spread spectrum sequence is similar to the distribution of the cross-correlation function of Gold sequence. Its maximum cross-correlation value is larger than that of Gold sequence of the same length, but its occurrence times of maximum cross-correlation is less than that of Gold sequence.

Like m sequence, Gold sequence and other pseudo-random code sequence, the chaotic sequence has good autocorrelation. Therefore, the chaotic sequence has pseudo randomness, which is similar to the traditional pseudo-random code. The length of the chaotic sequence is random and aperiodic in theory. Even though truncation is used to make the sequence valuing periodic, its period can be easily up to 2^{32} to generate a super-long sequence to improve the anti-interference capability. For a direct spread spectrum system, under certain period requirements (e.g., the non-ambiguity range for ranging) increasing the length of the sequence can improve the frequency of code clock to directly spread the wideband and to improve the spread spectrum gain so as to improve the anti-interference capability. Unlike the pseudo-random code sequence, its cross-correlation function has no such numerous spike pulses, which helps to improve the anti-multipath interference and anti-multiuser interference capability.

The results of numerical calculation and simulation experiment show that a chaotic sequence with a limited period and limited accuracy often has a larger maximum cross-correlation value than that of Gold sequence. However, the diversity of chaotic sequence provides conditions for selecting the sequences of good performance. The performance of the chaotic sequences obtained by screening is better than or equivalent to that of the preferred pseudo-random code sequence.

Such sequences are of great number and all these sequence groups can be generated by only one mapping formula and initial value, so the efficiency is extremely high. While, the number of pseudo-random code sequence is limited and the preferred sequence becomes fewer after screening. This is a shortcoming of pseudo-random code sequence when compared with the chaotic sequence. For example, when using an n-level shift register to generate the Gold sequence, the number of the generated sequence will be $(2^n + 1)$ (when $n = 10$, the number will be 1,025), while for the chaotic sequence, its number is much larger. For the aforesaid improved Logistic chaotic sequence, 2×10^5 different sequences can be generated when the initial value interval is 10^{-5}.

It is difficult to derive the iterative equation from the chaotic sequence after "binaryzation" due to its sensitivity to the initial value and the dependence on the parameter of non-linear system, so the chaotic sequence is secure and not easily detected. On the one hand, the long-term prediction of the chaotic sequence is impossible and even for the short-term prediction, there are great errors. On the other hand, it is impossible to derive the initial conditions of the system from the limited length of the sequence. Even the enemies have obtained one or several segments of plaintext-ciphertext, they cannot decipher the whole system for they do not know the model, parameters and initial conditions of the chaotic sequence. Therefore the purposes of security and anti-interception are achieved. The encryption mode is effective and simple, and it can be changed only if the mode, parameter or initial condition of the chaotic sequence is changed.

The sensitiveness of the chaotic system to the initial condition makes the long-term prediction impossible. However, it is possible to conduct short-term prediction by using observational data because the motional orbit of the system diverges a little in a short term. The so-called chaotic sequence prediction refers to use of the partial prior knowledge or information, for example part of the received chaotic signals or the known structure knowledge about the chaotic sequence, to predict the future possible value of the chaotic sequence for the purpose of deciphering the intercepted signals. There are some prediction methods to predict the chaotic sequences of low complexity and hierarchy with little errors, so the resistance of the sequence to prediction must be considered when such chaotic sequences are applied to secure communication, or the security is difficult to be achieved.

The reliable and effective operation of the chaotic sequence spread spectrum communication system depends on the synchronous capability of the system to a great extent. The synchronous methods of the chaotic sequence are as follows:

(1) Synchronous method similar to the traditional pseudo-random code. If the chaotic sequence synchronization can be achieved by using the traditional pseudo-random code synchronization technology, for example slippage correlation method, Δ DLL and parallel acquisition, after truncation, binaryzation, and limited length taking are applied to the chaotic sequence.
(2) Using synchronization channel to deliver synchronization information. The information transmitting end transmits the synchronization information, for example the chaotic driving variable and synchronization code, to the receiving end through the special synchronization channel, to synchronize the chaos generation mechanism of the receiving end with the transmitting end.

(3) Using synchronization pulse to deliver synchronization information. The information transmitting end transmits the synchronization pulse to the receiving end through a single channel (communication channel) to synchronize the chaos generation mechanism of the receiving end with the transmitting end.
(4) Chaos synchronization based on phase control. When the phase trajectories at the transmitting and receiving ends arrive at the pre-set Poincare section in phase space, sample the phase difference and state difference at the two ends of the system, and then adjust the phase and state at the two ends of the system to achieve the synchronization of the whole chaotic synchronization system.
(5) Deadbeat synchronization of discrete chaotic system. It is a technology based on the observer and using a scalar signal to achieve the discrete chaotic system synchronization.
(6) Prediction-based synchronization. The analysis of the short-term prediction methods of the chaotic system may help the research of sequence synchronization.

4.7.2 Type, Selection, and Generation of Chaotic Sequence [6]

4.7.2.1 Type of Chaotic Sequence

The chaotic sequence is usually generated by the numerous iterations of a non-linear equation, i.e., different code words will be generated according to different initial values. The chaos theory develops very fast, and a variety of chaotic iteration algorithms are proposed such as Chua's circuit, SB mapping, VB mapping, IB mapping, composite mapping, two-dimensional mapping, cascade chaotic sequence, embedded chaotic sequence, and spatiotemporal chaotic sequence. They can be classified according to the iterative formula.

The most common chaotic iterations are as follows:

(1) Tent-Map

The definition of Tent-Map is: (where $0 < a < 1$)

$$X_{K+1} = f(X_K) = \begin{cases} X_K/a & (0 < X_K < a) \\ (1 - X_K)/(1 - a) & a \leq X_K \leq 1 \end{cases}. \tag{4.31}$$

The statistical characteristics of Tent-Map sequence are:

Tent-Map sequence is evenly distributed within (0,1)

$$p(x) = \begin{cases} 1 & (0 < x < 1) \\ 0 & \text{others} \end{cases}. \tag{4.32}$$

Mean value of the sequence is $E\{x\} = 0.5$.

Autocorrelation function of the sequence is:

$$\begin{aligned} r(m) &= \lim_{N\to\infty} \frac{1}{N}\sum_{K=0}^{N-1}\left(X_K - \bar{X}\right)\left(X_{K+M} - \bar{X}\right) \\ &= \frac{1}{12}(2a-1)^{|m|}. \end{aligned} \tag{4.33}$$

(2) Chebyshev map

w-order Chebyshev map is defined as: (where $-1 \le X_K \le 1$)

$$X_{K+1} = f(X_K) = \cos\left(w \cos^{-1} X_K\right). \tag{4.34}$$

Its probability density is: (where $-1 \le x \le 1$)

$$p(x) = \begin{cases} \dfrac{1}{\pi\sqrt{1-x^2}} . \\ 0 \quad \text{others} \end{cases} \tag{4.35}$$

(3) Logistic-Map

Logistic-Map is defined as:

$$X_{n+1} = rX_n(1 - X_n) \quad 0 < X_n < 1, \tag{4.36}$$

where $1 \le r \le 4$, r is called fractal parameter. When $3.5699\ldots < r \le 4$, the system operates in chaotic state.

The statistical characteristics of Logistic-Map sequence are:
Its probability density function is: (where $0 < x < 1$)

$$p(x) = \begin{cases} \dfrac{1}{\pi\sqrt{x(1-x)}} . \\ 0 \quad \text{others} \end{cases} \tag{4.37}$$

Mean value of the sequence is $E\{x\} = 0.5$.
Autocorrelation function of the sequence is:

$$\begin{aligned} r(m) &= \lim_{N\to\infty} \frac{1}{N}\sum_{K=0}^{N-1}\left(X_K - \bar{X}\right)\left(X_{K+M} - \bar{X}\right) \\ &= \begin{cases} 0.125 & m = 0 \\ 0 & m \ne 0 \end{cases}. \end{aligned} \tag{4.38}$$

(4) Improved Logistic-Map

Improved Logistic-Map is defined as:

$$x_{n+l} = 1 - 2(x_n)^2 \quad -1 < x_n < 1. \tag{4.39}$$

The statistical characteristics of Logistic-Map sequence are:
Its probability density function is: (where $0 < x < 1$)

$$p(x) = \begin{cases} \dfrac{1}{\pi\sqrt{x(1-x)}} \\ 0 \quad \text{others} \end{cases}. \tag{4.40}$$

Mean value of the sequence is $E\{x\} = 0.5$.
Autocorrelation function of the sequence is:

$$\begin{aligned} r(m) &= \lim_{N\to\infty} \frac{1}{N}\sum_{K=0}^{N-1} X_K X_{K+M} - \bar{X} \\ &= \begin{cases} 0.5 & m = 0 \\ 0 & m \neq 0 \end{cases}. \end{aligned} \tag{4.41}$$

Cross-correlation function of the sequence is:

$$\begin{aligned} R_{12}(m) &= \lim_{N\to\infty} \frac{1}{N}\sum_{K=0}^{N-1} X_{1K} X_{2(K+M)} - \bar{X}^2 \\ &= \int_0 \int_0 x_1 f^m(x_2) P_{X_1}(x_1) P_{X_2}(x_2) dx_1 dx_2 - 0 = 0 \\ & f^m(x_2) = \underbrace{f(f \ldots f(x_2) \ldots)}_{m}. \end{aligned} \tag{4.42}$$

Besides, spatiotemporal chaotic sequence, composite mapping, two-dimensional mapping, etc. were proposed in recent years.

In engineering application, it is required to select a proper chaotic sequence type for use by comparing the balance, autocorrelation, cross-correlation, difficulty of generation, maturity of research, etc. of the chaotic sequences according to engineering requirements.

4.7.2.2 Selection of Chaotic Sequence

At present, the commonly used chaotic sequences are four types: Logistic chaotic mapping, Chebyshev chaotic mapping, Tent chaotic mapping, and improved

Logistic mapping. The results of the comparative study of the chaotic sequences generated by the four chaotic mapping are as follows [7]:

(1) Balance

For the generated code word, calculate the quantity of the balanced code under each kind of chaotic mapping and code length according to the balance principle, i.e., the quantity difference between "1" and "−1" in the code word is 1. The calculation results show that:

When the code length and quantity of the chaotic sequence are the same, Tent chaotic sequence has fewer balanced code, while there is no obvious difference among the quantities of balanced codes generated by the other three chaotic mappings. For chaotic sequences generated by all chaotic mappings, the quantity of the code meeting the balance requirements decreases with increase of the code length.

(2) Correlation

Calculate the autocorrelation function of the balanced code generated by different chaotic mappings with different code length. The calculation formula of correlative is as follows in simulation:

Autocorrelation function of the sequence:

$$R_{\mathrm{ac}}(m) = \frac{1}{N}\sum_{i=0}^{N-1} x_i x_{(i+m)\mathrm{mod}N} \quad 0 \le m \le N-1. \tag{4.43}$$

Root-mean-square value of autocorrelation sidelobe:

$$\sigma_{\mathrm{ac}} = \sqrt{\frac{1}{N-1}\sum_{m=1}^{N-1} R_{\mathrm{ac}}^2(m)}. \tag{4.44}$$

Cross correlation function of the sequence:

$$R_{\mathrm{cc}}(m) = \frac{1}{N}\sum_{i=0}^{N-1} x_i y_{(i+m)\mathrm{mod}N} \quad 0 \le m \le N-1. \tag{4.45}$$

Root-mean-square value of cross correlation:

$$\sigma_{\mathrm{cc}} = \sqrt{\frac{1}{N}\sum_{m=0}^{N-1} R_{\mathrm{cc}}^2(m)}, \tag{4.46}$$

where N is the length of the sequence.

The average values of all correlation root-mean-square values have been calculated. The calculation results are shown in Tables 4.1 and 4.2 respectively [7].

Table 4.1 Statistics of autocorrelation and cross correlation for code length of 1,023

	Logistic	Chebyshev	Tent	Improved logistic
Maximum sidelobe value of autocorrelation	0.1437	0.1437	0.1241	0.1437
Maximum root-mean-square value of autocorrelation sidelobe	0.0355	0.0359	0.0305	0.036
Minimum root-mean-square value of autocorrelation sidelobe	0.0277	0.0273	0.0282	0.0278
Peak value of cross correlation function	0.1984	0.1828	0.1652	0.175
Maximum root-mean-square value of cross correlation	0.0343	0.0345	0.0336	0.0341
Minimum root-mean-square value of cross correlation	0.0287	0.0285	0.0294	0.0286

Table 4.2 Statistics of autocorrelation and cross correlation for code length of 4,095

	Logistic	Chebyshev	Tent	Improved logistic
Maximum sidelobe value of autocorrelation	0.0716	0.0779	0.0662	0.0735
Maximum root-mean-square value of autocorrelation sidelobe	0.0166	0.0166	0.0163	0.0164
Minimum root-mean-square value of autocorrelation sidelobe	0.0147	0.0146	0.015	0.0148
Peak value of cross correlation function	0.0852	0.0857	0.0745	0.0872
Maximum root-mean-square value of cross correlation	0.0162	0.0163	0.0161	0.0162
Minimum root-mean-square value of cross correlation	0.0149	0.015	0.0153	0.0149

According to the above tables, Tent has the best cross correlation performance and the minimum autocorrlation sidelobe. The cross correlation performances of the other three chaotic sequences have no obvious difference. The correlation performance of all chaotic sequences will be improved with the increase of the sequence length.

4.7.2.3 Generation of Code Library

As infinite chaotic sequences are non-periodic, there are many problems in their synchronization and ranging. However, if they are under truncation and periodization, the synchronization and ranging of them can be similar to those of PN codes. Although the sequences' anti-interception performance and confidentiality will be weakened, they are difficult to be reconnoitered because such different sequences have different formulas, formula parameters, initial values, binaryzation, and sequence lengths. Therefore, it is feasible to constantly change chaotic codes whose amount is so huge to obtain anti-interception and anti-jamming performance.

In peacetime, PN codes are used, and in wartime, chaotic codes are adopted. As a result, a code library with a huge number of chaotic codes needs to be established. However, a chaotic sequence is different from another in performance, and not all chaotic sequences meet requirements. For this reason, proper sequences shall be selected to meet specific application requirements. The following are common principles in sequence choice:

(1) Balance principle

Sequences generated in huge number are selected according to balance E threshold value. If an absolute balance quality threshold value is used, the difference between the numbers of "1" and "0" in the whole code word shall not be larger than 1. That is, $E \leq 1/N$, N is the code length.

Usually, balance is related to the code length. In actual cases, a relative balance quality can also be used, namely, E value is kept the same, and at a code word where $E = 0.001$, the difference between numbers of "1" and "0" is made to be less than or equal to $E \times N$.

(2) Autocorrelation criterion

In the selected balance codes, the autocorrelation functions of all balance code words are calculated, and the largest sidelobe values $R_{\max}$ of such functions are worked out. This $R_{\max}$ is then compared with R_{mT} (the largest set autocorrelation sidelobe threshold value). If $R_{\max}$ is larger than R_{mT}, such code word is given up. Next, the mean square value of the autocorrelation function sidelobe of a code which meets requirements for the largest autocorrelation function sidelobe value, then such mean square value is contrasted with $R_{\sigma T}$ (the threshold value of the set autocorrelation sidelobe mean square value). If the autocorrelation sidelobe mean square value is larger than the threshold value, such code is abandoned.

In the case a chaotic sequence is contrasted with a Gold sequence whose length is the same as the chaotic sequence, all the largest sidelobe values of chaotic sequence autocorrelation functions are larger than those of the Gold sequence. However, the mean square values of chaotic sequence autocorrelation functions are similar to those of the Gold sequence (in the case, the code length is 1,023). For these reasons, it is proper to set the largest sidelobe value of the autocorrelation function $R_{mT} = 0.1$, rather than that of 0.0635 of the Gold sequence. However, the selected threshold value of the autocorrelation sidelobe mean square value can be based on that of the Gold sequence, namely, $R_{\sigma T} = 0.0346$.

(3) Cross-correlation criterion

Multi-site interference is a main factor which affects the spread spectrum TT&C system. Multi-site interference affecting a user in the spread spectrum system is mainly dependent on the square of the cross-correlation function of this user with other users in the system (or the root mean square value of other users' interference). It is the mean square value of the cross-correlation function that decides the anti-multisite interference performance of a chaotic spread spectrum system; the peak value just shows the worst situation, but it can cause false lock.

In a chaotic code set where balance and autocorrelation properties meet respective requirements, pairwise cross-correlation of the sequences is calculated. The threshold value of cross-correlation function peak value is supposed to be C_{mT}, and the mean square threshold value to be $C_{\sigma T}$. Searching cross-correlation functions is an arduous task. One reason is that the operand is huge (unreasonable design usually leads to computer memory crash); the other reason is that according to pairwise correlation values, it is hard to some extent to choose code words with all of their cross-correlation properties meeting requirements. In the study in respect to this aspect, researchers have put forward the connected node set searching model. Namely, sequences (supposed to be Q sequences) that have passed balance and autocorrelation screening are treated as nodes of a flow diagram, and the nodes which meet cross-correlation criterion are mutually connected. At first, pairwise connection of the Q sequences are calculated, then the recursive algorithm is used to work out the largest connected node set in the diagram.

According to the foresaid criteria, the code library selection process and computer procedures can be designed to select a good code library. After analyzing a good code library selected by chaotic sequences, a few useful conclusions can be made as follows:

1) Both the autocorrelation and the cross-correlation properties of chaotic sequences are not too bad, and as code length increases, related properties are better. It should be noted that the code length calculated in the case of multisite interference shall be the one that is contained in the code element width of information code. Because the optimum reception filter for information codes is an "integral – zero clearing" filter which integrates within information code element width, the integration time of corresponding cross-correlation functions shall also be the same time. Therefore, a too long integration time will not help to obtain optimum signal to noise ratio (SNR).
2) Both the largest autocorrelation sidelobe value and the largest cross-correlation value of a chaotic sequence are worse than a Gold sequence. However, when the code length is 1,023, its autocorrelation sidelobe root-mean-square value and the cross-correlation root-mean-square value are similar to those of the Gold sequence; and when the code length is larger than 1,023, they are better than those of the Gold sequence.
3) If the largest values of the autocorrelation sidelobe and the cross-correlation of the Gold sequence are chosen to be the selection criteria, there are no chaotic sequences that are better than the Gold sequence; if the root-mean-square values of the autocorrelation sidelobe and the cross-correlation of the Gold sequence are chosen to be the selection criteria, there are a vast amount of chaotic sequences that are better than the Gold sequences.

(1) The generation method of a truncated and periodic chaotic sequence

A possible way to produce a periodic chaotic sequence with a truncated chaotic sequence is shown in Fig. 4.29.

In Fig. 4.29, an n-bit truncated chaotic code taken from its library is registered in advance at the n-bit register; and under control of the f/n code clock, it is taken out

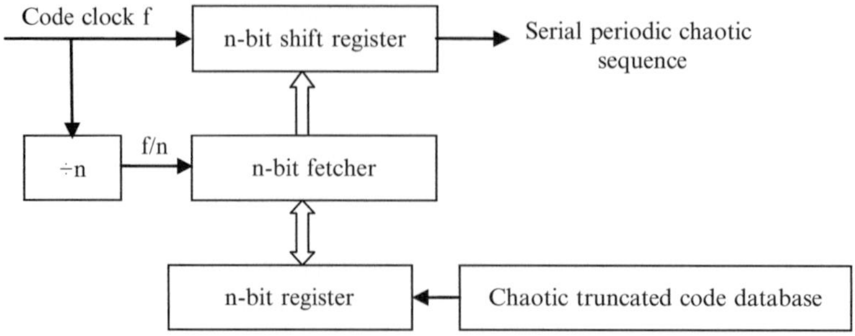

Fig. 4.29 Periodic chaotic sequence generator

and placed into an n-bit shift register by a fetcher. The placing happens once every n/f time interval. Driven by the code clock, the shift register produces a serial n-bit chaotic sequence. After every n-bit is shifted (amount to n/f time), the shift register will place an n-bit chaotic sequence once again, and go on to produce an n-bit serial sequence. Just like that, this process runs in a circle, forming a periodic chaotic sequence.

(2) The method to produce an infinite, non-periodic chaotic sequence

A production source of an infinite chaotic sequence may be an analog circuit or a digital circuit, namely:

1) by using a physically implementable circuit to produce chaotic oscillation, such as Chua's circuit;
2) by using high-speed DSP or FPGA to produce digital chaotic sequence through a recursion relation.

The first method is hard to implement, because the parameter change and the thermal noise influence of a non-linear circuit cannot be controlled artificially. It is relatively easy to implement a discrete chaotic system of the second method. A discrete chaotic sequence can be generated in a comparatively reliable way due to a relatively small influence from circuit parameters and noises. It is usually produced by the chaos iteration algorithm, which is quite easily implemented with the FPGA chip. The following is an example of Logistic-map to illustrate the digital implementation of chaos.

For Logistic chaotic mapping, the following is the expression of such mapping in the case of $r = 4$:

$$x_{k+1} = 4x_k(1 - x_k) = -4(x_k - 0.5)^2 + 1. \tag{4.47}$$

One way to implement chaotic mapping of the above expression is to use floating point arithmetic. Actual calculations suggest that the result of floating point arithmetic with single precision (32 bit) is separate from the chaotic state, and the arithmetic result with double precision (64 bit) is close to the present theoretical value. In actual application, however, equipment of floating point arithmetic with 64 bit double precision is expensive, and inconvenient for digital hardware implementation. For this reason, the above expression about Logistic-map can be implemented with a type of integer arithmetic. As both imputing and outputting of Logistic-map distribute within the range (0, 1), decimals x within such range can be presented as a binary number expression:

$$x = \sum_{v=0}^{\infty} a_v 2^{-(v+1)} = (a_0 a_1 a_2 \ldots), a_v = 0 \text{ or } 1. \tag{4.48}$$

If the front L bits are used to express x, namely, bits behind are omitted, we can get:

$$\begin{aligned} x &= \sum_{v=0}^{L} a_v 2^{-(v+1)} = (a_0 a_1 a_2 \ldots a_{L-1}) \\ &= 2^{-L} \sum_{v=0}^{L-1} a_v 2^{(L-1)-v} = 2^{-L} X. \end{aligned} \tag{4.49}$$

Such approach causes round-off error, which is unavoidable in any digital quantification. In the above expression:

$$X = \sum_{v=0}^{L-1} a_v 2^{(L-1)-v}. \tag{4.50}$$

It is an integer expressed by L bits binary numbers and has a one-to-one corresponding relationship with decimals x. If x_{k+1} and x_k are written in the form of Expression (4.48) and substituted into Expression (4.47), the results will be:

$$\begin{aligned} X_{k+1} &= 4X_k\left(2^L - X_k\right)/2^L \\ X_0 &= \left[2^L x_0\right], \end{aligned} \tag{4.51}$$

which is the integer expression of Logistic-map in the case of $r = 4$.

To implement the above expression, the complement of X_k can be calculated to work out $2^L - X_k$. Both $(2^L - X_k)$ and X_k are integers of L bits. Their product is $2L$ bits, and multiplying 4 amounts to the fact that the $2L$ bits product is moved leftwards 2 bits. Finally, being divided by 2^L means high-order L bits of the $2L$ bits integers are taken as output results. Therefore, Expression (4.51) can be implemented through digital hardware, which is shown in Fig. 4.30.

If L bits integers output as shown in Fig. 4.30 are multiplied by weight 2^{-L}, $\{x_k : k = 0, 1, 2, \ldots\}$ produced by Expression (4.26) can be obtained.

Selection of Expression (4.51) parameter L is limited. If the value is too small, calculation accuracy cannot be ensured, and the obtained sequence will be separate from chaotic state. Actual calculation shows that when $L = 32$, a sequence generated by Expression (4.51) is still at chaotic state. However, under the same 32 bit condition, a chaotic sequence cannot be obtained through floating point arithmetic. It is obvious that adopting an integer arithmetic expressed by Expression (4.51) is better than using floating point arithmetic. The former requires less calculation accuracy, leading to more simplified circuit and faster calculating speed.

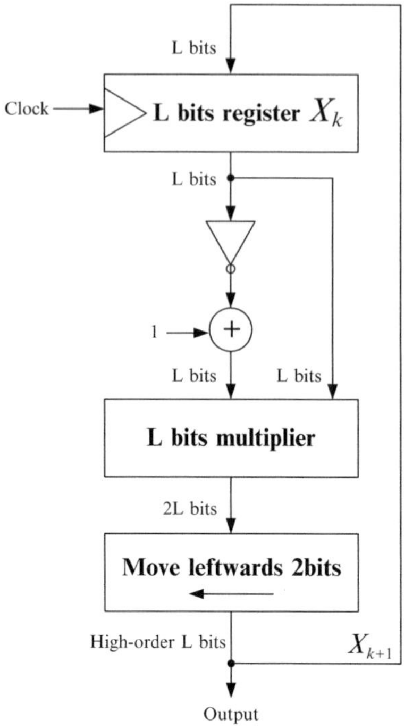

Fig. 4.30 Block diagram of logistic-map hardware implementation

4.7.3 Synchronization and Ranging of Chaotic Spread Spectrum Signals

The main differences between chaotic sequence spread spectrum C&T and pseudo-code spread spectrum C&T are production modes and synchronization methods of spread spectrum sequences. The rest parts of a system are similar to those of the existing spread spectrum C&T system, but synchronization of chaotic sequences is a difficult problem. Schemes under discussion to solve it are in abundance. The following are some of the schemes related to C&T.

4.7.3.1 "Chaotic Code Hopping"-Based Synchronization Method of an Infinite Chaotic Sequence – "Following Water Access Method"

(1) Basic principles

Based on "code hopping" and its synchronization principle introduced in Sect. 4.3.5 of this Book, a truncated chaotic sequence is stored in the memory unit under this scheme, and fixed storage is changed to following water storage for infinite chaotic codes, whose synchronization principles are:

1) At the transmitting end: driven by a code clock in transmitter, an infinite chaotic sequence is stored in circle in a following water way. Every storage unit stores

one code element. As the production source and its following water are infinite, the stored chaotic sequence is infinite.

2) At the receiving end: After code loop synchronization, driven by the code clock in receiver, the chaotic sequence stored in circle is extracted in a constant following water way as local codes, to implement its synchronization with the received infinite chaotic sequence.
3) The "chaotic code hopping" scheme is still adopted in capture and tracking of chaotic codes, whose "code hopping identification code" is changed to "following water opening code".
4) Memory's capacity shall be larger than or equal to the code length with the longest operating range, when:

- The target is static: Its speed is zero, so the code clock in transmitter and the code clock in receiver are in the same state. As a result, the flowing-in amount of following water memory is equal to that of the flowing-out amount, leaving an unchanged number of codes in the memory. It is equal to such static range. Fine measurement is determined by the code clock phase.
- The target moves away: Doppler frequency shift of the code clock in receiver is negative, making the flowing-in amount of following water memory larger than the flowing-out amount; and when the furthest range is reached, the storage amount tops. It corresponds to the longest operating range (the code clock is special for fine measurement). If a longer operating range is required, more storage capacity shall be added.
- The target moves closely: Doppler frequency shift of the code clock in receiver is positive, making the flowing-in amount of following water memory smaller than the flowing-out amount; and when the closest range is reached, the flowing-in amount stops decreasing. Its storage amount corresponds to the shortest operating range (the code clock is special for fine measurement).

(2) Production of an infinite chaotic sequence

The principle block diagram is as shown in Fig. 4.31.

In Fig. 4.31, the memory storage bit number $n \geq$ the code element number corresponding to the longest operating range. Each storage unit stores a code element. Chaotic codes are stored in order of storage bits 1, 2, 3..., which are started by the "following water driver code" and driven by the code clock at the transmitting end. Fetching of codes stored in memory is carried out in order of storage bits 1, 2, 3...n, which is started when the "following water startup code" is received by the receiving end. As the total storage bit number n is larger than or equal to the longest operating range, when the code element at $(n+1)$ bit is to be stored, the first storage unit has long been vacated, therefore, the code element at $(n+1)$ bit can be stored back to the first bit of storage unit, and the element at $(n+2)$ bit will be stored to the second bit of storage unit. Just like that, accessing is in a circle. So following water access of an infinite chaotic sequence is implemented.

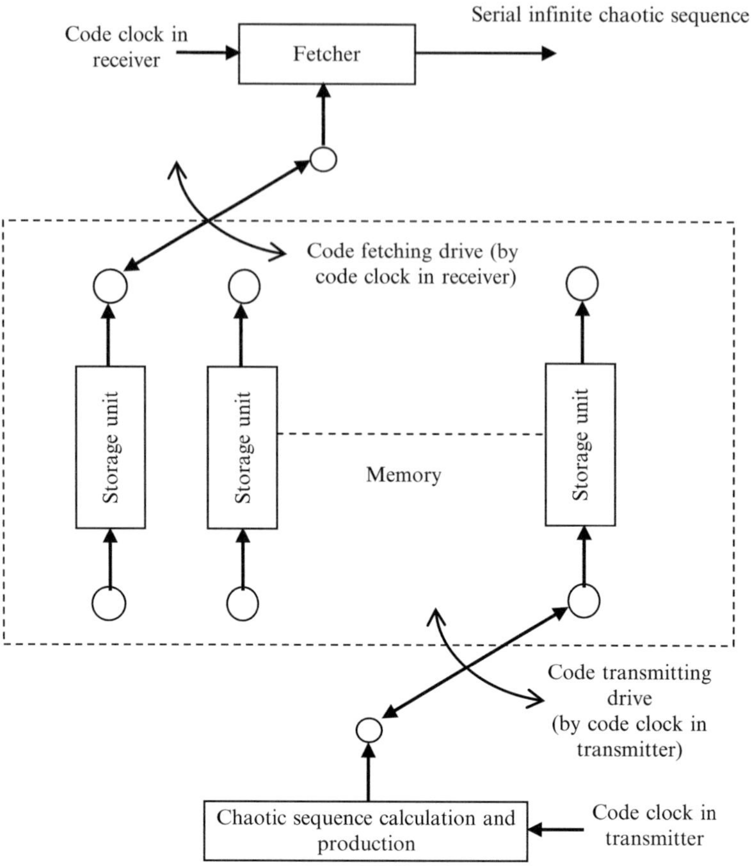

Fig. 4.31 Production principle of an infinite chaotic sequence

(3) Synchronization of an infinite chaotic sequence

Its basic principles are similar to synchronization of code hopping, which are:

1) Using a periodic leader sequence to implement capture, and after that, following water access is carried out to help the formation of an infinite chaotic sequence;
2) A certain chaotic code immediate to the leader sequence acts as "following water startup mark", which is known and can be solved through correlation reception.
3) At the transmitting end, startup of following water storage of the infinite chaotic code is controlled by a "following water startup code"; when such startup code is received by the receiving end, following water fetching at the receiving end will be started, whose following water rule is the same as the transmitting end, hence the synchronization.
4) The code extractor at the receiving end is driven by clock frequency with Doppler at the receiving end.

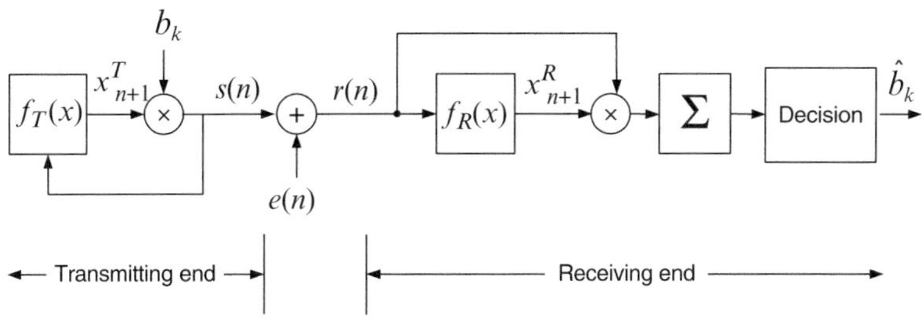

Fig. 4.32 A synchronization scheme using received spread spectrum sequence as a chaotic initial value

Figures 4.14 and 4.15 can be referred to for the structure and synchronization principles of an infinite chaotic sequence. It is just a little matter of changing "starting hop recognition" into "following water startup recognition".

4.7.3.2 Synchronization Scheme by Taking the Received Spread Spectrum Sequence as the Initial Value of Iteration at the Receiving End

Such scheme is shown in Fig. 4.32.

In Fig. 4.32, $s(n) = r(n)$ is used as the initial value of chaotic iteration at the receiving end; $f_T(X)$ and $f_R(X)$ are iteration formulas with different expressions for different chaotic types. In the figure, the input signal of $f_T(X)$ at the transmitting end is $s(n)$. After iteration, the output signal is $X^T{}_{n+1}$, while $s(n)$ is the resulted signal of $b(k)$ after being spread by the previous iteration computation $X^T{}_n$. After $s(n)$ is received by the receiving end, it is taken as the input of $f_R(X)$ at the receiving end, and after iteration, the receiving end output $X^R{}_{n+1}$ is generated. As $f_T(X)$ and $f_R(X)$ are the same iteration formulas, and their inputs are the same: $s(n) = r(n)$ (without regard to noise $e(n)$), their outputs $X^T{}_{n+1}$ and $X^R{}_{n+1}$ are the same, namely, transmitting and receiving are synchronized. When $X^R{}_{n+1}$ is used to de-spread the output signal $r(n)$, information data $b(k)$ can be restored. Although the above procedure involves spread spectrum of $b(k)$, a chaotic iteration arithmetic is still implemented as the processing result. The analysis is shown below.

Because:

$$s(n) = r(n)$$

while:

$$s(n) = X^T(n)b_k$$
$$r(n) = X^R(n).$$

Therefore,

$$X^T(n)b_k = X^R(n). \tag{4.52}$$

According to the above shown principle block diagram, there is:

$$X_T(n+1) = f[X_T(n)b_k] \tag{4.53}$$

$$X_R(n+1) = f[X_R(n)]. \tag{4.54}$$

By substituting Expression (4.52) into Expression (4.53) we can obtain:

$$X_R(n+1) = f\left[X^T(n)b_k\right]. \tag{4.55}$$

Comparing Expression (4.53) with Expression (4.55) shows that:

$$X_T(n+1) = X_R(n+1),$$

which means the synchronization is achieved.

The mapping expression of an improved Logistic-Map sequence is:

$$x_{n+1} = 1 - 2(x_n)^2 \quad (-1 < x_n < 1). \tag{4.56}$$

Under a condition when channel noises are omitted, if the transmitted information is assumed as $b(k) \in \{\pm 1\}$, and every information bit is distributed with N chips, there is comparison expression as follows:

$$s_n = r_n = x_n^T b(k)$$

$$x_{n+1}^T = f\left[x_n^T b(k)\right] = 1 - 2\left[x_n^T b(k)\right]^2 = 1 - 2\left(x_n^T\right)^2 \tag{4.57}$$

$$x_{n+1}^R = f\left(x_n^R\right) = 1 - 2\left[x_n^T b(k)\right]^2 = 1 - 2\left(x_n^T\right)^2. \tag{4.58}$$

It is clear to see that iteration arithmetic of Expression (4.56) is implemented, and the transmitting and receiving ends are totally synchronized.

4.7.3.3 Differential Chaos Shift Keying (DCSK) Scheme

In DCSK, a simple differential coherent receiving technique can be used to carry out demodulation, demodulation reference signals produced locally by chaotic synchronization circuit is not needed.

The block diagram of DCSK signal and modulation processing is as shown in Fig. 4.33. The front section of signals is for reference, and the back section is for information bearing. Information comprises symbols of “0” or “1”. In the information bearing section, the chaotic sequence is added by the symbol “modulo-two”

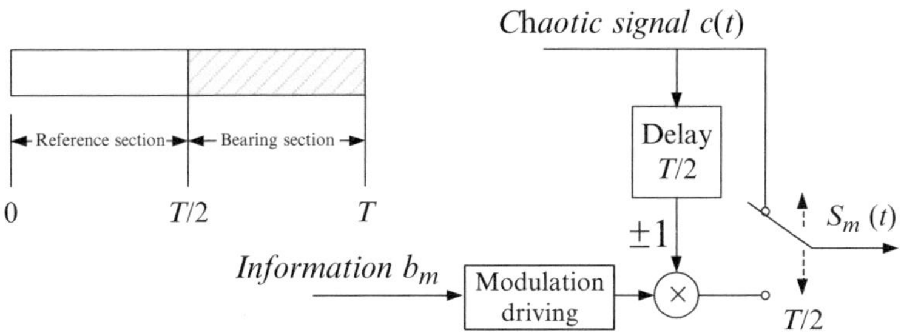

Fig. 4.33 Block diagram of DCSK signals and modulation processing

(spread spectrum); and according to whether the transmission symbol is "0" or "1", the output will be respectively either in-phase or out-phase with respect to the reference signal.

The modulation driving turns input symbols into different polarity and outputs them, and at $t=T/2$, it drives the switch to transmit corresponding signals of the information bearing section.

In the case the symbol is "+1" or "−1", transmitted signals are separately expressed as:

$$\text{“}+1\text{”}: \quad s(t)=\begin{cases} +c(t) & 0\le t<T/2 \\ +c(t-T/2) & T/2\le t<T \end{cases} \tag{4.59}$$

$$\text{“}-1\text{”}: \quad s(t)=\begin{cases} +c(t) & 0\le t<T/2 \\ -c(t-T/2) & T/2\le t<T \end{cases}. \tag{4.60}$$

In the above two expressions, $c(t)$ means chaotic synchronization signals.

At the receiving end, the differential coherent demodulation method is used to correlate the reference signal section with the information bearing section, as shown in Fig. 4.34. At first received reference signals are delayed for half symbol period, then they are correlated with the information bearing section to obtain decision observations z_m.

$$z_m=\int_{T/2}^{T} s(t)s(t-T/2)\mathrm{d}t-\begin{cases} \int_{T/2}^{T} c^2(t)\mathrm{d}t>0, & \textit{if } \text{“}+1\text{” is transmitted} \\ -\int_{T/2}^{T} c^2(t)\mathrm{d}t<0, & \textit{if } \text{“}-1\text{” is transmitted} \end{cases}. \tag{4.61}$$

In the above expression, $c^2(t)>0$.

In chaotic digital modulation using chaotic signals as carriers, since chaotic signals are non-periodic, observations z_m approaches the mean value $E_b/2$ as integral time increases only in statistical sense (E_b is the average energy of symbol period chaotic signals). For each symbol, z_m varies. Such change of observations

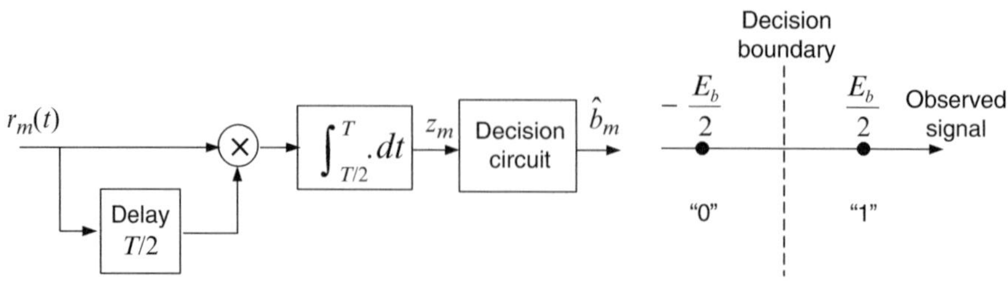

Fig. 4.34 Principle block diagram of DCSK demodulation processing

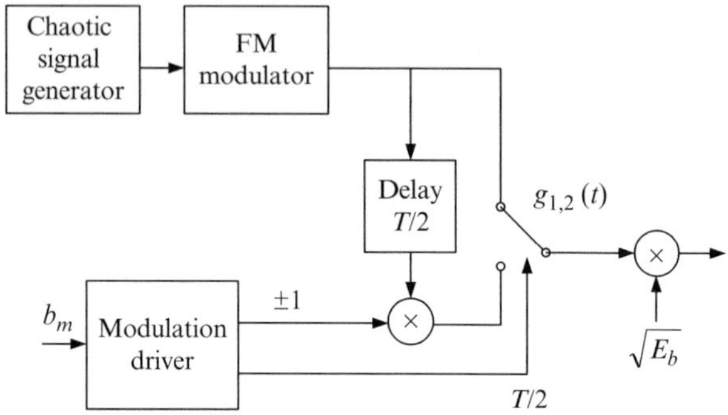

Fig. 4.35 Structure block diagram of FM-DCSK modulation

introduces variance accumulation of observations, resulting into increased bit error rate. To overcome such change of symbol energy, signals with constant power can be used as carriers to change waveforms of chaotic signals into those with chaotic phase and constant power, and FM-DCSK modulation is used here, as shown in Fig. 4.35.

The bit error rate formula in FM-DCSK modulation is:

$$\mathrm{BER} = \frac{1}{2}\mathrm{erfc}\left(\frac{|E[z_m]|}{\sqrt{\mathrm{Var}[z_m]}}\right) = \frac{1}{2}\mathrm{erfc}\left(\sqrt{\frac{1}{4\frac{N_0}{E_b} + 4L\left(\frac{N_0}{E_b}\right)^2}}\right), \tag{4.62}$$

where L is the number of reference section signals transmitted in each signal period.

The advantage of the above-mentioned DCSK is that it is easy to implement, but once $T/2$ is reconnoitered, information symbols can be demodulated. Therefore, it is not a good scheme in terms of confidentiality and anti-interception.

4.7.3.4 A Ranging Scheme for the Infinite Chaotic Sequence

When the chaotic sequence is used to carry out modulations to the suppressed carrier or the residual carrier, the carrier recovery loop or the residual carrier phase-locked loop can still be used for carrying out Doppler velocity measurement. Direct phase discrimination with sum and difference signals can be used to implement single pulse angle tracking. What is different is that an infinite chaotic sequence is non-periodic, that is the reason that the code loop cannot be used for ranging. One possible solution is the "open loop measurement" method, in which coherent integration takes the place of code loop filtering. One possible way to implement this method is to adopt an approach that is similar to "open loop ranging" in "sequence ranging" mode used in deep space TT&C. Namely, to obtain a set of ranging data at each transmission of a section of chaotic sequence, and to get K sets of ranging data in K times of periodic circulated transmission. Strictly speaking, it is a kind of discrete ranging data, but it is generally acceptable in ranging. A circulation period depends on data rate requirements and the unambiguous range; and the length of a chaotic sequence is dependent on ranging accuracy, namely, the integral time T. A section of a chaotic sequence is truncated to be the "ranging code"; and the delay between "transmitted ranging codes" (amounting to transmit "1") and "received ranging codes" (amounting to receive "1") is used for ranging. Such delay can be measured by a matching filter which is used to obtain optimum SNR and to conduct coherent integral filtering. This kind of filtering is in place of loop filtering to reduce the effect of thermal noises to capture and ranging accuracy.

Such scheme is shown in Fig. 4.36 [7]: First, the transmitting end constantly sends the modulated chaotic sequence. Then when each ranging datum is needed, a code selection pulse is used to choose N chaotic codes as "transmitted ranging codes"; and at each interval of M transmitted chaotic codes, the first N of such M codes are stored for one time of ranging. Therefore, its ranging data are discrete. It is similar to sequence ranging in deep space ranging where the matching filter structure is applied to capture such N returned chaotic codes to obtain the coarse range.

Its operating procedure is described as follows: Chaotic sequences are produced continuously driven by the code clock, and after binaryzation they generate binary sequences with amplitude being "+1" or "−1". The code clock produces a code selection pulse through the code selection pulse generator circuit, as shown in Fig. 4.37 below. Width of such a code selection pulse is $T_B = NT_c$, its period $T_M = MT_c$, generally $M \gg N$, and M is determined by the unambiguous range. The falling edge of the code selection pulse latches the shift register, in this way, the ranging sequence comprising N chaotic codes contained in the code selection pulse at every period is latched (namely, only filtering of selected N codes is matched, rather than all codes). The ranging sequence is stored once again at every period T_M. N chaotic codes in storage are used for matched receiving. The falling edge of such code selection pulse is used to trigger the T flip-flop, then it triggers the T flip-flop again through the NOT gate. Output of such T flip-flop is as shown in Fig. 4.37 below. At first chaotic signals are modulated and transmitted, then they reach the ranging station after having been forwarded by the target (transparent forwarding). When all transmitted chaotic codes (the same as the stored N chaotic codes) arrive at the receiving shift register, relevant peak values

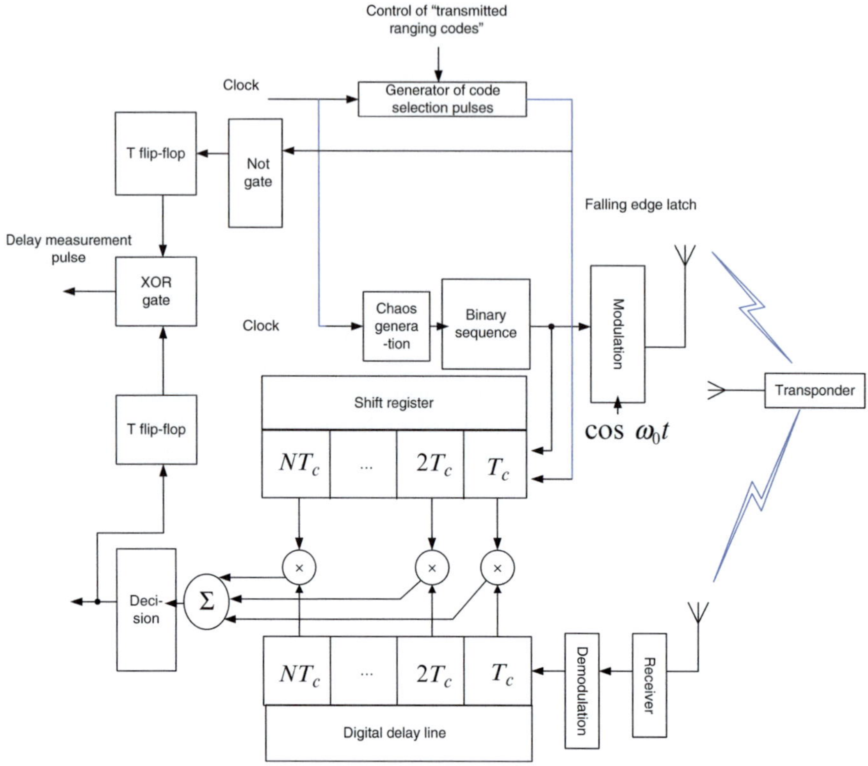

Fig. 4.36 Capture and coarse ranging scheme for infinite chaotic sequences

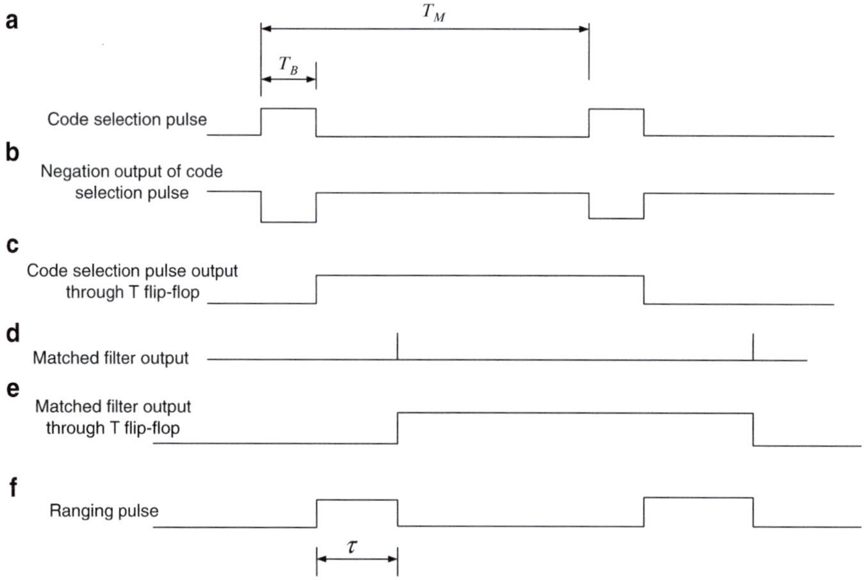

Fig. 4.37 Sketch map of ranging waveforms of an infinite chaotic sequence

are gained. As shown in Fig. 4.37 below. Such related peak value corresponds to triggering the T flip-flop, whose output waveform is as shown in Fig. 4.37 below. Output of the two T flip-flops is processed by XOR to get a period of delay measuring data τ, as shown in Fig. 4.37 below. K delay measuring data can be gained through K periodic circulated transmission to implement "sequence ranging". In such ranging method with matched filters, transmitted chaotic ranging codes are stored, and such a priori information is fully used in related tests. As a result, testing received quasi-random chaotic ranging codes can be implemented. However, the reconnaissance performer is without such information, equipping this scheme with anti-interception performance and confidentiality.

Under the condition of 1 detection probability, the shortest capture time is NT_c, the longest capture time is 2 NT_c, and the average capture time is 1.5 NT_c. However, in such application case, there is no situation under which receiving starts from within the codes. For such reason, the average capture time is NT_c.

In an ordinary range capture, for a digital matched filter, the worst range resolution is $T_c/2$. Its corresponding ranging resolution is:

$$d_e = \frac{1}{2} c \frac{1}{2} T_c = \frac{1}{4} c\, T_c. \tag{4.63}$$

It is plain that ranging accuracy of this scheme is not desirable. One improved scheme combines "coarse ranging" and "fine ranging", in which "coarse ranging" means the above-mentioned open loop "matched filter method". Namely, the range delay chip number at capture is used as "coarse range", and the error properties showed by chaotic codes are utilized to obtain "fine range", whose principle is as shown in Fig. 4.38.

In Fig. 4.38, early codes and late codes are unnecessary at capture. They are only needed in fine ranging. An instant code is needed at capture (i.e., the middle code

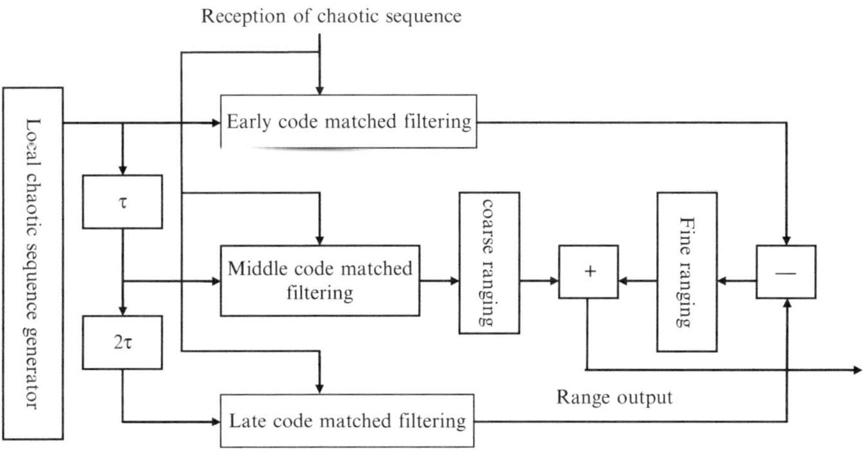

Fig. 4.38 Principle block diagram of chaotic sequence open loop ranging

aimed at received chaotic sequence), which produces the largest correlation value at alignment. For that reason, it is used to implement capture, and code clock counting is used to implement coarse ranging.

As bandwidth of a receiver is limited, its amplitude/phase frequency properties make the correlated triangle asymmetrical. As every edge of such a triangle is not an ideal deflected ascent or descent, this issue will cause shift of the discriminator zero cross point under some conditions, leading to ranging error. At the same time, noises will jitter its zero cross point, reduce the gradient, and weaken error identification properties, which will impose adverse influence on ranging accuracy.

References

1. Tian Ricai (2007) Spread spectrum communication. Tsinghua University Press, Beijing
2. Ziemer RE et al (2005) Introduction to digital communication (trans: Yin Changchuan, Hao Jianjun). China Machine Press, Beijing
3. Kaplan ED (2007) Understanding GPS: principles and applications (trans: Qiu Zhihe, Wang Wanyi). Publishing House of Electronics Industry, Beijing
4. Liu Jiaxing (2010) Spacecraft C&T engineering. National Defence Industry Press, Beijing
5. Liu Jiaxing, Wen Ji (2009) Conception for a Ka-band chaotic spread spectrum TT&C system [J]. Telecommun Eng 49(5):33–37
6. Liu Jiaxing, He Shibiao (2011) Technical considerations on an integrated space-air-ground TT&C and communication network for air and space vehicles [J]. J Spacecraft TT&C Technol 30(1):1–5
7. He Shibiao et al (2009) Applications of chaotic sequences in C&T. J Chongqing Commun Inst

Chapter 5
Special Issues on Radio Transmission Channel in C&T

The channel referred to in this chapter is the medium for information transmission. Radio transmission channel refers to the channel transmitting electromagnetic signals in free space (in a broad sense, it may include related components and links). Radio transmission channel may bring in many adverse impacts on C&T signals, such as orbit determination error and signal attenuation caused by radio wave transmission characteristics, multipath interference caused by reflected wave, various types of noise and interference introduced by the channel, large Doppler frequency shift in high-dynamic radio channel, and the non-real time caused by the channel with time delay. The impacts of the last three transmission characteristics have been addressed in the early chapters.

The impact of radio transmission channel on C&T varies with its operating frequency band, so the selection of operating frequency band for TT&C is the first issue in the design of C&T system.

5.1 C&T Frequency Band Developing to Ka-Band and Optical Bands

5.1.1 Principle for Selecting C&T Frequency Band [1]

Restricted by many factors, TT&C frequency band should be selected in light of the regulation's restrictions applicable to the frequency, propagation characteristic, noise effect, interference from other systems, etc. Main factors to be considered include:

(1) Compliance with SRRC and ITU regulations. If there is no express provision, an application should be submitted to SRRC for approval. This is a regulatory restriction. The spacecraft has a borderless flight, so compliance with ITU

J. Liu, *Spacecraft TT&C and Information Transmission Theory and Technologies*,
Springer Aerospace Technology, DOI 10.1007/978-3-662-43865-7_5

Table 5.1 Allocation of radio frequency bands

S/N	Frequency band name	Frequency range	Wave band name		Wavelength range
1.	Extremely low frequency (ELF)	3–30 Hz	Extremely long wave		100 Mm–10 Mm
2.	Super low frequency (SLF)	30–300 Hz	Superlong wave		10 Mm–1 Mm
3.	Ultralow frequency (ULF)	300–3,000 Hz	Ultra long wave		1,000 km–100 km
4.	Very low frequency (VLF)	3–30 kHz	Very long wave (myriametric wave)		100 km–10 km
5.	Low frequency (LF)	30–300 kHz	Long wave (kilometric wave)		10 km–1 km
6.	Medium frequency (MF)	300–3,000 kHz	Medium wave (hectometric wave)		1,000 m–100 m
7.	High frequency (HF)	3–30 MHz	Short wave (decametric wave)		100 m–10 m
8.	Very high frequency (VHF)	30–300 MHz	Ultrashort wave (metric wave)		10 m–1 m
9.	Ultrahigh frequency (UHF)	300–3,000 MHz	Microwave	Decimeter wave	100 cm–10 cm
10.	Superhigh frequency (SHF)	3 Hz–30 GHz		Centimeter wave	10 cm–1 cm
11.	Extremely high frequency (EHF)	30 Hz–300 GHz		Millimeter wave	10 mm–1 mm
12.	Tremendously high frequency (THF)	300 Hz–3,000 GHz		Decimillimeter wave	1 mm–0.1 mm

Note: The frequency range includes upper limit but excludes lower upper; the wavelength range includes lower limit but excludes upper limit

regulations on frequency band allocation can protect against RF interference and is helpful for international cooperation. Moreover, the selection of frequency band should meet demands of aerospace industry development and combine the frequency setting suggestions in the development planning of this discipline.

(2) Radio wave propagation characteristic. Radio wave at different frequency will have different physical phenomena in propagation via different media (such as ground, water, troposphere, and ionosphere). Radio wave band and corresponding frequency band can be allocated in many methods. According to the allocation method referred to in the Encyclopedia of China, 12 frequency bands are allocated, as listed in Table 5.1.

When radio wave propagation space is clear of any refractive, reflective, scattering, or absorbing substances, it is deemed that radio waves are propagated in free space. Energy dispersion loss (also called path loss L_r) is proportional to the square of propagation distance and inversely proportional to the square of wavelength. However, not all of waves of TT&C system are propagated in free space. Waves may be absorbed by free electrons and ions in the ionosphere and absorbed and scattered by oxygen molecules, water vapor molecules, cloud, fog, rain, and snow in the troposphere, thus leading to a loss. Such loss is closely related to radio wave frequency, beam elevation, and weather conditions. In microwave band, the higher the frequency is, the larger the attenuation caused by cloud, fog, rain, and snow will be. Atmospheric loss under the frequency less than 1,000 MHz mainly results from absorption loss of ionosphere. The lower the frequency is, the more the loss will be. In space propagation, radio wave may be affected by radio noise from the atmosphere, cosmos, and the Sun. In a word, radio wave at a frequency band of 300 MHz to 10 GHz has less atmospheric loss and is applicable to radio wave propagation through the atmosphere.

Atmospheric refraction error is another factor to be considered when selecting the frequency band. The change of radio wave refraction index in troposphere makes the radio wave propagation velocity different from the velocity of light and results in ray bending and velocity-measuring and ranging error. When the frequency is lower than 100 GHz, the troposphere can be deemed to be nondispersive medium and the refraction index does not vary with the frequency; the ionosphere is dispersive medium and velocity-measuring and ranging error is inversely proportional to the square of frequency. So the propagation error may decrease with the increase of the frequency.

(3) Blackout and flame attenuation. When the flight vehicle moves at a speed higher than dozen Mach in the atmosphere, the plasma sheath will be formed (e.g., missile reentry and near space vehicle). When passing through this sheath, radio signal will be seriously attenuated and signal phase will fluctuate. Rocket motor exhaust plume is also a kind of uneven plasma with electron concentration and collision frequency much higher than those in the ionosphere. In upper air, a plasma zone several times larger than the missile body will be formed. The effect of exhaust plume on radio signals is related to motor power, fuel type, and length and position of passing through the flame zone as well as the operating frequency of radio signal. The higher the frequency is, the less the effect will be. In C-band, the exhaust plume generally attenuates radio signals by 5–10 dB. In P-band, attenuation may reach 20 dB or so. Solid-fuel motor exhaust plume attenuates radio signal more. Higher frequency (e.g., Ka-band) cannot only reduce the effect of the exhaust plume but also improve the capability of penetrating plasma zone.
(4) Effect of different frequency bands on the equipment performance, such as data transmission rate, operating distance, measurement accuracy, and anti-interference capability. Both high reliability and real-time TT&C operation and high-speed data transmission of sharply increased mission data should be met. If data transmission rate is higher, a higher operating frequency band should be selected.
(5) Manufacturing level of HF devices and equipment. In order to meet certain tracking distance, operating frequency and antenna size should be increased to improve antenna gain. However, the increase of operating frequency and antenna size may increase the cost and be restricted by the level of processing. Moreover, the performance level of HF high-power transistor and HF low-noise amplifying and receive devices would also become a limiting factor for improving operating frequency.

(6) The selection of TT&C frequency band is also limited by such factors as miniaturization level of missile-borne equipment, tracking performance, and carrier frequency bandwidth of ground equipment. TT&C frequency band should coordinate with spacecraft transponder. In order to simplify spacecraft equipment, multiple TT&C systems should use the same type of transponder.
(7) Electromagnetic compatibility. It should be electromagnetically compatible with other equipment in the system, that is, it has the capability of neither bringing interference to others nor being interfered by others. The unification of frequency band with the Earth observation satellite services should be taken into account.
(8) Economic factors such as equipment cost.

With the development of TT&C and information transmission technology, higher and higher requirements are imposed on the information transmission rate, orbit determination accuracy requirement and anti-interference capability, especially future requirements for high-speed data transmission above 1 Gb/s. This makes the improvement of operating frequency band imperative. Under such a development situation, the International Telecommunications Union (ITU) makes a new allotment of TT&C and information transmission frequency bands.

The ITU Radio Regulations provide corresponding frequency band for different services of the spacecraft. TT&C often uses the frequency band of space operation and space research service. The following frequency bands are provided for near Earth (2×10^6 km below) and deep space (2×10^6 km above), as listed in Table 5.2.

5.1.2 Development Trend

Development toward millimeter wave and optical bands is the overall trends. In the space segment, there is no atmospheric effect. Therefore, optical communication has great advantages. However, in the space-ground segment, due to atmospheric loss, some issues are to be solved. Millimeter wave is less affected by atmosphere, especially Ka-band at the lower end of millimeter wave. At present, Ka-band has been in practical stage. This book will focus on Ka-band C&T technology.

Table 5.2 Space research frequency bands allocated by ITU

	Deep space		Near Earth	
Frequency band	Uplink (GHz)	Downlink (GHz)	Uplink (GHz)	Downlink (GHz)
S	2.11–2.12	2.29–2.3	2.025–2.11	2.2–2.29
X	7.145–7.19	8.4–8.45	7.19–7.235	8.45–8.5①
Ka	34.2–34.7	31.8–32.3	22.55–23.55③	25.5–27②

① Near-Earth exploration service band: 8.025–8.4 GHz
② TDRS satellite inter-satellite backward link band, can be used for downlink for near-Earth C&T
③ TDRS inter-satellite forward link band, which is applied by US to ITU for near-Earth C&T downlink

5.1.2.1 Increase to Ka-Band

Due to a long C&T distance, increasing strength of signals arriving at the Earth is an eternal topic for the downlink channel. In theory, when antenna aperture, noise temperature, and transmitting power of the space/ground system are constant, the receiving level on ground will be increased by 20Log*N* by increasing frequency by *N* times. This effect is quite attractive. But there are several factors restricting this advantage. First, rainfall and water vapor in the Earth atmosphere would significantly attenuate Ka-band; second, the surface accuracy of Ka-band antenna must be very high; third, due to narrow beam, antenna pointing should be more accurate, so angle acquisition is more difficult; fourth, compared with lower frequency bands, Ka-band front-end LNA noise is high and transmitter power is low; and fifth, phase noise increase will affect the data transmission. However, with the technological development, now more efficient spacecraft solid-state power amplifier (SSPA) has been produced. Waveguide beam antenna used in the ground station makes antenna pointing more accurate and improves antenna efficiency. The noise temperature of Inp HEMT cryogenic low-noise amplifier has lowered down to 11 K, which has become a very minor contribution to system noise temperature. These improvements give more obvious advantages of Ka-band.

The biggest problem of Ka-band is rain attenuation. A better solution is to establish two or more stations at different place to realize space diversity for preventing rain attenuation. These stations can also be used to arrange antenna arrays and conduct interferometer angle measurement. The station should be located at an area of low precipitation (decrease rain attenuation and improve availability), high altitude (reduce atmospheric loss in Ka-band), and low environmental noise (reduce external interference). In addition, some rain attenuation compensation technology should be used, for example, automatic power compensation and automatic adjustment of code rate and error correction coding method, etc.

The design idea of uplink channel is different from that of downlink channel. In order to ensure command reliability, receiving antennas of a probe should be wide beam. Under the condition of a fixed beam width, if the frequency increases, the gain of received signals could not be obtained. Uplink band of deep-space TT&C still uses S-band and X-band, because they have small rain attenuation and wide beam. In this scheme, probe receiving level is improved by increasing ground transmitting power. In the ground station, the use of a superpower transmitter is available and will not be limited by size or weight like the space probe.

5.1.2.2 Optical Communication Technology

According to the ideas of increasing operating frequency band, the further development is to use optical communication. The information is transmitted via laser or telescope. In such case, optical beam is very narrow, energy is more centralized, and the frequency band is wider. Therefore, transmitted information rate is higher, and optical communication equipment on the flight vehicle can have smaller size. However, optical communication has a disadvantage of large atmospheric loss.

Future deep-space exploration will adopt the method of constellation and networked communication. Interstellar scheme is also under progress, that is, a relay satellite is placed in lunar orbit, on which an optical telescope is used to receive optical signals from the deep-space probe and then these signals will be retransmitted to the ground deep-space station. Since there is no atmospheric attenuation here, C&T capability in optical band will not be affected. Besides the Moon, Mars and the other planets can also be used to constitute a planetary domain network.

Ground terminal of space-ground link in optical band can be either ground-based or space-borne. Currently, ground-based solution is preferable. Several 10 m-diameter telescopes are used to receive deep-space signals to realize space diversity against atmospheric attenuation. Furthermore, performance requirements for optical communication telescope are much lower than those for imaging telescope. Thus, the cost is much lower. In a direct detection method with pulse code modulation, it is only to determine the time of arrival of photons.

In the ground-based solution, the telescope can be deployed in two methods. One is to arrange 6–8 optical telescopes at an equal interval around the Earth; the other is to arrange three 10 m-diameter optical telescopes in each station, three stations in total.

The space-borne solution is to deploy optical telescopes in the low Earth orbit (LEO) and high Earth orbit (HEO). Since the station is distributed in space, atmospheric attenuation decreases and the diameter of the optical telescope can decrease to about 7 m. However, compared with the ground-based station, the cost of space-borne station is much higher, and the space-borne station can only support one target at the same time.

5.1.3 Background of Developing Ka-Band C&T System [2]

Currently, the demands to Ka-band C&T system mainly come from following eight aspects:

(1) Requirements for high-speed data transmission and high orbit determination accuracy of the Earth-orbiting spacecraft:

By using Ka-band, the available frequency band gets broadened, allowing for data transmission at a higher rate. In comparison with the "elbow" transmission of space-based TDRSS, the transmission path loss of space-to-ground "through" Ka-band ground C&T system will be reduced by about $20\log\frac{40,000\ \text{Km}}{400\ \text{Km}} = 40\,\text{dB}$, and the gain of ground 12 m antenna will be $20\log\frac{12\ \text{m}}{4.9\ \text{m}} = 8\,\text{dB}$ higher than that of TDRS 4.9 m antenna. Hence, the gain of space-to-ground "through" system will be 48 dB higher than that of the "elbow" system. Moreover, the use of Ka-band can also improve velocity-measuring accuracy and angle-measuring accuracy.

(2) Requirements for high-resolution Earth observation and remote-sensing reconnaissance. The data transmission rate for these applications is required to reach several Gb/s. Such a high data transmission rate could not be accomplished even by future's upgraded TDRSS. So Ka-band C&T system in the space-to-ground "through" transmission method is to be used. For the sake of confidentiality in military and reconnaissance applications, a store-and-forward method is often used, with the stored data being retransmitted in a high data rate when the flight vehicle is flying over own territorial air space. In civil and scientific research, Ka-band space-to-ground "through" ground C&T system is more suitable for high-resolution observation of own territory.

(3) Requirements for interconnection, intercommunication, and complementary advantages of space-borne/ground-based networks. TDRSS space-borne C&T system has many advantages including high-coverage real-time transmission and multi-target C&T. However, with a sharp increase of land, sea, air, and space users, resources provided by TDRSS are obviously insufficient. In order to cover the shortage of TDRSS and meet requirements of lots of high-speed data transmission users, it is required to establish the Ka-band C&T network and gradually adapt to the requirements of space-to-ground integrated C&T network development.

(4) Requirements for deep-space C&T. Under the condition that the apertures of satellite antenna and ground antenna are not changed, by increasing the operating frequency band, unidirectional antenna gain can get increased by $({}^{c}/_{11})$, and so deep-space detection range is increased. Furthermore, the increase of the operating frequency band can improve velocity-measuring and angle-measuring accuracy and reduce the effects of ionosphere and the Sun plasma zone, thus increasing orbit determination accuracy.

(5) For solving the "blackout" issue in the C&T of near-space flight vehicle, reentry vehicle, and antimissile system, the operating frequency band needs to be increased to Ka-band or above.

(6) Booming Ka-band communication satellite requires the support of Ka-band C&T system.

(7) Requirements from military TT&C. The current USB C&T network is a civil network in the international standard system. S-band is an internationally used TT&C band. It is poor in security protection and is incapable of anti-interference and anti-interception. In order to guarantee C&T security of military satellites and other flight vehicles, new explorations must be performed in carrier frequency and modulation scheme. Ka-band is a choice better than S-band.

(8) Inter-satellite link requirement. Due to no atmosphere effect now, inter-satellite link in Ka-band is used increasingly and is developed toward to V-band.

5.1.4 Characteristics of Ka-Band C&T System

Ka-band C&T system is characterized by:

(1) Wide frequency band with data transmission rate of 1.6 Gb/s above

The frequency band allocated for Ka-band C&T is 25.5–27 GHz, with the bandwidth of 1.5 GHz. Even if transmission bandwidth B is to be selected based on 1.5 R_c (R_c is transmitted channel symbol rate. Generally, the bandwidth is equal to 1.2–1.5 times of such rate), for BPSK modulation, transmissible symbol rate is 1 Gb/s; for QPSK, the rate is 2 Gb/s. Moreover, the increase of carrier frequency f_0 decreases relative bandwidth B/f_0, so in the same B, phase-frequency characteristic linearity and amplitude-frequency characteristic uniformity of Ka-band RF front-end are improved to help reduce the bit error rate of data transmission.

(2) Higher orbit determination accuracy

1) Velocity-measuring error. The velocity-measuring error $\delta_{\dot{R}n}$ caused by thermal noise shall be determined by the following expression:

$$\delta_{\dot{R}n} = \frac{c}{4\pi f_0 \tau}\sqrt{\frac{N_0 B_L}{P_S}} \quad \text{(in case of residual carrier velocity measurement)}$$

$$\delta_{\dot{R}n} = \frac{c}{4\pi f_0 \tau}\sqrt{\frac{N_0 B_L}{P_S}\left(1 + \frac{1}{2T(P_s/N_0)}\right)} \quad \text{(in case of spread spectrum TT\&C)} \tag{5.1}$$

where f_0 is carrier frequency, τ is integration time, B_L is PLL bandwidth, and P_s/N_0 is signal-noise power spectral density ratio under thermal noise.

It is obvious that f_0 increase can efficiently reduce the velocity-measuring error caused by thermal noise. The physical cause is that the increase of f_0 increases corresponding Doppler frequency offset, equivalently, increase speed "signal," thus improving velocity-measuring SNR.

2) Angle-measuring error. Angle-measuring error $\delta_{\theta n}$ caused by thermal noise shall be determined by the following expression:

$$\delta_{\theta n} = \frac{\theta_{3\,\text{dB}}}{K_m\sqrt{\frac{S}{N}\cdot\frac{Bn}{\beta n}}} = \frac{70\lambda}{DK_m\sqrt{\frac{S}{N}\cdot\frac{Bn}{\beta n}}} \tag{5.2}$$

where λ is the operating wavelength; K_m is normalized difference slope, generally 1.2–1.6; D is antenna aperture; Bn is IF equivalent noise bandwidth; S/N is sum channel signal-noise ratio within the bandwidth Bn; and βn is servo bandwidth.

It is obvious that the higher the operating frequency, the less the angle-measuring error caused by thermal noise. The physical cause is that the narrower the beam, the higher the sensitivity of antenna to angle error, that is, angle error signal is increased, thus improving angle-measuring *S*/*N*.

Angle-measuring error caused by multipath effect in the elevation branch is

$$\delta_m = \frac{70\lambda\rho}{DKm\sqrt{2(G_\Sigma/G_{s2})_{rms}}} \quad (5.3)$$

where ρ is multipath reflection coefficient, G_Σ is sum beam gain, and G_{s2} is difference beam gain when deviating from the electric axis twice elevation.

When D is invariant, the frequency increases, the beam will become narrow. Accordingly, side lobe value G_{s2} deviating from twice of elevation becomes smaller. It is obvious that the increase of frequency will decrease λ and increase G_Σ/G_{s2}. Therefore, angle-measuring error caused by multipath reflection is reduced.

3) Ranging error. Ranging error shall be determined by ranging signal. When the multitone ranging system is used, the error is determined by fine tones; when pseudocode and spread-spectrum ranging is used, it is determined by the code clock; when carrier signal ranging is used, it is determined by carrier signals. If f_R is used to represent the said ranging frequency, the ranging error δ_{Rn} caused by thermal noise shall be determined by the following expression:

$$\begin{aligned} \delta_{Rn} &= \frac{C}{18f_R}\sqrt{\left(\frac{N}{S}\right)_{B_L}} \quad \text{(in case of ranging tone)} \\ \delta_{Rn} &= \Delta\sqrt{\frac{B_L(1+2/T(S/N_0))}{2S/N_0}} \quad \text{(in case of spread spectrum TT\&C)} \end{aligned} \quad (5.4)$$

where B_L is equivalent noise bandwidth of ranging signal PLL, $(S/N)_{B_L}$ is equivalent signal-noise ratio in B_L, and T_c is symbol width, namely, $T_c = 1/f_R$°

It can be seen from Expression (5.4) that when modulation signal ranging is used, thermal noise random error is unrelated to carrier frequency. Then, an important measure to improve ranging accuracy is to use a high f_R. However, if high-accuracy carrier ranging is used, the higher the carrier, the less the ranging error by thermal noise.

4) Decrease of radio wave refraction error. Ranging value including ionospheric refraction is

$$R_e = R_a + (A_1/f^2) + (A_2/f^3) + (A_3/f^4) + \cdots \quad (5.5)$$

where the second item and the following refer to ranging error induced by ionospheric refraction; A_1, A_2, A_3, ... refer to ionospheric structure coefficient;

and f is operating frequency. It can be seen that the increase of operating frequency can reduce this error. For velocity measuring and angle measuring, similar results are also obtained. It is possible to increase operating frequency to reduce refraction error.

(3) Higher S/N

In theory, when the aperture, noise temperature, and transmitting power of the space-ground system are invariant, ground receiving level will increase by 20LgN by increasing frequency of N times. This result is quite attractive. By collecting statistics on effects of weather factors on G/T value, documents showed that compared with optimized X-band link, optimized Ka-band link can obtain 5.9–7.2 dB average gain. S-band is even 11 dB higher than that of X-band.

(4) Wideband frequency hopping, anti-interference, and anti-interception capabilities

Since the bandwidth of operating frequency band is up to 1.5 GHz, hopping in such a wide band is available. This greatly increases anti-interference, anti-interception, anti-multipath, and anti-flicker capabilities. $N = \Delta f/B, P_e = J/N$, where Δf is available operating bandwidth, B is information bandwidth, N is the number of hopping points, J is the number of hopping points where interference power is greater than signal power, and P_e is the error probability of hopping system without an increase of redundancy. It is obvious that the increase of Δf can increase N and thus reduce the error probability of transmitted information to improve system anti-interference performance.

If the operating bandwidth increases, the frequency of spread-spectrum code clock can be increased to realize wideband direct spread, gain a very high spread-spectrum gain, enhance anti-interference capability, and improve ranging accuracy.

At present, S-, X-, and Ku-bands are overcrowded and are difficult to provide a wide operating bandwidth, so we should transfer to Ka-band.

(5) High anti-interference capability

It has narrow beam and better space selectivity to improve anti-interference capability.

Beam width $\theta_{3\ \mathrm{dB}} = 70\lambda/D$, where D is antenna aperture. With respect to a same antenna aperture D, the higher the operating band, the narrower the beam. When the operating band increases from 2.1 GHz in S-band to 26 GHz in Ka-band, the beam will narrow 12.4 times, greatly reducing the susceptibility to interference in other directions and anti-multipath interference. On the other hand, beam narrowing will lead to the difficulty in acquisition and tracking of the angle-tracking system.

(6) Smaller antenna used in the ground station

Vehicular minimum-sized station is designed for more mobility and concealment to help anti-destruction and enhance survivability.

As mentioned above, since S/N is increased, it is possible to reduce antenna aperture in the ground station.

It is known that antenna gain is determined by $G = \eta(\pi D/\lambda)^2$. When G is invariant, a decrease of λ can decrease D in the same times. In practical design, it is required to consider the increase of path loss caused by the decrease of λ. Therefore, antenna aperture in the ground station should be reduced in integrated space-ground design.

(7) Compatibility of ground-based network and space-based network (TDRSS)

Due to the same band being used both in ground-based C&T network and the space-space link of TDRSS, their compatibility is possible. Compatibility involves three aspects:

1) Systems are identical in operation system, signal format, and operating band.
2) Compatibility with spacecraft transponder. Besides the design of same operating frequency (S/Ka dual mode), the antenna should be compatible with two operating modes, namely, "pointing up to TDRSS" and "pointing down to TT&C ground station". Antenna switch may be used.
3) Ground C&T station and TDRSS ground terminal station implement partial compatibility. A same IF (e.g., 1.2 GHz) should be used in the design. If the condition allows, data can be transferred between them via fiber cable, and can be processed in the same baseband device.

(8) Reducing size and weight of the transponder

Using Ka-band, the size and the weight of waveguide are greatly reduced, compared with those of S-band, which is helpful for developing compact and light weight transponder.

(9) Better EMC

Ka-band is a new band. Electronic equipment in this band is relatively few. It is less likely to be disturbed by other equipment or interfere with other equipment (allowable flux density as specified by ITU is a bigger value). The power of jammers in this band is also relatively low.

(10) Reduction of "blackout" effect for near-space vehicle and reentry TT&C

"Near-space vehicle" refers to a flight vehicle at an altitude of 20–100 km. When a high-speed vehicle flies at such an altitude and reenters the atmosphere, the plasma sheath will be formed to attenuate and refract radio waves. The attenuation constant of radio wave propagation is

$$\alpha = \frac{8.68\omega}{c\sqrt{2}}\left[\sqrt{\left(1-\frac{\omega_p^2}{\omega^2+\nu^2}\right)^2+\left(\frac{\omega_p^2}{\omega^2+\nu^2}\cdot\frac{\nu}{\omega}\right)^2}-\left(1-\frac{\omega_p^2}{\omega^2+\nu^2}\right)\right]^{\frac{1}{2}} (\mathrm{dB/m}) \tag{5.6}$$

where α is an attenuation constant (dB/m), ω is angular frequency of radio wave (rad/s), c is the velocity of light, ω_p is angular frequency of plasma (rad/s), and v is plasma collision frequency (1/s).

The integral of α on plasma thickness in radio wave transmission direction is the overall attenuation.

It can be seen that the increase of operating frequency ω can reduce the attenuation.

5.1.5 Main Technical Issues of Ka-Band C&T System

(1) Rain attenuation

The biggest problem for the use of Ka-band is rain attenuation. Currently, some countermeasures against rain attenuation have been taken for Ka-band satellite communication, Ka-band TDRSS, and Ka-band deep-space TT&C to make these systems for practical applications. In contrast, Ka-band TT&C system is less impacted by rain attenuation, because the ground TT&C station spends a very short time (about 20 min) for each TT&C of spacecraft in near-Earth orbit and can only cover a very short section of the orbit. Overall geometric coverage should be implemented by multiple stations. So rain attenuation effect on one station gives a little impact on overall coverage. For example, when 12 stations are used to implement 14 % overall coverage, if average rain attenuation availability of each station is 99 %, overall coverage will only decrease to $99\% \times 14\% = 13.86\ \%$. It is obvious that the effect is very small.

Some calculation results of Ka-band rain attenuation are listed in Table 5.3.

Overall, advantage of Ka-band overrides the disadvantage brought by rain attenuation in Ka-band. Therefore, many applications such as wideband satellite communication are using Ka-band increasingly.

(2) Angle acquisition and tracking of narrow-beam antenna

When the frequency increases to Ka-band, beam will become narrow. In such case, to acquire and track high-dynamic target in low orbit will be a new technological difficulty. When the frequency increases to Ka-band, machining accuracy of antenna surface should be higher. The effect of antenna surface accuracy on angle-measuring accuracy is also an issue to be paid attention on.

The acquisition of narrow-beam antenna has several solutions:

1) Wide-beam guidance and narrow-beam tracking.
2) Improve spacecraft orbit forecast accuracy and antenna pointing accuracy and gradually correct mathematical model of antenna pointing by angle-tracking data.
3) Multi-beam guidance and narrow-beam tracking. It includes several schemes such as "feeder focus-deviated multi-beam," "feeder phased array multi-beam," and "plane phased array."

Table 5.3 Predicted values of rain attenuation in Ka-band in Beijing, Guangzhou, and Xi'an

Suppose satellite positioning 100.5°E, operating frequency 30 GHz, and radio wave: horizontal polarization					
Average annual percent/%	Percent in the worst month/%	Attenuation/dB	XPD/dB	Average annual break period/h	Break period in the worst month/h
Beijing, 39.92°N, 116.42°E; elevation, 40.96°					
99.990	99.948	35.90	14.59	0.877	0.379
99.950	99.790	18.74	19.99	4.383	1.537
99.900	99.615	13.74	22.45	8.766	2.809
99.500	98.440	6.25	28.28	43.830	11.393
99.000	97.150	4.32	30.93	87.660	20.822
98.000	94.790	2.93	33.61	175.32	38.056
Guangzhou, 23.33°N, 113.5°E; elevation, 59.0°					
99.990	99.948	58.34	18.93	0.877	0.379
99.950	99.790	30.45	24.18	4.383	1.537
99.900	99.615	22.33	26.57	8.766	2.809
99.500	98.440	10.15	32.24	43.830	11.393
99.000	97.150	7.01	34.81	87.660	20.822
98.000	94.790	4.76	37.43	175.32	38.056
Xi'an, 34.27°N, 108.90°E; elevation, 49.16°					
99.990	99.948	17.97	22.55	0.877	0.379
99.950	99.790	9.38	27.67	4.383	1.537
99.900	99.615	6.88	30.00	8.766	2.809
99.500	98.440	3.13	35.53	43.830	11.393
99.000	97.150	2.16	38.05	87.660	20.822
98.000	94.790	1.47	40.61	175.320	38.056

The above schemes have been described in detail in Chap. 2. Please read relevant chapters and sections.

(3) Uplink and downlink (or Ka/X) dual-band C&T system

As previously mentioned, downlink band uses Ka-band to adapt the requirement of high-speed data transmission. The design idea of uplink channel is different from that of downlink channel. In order to ensure command reliability, receiving antennas of the transponder should be of wide beam. Under the condition of a fixed beam width, if the frequency increases, the gains of received signals could not be obtained. With respect to real-time and high-reliability TT&C, it is better to use S- or X-band TT&C band below 10 GHz, because its rain attenuation is less and the beam is wider. With a view to anti-interference and high-speed data transmission, the whole TT&C station should use Ka/S (or Ka/X) dual-band system. In particular, it is better to use Ka-band for downlink channel. The uplink channel may use S-band (or X-band). If there is some interference in S-band (or X-band) or uplink code rate is high, the uplink channel can use Ka-band.

The use of dual-band system has some advantages:

1) Low band and wide beam are used to guide Ka-band narrow-beam tracking.
2) Capability of dual-band hopping anti-interference.
3) Small low-band rain attenuation to enhance anti-rain attenuation capability of the system.
4) Compatible with existing USB system.
5) Correcting effects of radio wave propagation on measurement accuracy by dual-band receiving.

(4) Dual-mode transponder compatible with space-ground-based network

Researches of dual-mode transponder compatible with space-ground-based network include:

1) Capability to answer C&T signals of space-borne network or ground-based network by time sharing.
2) The antenna has time-sharing "up-look" or "down-look" function.

(5) Vehicular Ka-band phased array antenna

As previously mentioned, backward high-speed data transmission is a difficulty of the system. One key technology is to improve EIRP of the flight vehicle transponder. This requires improving transmitting power and antenna gain. Phased array antenna is a better method. Its advantages include:

1) Active phased array can be used to increase overall transmitting power.
2) In case the antenna has no mechanical movement, space can be more efficiently used and antenna area will be increased to enhance gain. If the conformal phased array antenna succeeds, antenna area can be further expanded.
3) Non-movement of antenna reduces disturbance on the flight vehicle.

5.2 Rain Attenuation and Atmospheric Attenuation in Signal Transmission Channel [5]

5.2.1 Significance of Rain Attenuation Study

If the operating band increases to Ka-band (18–40 GHz), some issues will be brought about, such as increase of rain attenuation and atmospheric propagation loss, tropospheric scintillation, depolarization, and noise increase. Rain attenuation is the main factor. For example, in 30 GHz, when antenna elevation is 10°, atmospheric attenuation is about 4 dB and the variance of scintillation attenuation is about 0.3 dB, while rain attenuation in Haikou and other regions may reach 60 dB and above (in 99.99 % availability). It is obvious that rain attenuation study is very important. The abovementioned factors decrease received C/N_0, increase bit error rate, affect system availability, and limit the increase of transmission code rate.

Taking TDRSS band increase from Ku-band to Ka-band as an example, after Ku-band is changed to Ka-band, return frequency in space-ground section increases from 14 to 18 GHz, antenna gain increases by 2.3 dB, and rain attenuation also increases. For example, in a heavy rain (16 mm/h) in some regions, rain attenuation increases from 0.5 dB/km to 1.2 dB/km, namely, an increase of 0.7 dB/km. If the path is 4 km in heavy rain area, total attenuation increases by 2.8 dB. It is equivalent to the increment of antenna gain. Therefore, above heavy rain, total gain of Ka-band is negative; below heavy rain, total gain is positive. When ground antenna aperture is same, under most climatic conditions, Ka-band is better than Ku-band. For forward frequency 30 GHz in ground-space section, rain attenuation is much bigger. However, the code rate of forward transmission is lower than that of return transmission, and forward transmitting power is on the ground and can have a larger reserve. Measured rain attenuation value or received C/N_0 value will be used by the ground for open-loop power control to realize rain attenuation compensation.

5.2.2 Characteristics of Rain Attenuation

The cause for rain attenuation is that when radio waves pass through rainfall area, raindrops will absorb and scatter radio waves, thus causing attenuation of radio waves during propagation. The radium of a raindrop is generally 0.025–0.3 cm, while the wavelength of Ka operating band is about 1–1.7 cm. Main rain attenuation is absorption attenuation. Most of it is thermal loss. Rain attenuation has the following characteristics:

(1) Rain attenuation is related to the ratio of raindrop diameter to radio wavelength. The bigger the raindrop diameter, the higher the rain attenuation. The more the rainfall, the bigger the raindrop diameter and the higher the rain attenuation.
(2) The higher the frequency, the bigger the ratio of raindrop diameter to radio wavelength, so the higher the rain attenuation.
(3) When the elevation of ground antenna to the target is lower, the path of radio wave passing through the rain area is longer and rain attenuation is higher.
(4) Rain attenuation in summer is higher than that in winter.
(5) The denser the raindrop, the higher the rain attenuation caused by scattering.
(6) Nonspheric raindrop would change polarization direction of radio wave.
(7) Rain attenuation would cause the increase of received noise temperature.

5.2.3 Calculation of Rain Attenuation

Since rainfall differs from regions and seasons and is inhomogeneous in time and space, long-term measurement must be performed to obtain systematic characteristics of the rainfall. However, stations are not provided with these conditions. In general, accessible weather data such as rainfall capacity and rainfall time are used to forecast rain attenuation value. Currently, common forecast methods are ITU-R method, Lin method, SAM method, Crane method, and Bothias method. ITU-R method is very simple, requiring fewer parameters but giving higher accuracy. It is widely used internationally.

With respect to the relationship between rainfall intensity R_P (in mm/h) and rain attenuation A_{R}, ITU-R recommends an approximate representation:

$$A_P = aR_P^b L_e \tag{5.7}$$

where L_{e} is the effective path length of radio wave passing through rain area and a and b are coefficients related to frequency and temperature, respectively.

According to the estimation methods given by ITU, rain attenuation can be calculated as follows:

(1) Estimation of rain attenuation coefficients a and b

Typical values are listed in Table 5.4. The subscript H refers to horizontal polarization wave; V refers to vertical polarization wave.

Table 5.4 only gives some data. Other frequency data can be obtained by logarithmic scale insertion method. The expression is

$$\begin{aligned} a(f) &= \lg^{-1}\left\{\lg\left[\frac{a_2}{a_1}\right] \bullet \left[\frac{\lg(f/f_1)}{\lg(f_2/f_1)}\right] + \lg a_1\right\} \\ b(f) &= (b_2 - b_1) \bullet \left[\frac{\lg(f/f_1)}{\lg(f_2/f_1)}\right] + b_1 \end{aligned} \tag{5.8}$$

where a_1, b_1, a_2, and b_2 are rain attenuation coefficients in case of f_1 and f_2.

It can be seen from Table 5.4 that when the frequency is lower than 40 GHz, b is approximately equal to 1, especially for Ka-band. When the frequency increases from 2 GHz (Ku-band) to 20 GHz (Ka-band), a_H increases four times and corresponding rain attenuation increases four times. In 20–30 GHz, a_H has little change, while up to 40 GHz it increases four times. It is obvious that ITU has considered this factor when allocating 25.5–27 GHz to satellite-ground observation service.

Table 5.4 Coefficients for calculating rain attenuation

	Coefficient			
Frequency/GHz	a_{H}	a_{V}	b_{H}	b_{V}
4	0.00065	0.000591	1.121	1.075
6	0.001750	0.00155	1.308	1.265
12	0.0188	0.0168	1.217	1.2
20	0.0751	0.0671	1.099	1.065
30	0.087	0.167	1.021	1.0
40	0.350	0.31	0.039	0.929

(2) Estimation of rain attenuation intensity

Rain attenuation index means how much time in each year the rain attenuation value is exceeded. "Time" here means average statistical value within 1 year. The ratio of such average time to total time in one year is the time probability $P\%$ exceeding the rain attenuation value A in 1 year. Therefore, one precondition for rain attenuation index is the time probability $P\%$. $(1 - P\%)$ is the availability. For example, if the availability is required to be 99.5 %, rain attenuation time probability $P\% = 1 - 99.5\ \% = 0.5\ \%$. According to $P\%$, rainfall is obtained to solve rain attenuation value A_R.

Rainfall intensity R_P refers to the rainfall per hour in case of time probability of $P\%$. Rain attenuation per unit path in rain area is

$$A_P = aR_P^bL(\mathrm{dB/km}) \tag{5.9}$$

Rainfall may be heavy or light. The probability of heavy rain is low while the probability of light rain is high. Annual time probability of rainfall within one year is $P\%$, which can be referred to Table 5.5 as given by ITU.

As shown in Table 5.5, C, E, F, K, and N are rain areas as divided by ITU. Regions and areas in China belong to five rain areas as listed in the above table, respectively, which can be found out in the rain area distribution map for Far East areas as provided by ITU, as shown in Fig. 5.1.

Countries and regions have their own situations, so rainfall data given by their own meteorological departments will be more accurate. Reference [3] gives rainfall intensity of main cities in China based on the data given by Chinese meteorological departments and ITU rain areas, as shown in Table 5.6.

Rainfall intensity listed in Table 5.6 is measured by the meteorological department at each 10 min, while rainfall calculation model recommended by ITU is the rainfall intensity per minute. Since heavy rain and rainstorm last for a short time, rainfall intensity varies within a time period. Measured average values in 1 min and 10 min are different. Therefore, it is required to convert rainfall intensity in 10 min into rainfall intensity in 1 min, namely,

$$R_1(P) = \alpha P^{\beta} R_{10}(P) \tag{5.10}$$

Table 5.5 Relationship between rainfall intensity R_P and annual time probability $P\%$ in rain areas in China

$P\%$	B	C	D	E	F	G	K	M	N
0.001	32	42	42	70	78	65	100	120	180
0.003	21	26	29	41	54	45	70	95	140
0.01	12	15	19	22	28	30	42	63	95
0.03	6	9	13	12	15	20	23	40	65
0.1	3	2.8	8	2.4	4.5	12	4.2	22	15
1.0	0.5	0.7	2.1	0.6	1.7	3	1.5	4	5

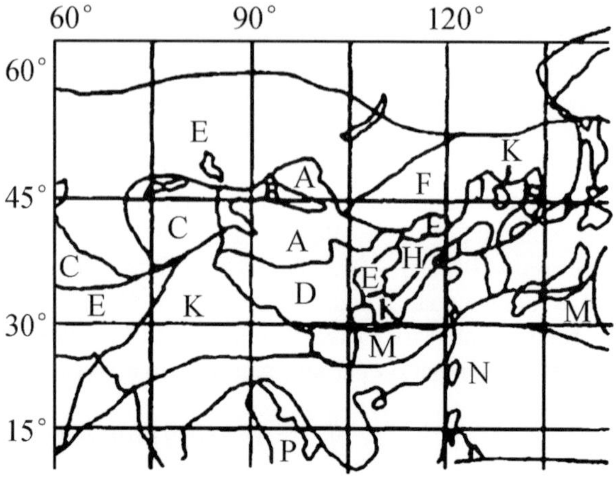

Fig. 5.1 Distribution in Far East areas

Table 5.6 Relationship between rainfall intensity R_p and time probability $P\%$ in some cities of China

City (ITU rain area) \ R_p \ P%	0. 001	0. 003	0.01	0.03	0.06	0.1	0,3	0.6	1
Beijing (K)	96	71	48	28	17	11	4.8	2.1	1.1
Xi'an (G)	58	37	20	11	8.2	6.1	3.5	2.1	1.3
Taiyuan (F)	78	51	30	15	10	7	3.2	1.7	1.1
Lanzhou (E)	51	32	18	10	7	5	2.3	1.3	1.1
Urumchi (B)	15	11	8	5.2	4.1	3.2	1.8	1.0	0.6
Lhasa (D)	39	26	17	11	8.2	6.5	3.4	2.1	1.2
Haikou(N)	120	108	87	62	54	35	13	6.6	3.2
Kunming (K)	82	67	48	30	21	16	7.8	3.1	2.6
Qingdao (M)	100	800	59	32	21	13	5.9	2.3	1.3
Changchun (K)	97	70	45	22	13	10	4.1	2.1	1.2
Harbin (K)	79	55	32	17	11	8	3.3	1.8	1.1
Fuzhou (M)	105	89	62	41	30	21	10	6.1	4.1
Chengdu (M)	100	78	53	31	21	15	6.2	3.1	2.0

Table 5.7 List of α and β parameters of 10 min rainfall intensity conversion

Rain area	A, B	D, E	F, G, H, J, K	M	N, P
α	0.796	0.863	0.847	1.068	1.050
β	−0.0745	−0.0736	−0.0820	−0.0353	−0.0587

where $R_1(P)$ and $R_{10}(P)$ refer to 1 min rainfall intensity and 10 min rainfall intensity, respectively; P is time probability; and α and β are conversion parameters for rainfall intensity, determined by Table 5.7.

1 min rainfall intensity in Haikou and Fuzhou (N and M areas as specified by ITU) can be calculated by using the above expression and table. Comparing it with ITU results (Table 5.8), it is obvious that there are some differences.

Table 5.8 Difference in rainfall intensity given by meteorological department and ITU

Rain area	M			N		
Time probability	0.01	0.1	1	0.01	0.1	1
Rainfall intensity given by Chinese meteorological department	77.87	24.32	4.38	119.7	42.07	3.36
Rainfall intensity given by ITU	63	22	4	95	35	5

Causes for such differences are that rainfall intensity is randomly distributed in one year and changes with the years. In one rain area, the rainfall intensity also changes. The measured value of such a random variable will be different with different measurement time and measurement method. Measuring results are different, so rain attenuation value is only a predicted value and is not entirely accurate.

Since the calculation method given by ITU is simple and is a world standard, it is recommended to use ITU method to estimate rain attenuation. Accurate estimation requires accurate rainfall intensity data given by the meteorological department of the region where the ground station is located.

(3) Estimation of effective length L_e of rain area path and rain height

L_e is effective rain area length of electromagnetic wave passing through the rain area. Its calculation expression is

$$\mathrm{L_e} = \gamma_\mathrm{p} \times \mathrm{L_s} \tag{5.11}$$

L_s is the actual length of radio wave passing through the rain area. It is related to rainfall layer height h and antenna elevation θ, namely,

$$\mathrm{L_s} = (h_\mathrm{R} - h_0)/\sin\theta(\text{in case of } \theta \geq 5^\circ) \tag{5.12}$$

where h_0 is the altitude of the antenna, and h_R is rain layer height. Expression (5.12) means that there are raindrops within a height of $(h_\mathrm{R} - h_0)$ and it rains within the ground area of $L_\mathrm{G} = L_s\cos\theta$. If the range of ground rain area is less than this value, it is indicated that the actual length of radio wave passing through the rain area is less than L_s. It should be corrected by reducing L_e, namely, multiplying a shortened factor γ_p which can be estimated according to the following expression:

$$\gamma_p = 1/(1 + L_\mathrm{G}/L_0) \tag{5.13}$$

where $L_0 = 35\exp(-0.015\,R_p)$. When R_p is more than 100, 100 is taken.

h_R is rain layer height, namely, isothermal height at 0 °C as given by the meteorological observatory. However, actual rainfall height is always lower than 0° isothermal height. Therefore, rain attenuation calculated here is more than actual value.

If 0° isothermal height could not be obtained from the meteorological department, rain layer height mode recommended by ITU may be used. The approximate expression of rain top height h_R is

$$h_R(\mathrm{km}) = \begin{cases} 4 - 0.075(\Phi - 36^\circ) & \Phi \geq 36^\circ \\ 3 + 0.028\,\Phi & 0^\circ < \Phi < 36^\circ \end{cases} \tag{5.14}$$

where Φ is the latitude of the region where the observation station is located.

(4) Estimation of rain attenuation value

Rain attenuation calculation expression is

$$A_R = a(R_p)^b L_e = r_R L_e \tag{5.15}$$

Availability calculation expression is

$$\mathrm{A} = (1 - P\%) \tag{5.16}$$

According to Expressions (5.15) and (5.16), rain attenuation values under a certain availability can be obtained.

The abovementioned calculation method is to calculate rainfall time probability based on system availability requirements, search out corresponding rainfall intensity and then calculate rain attenuation value. However, if it is unable to search out the rainfall intensity for the rainfall time probability and only rainfall intensity at 0.01 % rainfall time probability is known, the following method can be used.

1) Calculate rain attenuation value $A_{0.01}$ at 0.01 % rainfall time probability, namely,

$$A_{0.01} = r_R L_s \gamma_{0.01} \tag{5.17}$$

2) Calculate rain attenuation value A_p at any rainfall time probability $P\%$ by using $A_{0.01}$, namely,

$$A_p = A_{0.01}\left[0.12\mathrm{P}^{-(0.546+0.043\log\mathrm{P})}\right] \tag{5.18}$$

Reference [4] provides rainfall rate per hour $R_{0.01}$(mm/h) in China. Accordingly, we can estimate isoline distribution curves of rain attenuation when the geosynchronous satellite is located at 92°E and the frequencies are 30 and 20 GHz, as shown in Figs. 5.2 and 5.3.

Calculated results show that in China, areas having the largest rain attenuation in Ka-band are coastal areas, such as Fujian, Guangdong, and Hainan; the second

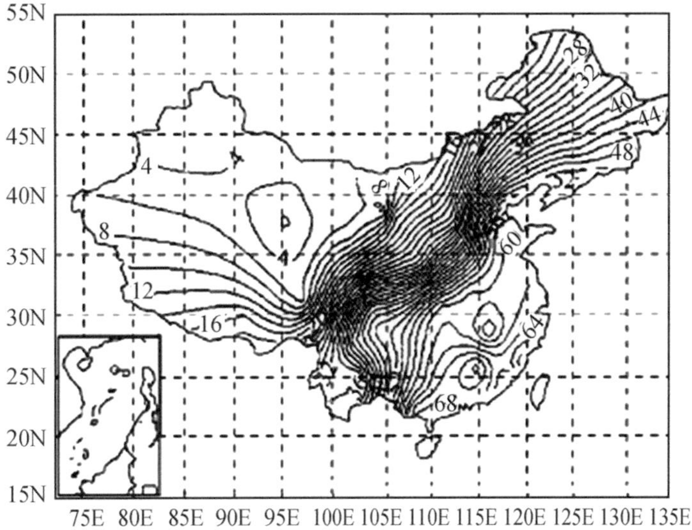

Fig. 5.2 Isoline distribution diagram of rain attenuation in case of 30 GHz and 0.01 % time probability (satellite: 92°E)

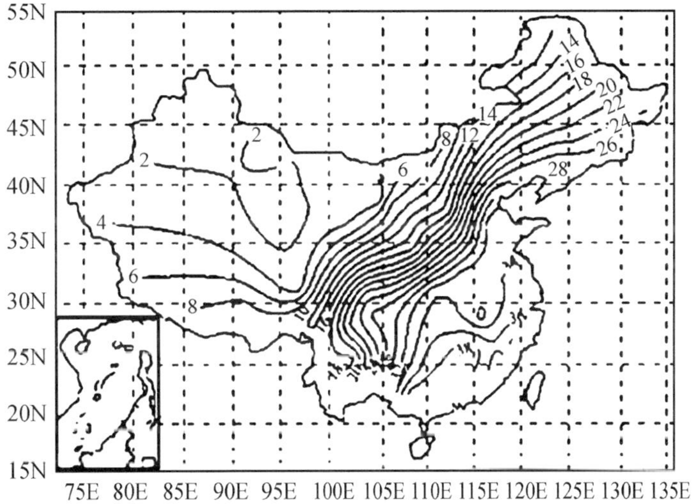

Fig. 5.3 Isoline distribution diagram of rain attenuation in case of 20 GHz and 0.01 % time probability (satellite: 92°E)

largest is Liaodong peninsula, Jiaodong peninsula and coastal areas in Shandong, then North China and Northeast China Region; the last is northwest region. Factors affecting rain attenuation include rainfall intensity of an area, altitude and antenna elevation of a ground station, and operating frequency of the ground station.

5.2.4 Increment in System Noise Temperature Caused by Rainfall

Rain layer operates as a lossy waveguide, and liquid water generally has a temperature of 273 K, the increment ΔT_a in system noise temperature caused by rain attenuation is calculated according to the formula below:

$$\Delta T_A = T_r\left(1 - \frac{1}{A_R}\right) \tag{5.19}$$

where T_r means temperature of raindrop, generally being 273 K, and A_R represents rain attenuation, being expressed by multiples.

According to Expression (5.19), the antenna noise temperature will increase by 136.5 K when rain attenuation is doubled (namely, 3dB).

5.2.5 Rain Attenuation Countermeasure Technology

Rain attenuation above Ka-band is very large, which makes difference of signal level between fine days and rainstorm to be 16 dB or higher. If the rainfall reserve of the system is used to overcome impact caused by rain attenuation, serious power waste will be caused in fine days. Furthermore, appropriate compensation can not be made in case of large rain attenuation; as a result, the system performance will deteriorate. Therefore, adaptive anti-attenuation measures are required. Adaptive anti-attenuation measure may vary with respect to different weather and climate conditions, thereby maintaining the receiving signal level at a normal level and reducing impact upon other performances of the system. Adaptive anti-attenuation measure comprehensively considers reliability and economy of the system. The core idea is to share adaptive resources which include bandwidth, time slot, and power. In general, adaptive anti-attenuation technologies consist of adaptive modulating, adaptive coding, adaptive power control, and adaptive TDMA. The foresaid technologies have their own advantages. In Ka-band, application of only one technology can not effectively deal with attenuation of the entire link. Forward and backward transmission of Ka-band is of asymmetry and has large frequency difference; therefore, different anti-attenuation measures can be employed to uplink and downlink of the ground terminal. The main measures are as follows:

(1) Attenuation detection and power automatic compensation technology

Downlink rain attenuation value can be measured and uplink rain attenuation value can be calculated. Adaptive technologies like power automatic compensation, bit rate and error-correction coding methods, as well as appropriate availability design are used to minimize impact upon the availability specifications of the system.

For power compensation technology, the core is to estimate attenuation of uplink. For uplink, the satellite beacon signal (such as telemetry signal) received by downlink is usually employed to estimate beacon strength in real time, which is free from impact of flick, rain attenuation, and noise and is called as blue sky beacon reference value. Since the power spectrum of flick component does not overlap the power spectrum of rain attenuation component, filtering method can be used to separate the flick component of the real-time beacon, and then beacon level (S/N, C/N_0) received when it rains subtracts that in fine days to obtain the rain attenuation component in slow change. After flick component and rain attenuation component respectively undergo frequency conversion, add them together to get total attenuation of the uplink signal. For compensation of power control algorithm to slowly changing attenuation, the slowly changing attenuation caused by rainfall and dense clouds is mainly compensated. Sometimes, influence of flicker component can be ignored. The attenuation estimation methods include:

1) Frequency conversion factor method. Its results have little influence upon the power control accuracy, and the frequency conversion factor of rainfall attenuation is

$$\frac{\mathrm{A_u}}{\mathrm{A_d}} = \frac{\phi(f_u)}{\phi(f_d)} \tag{5.20}$$

where

$$\phi(f) = \frac{f^{1.72}}{1 + 3 \times f^{3.44}}$$

Frequency conversion factor of flick component is

$$\frac{\dot{S}_{\mathrm{u}}}{S_{\mathrm{d}}} = \left(\frac{f_{\mathrm{u}}}{f_{\mathrm{d}}}\right)^{\frac{7}{12}} \tag{5.21}$$

Aggregate signal attenuation:

$$A = A_{\mathrm{u}} + S_{\mathrm{U}} \tag{5.22}$$

where uplink rain attenuation A_{u}, downlink rain attenuation A_{d}, uplink scintillation attenuation S_{u}, downlink scintillation attenuation S_{d}, and aggregate attenuation A are expressed by dB, and the unit of uplink frequency f_{u} and downlink frequency f_{u} is GHz.

2) Direct measurement method. If the foresaid downlink rain attenuation has been measured, and then the sum of downlink and uplink rain attenuation is measured, thus the difference between the foregoing two attenuations is the uplink rain attenuation value, thereby achieving the purpose by directly adjusting uplink power.

For closed-loop power compensation technology, uplink rain attenuation is required to be measured on satellite and sent to ground station by telemetry for direct adjustment of the uplink power. For downlink power compensation, since it is difficult to implement on-satellite measurement, several fixed powers can be switched.

(2) Adaptive error-correction coding

For the convenience of discussion, take TDRSS for instance below: The basic principle of adaptive forward error-correction coding (AFEC) is that with the change in attenuation of relay satellite downlink, the transponder of user satellite changes coding modes and code rate of the transmitting signals, and the receiving end of ground terminal changes decoded mode accordingly, so as to obtain coding gain with different depth. Furthermore, decrease of code rate may lower C/N_0 thereby compensating different rainfall attenuation to ensure the channel transmission quality. AFEC anti-attenuation measure is to conduct adaptive switch according to rain attenuation degree with a series of coding methods with different code rates or modes, thereby compensating for different attenuation. Since additional symbols are required to serve as monitoring symbols when error-correction coding is employed, so the bit rate of channel will increase. When the channel bandwidth is fixed (such as satellite transponder bandwidth), the information rate should be adaptively adjusted downward to make it match the channel bandwidth. Among the users, the bit rate will be assigned adaptively to ensure key users.

Adaptive error-correction coding anti-attenuation system plan is implemented as follows: the receiving station determines its own attenuation degree through monitoring bit error rate of downlink signal or receiving power level. According to the results, the ground terminal will control the variable rate encoder of the user responder to select appropriate error-correction coding mode; furthermore, the ground terminal can control the variable rate encoder of the local station to be corresponding to the user responder. Error-correction coding mode varies against attenuation degree of the receiving signal, in order to improve transmission capacity of effective information while link transmission quality is ensured. Therefore, the variable rate coding plan shall be of adaptability. In case of application in TDRSS, uplink remote controlling shall be in normal operation; namely, rain attenuation compensation shall be made to uplink to ensure normal uplink receiving, thereby controlling coding mode of the user responder. Presently, it is difficult to control the coding mode and code rate of the user spacecraft in terms of implementation and application; therefore, it will be better that the user responder shall be designed with error-correction coding mode; meanwhile, the code rate shall meet the user requirements, so as to improve the rain attenuation margin of the system. Compared with transmission capacity of fixed gain coding plan, the adaptive coding plan has higher transmission capacity for effective information.

(3) Storage and retransmission

Storage and retransmission technology is of practicability and also called as time diversity technology. Since much of high-speed mission data in C&T is not required to be transmitted in real time, "storage in rain area" and "retransmission in rainless area" is allowed.

(4) Reservation of sufficient rain attenuation margin

1) Apply error-correction coding to lower E_b/N_0 requirements.
2) Lower C/N_0 requirements by adaptively reducing code rate.
3) Increase G/T value.
4) Increase EIRP value.
5) Reasonably select station sites.

Appropriate station sites shall be located in the place featuring small rainfall (decrease rain attenuation and improve availability) and high elevation (reduce atmospheric loss at Ka-band). Weather forecast shall be further improved and the antenna shall be arranged to operate at the high elevation.

(5) Application of diversity technology

1) Space diversity. Also called as "position diversity," it is an effective rain attenuation countermeasure technology. By applying such technology, two antenna heads shall be established on two points (which should be 20 km away from each other and cannot be both located in a same rainstorm area). The output of the two antennas shall be connected with a wideband HF receiver and D/C with same IF. IF output shall be transmitted to a common terminal by optical cables, as shown in Fig. 5.4. By employing such technology, several functions can be implemented, including space diversity anti-attenuation, connection terminal interferometer, multi-station position locating, and antenna arraying.

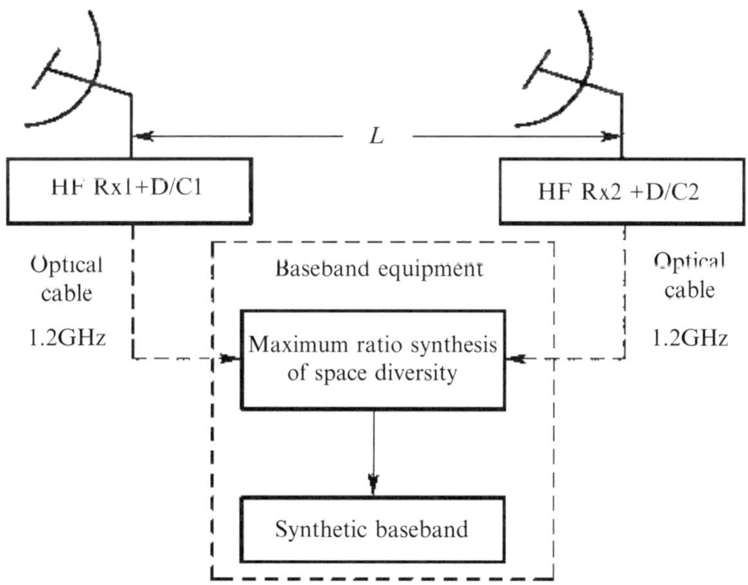

Fig. 5.4 Principle block diagram of space diversity synthesis for rain attenuation countermeasure

2) Employ polarization diversity to decrease change of circularly polarized axial ratio and impact caused by depolarization.
3) Employ dual-band receiving antenna, which is also called as frequency diversity technology. In case of occurrence of rain attenuation in high band, the antenna shall operate at low band to obtain rain attenuation margin.

(6) Adjusting gain and corresponding beam width of satellite antenna

It is also called as "beam diversity technology." By using such technology, the antenna gain will be adjusted and increased to compensate rain attenuation; at this time, the antenna beam width will become narrow and only key service area will be protected.

5.2.6 *Atmospheric Attenuation in Signal Transmission Channel*

In the atmosphere, electromagnetic wave is mainly absorbed by oxygen and water vapor at 22.2 GHz, 60 GHz(O_2), and 118.8 GHz(O_2). The empirical formula of synthetic attenuation γ_a (including absorption by oxygen and water vapor) of atmospheric loss is

$$\gamma_a = a + b\rho_0 - cT_0 \text{ (dB/km)} \tag{5.23}$$

where a, b, and c are empirical coefficient and correlated with frequency; ρ_0 represents mean density of surface water vapor of local area; and T_0 represent mean temperature (°C) of local surface. The empirical coefficients are listed in Table 5.9.

In addition to frequency as specified in Table 5.9, the coefficients on other frequency can be obtained from interpolation on logarithmic scale. Table 5.9 shows that the area with the higher RH will result in larger atmospheric attenuation.

ITU-R provides a more accurate calculation method. By using such method, the foresaid synthetic attenuation falls into dry air attenuation and water vapor attenuation which will be calculated respectively and the sum of them will be the aggregate attenuation.

(1) Equivalent height

Equivalent height of dry air is

$$\begin{aligned} h_0 = & 5.386 - 3.32734 \times 10^{-2} + 1.87185 \times 10^{-3} f^2 \\ & - 3.52087 \times 10^{-5} f^3 + \frac{83.26}{(f-60)^2 + 1.2} \end{aligned} \tag{5.24}$$

Table 5.9 Coefficient for calculating atmospheric attenuation

Frequency/GHz	a	b	c	α	β	ε
4	0.00802	0.000141	0.0000850	0.0397	0.000276	0.000176
6	0.00824	0.000300	0.0000895	0.0404	0.000651	0.000196
12	0.00898	0.00137	0.000108	0.0436	0.00318	0.000315
15	0.00953	0.00269	0.000125	0.0461	0.00634	0.000455
16	0.00976	0.00345	0.000133	0.0472	0.00821	0.000536
20	0.0125	0.0125	0.000101	0.0560	0.0346	0.00155
22	0.0181	0.0221	0.000129	0.0760	0.0783	0.00310
24	0.0162	0.0203	0.0000563	0.0691	0.0591	0.00250
30	0.0179	0.0100	0.000280	0.0850	0.0237	0.00133
35	0.0264	0.0101	0.000369	0.1230	0.0237	0.00149
41	0.0499	0.0121	0.000620	0.2370	0.0284	0.00211
45	0.0892	0.0140	0.00102	0.4260	0.0328	0.00299
50	0.267	0.0171	0.00251	1.27	0.0392	0.00572
55	3.93	0.0220	0.0158	24.5	0.0490	−0.00121
70	0.499	0.0319	0.00443	2.14	0.0732	0.0104
80	0.160	0.0391	0.00130	0.705	0.0959	0.00586
90	0.113	0.0495	0.000744	0.458	0.122	0.00574
94	0.106	0.054	0.000641	0.417	0.133	0.00594

Dry air: RH, 10 %, $\rho_0 = 0.00l$ g/m^3
Moderate weather: RH, 42 %, $\rho_0 = 7.5$ g/m^3
Humid weather: RH, 60 %, $\rho_0 = 18.0$ g/m^3

Equivalent height of water vapor is

$$h_0 = 1.65\left\{1 + \frac{1.61}{(f-22.23)^2 + 2.91} + \frac{3.33}{(f-183.3)^2 + 4.58} + \frac{1.90}{(f-325.1)^2 + 3.34}\right\} \tag{5.25}$$

(2) Atmospheric attenuation coefficient

The calculation formula recommended by ITU-R is as follows:

1) The calculation formula of dry air attenuation coefficient γ_o (dB/km) (frequency 54 GHz below) is as follows:

$$\gamma_o = \left[\frac{7.34 r_p^2 r_t^3}{f^2 + 0.36 r_p^2 r_t^2} + \frac{0.3429 b \gamma_o'(54)}{(54-f)^a + b}\right] f^2 \times 10^{-3} \tag{5.26}$$

where f represents frequency (GH$_Z$); $r_p = p/1{,}013$; $r_t = 288/(273+t)$; p represents pressure (hP$_a$); t represents temperature (°C); and $\gamma'_0(54)$, a, and b mean the parameters obtained through intermediate calculation.

2) The calculation formula of attenuation coefficient γ_o (dB/km) of water vapor transmitted below 350 GH_Z is as follows:

$$
\begin{aligned}
\gamma_w = \Bigg\{ & 3.13\times 10^{-2} r_p r_t^2 + 1.76\times 10^{-3}\rho r_t^{8.5} + r_t^{2.5}\Bigg[\frac{3.84\xi_{w1}g_{22}\exp(2.23(1-r_t))}{(f-22.235)^2+9.42\xi_{w1}^2} \\
& + \frac{10.48\xi_{w2}\exp(0.7(1-r_t))}{(f-183.31)^2+9.48\xi_{w2}^2} + \frac{0.078\xi_{w3}\exp(6.4385(1-r_t))}{(f-321.226)^2+6.29\xi_{w3}^2} \\
& + \frac{3.76\xi_{w4}\exp(1.6(1-r_t))}{(f-325.153)^2+9.22\xi_{w4}^2} + \frac{26.36\xi_{w5}\exp(1.09(1-r_t))}{(f-380)^2} \\
& + \frac{17.87\xi_{w5}\exp(1.46(1-r_t))}{(f-448)^2} + \frac{883.7\xi_{w5}g_{557}\exp(0.17(1-r_t))}{(f-557)^2} \\
& + \frac{302.6\xi_{w5}g_{752}\exp(0.41(1-r_t))}{(f-752)^2}\Bigg]\Bigg\} f^2\rho\times 10^{-4}
\end{aligned}
\tag{5.27}
$$

Figure 5.5 [6] shows the absorption attenuation curve of water vapor passing through troposphere under typical atmosphere condition (water vapor density is 7.5 g/m^3, $N_0 = 313$).

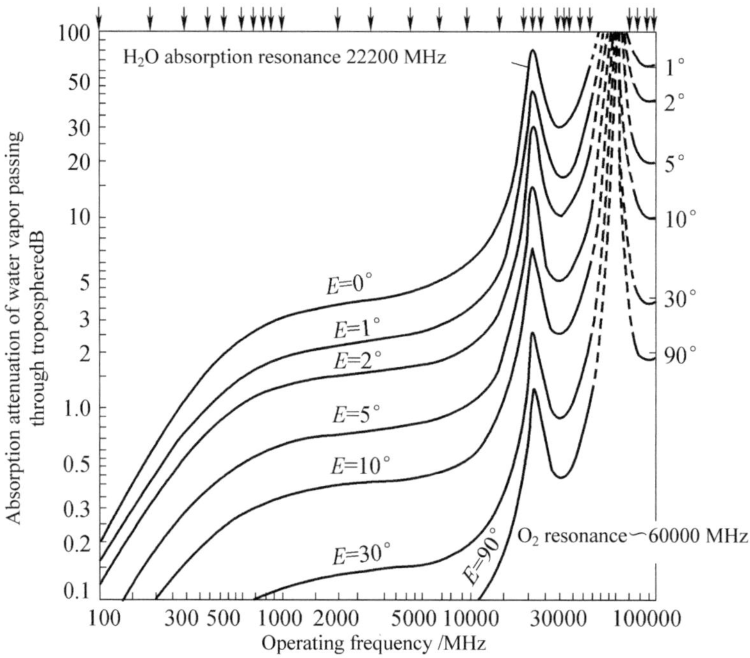

Fig. 5.5 Attenuation coefficient curve

(3) Aggregate atmospheric attenuation value

ITU-R Recommendations provide two approximate calculation methods when operating frequency is 350 GHz below.

1) Atmospheric attenuation A(dB) (elevation angle is $5° < Ø < 90°$):

$$A = \frac{A_0 + A_W}{\sin \Phi} \tag{5.28}$$

where

$$\mathrm{A}_0 = h_0 \gamma_0, \ A_W = h_W \gamma_W$$

where h_0 represents equivalent height of dry air, h_W represents water vapor attenuation coefficient, and γ_W is calculated based on the density of water vapor on Earth surface.

2) Calculation formula of A:

$$A(p) = \frac{A_0 + A_W(p)}{\sin \Phi} \tag{5.29}$$

where

$$A_W(p) = \frac{V_t(p) \cdot \gamma_w(\rho)}{\rho} \tag{5.30}$$

where $V_t(p)$ represents water vapor content (kg/m^3) of air column per unit area when time percentage is p, ρ means water vapor density of the Earth surface (g/m^3) on the Earth surface, γ_W (dB/km) represents water vapor attenuation coefficient which is calculated based on Expression (5.27), and Ø is the elevation of the antenna.

$V_t(p)$, t, and ρ can be found out from ITU-R Recommendations. There is little difference between the calculation results from the foregoing calculation methods. The calculation results show that the atmospheric attenuation value is about 1 dB in South China with muggy climate when operating frequency is 30 GHz and elevation angle is higher than 35°. The attenuation value may reach 3–4 dB in case of low-elevation angle (such as 10°).

5.3 Influence of Multipath Transmission [7]

When C&T is made for low-altitude flight vehicle and long-range vehicle (such as cruise missile, UAV, etc.), the antenna of the ground C&T station operates at low-elevation angle, as a result, the multipath reflection signals will cause serious multipath effect, which has become a key problem. In C&T, it has the following features:

(1) The flight vehicle flies at a high speed, as a result, the attenuation feature changes dynamically and the Doppler frequency-shift is relatively large.
(2) It is a frequency selective-fading channel because the downlink data transmission rate is higher than that of mobile communication; it is a flat fading channel because uplink data transmission rate is relatively high.
(3) Flat fading channel generally has ranging, angle-measurement, and velocity-measurement functions.
(4) Compared with mobile communication, most of C&T communication stations are located in open areas and the multipath environment is not complicated, so the multipath reflection signals are less than those generated from mobile communication. Moreover, the antenna is of narrow beam; it has better space filtering effect compared with wide-beam antenna.
(5) The equipment on the flight vehicle shall have small volume, light weight, and little power consumption. Antenna installation position is restricted, which is unfavorable for employing many anti-multipath measures.
(6) The power amplifier on the flight vehicle is required to operate under saturation status to improve efficiency and power, so the signals to be transmitted shall be in constant envelope.

The radio wave which is generally transmitted between receiving antenna and transmitting antenna mainly contain the following components:

(1) Direct wave: direct transmission within line of sight, being the most powerful component.
(2) Multipath reflected wave: it reaches the receiving antenna after being reflected by different objects; the component intensity is inferior to direct wave.
(3) Diffracted wave: it reaches the receiving antenna after being diffracted and blocked by large-size object on the transmission channel. Its intensity is similar to reflected wave.
(4) Scattered wave: it is the radio wave received by the receive antenna, which is caused by diffuse reflection due to secondary transmitting of the ion in air after being excited.

The foregoing radio waves will generate three different kinds of losses during transmission: the direct wave generates radio wave diffusion loss, which is also called as path loss. Diffracted wave will generate slow fading loss and even forms semi-blind zone due to shadow effect arising from blocking of the objects. Multipath effect will generate fast fading loss which greatly influences receiving quality. The section below will focus on it.

5.3.1 Three Types of Fast Fading Caused by Multipath Effect

In general, fast fading means fast decrease or fluctuation of the signal amplitude (but the values may change slowly). Such fading includes frequency selective-fading channel, space selective fading and time selective fading. Selective fading means that the fading features vary at different frequency, in different space, and at different time. The antenna may receive various selective fading signals independently or concurrently.

5.3.1.1 Frequency Selective Fading

Frequency selective fading means that different fading features will be caused by multichannel transmission at different frequency. It can be described from "frequency domain" or "time domain." In terms of frequency domain, there is a path difference ΔR between the transmission path of the multipath reflected signal and that of direct-path signal; accordingly, there is phase difference which varies from different frequency. In this case, the direct wave and multipath reflection wave on certain frequency will be subject to in-phase addition or deduction, thereby forming frequency selective fading. The bigger the path difference is, the higher the scintillation fading will be. In terms of time domain, it is reflected by the phenomenon that a serial of pulses with different time delay will be received due to different lengths of transmission path after a narrow pulse is transmitted. Such pulses will combine and form a wide pulse, thereby generating "time delay extension." The relationship between "frequency domain" and "time domain" is $B_c = 1/2\pi\sigma_c$, where Bc represents correlated bandwidth, its physical significance. If two carriers separated by Bc pass through the channel with frequency selectivity, they would have a correlation function value of 0.5 [8]; σ_c represents "time delay diffusion," indicating r.m.s. value of multipath distributed power spectrum. Compared with mobile communication, σ_c value is 0.5 ms in flat rural area, 20 ms in hilly land, and 0.1 ms in house. The physical model formed by such fading is as shown in Fig. 5.6.

In this model, it is assumed that the input signal is constant amplitude-frequency sweep signal. Such signal will become envelope scintillation frequency sweep signal after multipath interference by multipath transmission channel, which can be

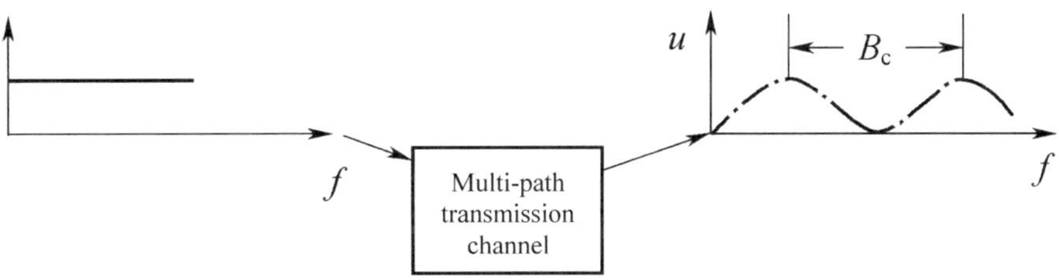

Fig. 5.6 Frequency selective-channel model

considered as filter effect and also called as "filtering effect of multipath interference." As an important fading type, the amplitude/phase feature of fluctuation will cause signal distortion and error code. In Fig. 5.6, B_L means interval between peak values of amplitude/frequency feature fluctuation of equivalent filter of multipath interference.

5.3.1.2 Space Selectivity Fading

Space selectivity fading means that the fading characteristics caused by multipath transmission in different space and locations (usually different heights and angles are under discussion) are different. It is also called Rayleigh fading, with its physical model shown in Fig. 5.7.

In this model, one point beam antenna pointing at θ_o degree is used to transmit single frequency sine wave.

After multipath interference by multipath transmission channel, the amplitude of sine wave received by the same antenna on different spatial stations A(S1), B(S2), and C(S3) (S means of the coordinates of the spatial stations, and different heights are given in Fig. 5.7) are different. Namely, the fading in different places at the same time are different from each other. In the figure above, the signal amplitude (U) received at corresponding S point is different. The amplitude fluctuation period is T_1 which is determined by the following formula:

$$T_1 = \frac{\lambda}{\Delta\theta} \tag{5.31}$$

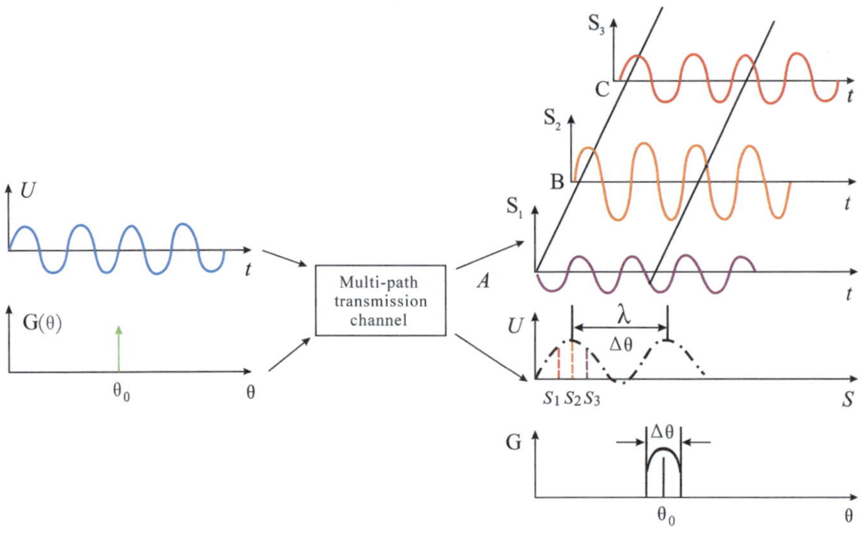

Fig. 5.7 Space selectivity channel model

where λ represents wavelength of the signal; $\Delta\theta$ represents diffusion width of the angle, its meaning is at the angle deviating from the beam pointing θ_0 of the transmission point, the multipath reflected signals can also be received. It means that spread width is incurred by the beam, with diffusion width shown as $\Delta\theta$ in the figure. The space selectivity fading period caused by it is

$$T_1 = \frac{\lambda}{\Delta\theta} \tag{5.32}$$

Space selectivity fading is generated as a result of multipath interference caused by the surrounding objects.

5.3.1.3 Time Selectivity Fading

Time selectivity fading means the fading features caused by multipath transmission at different time are different. Flight vehicle C&T communication is featured by time-varying characteristics. The physical model of such fading formation is as shown in Fig. 5.8.

In this model, the input signal is a single-frequency sine wave with a frequency of f_0. After multipath interference is imposed by multipath transmission channel, the output signals are carriers with envelope fluctuation as shown in the figure. When the receiving antenna is moving, the direct wave and multipath reflected wave will be subject to in-phase addition at a certain point. However, when the receiving antenna is moving toward another point, the direct wave and multipath reflected wave will be in reversed-phase subtraction; therefore, with movement of receiving antenna, the synthesis signal amplitude will change accordingly, thereby generating a fluctuated envelope with period being π/B, where B represents relative frequency shift between direct wave and multipath reflected wave, Δf means absolute Doppler frequency shift of carrier, and $(f_0 + \Delta f)$ refers to center frequency of the carrier. The higher f_0 is, the bigger Δf is, so the AGC bandwidth eliminating such amplitude fluctuation shall be wider, and the adjustment velocity shall be

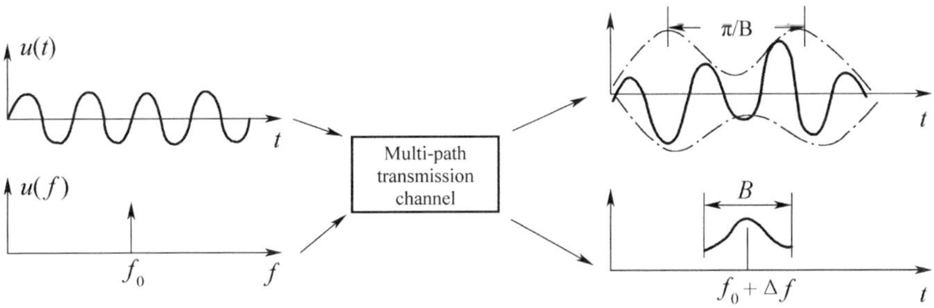

Fig. 5.8 Time selectivity channel model

higher. Similarly, the bandwidth of phase-locked loop deleting corresponding frequency fluctuation shall be wider, or loss-of-lock may be caused.

Figures 5.6, 5.7, and 5.8 show the amplitude change of multipath signals along the direction of three horizontal coordinates (frequency axis f, space axis s, and time axis t). Such change indicates the multipath fading status on frequency domain, space domain, and time domain. C&T communication of flight vehicles is subject to the three types of fading individually or simultaneously, which can be deemed as an important feature.

5.3.2 *Nature of Reflection Coefficient*

The actual amplitude value ρ of reflection coefficient can be considered as the product of three independent factors (i.e., ρ_o, ρ_s, and D). ρ_o is electromagnetic reflection coefficient of the mirror reflecting surface material; ρ_s (roughness coefficient) means the factor of mirror reflection component of reflected wave which is reduced due to surface roughness; D refers to diffusion factor resulting from curvature of reflecting surface.

5.3.2.1 Electromagnetic Reflection Coefficient

If the reflecting surface is very smooth (mirror reflection), the only factor that causes change in amplitude and phase (compared with incident wave) of reflected wave is electric characteristic of the surface material. Such relationship can be expressed by plural electromagnetic reflection coefficient Γ, namely,

$$E_r = \Gamma E_i \tag{5.33}$$

where E_r represents the electric field intensity of reflected wave vector (at the moment upon occurrence of reflection) and E_i means that of the corresponding incident wave.

The reflection vector expressed by plural is

$$\Gamma = \rho_0 e^{-j\phi} \tag{5.34}$$

where ρ_0 is amplitude value of Γ and ϕ means its phase angle.

$$\rho_0 = |E_r/E_i|, \text{so } 0 \le \rho_0 \le 1.$$

Phase angle ϕ is a phase change which occurs during reflection; its value ranges between 0 and $\pm\pi$rad; if ϕ is positive, E_r lags behind E_i in terms of phase.

Reflection coefficient Γ is dependent upon the electric characteristics of the material, particularly upon complex dielectric constant ε_c; in addition, it is relevant to the grazing angle Ψ and polarization (horizontal or vertical) of the wave. Employ Γ_h and Γ_v to express horizontal and vertical component of Γ; its numerical expression can be derived from Maxwell Equation, namely,

$$\begin{aligned} \Gamma_h &= \rho_{0(h)} e^{-j\phi h} = \frac{\sin\psi - \sqrt{\varepsilon_c - \cos^2\psi}}{\sin\psi + \sqrt{\varepsilon_c - \cos^2\psi}} \\ \Gamma_v &= \rho_{0(v)} e^{-j\phi v} = \frac{\varepsilon_c \sin\psi - \sqrt{\varepsilon_c - \cos^2\psi}}{\varepsilon_c \sin\psi + \sqrt{\varepsilon_c - \cos^2\psi}} \end{aligned} \tag{5.35}$$

Expression (5.35) shows that the reflection coefficient ρ_o varies from different dielectric constants of the same smooth reflection surface. The vertical polarization wave which is not fully perpendicular to the ground can be divided into a horizontal component and a vertical component. In this case, the coefficient Γ_h is only applicable to the vertical component of such wave. The reflection coefficient of horizontal component is $-\Gamma_{v.}$

It is worth noting that $\Gamma_h \cong \Gamma_v \cong -1$ when the grazing angle is very small ($\psi \cong 0$), which is attributed to the fact that the path difference δ is null when $\psi = 0$, namely, in this case, the effect of the reflected wave will be accurately cancelled from the view of flat ground interference theory. The experimental results prove the foregoing content. The grazing angle of reflected wave of the radar target can not be detected when it approximates to null. However, the filed intensity of the null grazing angle is not equal to null, because (1) the ray theory is false in case of null grazing angle, and (2) reflection arises out of curve ground surface other than flat ground, and the diffusion factor will weaken reflected ray. The result shows the field intensity is not null but a very small value.

The real part of complex dielectric ε_c is common dielectric constant ε_r, and its imaginary part is relevant to electric conductivity and frequency (or wavelength). When MKS system is used, we can obtain

$$\varepsilon_c = \varepsilon_r - j\varepsilon_i = \varepsilon_r - j60\lambda\sigma \tag{5.36}$$

where ε_r is common dielectric constant (real part of ε_c), λ is wavelength (m), and σ is electric conductivity (Ω/m).

Reference [9] provides reflection coefficient ρ_0 and the relationship between its phase and radio wave polarization and operating band. Refer to Figs. 5.9, 5.10, 5.11, and 5.12 for such relation.

Figure 5.9 shows, when the reflection surface is sea level, the horizontally polarized reflection coefficients at zero grazing angle are all equal to 1 for all frequencies and slightly decreased with the increase of the grazing angle. Within the plotted frequency range and grazing angle range, $\rho_{0(\mathrm{H})}$ is not lower than 0.95. Within the range of such frequency and grazing angle, ϕ_{H} almost equals to πrad; but within the whole grazing angle range 0–90°, ϕ_{H} just increases by 4°. For all practical applications, it can be considered that $\phi_{\mathrm{H}} = \pi$ or 180°.

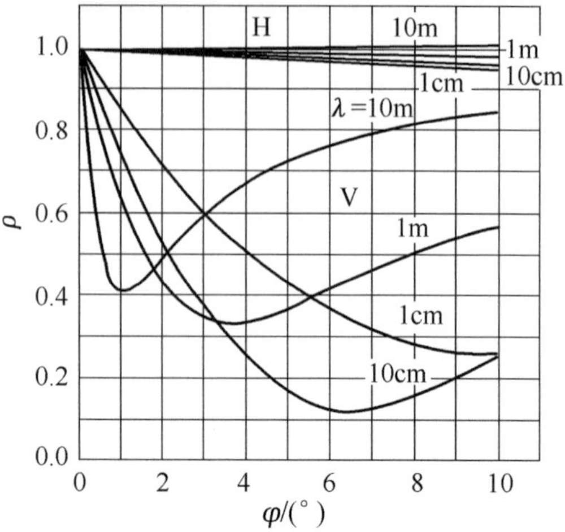

Fig. 5.9 Rating curve between seawater ($T = 10$) reflection coefficient and grazing angle

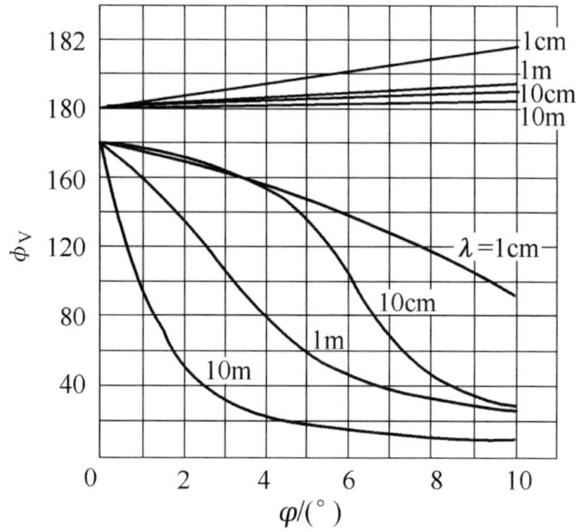

Fig. 5.10 Rating curve between seawater ($T = 10$) reflection coefficient phase and grazing angle

Refer to Fig. 5.9 for vertical polarization. In case that grazing angle is 0°, $\rho_0 = -1$. However, with increment of grazing angle, ρ_o decreases accordingly to the minimum value when a particular grazing angle value is achieved. For perfect dielectric, the minimum value is zero and very sharp. In terms of optics, the definite angle is called as Brewster angle. For imperfect dielectric, the minimum angle has not definite boundary and sometimes is called as quasi-Brewster angle.

The relative dielectric constant ε_r and electric conductivity σ of the land is less than the values of ε_r and σ of the water but has similar change trend. ε_r changes between 2 and 30, and σ changes between $10^{-4}\ \Omega/\text{m} - 10^{-1}\ \Omega/\text{m}$. The values of ε_r and σ of dry land are relatively high. On the land, except for small grazing angle, the effect caused by surface roughness usually exceeds that caused by surface electromagnetic characteristics. The effect from the two aspects determines the final reflection coefficient ρ.

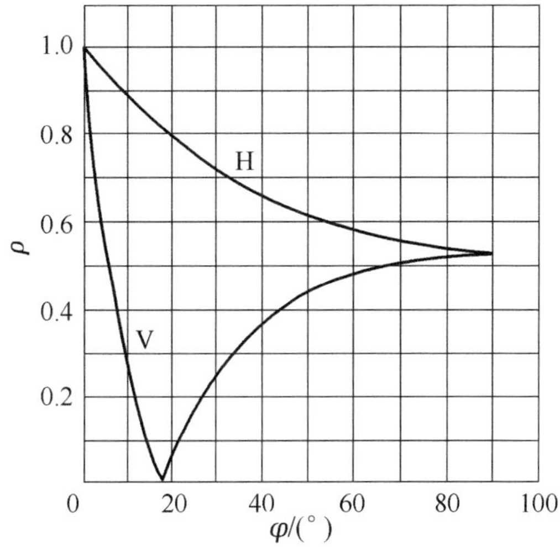

Fig. 5.11 Rating curve between common land ($\varepsilon_c = 10$, $\sigma = 1.6 \times 10^{-3}$ Ω/m) reflection coefficient and grazing angle

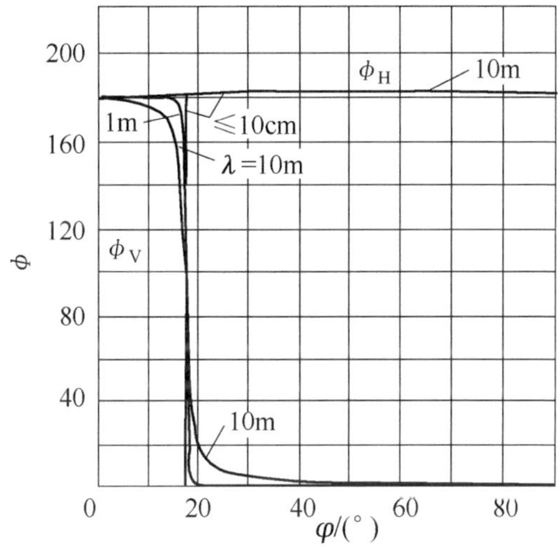

Fig. 5.12 Rating curve between common land ($\varepsilon_c = 10$, $\sigma = 1.6 \times 10^{-3}$ Ω/m) reflection coefficient phase and grazing angle

5.3.2.2 Circular Polarization Reflection Coefficient

It is generally known that circular polarization wave is considered to be constituted by a horizontal polarization wave and a vertical polarization wave with 90° phase difference. Right or left hand is dependent upon the sign of phase difference.

Obviously, if the horizontally polarized reflection coefficient is the same as that of vertically polarized reflection coefficient, the reflection of the two components is the same. If they are in the same phase change, the circularly polarized reflection coefficient will be the same as horizontally/vertically polarized reflection coefficients.

When the grazing angle is not zero, the horizontal reflection coefficient does not equal to vertical reflection coefficient; therefore, the reflected wave will not be purely circularly polarized wave.

When circularly polarized wave is reflected from the direction of vertical incidence, the phase of vertically polarized reflected wave is 0°, so the phase of the horizontally polarized reflected wave is 180°; the reflected wave turns reversely. Therefore, in order to receive the signal of reflected wave with full intensity, the receiver shall rotate reversely but the perfect corotation receiver will receive no signal. Relying on such theory, multipath interference can be rejected at high-elevation angle. However, at low-elevation angle, the phase of the vertically polarized component reflected wave approximates to 180°, thereby resulting in poor anti-multipath effect.

5.3.2.3 Reflection Coefficient of Rough Surface

Reflection on smooth surface is called a mirror reflection, which can be analyzed by reflection ray theory. However, if the surface is rough and has irregular "sags and crests," the wavefront shape of the reflected wave is irregular, and the phase and direction of wavefront cannot be determined early. In this case the reflection is called a diffuse reflection.

If the surface is not rough, the reflected wave can be divided into a mirror component and a diffuse reflection component. The diffuse reflection component occupies a part of total power of incident wave, so the reflected wave is weaker than reflected wave on smooth surface. The decrement is generally expressed by roughness coefficient and marked as $\rho_{s.}$ If the surface is flat but rough, the ratio of reflection component amplitude to incident wave amplitude is $\rho_s\rho_0$. Since $\rho_s\rho_0 \leq \rho_{0,}$ then $0 \leq \rho_s \leq 1$. The coefficient is a real number. In other words, when reflected mirror component is considered as such reflection from an even or smooth surface, the roughness will not cause any phase change.

The reflected wave component from diffused reflection on rough surface will have an amplitude fluctuation with Rayleigh probability density and phase fluctuation with an even density with 27πrad. Diffused reflection power and resultant field of direct wave field and mirror field are in noncoherent addition to form a total field. Its mean power is slightly higher than the power without diffused reflection component (assumed that direct component and mirror component are subject to no change). Total fluctuation of the field is described by Rice probability density function. In principle, change in fluctuation and power will impose impact to detection performance. Diffused reflection field will cause fluctuation of the signals.

Diffused reflection may improve the mean power of the signal; however, it is difficult to estimate the improvement degree. Mirror reflection is usually in reverse proportional to diffused reflection. In case of increase in roughness, mirror reflection will decrease and diffused reflection increase accordingly. However, it is difficult to get quantitative relation between them in theoretically.

At the receiving terminal of diffused reflection path, the r.m.s. value of diffused reflection field is expressed by diffused reflection coefficient $\rho_{d.}$ Power of diffused

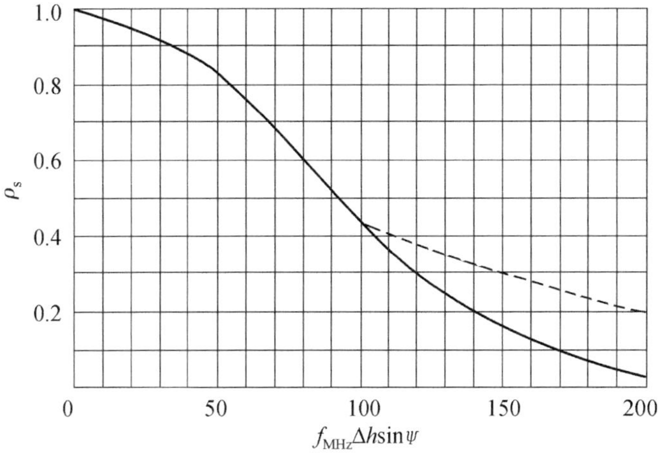

Fig. 5.13 Reflection coefficient ρ_s of rough surface calculated by Ament formula (*Dotted line* represents experimental results of Beard-Katz-Spetner)

reflection is noncoherently added to the synthesized component of direct wave path and mirror reflection; therefore, for dual-way path, the total mean power is increased to about $1 + 2\rho_d^2$ times of the original power.

As a conclusion, reflection coefficient p is determined by the following method: according to Expression (5.34) or Fig. 5.9 through Fig. 5.12, solve ρ_o for different planar media ε_c in case of smooth surface. In case of rough surface (such as macadam, hack, grassland), solve roughness coefficient ρ_s according to Fig. 5.13 [9], in order to solve reflection coefficient of rough surface $\rho = \rho_s\rho_0$.

In Fig. 5.13, Δh refers to the standard deviation of surface fluctuation height distribution.

5.3.3 *Path Loss and Scintillation Fading in Case of Multipath Propagation*

5.3.3.1 Path Loss in Case of Multipath Propagation

Firstly discuss flat fading. As is well-known, in free space, we can obtain

$$P_r(d) = P_t G_t G_r \left(\frac{\lambda}{4\pi d}\right)^2 \tag{5.37}$$

where P_t represents transmitting power (W), G_t means gain of transmitting antenna in the direction of receiver, G_r represents gain of receiving antenna in direction of transmitter, $\lambda(m)$ is the carrier length, and $d(m)$ represents the distance between transmitter and receiver.

Path loss L_F is defined to be the ratio of receiving power to transmitting power when perfect omnidirectional antenna is employed, i.e., $G_t = G_r = 1.0$.

$$L_F = \frac{P_r(d)}{P_t} = \left(\frac{\lambda}{4\pi d}\right)^2 = \left(\frac{c}{4\pi df}\right)^2 \tag{5.38}$$

where the subscript F represents free space and wavelength is expressed as a function of carrier f and light velocity c. Thus, it can be seen that the path loss in free space is reversely proportional to the square of distance; its dB is $L_f = -101\ \mathrm{g}L_F$, namely,

$$L_F = -10\lg L_F = +32.44 + 20\lg d_{km} + 20\lg f_{MHz}\ \mathrm{dB} \tag{5.39}$$

where the unit of d_{km} is km; that of f_{MHz} is MHz.

Refer to Fig. 5.14 for the propagation model of the signal from transmitting antenna to receiving antenna and reflected by smooth surface. The smooth surface may be the media plane of various dielectric constants. Here, only electromagnetic reflection coefficient ρ_o is considered, but neither the decrement factor ρ_s nor diffusion factor caused by reflective convex curvature is taken into consideration. The figure shows that the proximate analysis method for geometric optics is used. Based on the common optics ray principle, A represents the mirror image of receiving antenna against the plane; h_t and h_r, respectively, represent the heights of transmitting antenna and receiving antenna; and d means distance between the antennas. There are two possible ray propagation paths between the transmitter and receiver: one is LOS path, the other is the path reaching the receiving antenna after reflection; the signals received by the receiving antenna are the vector sum of LOS and reflected signals. Assuming h_t and $h_r \ll d$, the length of LOS path is approximately equivalent to that of the reflection path; therefore, attenuation of the two paths is also equivalent approximately. Receiving power of LOS is obtained from Expression (5.37). The root-mean-square value of the power is

$$|E| = \sqrt{P_t G_t G_r}\left(\frac{\lambda}{4\pi d}\right) \tag{5.40}$$

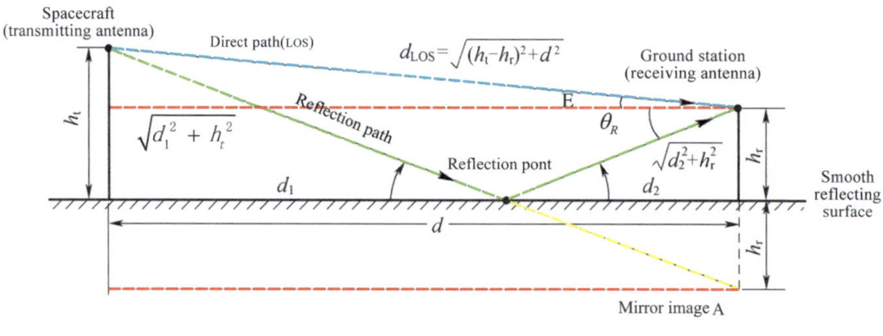

Fig. 5.14 Model of "mirror image ray" reflected by smooth surface

Reflection signal is expressed by the same expressions.

Different phase of two signals is caused by different path lengths. The path different is very small in terms of absolute value, but it will cause obvious phase difference with respect to carrier wavelength.

To calculate path loss, the vector sum of the two signals shall be solved. For this, the carrier phase of the two paths shall be available. The distance of transmitting antenna away from the reflection point is d_1, and distance of the reflection point away from receiving antenna is d_2, and total length of the reflection path is d_R; through geometric relationship, we can obtain the following:

The distance d_2 of reflection point away from ground station is

$$d_2 = \frac{h_r}{h_r + h_t} d \tag{5.41}$$

The difference of path lengths $\Delta R = d_R - d_{COS}$ is

$$\Delta R = d\sqrt{1 + \left(\frac{h_t + h_r}{d}\right)^2} - d\sqrt{1 + \left(\frac{h_t - h_r}{d}\right)^2} \tag{5.42}$$

The relationship between elevation angle θ_R of reflected ray and elevation angle E of flight vehicle is

$$\frac{tg\,\theta_R}{tgE} = -\frac{h_t - h_r}{h_t + h_r}$$

In case of low-elevation angle, $tg\theta_R \approx \theta_R$, $tgE \approx E$; therefore,

$$\frac{\theta_R}{E} = -\frac{h_t - h_r}{h_t + h_r} \tag{5.43}$$

When $h_t \gg h_r$,

$$\theta_R \approx -E \tag{5.44}$$

The phase difference between direct and reflection vectors (rad)

$$\theta = \frac{2h_t h_r}{d} \bullet \frac{2\pi}{\lambda} \tag{5.45}$$

Before vector addition is completed, the change of signal amplitude shall be considered when the signal is reflected from flat surface. As said above, it shall be multiplied by plural reflection coefficient $\rho\cdot$. $\rho\cdot$ is the function of incidence angle, ground insulation constant of reflection point and conductivity. The research also reveals that $\rho = -1$ when the incidence angle is small. According to module value of the vector (|E|) given in Expression (5.40), the vector sum is

$$E_{\Sigma} = |E|\left(1 + \rho e^{j\theta}\right) = \sqrt{P_i G_i G_r}\left(\frac{\lambda}{4\pi d}\right)(1 - \cos\theta - j\sin\theta) \tag{5.46}$$

The modulus of vector sum is

$$\left|E_{\sum}\right| = \sqrt{4P_t G_t G_r}\left(\frac{\lambda}{4\pi d}\right)\sin\left(\frac{\theta}{2}\right) \tag{5.47}$$

Substitute θ in Expression (5.45) to obtain

$$\left|E_{\sum}\right| = \sqrt{4P_t G_t G_r}\left(\frac{\lambda}{4\pi d}\right)\sin\left(\frac{h_r h_t}{d} \bullet \frac{2\pi}{\lambda}\right)$$

When d is large, the sinusoidal function can be approximated by $\sin x = x - (x^3/3!) + (x^5/5!) - \cdots\cdots$; only the first item is taken, thereby obtaining

$$\sin\left(\frac{h_r h_t}{d} \bullet \frac{2\pi}{\lambda}\right) \approx \frac{h_r h_t}{d} \bullet \frac{2\pi}{\lambda}$$

then,

$$\left|E_{\sum}\right| = \sqrt{4P_t G_t G_r}\left(\frac{h_r h_t}{2d^2}\right) \tag{5.48}$$

The receiving power is proportional to the square of the receiving voltage value, therefore,

$$P_r = 4P_t G_t G_r\left(\frac{h_r h_t}{2d^2}\right)^2 \tag{5.49}$$

When gain of transmitting or receiving antenna is that of omnidirectional element antenna, the path loss is

$$L_{\mathrm{PE}} = \frac{P_r}{P_t} = \frac{(h_r ht)^2}{d^4} \tag{5.50}$$

According to this, the path loss of flat ground propagation is in reverse proportional to the fourth power of the distance. The path attenuation expressed by dB is

$$L_{\mathrm{PE}} = -20\lg(h_r h_i) + 40\lg d_{\mathrm{km}} + 120 \tag{5.51}$$

Compared with the free space propagation path loss formula given in Expression (5.38), the Expression (5.38) is a function of carrier frequency, but Expression (5.50) is uncorrelated with carrier frequency. Another difference is that the path attenuation will vary from distance. d_{km} in Expression (5.38) increases by 1 time, and the path attenuation will increase by 6 dB. But in Expression (5.50), d_{km} increase by 1 time, and the path attenuation will increase by 12 dB. It is revealed from Expression (5.50) that the path loss is a function of the height of transmitting and receiving antenna in case of reflection by perfect flat surface. The higher the antenna is, the phase angle θ between direct signal and reflection signal larger. When it changes toward the in-phase addition, the synthesized signal amplitude will be larger. Therefore, the system designer may control total path attenuation by controlling height of the antenna.

5.3.3.2 Scintillation Fading Caused by Multipath Propagation

Other factors influencing path attenuation are not considered for the path loss for perfect flat surface, including types and density of the buildings, plants, and altitude between the two antennas. If these factors are considered, the analysis may become very complicated. In fact, the reflection coefficient ρ is not -1 but a plural $\rho e^{j\varphi}$. They change with different objects, grazing angle ψ, and operating frequency ω_o; its function is $\rho(\omega_o\psi)$. At this time, Expression (5.46) becomes

$$E_{\Sigma} = |E|\left[1 + \rho(\omega_0\psi)e^{j\theta+\varphi}\right] \tag{5.52}$$

The phase of reflection signal given in Expression (5.52) includes phase φ of reflection coefficient and phase θ introduced by path difference. After the foregoing factors are considered, the synthesized signal amplitude E_E and phase will be fluctuated with the change of these factors. Such scintillation fading is manifested by multipath interference.

The section above discusses the case in which perfect omnidirectional antenna $G_r = 1$. In fact, the antenna gain at direct signal angle cp_d is $G(\varphi_d)$; the antenna gain at reflection signal angle φ_r is $G(\varphi_r)$. Therefore, the synthesized signal at the output terminal of receiving antenna is

$$E_{\Sigma A} = |E|\left[\sqrt{G(\varphi_d)} + \sqrt{G(\varphi_r)} \cdot \rho(\omega_0\psi)e^{j(\theta+\varphi)}\right] \tag{5.53}$$

When antenna output includes two kinds of polarized wave, Expression (5.53) shall be used to respectively calculate polarized wave once. According to polarization features of transmitting wave and the proportion of the two polarized waves, adaptive weight is conducted for each type of polarized wave; the sum of the squares of the two polarized signals is the total receiving power.

Expression (5.53) considers several factors and shows that the signal fading caused by multipath effect is correlated with the nature of the reflector ($\rho e^{j\varphi}$), path difference (θ), frequency (ω_o), and antenna pattern (G).

5.3.4 *Orbit Determination Error Caused by Multipath Propagation*

Ranging, velocity-measuring, and angle-measuring accuracy will be influenced by the interference of multipath signal. For ranging, ranging error will be caused by the phase change of the composed signal due to the superposition of multipath interference signal and direct signal. For angle tracking, an angle interference error signal will be generated and make the antenna direction deviate due to the directional difference between multipath interference and direct signal. For velocity measuring, the Doppler frequencies between multipath interference signal and direct signal are different due to their directional difference. The Doppler frequency shift after synthesis will change.

5.3.4.1 Effect of Multipath Interference on Velocity-Measuring Accuracy

The phase and amplitude of the multipath reflected signal is different from direct signal, so there will be a phase shift $\Delta\Phi$ between the composed signal after superposition and the direct signal without multipath interference, and the change of $\Delta\Phi$ will introduce Doppler frequency error, as shown in Fig. 5.15.

In Fig. 5.15, ρ is the reflection coefficient of multipath signal and Φ is the sum of phase difference generated by direct/reflected-wave path difference $\triangle R$ and the phase angle of the reflection coefficient. The latter is as shown in Figs. 5.9, 5.11, 5.10, and 5.12. G_{sr} is the ratio between the antenna gain at the direction of direct wave and the antenna gain at the direction of multipath reflected wave. By using the geometric relationship between flight vehicle and ground station, we can obtain

$$\Delta\Phi \approx \rho/\sqrt{G_{sr}}\sin\left[\left(4\pi\frac{h}{\lambda}\right)\cdot\sin E + \Phi\right] \tag{5.54}$$

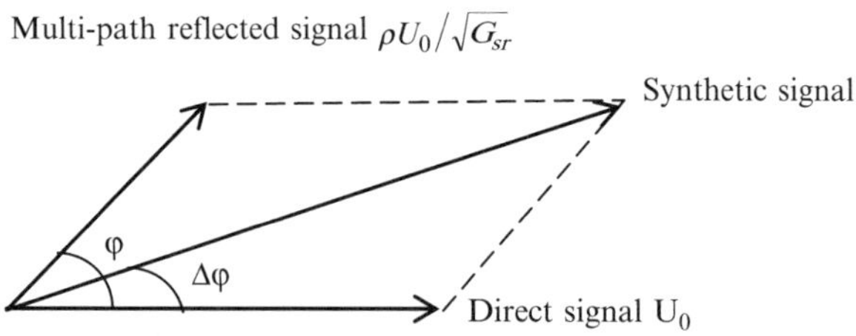

Fig. 5.15 Phase and amplitude relationship between multipath reflected signal and direct signal

where h is antenna height (m), λ is signal wavelength (m), and E is antenna elevation (°). It can be seen that any change of ρ, Φ, E, h, G_{sr} will cause the change of $\Delta\Phi$, while the change of $\Delta\Phi$ will cause the change of Doppler frequency shift, which is equal to

$$\Delta f_d = \frac{d(\Delta\Phi)}{dt} = \frac{\sqrt{2}\rho h\dot{E}}{\lambda\sqrt{G_{sr}}} \quad (\mathrm{rad/s}) \tag{5.55}$$

The velocity-measuring error caused by Δf_d is

$$\delta_v = \frac{\sqrt{2}\rho h\dot{E}}{\sqrt{G_{sr}}} \quad (\mathrm{m/s}) \tag{5.56}$$

where $\dot{E}$ is the change rate of antenna elevation.

δ_v is a randomly fluctuant variable due to the changes of radial reflection characteristics at multipath reflection point and the movement of the target.

5.3.4.2 Effect of Multipath Interference on Ranging Accuracy

Besides causing phase shift $\Delta\Phi$ to the phase of direct signal carrier, the multipath interference also causes the same phase shift $\Delta\Phi$ to the ranging signal modulated on the carrier. The phase shift of ranging signal will cause ranging error δ_R:

$$\delta_R = \frac{(\Delta\Phi/2\pi)c}{2f_R} = \frac{\rho\sin\left[\left(4\pi\frac{h}{\lambda}\right)\sin E + \Phi\right]c}{4\pi f_R\sqrt{G_{sr}}} \tag{5.57}$$

where f_R is ranging tone frequency and c is the velocity of light.

δ_R is also a randomly fluctuant variable.

5.3.4.3 Effect of Multipath Interference on Angle-Measuring Accuracy

Multipath reflection will affect the elevation angle-measuring accuracy since it may make the elevation of the tracking antenna jitter or even lose the target.

Make the main lobe of the antenna A aim at the target being tracked, and set the elevation angle as E, antenna height as h, and the image of the antenna on the ground as A′. As shown in Fig. 5.16, the reflected wave will enter the receiving antenna at an angle of $2E$ lower than the antenna axis (see Expression 5.44). The phase difference between the ground reflected wave and direct wave is

$$\phi = \frac{2\pi}{\lambda}2h\sin E + \phi = \frac{4\pi h}{\lambda}\sin E + \phi \tag{5.58}$$

where λ is operating wavelength, E is antenna elevation angle, and φ is the phase of ground reflection coefficient.

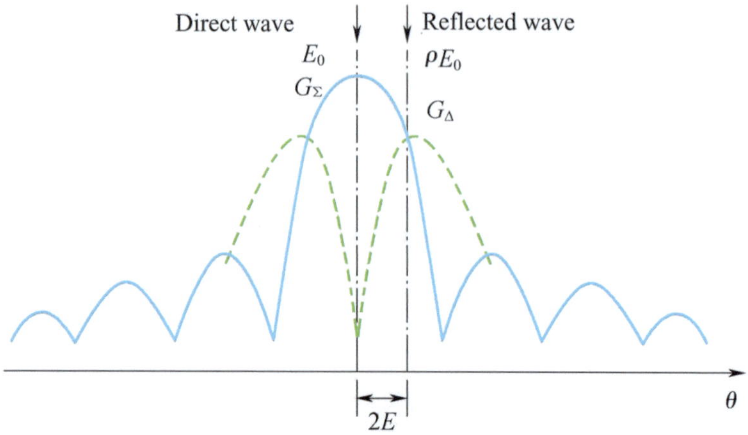

Fig. 5.16 Angle-measuring error caused by multipath reflected wave

Ground reflection affects both the sum channel and the difference channel of the antenna. When the ground reflected signal is much smaller relative to the direct wave, its effect on the sum channel can be omitted and only the effect on the difference channel is considered.

Set the voltage generated by the direct wave of the target as E_0, ground reflection as ρ, then the voltage of the reflected wave will be pE_0. If the reflected wave enters from a side lobe of the difference pattern of the antenna, the output of the reflected wave in the difference channel of the antenna is

$$E_A = \rho E_0 \sqrt{G_\Delta} \tag{5.59}$$

where C_Δ is the difference pattern gain at an angle of $2E$ away from the electric axis.

The sum channel output of the antenna is

$$E_\Sigma = E_0 \sqrt{G_\Sigma} \quad \text{(only the direct component is considered)} \tag{5.60}$$

where G_Σ is the sum pattern gain.

$$E_0 = \frac{E_\Sigma}{\sqrt{G_\Sigma}} \tag{5.61}$$

Substitute Expression (5.61) into Expression (5.59) to obtain

$$E_\Delta = \rho E_\Sigma \sqrt{\frac{G_\Delta}{G_\Sigma}} \tag{5.62}$$

In the difference channel, this ground reflected signal seems like an error signal. It can drive the servo system until it is balanced by the error component with equal value but opposite phase. At this moment the antenna can be stable and generate angle-measuring error. It should be appointed out that the phase between E_Δ and E_Σ is random; so is the amplitude of the difference signal. Its root mean square is $E_\Delta/\sqrt{2}$. This error component changes as the function of elevation angle E.

The existence of ground reflected signal E_Δ makes the antenna generate a disturbance angle (only for elevation angle) $\Delta\theta$,which generates an error voltage to balance with E_Δ. Based on the definition of K, there is

$$E_\Delta = E_0\sqrt{G_\Delta} = E_0\frac{d\mathrm{G}_\Delta(\theta)}{d\theta}(\Delta\theta) \tag{5.63}$$

where $\frac{dG_\Delta(\theta)}{d\theta}$ is the absolute slope of the difference lobe, and $K_\mathrm{m} = \frac{\theta_{3\,\mathrm{dB}}}{\sqrt{G_\Sigma}}\frac{d\mathrm{G}_\Delta(\theta)}{d\theta}$, where $\mathrm{K_m}$ is normalized error slope; $K_\mathrm{m} = 1.2 \sim 1.6$ in different antenna schemes; substitute it into Expression (5.63) to obtain

$$\Delta\theta = \frac{E_\Delta}{K_\mathrm{m}\frac{E_0}{\theta_{3\,\mathrm{dB}}}} \tag{5.64}$$

Substitute Expression (5.60) into Expression (5.63) to obtain

$$\Delta\theta = \frac{\rho E_\Sigma\sqrt{\frac{G_\Delta}{G_\Sigma}}}{K_\mathrm{m}\frac{E_\Sigma}{\theta_{3\,\mathrm{dB}}}} = \frac{\rho\theta_{3\,\mathrm{dB}}}{K_\mathrm{m}\sqrt{\frac{G_\Sigma}{G_\Delta}}} \tag{5.65}$$

Because the position at which the reflected wave enters the difference pattern side lobe is random, its root-mean-square value can be assumed as $G_\Delta/\sqrt{2}$. Consider the function of the abovementioned $E_\Delta/\sqrt{2}$; the root-mean-square value of $\Delta\theta$ is

$$\sigma_\mathrm{EL} = \frac{1}{2}\frac{\rho\theta_{3\,\mathrm{dB}}}{K_\mathrm{m}\sqrt{\frac{G_\Sigma}{G_\Delta}}} \tag{5.66}$$

If $K_\mathrm{m} = 1.4 = \sqrt{2}$, there is

$$\sigma_\mathrm{EL} = \frac{\rho\theta_{3\,\mathrm{dB}}}{\sqrt{8}\sqrt{\frac{G_\Sigma}{G_\Delta}}} = \frac{\rho\theta_{3\,\mathrm{dB}}}{\sqrt{8G_\mathrm{se}}} \tag{5.67}$$

where $\theta_{3\text{ dB}}$ is half-power beam width of antenna; ρ is ground reflection coefficient, $\rho < 1$; G_{se} is the sidelobe relative gain of antenna difference pattern at an angle of $2E$ lower than the electric axis; $G_{se} = G_{\Sigma}/G_{\Delta}$; G_{Δ} is the gain of difference pattern at an angle of $2E$ lower than the electric axis; and G_{Σ} is the maximum gain of sum pattern.

It is difficult to predict the accurate value of elevation angle error caused by multipath for the ground reflection coefficient, and its changes are different at different positions around the antenna. Expression (5.67) is only for approximate evaluation in engineering.

5.3.5 *Effect of Multipath Interference on Data Transmission Bit Error Rate*

The characteristics of multipath transmission channel are decided by the characteristics of multipath reflector (including the surface nature, position, direction, number, etc. of the reflector). Different reflector corresponds to different multipath channel transmission function. When a signal passes through the channel, it is equivalent to multiplying with the transmission function which changes the output signal. The effect of the same channel on different signal is different, so the bit error rate is different. For narrowband signal, it is indicated by flat fading, and for wideband signal, the frequency selective fading. Multipath narrowband fading channel means the situation when the bandwidth Bc of multipath channel frequency response is larger than signal bandwidth Bs (refer to Sect. 5.3.1.1 of this book). Here, in the range of Bs, frequency response of multipath channel is basically flat, so the spectrum and group delay of the narrowband signal is basically flat when it passes through multipath channel. Therefore, it is also called flat fading channel. When $B_c < B_s$, it is wideband fading channel. Here, in the range of B_s, the frequency response of multipath channel is fluctuant and expressed as frequency selective characteristics. The spectrum and group delay of a wideband signal when passing through the channel are fluctuant, so it is also called frequency selective fading channel. Therefore, "narrowband" or "wideband" in multipath fading channel is decided by the relationship between B_c and B_s. From Expression (5.53) we can see that the fluctuation and fading characteristics of multipath channel are in connection with path difference (θ), the nature of reflector ($\rho e^{j\varphi}$), frequency (ω_0), antenna pattern (G), as well as the position and motion state of the flight vehicle relative to the ground station. That is to say, B_c will vary according to the mentioned different reflection conditions, so the definition of "narrowband" or "wideband" of B_c varies. An approximate qualitative concept may be given to readers, i.e., it can be deemed that in satellite mobile communication, the low-speed data in the range of KHz is pure narrowband signal, and the high-speed data in the range from MHz to GHz is "wideband" signal [10]. Their corresponding measures against multipath interference are different.

5.3.5.1 Narrowband Fading Channel

When a signal is a single frequency or $B_s < B_c$ narrowband signal, multipath channel is a flat fading channel which causes an average fade on signal receiving power, but not the distortion of data transmission signal which can cause bit error. However, bit error can also be caused by the decrease of E_b/N_0 due to the fade of signal power. Therefore, it is required to determine the bit error at each E_b/N_0 and then use the probability distribution of E_b/N_0 to calculate the average bit error. Or in return, use the required average bit error rate to calculate the allowable E_b/N_0 distribution so as to obtain the required minimum E_b/N_0 and finally determine the required fading margin. Therefore, enough fading margin is the basic measure against multipath for narrowband fading channel.

For the measurement of bit error performance, there is another indicator, outage probability. It is the percentage that the bit error rate can not meet the required time. The minimum E_b/N_0 meeting the requirements of outage probability can also be a method to determine the margin of narrowband fading channel.

5.3.5.2 Wideband Fading Channel

When signal bandwidth $B_s > B_c$, multipath channel is a frequency selective fading channel. It will cause spectrum and group delay distortion of wideband signal; thus, bit error is generated. The causes of frequency selectivity are as shown in Fig. 5.17.

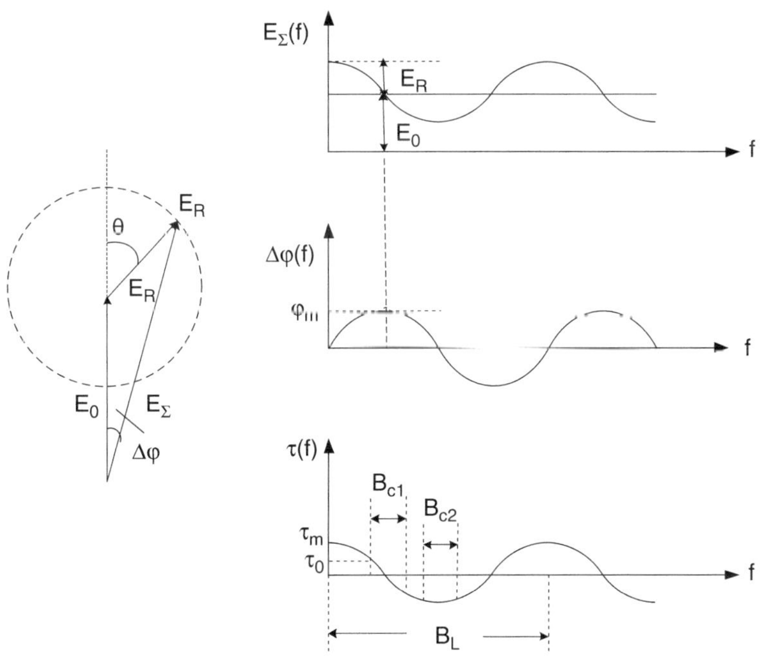

Fig. 5.17 Characteristics of frequency selective channel caused by multipath interference

In Fig. 5.17, E_0 is direct wave, E_R is reflected wave, E_Σ is composed wave, θ is the phase difference between E_R and E_o, $\Delta\varphi$ is the phase difference between E_Σ and E_o, and θ is caused by the path difference ΔR between E_R and E_o.

$$\theta = (2\pi f)\frac{\Delta R}{C} \quad (\text{rad}) \tag{5.68}$$

When f changes, θ, E_Σ, and $\Delta\varphi$ change too, causing frequency selective characteristics as shown in Fig. 5.16. It can be seen from this Figure that when θ changes 2π rad, the change of $E_\Sigma(f)$ fluctuation will be a circle, which is caused by the change of corresponding frequency from f to $f+B_L$, i.e.,

$$2\pi = (2\pi B_L)\frac{\Delta R}{C}$$

$$B_L = \frac{C}{\Delta R} = \frac{1}{\sigma_c}(\text{H}_\text{Z}) \tag{5.69}$$

where σ_c is the time delay between direct wave and reflected wave. In the bandwidth of B_L, the maximum relative value of amplitude-frequency characteristics fluctuation is

$$H_{\max} = \frac{E_R}{E_\Sigma} \tag{5.70}$$

When E_R is small, phase-frequency characteristics approximate to sine fluctuation:

$$\Delta\varphi(t) = \varphi_m \sin\left(\frac{2\pi}{B_L}f\right) \tag{5.71}$$

The corresponding group delay fluctuation is

$$\tau(f) = \frac{d[\Delta\varphi(f)]}{df} = \frac{2\pi}{B_L}\varphi_m \cos\left(\frac{2\pi}{B_L}f\right) \tag{5.72}$$

When E_R/E_O is small,

$$\varphi_m \approx \arctan(E_R/E_O)$$

The abovementioned $\tau(f)$, $E(f)$ fluctuation will cause data transmission bit error, and its simulation result is as shown in Fig. 3.50a. From this Figure, we can see that for QPSK modulation, when the ratio between the peak value τ_m of group delay fluctuation and symbol width is 0.9, E_b/E_O loss will be up to 9 dB. However, if the bandwidth of the signal is narrow and the corresponding operating bandwidth is also narrowed to B_c (refer to Fig. 5.16), the group delay fluctuation in the band will become approximate linearity (B_{c1}) or approximate parabola (B_{c2}). At this moment, its effect on bit error rate will be decided by Figs. 3.48 and 3.49. If $B_c = \frac{1}{2\pi}\frac{1}{\sigma_c}$, then

$\frac{B_c}{B_L} = \frac{1}{2\pi}$. For B_{c1}, $\frac{\tau_{c1}}{\tau_m} = 0.48$. Compared with the abovementioned $\frac{\tau_m}{T_s} = 0.9$, it will decrease to $\frac{\tau_0}{T_s} = 0.43$. Refer to Fig. 3.48; loss of E_b/N_0 decreases 0.5 dB and it is very small. It can be analyzed in a similar way that for B_{c2}, the loss of E_b/N_0 is small. Therefore, bandwidth B_c is generally set as the limit of frequency selective fading [8], i.e., when signal bandwidth $B_s > B_c$, the effect of frequency selective fluctuation on bit error rate should be considered. When $B_s < B_c$, the effect of frequency selective characteristics can be omitted and regarded as flat fading. This is the physical reason for defining $B_c = \frac{1}{2\pi}\frac{1}{\sigma_c}$.

Bit error of frequency selective fading is caused by distortion instead of small E_b/N_0, so the increase of transmission power will not decrease the bit error. This phenomenon is called error floor. Such bit error can only be solved by such measures as adaptive equalization and fading channel error correction coding. This is the significant difference from narrowband fading channel. In a narrowband fading channel, the increase of E_b/N_0 can decrease bit error. Besides frequency selective fading, the causes of error floor are time-selective fading, modulation mode, bandwidth of filter, sampling time and the shape of time delay profile, etc.

The above is based on frequency domain analysis. It can also be analyzed from time domain. If the delay time of multipath signal is so long as to cover a symbol width, the direct code element signal and the reflected code element signal will overlap and extend the code element. The duration of code element extension may cause interference to the next code element, called inter-symbol interference which will cause bit error. If the duration of such extension increases, the bit error rate increases too and finally it may break the allowable error floor.

In a same delay spread channel, the narrower the symbol width is, the easier the inter-symbol interference will be caused. Therefore, the higher the data transmission rate is, the greater the effect of frequency selective fading on bit error rate will be.

The above is only the analysis under ideal smooth surface and "two-path model." The multipath model in actual scene is much complicated, and it is required to research and establish all kinds of environmental models. For example, fuselage scattering is an important and complicated multipath environment. The difference of multipath environment will make different envelope distributions of multipath composite signal. For example, its envelope distribution is quasi-Gaussian distribution at a high elevation angle; Rice distribution at a low elevation angle, which may cause intermittent interruption of signals, and Rayleigh distribution at a negative elevation angle, which may cause frequent interruption of signals. It is very difficult to resist such kind of multipath interference.

5.3.6 Anti-multipath Interference Measures

The severity degree of multipath interference is relative to the gain at the directions of antenna beam pointing to ground, of which there are two situations, i.e., "sidelobe pointing to ground" and "main lobe pointing to ground." Different situation should take different anti-multipath interference measures. The latter is more difficult to deal with.

There are two approaches for anti-multipath interference measures: first, try to decrease the strength of the received multipath interference to increase the signal-to-jamming ratio (S/J), and then in low S/J situation, find the best way to extract signals.

5.3.6.1 Decrease Multipath Signal to Improve S/J

(1) Spatial filtering

Due to the different spatial direction between direct wave and reflected wave, spatial filtering can be used to improve S/J. It can be realized by the following methods:

1) Low sidelobe and narrow-beam antenna: the antenna pattern itself is a characteristic curve of spatial filter, as shown in Fig. 5.18.

In Fig. 5.18, φ_d is the spatial direction of direct wave and the corresponding antenna gain is $G(\varphi_d)$; φ_{r1} is the spatial direction of the reflected wave when the

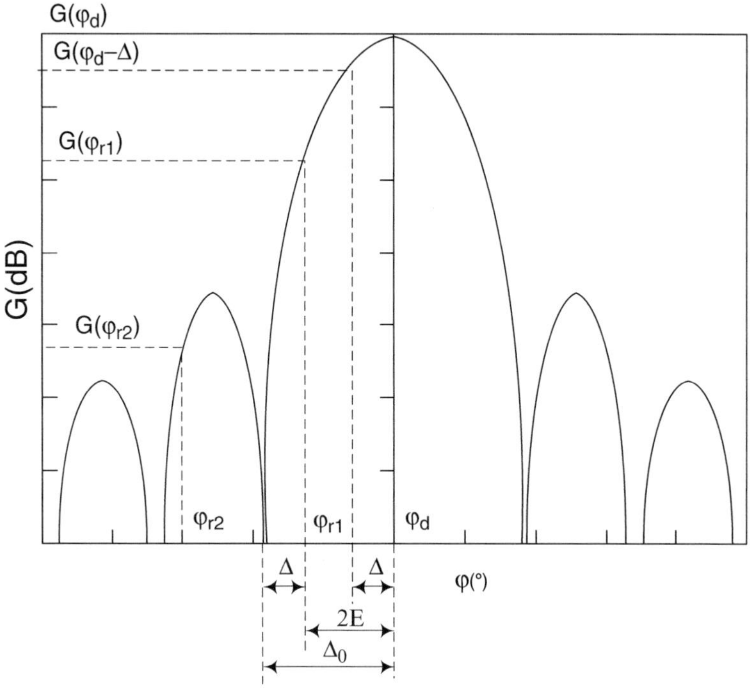

Fig. 5.18 Spatial filtering characteristics of antenna pattern

main lobe points to ground and the corresponding antenna gain is $G(\varphi_{r1})$. Antenna sidelobe gain is $G(\varphi_2)$ when the sidelobe points to ground. Therefore, the antenna pattern of narrow beam acts as the spatial filter. For "main lobe pointing to ground," S/J improves times of $G(\varphi_d)/G(\varphi_{r1})$. The narrower the beam is, the higher $G(\varphi_d)/G(\varphi_{r1})$ will be. Thus, the narrow beam can be used to improve the capability of anti-multipath interference. For "sidelobe pointing to ground," decreasing the sidelobe level can improve $G(\varphi_d)/G(\varphi_{r2})$, so low sidelobe antenna is used to improve the capability of anti-multipath interference. Therefore, the most direct and effective way for anti-multipath interference is to use low sidelobe and narrow-beam antenna [9].

2) Offset-axis tracking: according to the image theory of multipath rays, for low-elevation case, when the elevation of the target is E, the angle between the direct ray and multipath reflected ray is 2E (refer to Expression 5.44). What shown in Fig. 5.17 is such an angle relationship when the main lobe points to ground. From Fig. 5.17, we can see that if the electric axis of the antenna is offset downward for an angle of Δ, making the reflected ray to be located around the sidelobe zero, the multipath interference will be greatly decreased. Although it decreases $G(\varphi_d - \Delta)$ compared with $G(\varphi_d)$, S/J is improved a lot. "Offset-axis angle Δ" can be obtained according to Fig. 5.17:

$$\Delta = \Delta_0 - 2E \tag{5.73}$$

where Δ_0 is the width of the main lobe zero and E is the elevation of the target.

"Offset axis" can be achieved by using "closed loop" (offset-axis tracking) or "open loop" (offset axis). Closed loop can be achieved by Fig. 2.87, where the scanning voltage at the error voltage output of the tracking receiver needs to be changed to an error DC voltage ΔU of the loop:

$$\Delta U = K\Delta \tag{5.74a}$$

where K is angle error characteristic slope at the error output of the tracking receiver. This ΔU will make the electric axis of the antenna deviate Δ from the target. For achieving by open loop, the elevation servo loop is to be opened, and the azimuth servo loop is still closed. Because the multipath interference usually takes place at the elevation branch, the azimuth can be closed to realize automatic tracking. When the target is at low altitude and long distance, the elevation is of little change, so the target is still within the beam when the elevation branch is in open-loop status. At this moment, there is an error voltage output at elevation branch, and it can be used to estimate the elevation value of the target after being subject to dynamic tracking and filtering.

3) Adaptive nulling phased array antenna: it uses an array antenna to conduct the amplitude and phase weighting on signals output by each array element, making the antenna gain $G(\phi_r)$ at the direction of multipath reflected wave to be zero and the antenna gain $G(\varphi_d)$ at the direction of direct wave to be the maximum,

so as to form a spatial filtering at the spatial direction to improve S/J. Adaptive adjustment of amplitude/phase weighted value is required because φ_d and φ_r are dynamically varied with time due to the movement of flight vehicle. The principle of adaptation is making the zero point of the antenna deepest at the direction of multipath reflection and making S/J the maximum. The calculation methods of adaptation include minimum mean square error, minimum jamming-to-signal ratio, maximum likelihood method, etc. You can compare and select a proper one according to the actual engineering needs.

(2) Higher operating band

It has two functions: ① Improving operating frequency can narrow the beam under the same antenna aperture. As previously mentioned, narrow beam can improve the capability of anti-multipath interference. In other words, it can make the antenna operate at a lower elevation under the same multipath interference environment, so it has a longer operation range for the target at the same altitude. ② After the frequency band is improved, when the concave-convex size of reflecting surface is comparable with the wavelength, diffuse reflection will be caused, presenting as the increase of "electrical roughness" on the reflecting surface. The relationship between diffuse reflection coefficient ρ_s and frequency f is as shown in Fig. 5.13. Diffuse reflection coefficient will decrease with the increase of f. The total diffuse reflection coefficient $\rho = \rho_s \rho_0$, so ρ will decrease and the multipath interference signal will decrease too. The disadvantage of improving operating frequency is that the narrow beam will increase the difficulty of angle acquisition and cause the impact of rain attenuation. One potential solution is to use dual-band scheme with the high-frequency band (e.g., K_a or K_u) for achieving anti-multipath, high-speed data transmission, high accurate measurement, and strong EW capability, while the low-frequency band (e.g., UHF or S) for achieving wide-beam acquisition and guide and obtaining small rain attenuation. The low-frequency band can also be used for low-speed data transmission to ensure the transmission of command, telemetry, and GPS data or served as a backup channel. Its multipath fading is a flat fading which is easier to counter than the frequency selective fading.

(3) Reserve sufficient fading margin

For the flat fading of narrowband signal, fading margin can be used to make up the E_b/N_0 decrease caused by flat fading to ensure the requirement of bit error rate can be met. Generally, 10–20 dB fading margin is required.

(4) Anti-clutter net and clutter absorbing floor

For a ground station, installing anti-clutter nets and clutter absorbing floors around the multipath reflection points (calculated according to Expression 5.41) can decrease the multipath interference reflected wave. This scheme is useful for fixed station with a rather high cost. For airborne antenna, the solution is to paint absorbing coating or to sticker absorbing material on the multipath reflection area.

(5) Considerations of antenna polarization mode selection

1) Vertical polarization is better for low-elevation operation: from Figs. 5.9, 5.10, 5.11, and 5.12, we can see that for horizontal polarized wave, whatever the grazing angle is, ρ of the reflected wave is always close to 1 and phase angle is close to 180°, which means the multipath reflected wave will offset the direct wave and so cause a serious attenuation. When using vertical polarized wave, ρ of the reflected wave is less than 1 and phase angle will decrease with the increase of elevation, so the offset effect of the multipath reflected wave can get alleviated.
2) For high elevation, circular polarization can be used. Circular polarized wave is composed of vertical polarization component and horizontal polarization component with a phase difference of 90°. From Figs. 5.9, 5.10, 5.11, and 5.12, we can see that when the grazing angle changes, the phase angle of horizontal polarization component is kept around 180°, but the phase angle of the vertical polarization component is 180° at low elevation, i.e., its phase relative to horizontal polarization component remains unchanged, so the rotational direction of circular polarization between the reflected wave and the direct wave remains unchanged, and no polarization isolation is caused on the direct wave. But when the elevation increase gradually, the phase angle of vertical polarization component will change to 0° from 180°, which makes the rotational direction of circular polarization of the reflected wave gradually reversed. It gradually has a rotational direction opposite to that of the direct wave, thus polarization isolation is gradually caused. From Fig. 5.12, we can see that, for a common land, when the grazing angle is over 20°, the phase of the vertical polarized reflected wave will become 0°, the reflected wave will become elliptically polarized wave and anti-multipath function is possessed. When the grazing angle is 90°, it is a completely inversed circular polarized wave, and the polarization isolation between the direct wave and the reflected wave is the highest at this moment.

The multipath interference gets stronger at a low elevation, so theoretically if vertical polarization is selected, a certain anti-multipath interference effect can be obtained. The disadvantage of vertical polarization is that when the attitude of the flight vehicle changes, the polarization direction of the antenna will change too, which make it mismatch with the vertical polarization of the ground antenna and cause polarization mismatch loss. Therefore, when the attitude of the flight vehicle is stable, vertical polarization can be selected. If the attitude of the flight vehicle changes greatly, it is required to compare the vertical polarization antenna and circular polarization antenna in actual engineering design according to the changes of flight vehicle and realization difficulty of the antenna. A compromise proposal is that the ground station uses vertical polarization and the flight vehicle uses circular polarization antenna.

(6) Selection of antenna erection location

The basic principle of station distribution is to prevent the reflected rays shown in Fig. 5.14 from entering the antenna or to make the reflection coefficient ρ at the reflection point small.

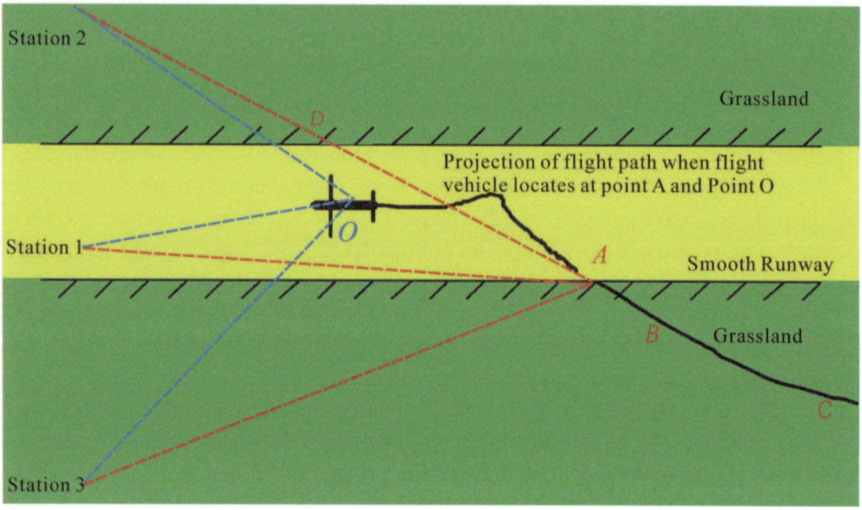

Fig. 5.19 Top view of multipath interference and antenna erection location

1) Select the installation location for the receiving antenna to make the reflection coefficient ρ at the reflection point small, including not selecting the smooth ground with strong reflection for station distribution. There should be no reflection surface with large reflection coefficient ρ between the receiving and transmitting antennas involved to generate multipath interface. For special usage, such measures as plowing at the reflection point of the multipath signal can be taken to reduce ρ.

What shown in Fig. 5.19 is the horizontal projection top view when UAV entering the runway and landing. CBAO is the UAV's flight path (as an example). Light and ray theory is used to draw the reflected ray diagram at point A and point O. The horizontal projection of the reflected ray is shown in dotted line (the long dotted line is the reflection at point A and the short dotted line is the reflection at point O). The multipath reflected signal generated by the part at which the reflected ray and the runway overlap with each other may enter the receiving antenna. From the figure we know that the overlap when station is arranged at station 1 is the maximum, station 2 the second place, and station 3 the minimum. Therefore, station 3 is the best and station 1 is the worst for station distribution. Figure 5.20 shows the front view of multipath reflected rays when the station is arranged at station 1.

If omnidirectional antenna is used by both the receiving and transmitting antenna, and the antenna has no spatial filtering function, when the target enters the sky over the runway from point A and flies at any altitude, the strong multipath reflected signal generated by the runway may enter the receiving antenna provided that the "mirror image" optical ray conditions shown in Fig. 5.20 are satisfied. The area between "mirror image" reflected rays is the multipath reflection generation area. When the station is arranged at station 2, its reflection area is

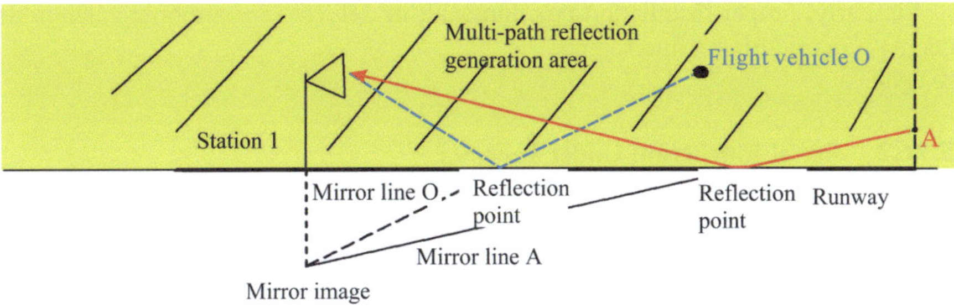

Fig. 5.20 Front view of multipath reflected rays when station is arranged at station 1

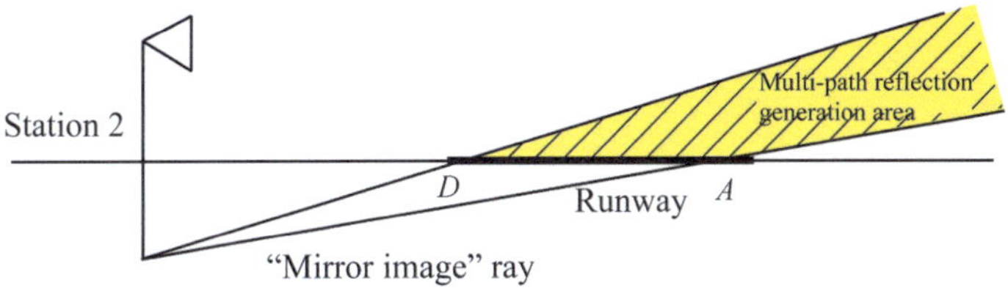

Fig. 5.21 Front view of multipath reflection when station is arranged at station 2

as shown in shadow area in Fig. 5.21. If the antenna is narrow beam, the multipath in some area may be attenuated by additional spatial filtering due to spatial selective action.

From Fig. 5.21 we can know that the strong reflected signal generated by the runway can enter the receiving antenna only when the target is at certain altitude range. This reflection area is at the same side as that of the flight path of the target as shown in Fig. 5.19. The target may pass through this reflection area so as to receive the strong signal reflected by the runway. However, compared with Fig. 5.20, this reflection area is much smaller, so the multipath interference probability is decreased.

The reflection area when station is arranged at station 3 is similar to that of Fig. 5.21, but the flight path of the target is at the other side. Therefore, the target will not pass through the reflection area to receive the multipath interference signal.

2) When erecting the antenna on a mountain, make sure the corresponding reflected signal can be reflected by the mountain so as to prevent it from entering the antenna. Besides, the more the antenna, the larger the angle between the mirror image ray and the horizontal line, that is to say, the larger the angle the reflection direction deviating from the antenna direction, the better the spatial filtering of the antenna.
3) Erect the antenna behind the natural shield (e.g., small hill, hillock, etc.) to prevent the reflected signal from entering the receiving antenna.

In a word, there are many similar schemes for station distribution and antenna erection. A proper scheme shall be selected according to local conditions and the "mirror image" ray reflection principle shown in Fig. 5.14.

Similarly, proper installation position shall be selected for airborne antenna to decrease the multipath interference.

5.3.6.2 Signal Extraction Under Low S/J

(1) Modulation/demodulation mode for anti-multipath interference

1) Narrowband signal: for narrowband signal, multipath interference is a flat fading channel. It is of such burst interference features as "interruption" and "discontinue" when the fading depth of the carrier is great. In this case it is better to use noncoherent demodulation mode. The reason is that PLL is required to be used to extract coherent reference signal in coherent demodulation. The instantaneous fading of carrier amplitude will introduce instantaneous losing lock of PLL. A period of time for PLL acquisition and stability is required after the carrier intensity is recovered. Long-time bit error, discontinuous losing lock, and even disable operation will be caused in such period of time. Noncoherent demodulation has no PLL, so it is free of such problems as discontinuous losing lock and reacquisition time. The SNR output by demodulation is low, and bit error is caused when the carrier is in deep fading. When the carrier intensity is recovered, it will immediately output demodulated signal. After demodulation, the means having strong anti-burst interference capability, such as RS error correction coding and interleaving, can be used to further decrease bit error. For additive white Gaussian noise at the same bit error rate, a higher E_b/N_0 is required by noncoherent demodulation compared with coherent demodulation, while noncoherent demodulation is the better choice when multipath interference is the principle contradiction. The common modulation/demodulation is noncoherent binary frequency shift keying (BFSK) for which only two filters and envelop detector are required to demodulate the code element signal. Another advantage of using frequency modulation mode is that it is insensitive to amplitude fluctuation and has better anti-interference capability. Besides, it is constant envelope modulation, and the power amplifier can operate under saturation state.
2) Wideband signal: for wideband signal, multipath interference is a frequency selective fading channel. At present, "multi-carrier modulation technique" and "single-carrier frequency domain equalization" (SC-FDE) are generally used for anti-multipath interference. In IEEE820.11 and 802.16d standard, SC-FDE and DFDM multi-carrier modulation technique are suggested to be used. The main difference between single-carrier modulated signal and multi-carrier modulated signal is that the former has a constant envelope and the latter has a nonconstant envelope. Therefore, the former is more suitable to be applied on flight vehicles and its power amplifier can operate under saturation state. The two modulation technologies are given as below:

① OFDM modulation technique: multi-carrier modulation techniques include OFDM, FDM, SQAM, SC-FDMA, MC-CDMA, etc., among which OFDM is the typical one and frequently applied.

The influence of multipath interference on the bit error rate is related to code rate. Figure 3.50 shows that for a certain group delay fluctuation, the lower the code rate is, the less the influence on the bit error rate. Therefore, the transmitted data will undergo serial-parallel conversion by OFDM technique and be distributed to several orthogonal subcarriers for frequency division multiplexing to divide the wideband signals into several parallel sub-channels. Because the bit error rate of sub-channel is greatly decreased, the frequency selective channel of wideband is decomposed of several parallel flat fading channels to decrease the bit error rate. Besides, cyclic prefix is used as a guard interval and the guard interval is made to be greater than the maximum delay spread so as to further eliminate the inter-symbol interference introduced by multipath delay spread. Another advantage of OFDM is using IFFT and FFT to modulate and demodulate signals. Its complexity depends on the complexity of FFT calculation and thus simplifying the complexity of realization. In addition, bit rate on all subcarriers is significantly decreased, so OFDM is used for high-rate data transmission.

The disadvantage of OFDM is that the signal has nonconstant envelope and has high peak-to-average ratio for it is a multi-carrier signal. It has high requirements on the linearity of the power amplifier of the transmitter. Besides, the orthogonal subcarriers formed in the transmitter are also complicated, so the volume, weight, power consumption, and cost of the transmitter are high. In addition, if the orthogonality of subcarriers in OFDM is poor, the signals in the sub-channel will interfere with each other, causing the decrease of signal to interference ratio and the increase of bit error rate. Therefore, OFDM has high requirements on frequency offset (including Doppler frequency shift), phase noise, and carrier synchronization loop. Thus, the complexity of the system and difficulty of realization are increased. Due to the abovementioned defects of OFDM, it is not suitable to be applied in the TT&C and information transmission system of flight vehicles.

② Single-carrier modulation mode: obviously, compared with "multi-carrier modulation," the most important feature of "single-carrier modulation" is that it can achieve constant envelope modulation and it is easy to be modulated. Because the transmitted signal is wideband signal and the corresponding multipath channel is a frequency selective fading channel, there is group delay characteristic fluctuation and amplitude-frequency characteristic fluctuation which will cause bit error (as shown in Figs. 3.50 and 3.52). Therefore, the channel equalization technology shall be correspondingly used when single-carrier modulation is used. "Single carrier" and "channel equalization" are closely integrated. The detailed introductions of the two are as follows:

(2) Adaptive equalization technique

Equalization technique includes "time domain equalization" (TDE) and "frequency domain equalization" (FDE). TDE is, in the time domain and for time response, making the waveform of the whole time domain impulse response including the equalizer meet the requirements of free from intersymbol waveform interference. For its realization, adaptive calculation and weighting are required to be applied on many filter taps, so the equipment is complicated, the convergence speed is slow, and extra noise is caused. If the algorithm with a fast convergence speed (e.g., RLS) is used, the calculation will be complex, and its complexity will increase rapidly with the multipath delay expansion, which will make the equipment more complicated and difficult to be used. The performance (especially the dynamic adaptive capability) of time domain equalizer is limited, and the equipment will be very complicated in order to meet the requirements of anti-multipath, so it is not suitable to be applied in the TT&C and information transmission system of flight vehicles.

FDE is used to conduct channel equalization on transmission function in frequency domain, as shown in Fig. 5.16 (2-path flat reflection is shown by the Figure). There is fluctuation in amplitude-frequency characteristic, phase-frequency characteristic, and group delay characteristic. Thus, it is required to connect in series an equalizer to make the product between the transmission function of equalizer and frequency selective transmission function a constant, so as to meet the requirements of distortionless waveform. In this way, the bit error caused by intersymbol interference can be avoided. For the TT&C and information transmission system of flight vehicles, there are ① changes of delay difference between the direct wave and the reflected wave; ② changes of complex reflection constant of the reflection point; ③ changes of the reflector, including concave-convex characteristics of the reflecting surface, position and direction of the reflector, number of reflection paths, etc.; ④ changes of Doppler frequency, especially the changes of Doppler frequency caused by the changes of path difference $\triangle R$. The transmission function of frequency selective channel is dynamic time varying due to such factors. The transmission function of the equalizer is required to be adjusted in real time to match with it. Therefore, adaptive equalization is used.

Adaptive equalization principle of SC-FDE is as shown in Fig. 5.22. First, at the transmitting end, the data to be transmitted are divided into blocks and then transmitted in form of "data blocks." Each data block contains user data and training sequences which are known. After being received at the receiving end, such sequences can be used to estimate the frequency selective characteristics of multipath channel which then are used as the reference signals to perform adaptive weighting on the coefficients of the equalizer and change its frequency response. Such continuously and automatically adjusting will make the total group delay characteristics and amplitude-frequency characteristics approach to constant and complete the adaptive equalization.

In Fig. 5.22, at transmitting end, the training sequence is inserted into the data block with a cyclic prefix. The cyclic prefix is not only used to transmit the training sequence but also work as guard interval. The length of the guard interval should be longer than that of the delay spread to decrease the intersymbol interference. Its disadvantage is that it will take some time, increase the overhead and decrease channel efficiency. However, by comprehensive balance, the benefits obtained are still the main points. The data demodulated at the receiving end will be subject to arrival of frame test and then stored in the memory according to the frame (a data block). The training sequence in one branch will be taken out for multichannel estimation, and the prefix in another branch will be removed for FFT conversion,

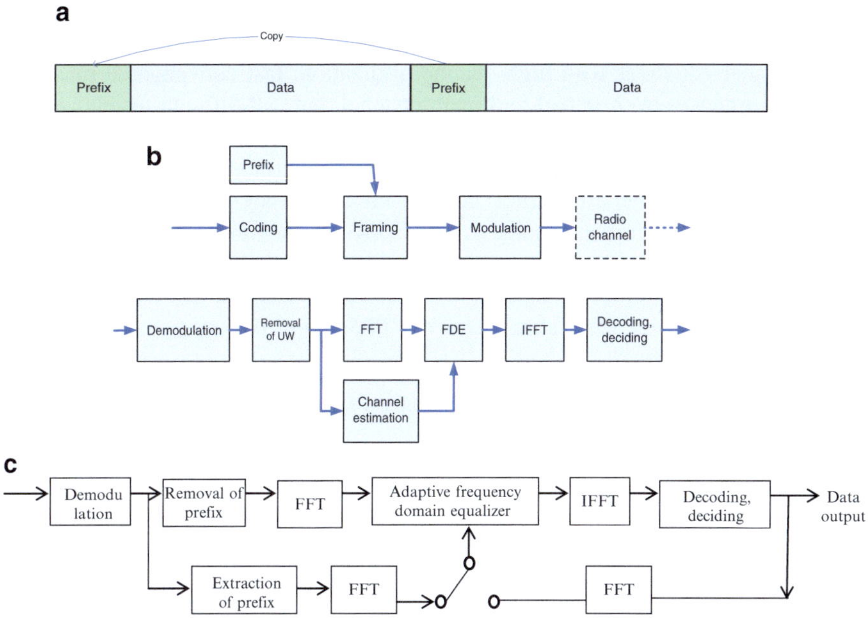

Fig. 5.22 Principle block diagram of SC-FDE. (**a**) Frame structure. (**b**) Block diagram of modulator. (**c**) Block diagram of demodulator

and then perform subtraction with the data of the training sequence after FFT to obtain the error signal. After that, use adaptive algorithm and change the weight coefficient of the equalizer to realize frequency domain adaptive equalization. Rapidly track the changes of multipath channel and finally use IFFT to convert back to the time domain waveform for decoding waveform decision. In this adaptive system, reference signal is the received training sequence. Adaptive criterion can be minimum mean square error (MMSE), least squares criterion (LS), etc. For adaptive algorithm, least mean square (LMS), recursive least squares (RLS), etc. can be used. The convergence rate of the latter is faster by one order of magnitude, and its steady state error is smaller, so it is more suitable for high-dynamic time-varying channel. However, its calculation complexity is much higher than LMS. Adaptive calculation is based on each data block; therefore, the shorter the data block is, the faster the convergence rate will be. In the observation time occupied by each data block, the less the changes of multipath channel characteristic are, the better the adaptive performances for dynamic time-varying characteristic will be.

In Fig. 5.22, the mode when the switch is placed at another endpoint is an operating mode for decision feedback equalizer. The time domain output waveform after decoding decision will be converted into frequency domain through FFT conversion, which is subtracted by the equalized signal in the current frame to generate an error signal. It will change the weight coefficients after adaptive equalization algorithm and the updated weight coefficients will play their role in the next frame. This kind of equalizer does not need to transmit the training sequence. It is a nonlinear equalizer with high channel utilization, fast convergence rate, and good performances even in bad multipath channel, but it is difficult to realize.

Compared with OFDM, there are influences on MC-FDM performances caused by frequency offset, phase noise, and carrier loop synchronization accuracy. However, they are reduced after IFFT conversion since they have been averagely divided into each symbol. Besides, MC-FDM can achieve constant envelope characteristics and fast adaptive convergence, and its equipment is simpler than that of OFDM. Therefore, it is more suitable to be used for the dynamic time-varying multipath wideband channel of the TT&C and information transmission system of flight vehicles.

(3) Improvement of receiver's dynamic performance

Multipath interference will cause strong amplitude fluctuation and frequency fluctuation. In this case, if the receiver cannot adapt to this kind of high-dynamic change, its normal operation will be interrupted (e.g., receiver losing lock). Therefore, in order to resist multipath interference, one of the basic measures is to improve receiver's dynamic performance, including quick AGC, PLL, and delay lock.

(4) "Memory" and "fast acquisition"

Lose lock of carrier loop, bit synchronization loop, range tracking loop, and frame synchronization out of synchronization, etc. may happen in short time when receiver's dynamic performance is not enough or SNR caused by scintillation fading is too low, leading to some possible serious consequences. For example, angle-tracking error caused by lose lock of carrier loop may result in antenna pointing away from the target; out of synchronization of bit synchronization and frame synchronization will give rise to frame drop and loss of telecommand and telemetry signals, in particular, loss of command signal may bring about disastrous consequences. For example, in an unmanned air vehicle (UAV) TT&C system, precise, exact, and reliable telecommand is crucial when the unmanned air vehicle is controlled to perform wheel landing over the runway. But when the UAV is about to land, the receiver may not receive command signals because of serious multipath reflective interference from the runway. In this case, the usual design is that the output command signal is null which is corresponding to the command of let the steering lever keep the UAV level flying. If in this case the command happens to be a command climbing, the steering lever will change from climbing to level flight, namely, pulse control of diving is generated, which will lead the UAV to dash onto the runway and break. One solution is that in the case signal lose lock or out of synchronization, use a memory circuit to "memorize" the output signal of the previous moment, then pulse interference of this type will not happen, and after the signal is locked, get rid of "memory" and restore the original normal output. As time for "memorizing" is limited, there is requirement for receiver fast acquisition when the signal becomes strong from brief fading.

As the phase-locked receiver comes with the shortcoming of "losing lock," phase demodulation method of "frequency discrimination + integral" or FM modulation/demodulation method can be adopted in resistance against multipath interference.

(5) Error correction coding for anti-multipath interference

Error correction coding will get coding gain to provide fading margin for multipath fading, but different from additive white Gaussian noise channel, fading channel has its own special problems. Following codes are usually used:

1) RS coding: As sudden signal fading caused by multipath interference will produce burst bit error, adoption of RS error correction coding capable of resisting sudden interference will generate certain anti-multipath effect.
2) Interleaved code: Multipath fading, if lasting relatively long, tends to cause a sheet of code element errors, and common error correction coding usually fails to correct such error. Therefore, if interweaving is used to disperse and randomize such burst errors and then is combined with other coding methods, average bit error rate can be decreased. However, without combination with other coding methods, interweaving alone cannot decrease such rate. Moreover, interweaving comes with relatively long transmission waiting time, which shall be noted in system design phase.
3) Concatenated codes: The one of most importance is "convolutional code + RS concatenated code." Under the condition of certain time and bit rate, fading time is much longer than the time that can be corrected by RS codes. In this case, if conduct convolutional coding at first and then perform interweaving to disperse burst errors, which are corrected by use of RS code, the outcome will be much more desirable.
4) Grid coding: According to the analysis, TCM owns relatively better error correction effect on bit error caused by frequency selective fading.
5) Space time coding: It combines space diversity, time diversity, and frequency diversity, improving anti-multipath performance to a great extent. But it generally requires multi-antenna, which makes system relatively complex.

(6) Use of spread-spectrum signal to deal with multipath interference and the rake receiver [11].

The anti-multipath interference capability of spread-spectrum signals falls into two conditions: Multipath signals that can be decomposed temporally and those that cannot be decomposed (temporally overlapping). For the anti-interference capability of those can be decomposed, there is an analysis as below: Let direct signal be $U(t)$, reflected signal be $U(t+\tau)$, then the receiver will receive the sum of $U(t)$ and $U(t+\tau)$ signals. Assume other conditions of these two signals are the same except there exists delay difference τ; then its power of composite signal is

$$\begin{aligned} P_{\Sigma} &= E(U(\mathrm{t}) + U(t+\tau)]^2 \\ &= E\big(U^2(t) + E\big(U^2(t+\tau)\big] + 2E[U(t)U(t+\tau)] = 2p_0[1 + r(\tau)] \end{aligned} \quad (5.74b)$$

where $[E\ldots]$ is mathematical expectation and $r(\tau)$ is the normalized correlation function of $U(t)$.

$$r(\tau) = \frac{R(\tau)}{R(0)} \tag{5.75}$$

where

$$R(\tau)] = E[U(\mathrm{t})U(t+\tau)] \tag{5.76}$$

where $R(\tau)$ is the autocorrelation function of $U(t)$ and $R(0)$ is the autocorrelation function of $U(t)$ when $\tau = 0$, namely, its average power.

$$R(0) = E\left[U^2(t)\right] = P_0 \tag{5.77}$$

When pseudocodes are used in spread spectrum, video PN codes (rather than carrier signals) share similar relevant parameters with white noises.

$$R_{\mathrm{cr}}(\tau) = \left\{ \begin{array}{ll} \frac{A^2}{T_{\mathrm{C}}}(T_{\mathrm{C}} - |\tau|) & |\tau| < T_{\mathrm{C}} \\ 0 & |\tau| \geq T_{\mathrm{C}} \end{array} \right\} \tag{5.78}$$

where A is pseudocode amplitude, T_{C} is the code element width of pseudocode, and τ is delay.

Substitute relevant functions into Expression (5.75) to obtain

$$P_{\Sigma} = 2P_0\left[1 + \frac{R_{cr}(\tau)}{R_{cr}(0)}\right] \tag{5.79}$$

Average power is

$$P_{\mathrm{cr}} = R_{\mathrm{cr}}(0) = A^2$$

Then there is

$$P_{\Sigma} = \left\{ \begin{array}{ll} 2P_0\left[1 + \left(1 - \frac{|\tau|}{T_c}\right)\right] & |\tau| < T_C \\ 2P_0\left[1 + \left(1 - \frac{|\tau|}{T_c}\right)\right] & |\tau| \geq T_C \end{array} \right\} \tag{5.80}$$

Expression (5.80) shows that for video signals, when $\tau \geq T_{\mathrm{c}}$ and direct and reflected signals can be decomposed, the sharp characteristics of pseudorandom codes make multipath interference completely independent. When multipath delay

$|\tau| < T_C$, reflected signals overlap with useful signals, making power of synthetic video signals take on fluctuation fading. The extent of fluctuation depends on the ratio between $|\tau|$ and T_c. By substituting the power into Expression (5.80), we see that it only takes on fluctuation fading when path propagation delay τ is smaller than code element width T_c of pseudocodes (this conclusion is made by contrasting with routine communication systems). For example, when $R_c = 10$ Mb/s ($T_c = 0.1$ μs) and the delay difference between path delay of reflected signals and direct signals is smaller than 0.1 μs, namely, path difference which is within 30 m occurs, then fluctuation fading happens. In this case, de-spread loss happens and produces partial de-spread and partial spread, which is the reason the anti-multipath interference capability decreases, but a part of such capability still remains. When $\tau = 0$, there exists no anti-multipath interference capability. But when $|\tau| \geq T_C$, $r(\tau) = 0$; at this time, DS spread-spectrum system deals with multipath reflected signals as interference noise. When the above $|\tau| < T_C$, multipath reflected signals overlap with direct signals; this kind of multipath signals are non-decomposable. When $|\tau| \geq T_C$, multipath reflected signals and direct signals are temporally separated; this kind of multipath signals are decomposable (decomposability, in the broad sense, can extend to each multipath reflected signals). As designable T_c is quite small when spread spectrum is adopted, the less $|\tau|$ overlaps with T_c, the stronger the capability of anti-multipath interference will be.

The spread-spectrum technique can not only resist multipath interference but compose each multipath reflected signal after decompose them, making them useful signals. This type of technique is called multipath diversity technique, a kind of implicit diversity technique. The way to realize it is using a rake receiver. As delay of multipath components is usually relatively large, in the case T_c of code type and chip is chosen properly, autocorrelation of pseudorandom sequence can ensure tiny correlation between multipath components. Therefore, when propagation delay between each path is equivalent to or longer than chip period, multipath signals can be treated as uncorrelated noise, which can be decomposed by using correlation of pseudorandom codes. In the case a frequency selective fading channel is used, on one hand, signals will be in severe distortion during transmission; on the other hand, when chip bandwidth is much larger than that of the frequency selective channel in the spread-spectrum system, the effect of frequency diversity, however, will happen. When transmission delay between multipath signals is longer than chip period, fading between multipath components is barely correlated. If a spread-spectrum receiver is used to discompose and then compose such barely related multipath components, signal-to-noise ratio of the received signals can be improved to implement the effect of time diversity. If chip width is W, then the time resolution will be $1/W$. In the case the largest delay spread caused by multipath is τ_M, the separable signal component number will be $\tau_M W$. According to the relationship of $\tau_M = 1/(\Delta f)_c$ between τ_M and relevant width $(\Delta f)_c$, the number of separable signal component can be expressed as $W/(\Delta f)_c$. In this way, use of broadband signal can be seen to gain $W/(\Delta f)_c$ frequency diversities.

Basic structure of a rake receiver is shown as in Fig. 5.23. The receiver comprises of a set of correlators. The delay difference dealt with by adjacent

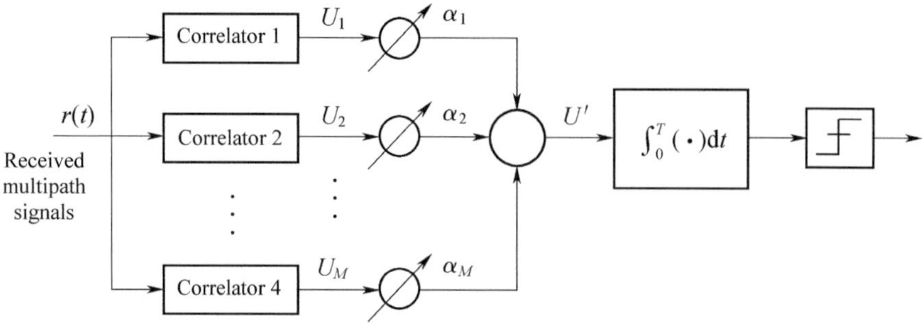

Fig. 5.23 Basic structure of rake receiver

correlators is integral multiples of $1/W$. Therefore, $1/W$ can be taken as a chip period T_c. The relative delay of multipath components, if shorter than a single T_c, cannot be distinguished. Such components are therefore treated as on the same single path which is actually formed by multiple paths. So there will be fading on this single path. Each correlator from the general received signals separates multipath signals that fall onto relevant single path delay and accumulate energy. Total M strongest multipath components are detected and combined optimally, and then are demodulated and decided.

A typical rake receiver contains M correlators and a weighting network, with the former used to capture M strongest multipath components and the latter used to carry out linear superposition and combination on correlator output. Correlator 1 has the same delay as multipath component $s(t-\tau_1)$ does. Correlator 2 has the same delay as multipath component $s(t-\tau_2)$ does, and its correlation with $s(t-\tau_2)$ is much strong, but its relation with $s(t-\tau_1)$ is quite weak. In this way when correlators are receiving corresponding signals, non-corresponding signals are filtered, in which way multipath separation effect is reached.

Output of M correlators is separately noted as $U_1, U_2 \ldots U_m$; their corresponding weights are $\alpha_1, \alpha_2 \ldots, \alpha_M$. Similar to maximal ratio synthesis, weighting coefficient can be determined by power output or signal-to-noise ratio of each correlator; if the power output or the signal-to-noise ratio of a correlator is comparatively small, the corresponding weighting coefficient will be allocated comparatively small values too. After combination, signal U' can be expressed as

$$U' = \sum_{m-1}^{M} \alpha_m U_m \tag{5.81}$$

where α_m is the normalization coefficient.

$$\alpha_m = \frac{U_m}{\sum_{m-1}^{M} U_m^2} \tag{5.82}$$

Comparing with signal estimation provided by single path component, output of each correlator gains composite signals through weighting, the transmitted signal estimation provided by which is more accurate. If there is only one correlator in the receiver and the signal is interfered severely, the receiver is probably unable to restore the original signal, which will increase the bit error rate. However, under the condition that the rake receiver is used, even though the output signals of a correlator are damaged, there is possibility that output signals of other correlators are unimpaired; after weighting, influence brought by damaged signals will be lowered. The rake receiver will conduct final decision on M statistical results acquired through independent reception and combination. This kind of diversity reception overcomes signal multipath fading with sound effects, thus the receiving performance of spread-spectrum signals is improved.

(7) Use of diversity techniques against multipath interference

The abovementioned method of using spread spectrum to realize multipath diversity is just one of the diversity techniques. Other such techniques, corresponding to each type of fading, are space diversity, frequency diversity, time diversity, polarization diversity, angle diversity, etc.

1) Space diversity and multi-antenna diversity

Place multiple receiving antennas at different space positions. When the distances between them are up to several wavelengths, diversity channels almost independent from each other are obtained, as paths, grazing angles of reflected waves, and reflection coefficients of their received signals are different. In this case, fading processes of their received signals are almost independent from each other. If such signals are dealt with by using the diversity combining technique, gain of diversity combining will be acquired, reducing signal fluctuation fading to a large extent. In this context, space diversity is realized by using multi-antennas which gives it another name "multi-antenna diversity." The higher the operating frequency band is, the less the physical separation distance between antennas required in space diversity will be. When S-band is used, space diversity can be realized with a distance of 10–20 cm. The advantages of space diversity combining are that diversity combining gain is obtained and fading is reduced by using multiple receiving antennas and receivers; the disadvantage is that the equipment is complicated. The more the number N of diversity antenna is, the more diversity synthesis gain will be. But, as N is becoming larger and larger, the growth rate of gain becomes less and less. In engineering, the case is usually be $N = 2$–4.

A relatively common space diversity combining method is height diversity. It means that antennas are placed at different heights. As paths, grazing angle ψ, and reflection coefficients of their received signals are not the same, total received signals can be strengthened after diversity combining. Horizontal and vertical polarization conditions are shown separately as in Fig. 5.24a, b (each curve represents a value of ψ). The composite signal formed due to mutual interference of direct signals and reflected signals is shown as in Fig. 5.24. It is obvious that at a certain ψ, the received composite signal fluctuates according to growing antenna

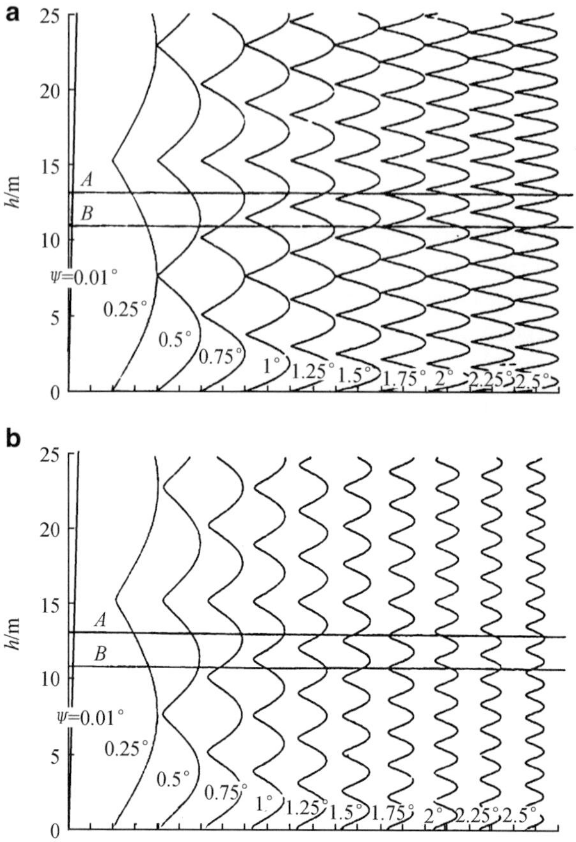

Fig. 5.24 Relation between received signals and h. (**a**) Horizontal polarization and (**b**) vertical polarization

height h. There are multiple composite valley points, but if two antennas are separately placed at the two points with $h = A$ and $h = B$, and are made at the same ψ, the probability of signal valley points appearing at the same time will be reduced considerably. In this case, effects of multipath fading can be mitigated by using the signals received by the two antennas to carry out diversity combining. Moreover, if sliding antennas are used, the valley points in the figure can be avoided by incessantly adjusting antenna height to change ψ. However, its realization is very difficult.

Draw 2 horizontal lines A and B in Fig. 5.24 at heights of the 2 antennas mentioned above. It is observed that when $\psi = 0.75°$, one of the two signals is close to valley bottom; the other is adjacent to the peak. Therefore, signals made by diversity combining are still relatively strong. As ψ is up to 1.75°, both antennas are next to valley points; the received signals are weakened a lot. When ψ is larger than 1.75°, both antennas are no longer at valley points simultaneously, and the total signals received become large again.

Figure 5.25 shows the relation between the total received energy and ψ under the condition that antenna heights are 13 m and 14 m, respectively. It can be seen that

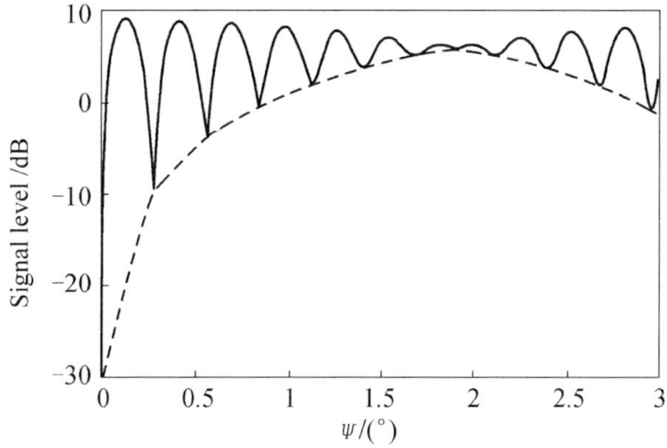

Fig. 5.25 Total energy received when antenna difference between both antennas is 1 m

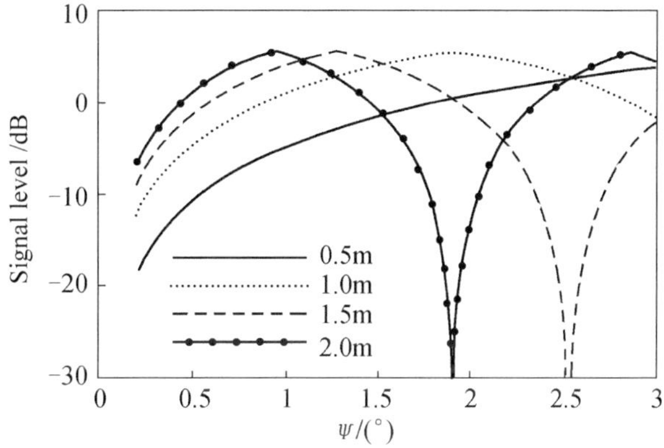

Fig. 5.26 Envelope curves of total energy received by a pair of antennas with different height differences

when $\psi > 1.5°$, relatively strong signals can be received, but as ψ drops, valley depth becomes large progressively. Figure 5.26 displays a set of envelope curves when antenna height difference varies from 0.5 to 2 m. It is clear that different antenna height differences bring on a different range of ψ value which leads to deeper and shallower valley. As what will be described in introduction of angle diversity below, height diversity can be used along with angle diversity.

Moreover, receiving antenna diversity can be equivalently moved to the transmitting terminal by using reciprocal theorem of the linear system, to realize diversity of multiple transmitting antennas.

2) Frequency diversity and frequency hopping

As previously mentioned, there is a kind of frequency selective fading in multipath interference, which means if different carrier frequencies are used for transmission, there will be different fading. In the case where carrier frequency interval is wider than channel correlation bandwidth (refer to relevant references to know that channel coherent bandwidth $\Delta F = 1/L$, L is width of delay power spectrum), fading on different carriers are independent from each other to a great degree. Therefore, when multiple signals at different carrier frequencies are transmitted, multiple diversity channels whose fading similar to be independent are existed. Fading of the composite signal can be reduced by adding the output signals to conduct a weighted average. A system adopting this sort of diversity method is called frequency diversity. An obvious shortcoming of this method is that signals with multiple frequencies shall be transmitted.

Through frequency hopping (FH) spread-spectrum modulation, frequency diversity is applied in some systems. Carrier frequency of the transmitter can be changed periodically by an FH-modulated system and by using a limited carrier frequency set; bandwidth sum of such carriers can be much large. In the case where changed bandwidth of FH carrier is larger than correlation bandwidth of the channel, frequency fading at each hopping will be independent from the adjacent two hopping in a significant measure. An effective FH diversity method is to combine forward error correction coding with interweaving and disperse output symbols of the encoder to multiple independent frequencies. Among such FH systems, only one carrier frequency is used at any time, therefore, frequency diversity of this type is, in fact, the combination of pure frequency diversity and time diversity. FH spread spectrum is a technique to good purpose. Periodical change of carrier frequency will not only realize frequency diversity but also at the same time has the effect of averaging interference, bringing much benefits for system capacity.

3) Time diversity

In the case the receiver and the transmitter are in relative movement, fading change can be presented as a function of time, whose change rate is directly proportional to the relative velocity of the receiver and the transmitter. If two samples of fading process are surveyed at different time, correlation between the samples decreases as their time interval increases. If the time interval of samples is long enough, such samples can be seen as almost independent from each other. The independence of long time interval samples can be used to obtain practically independent diversity channels. The way to realize it is that signals in waiting are sent repeatedly at certain intervals (larger than time correlation interval $\Delta T = 1/B$, where B is the spread width of Doppler shift), then independent diversity signals can be obtained at the receiving terminal and be combined. An effectively applied method is that information bits are coded and interweaved before transmission (of course, inverse processing shall be carried out at the receiver for restore). In coding process, information contents of single bit are dispersed into output symbols of multiple encoders. And interweaving ensures that output symbols of encoders, which are adjacent to each other temporally, are transferred at different time. Moreover, time intervals are long enough to be considered that independent samples of fading process are experienced. Although in many cases, use of forward error correction coding and interweaving can bring much performance gain, in some cases, such gain is limited. The performance relies on moving velocity, namely, performance in case of high velocity shall be better than that in case of low velocity. The reason of performance's dependence on velocity is that span (or length) of the interweaver is always fixed; therefore, there is always a certain moving velocity; if the velocity is lower than it, fading will become so slow that the interweaver is unable to separate symbols effectively by time. In the event that velocity is lower than the certain moving velocity, diversity gain will be reduced. With extremely low velocity or in still, the performance becomes the same as that of the system without diversity. In the process of interweaving and de-interweaving, delay will be introduced into the communication link, which should be controlled within a certain range. Moreover, similar effect will be obtained if accumulated decisions are sent repeatedly.

4) Polarization diversity

Let's discuss polarization diversity without multipath interference. Polarization characteristics of wide-beam transmitting antenna on a fight vehicle are different at different antenna angles. Normally, two kinds of orthogonal components are contained. In the direction close to flight vehicle metal surface, the single-polarized antenna tends to produce two types of orthogonally polarized radio waves. As the angle of flight vehicle antenna relating to the receiving antenna of the ground station is changing during vehicle flight, and the receiving antenna corresponds to the patterns of different directions of the flight vehicle transmitting antenna, the axial ratio of circular polarized wave of the received radio wave received by ground station receiving antenna is changing (as varying from circular polarized wave when the antenna is directly facing to vehicle surface to linear polarized wave when the antenna is parallel to the surface). Polarization of transmitting wave depends on polarization of transmitting antenna. As any attempts to make transmitting antenna polarization match with receiving antenna polarization is impossible, therefore, the solution to this problem is to use the receiving system to receive two types of orthogonally polarized radio waves and optimally combine received orthogonally polarized signals. That is, signals received by the dual-polarization receiving antenna are optimally combined to realize polarization matching of incident wave polarization.

In the case of multipath propagation, Figs. 5.9 and 5.10 show the relation between ρ value and ψ in sea surface reflection. It is obvious that for horizontal polarization H and all ψ angles, $|\rho|$ approximates 1, but the phase approximates 180°, leading to the result that for all ψ angles, peak values are added up by +6 dB while the subtraction effect has generated deep valleys, as shown in Fig. 5.27. The change relation between $|\rho|$ of vertical polarization V and ψ can also be known from Figs. 5.9 and 5.10. $|\rho|$ has a relatively large change scope, and its phase varies from $-180°$ ($\psi = 0°$) to 0°. This leads to the time of occurrence of the valley value (and the peak value) of multipath total composite signal in horizontal polarization is not the same as in vertical polarization. With ψ approximating 0°, it can be seen that two kinds of polarization start to correlate; when $\psi = 0°$, their peaks and valleys appear at the same time, as clearly shown in Fig. 5.27.

Multipath interference can be diminished by using the characteristics as analyzed above.

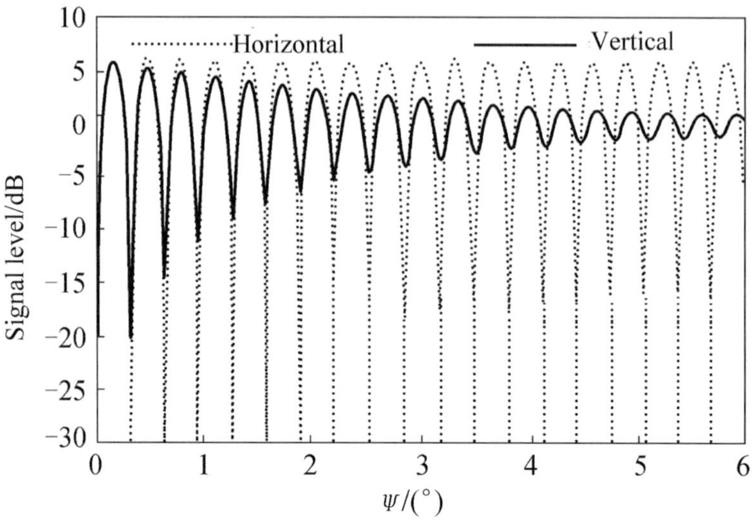

Fig. 5.27 Relation between received signals and grazing angles

- Vertical polarization is applied in transmitting and receiving; the valley point of vertical polarized signals is shallow, except for the grazing angle where it is very small.
- Two kinds of polarized signals are received and combined optimally. In the case $\psi > 6°$, combined signals are without valley points because valley points in horizontal and vertical polarization do not correlate; when $\psi < 6°$, valley depth will become large as the grazing angle tends to 0°. Since phases of both reflection coefficients are close to $-180°$, $|\rho|$ in vertical polarization approximates 1.
- Strive to enlarge the grazing angle, which increases as the height of the receiving antenna adds.

5) Angle diversity

Angle diversity is a means to strengthen received signals by changing antenna pointing. The elevation limitation method is just a good example. With this method, antenna elevation is adjusted near the horizon, making direct signals enter near the peak value of antenna main beam, while reflected signals enter near the first valley point of the main beam. When the direct and reflected signals at the antenna output are in opposite phases, the antenna output is equal to the value of direct signal level multiplied by the gain of main beam peak value subtracting the reflected signal level multiplied by the gain of antenna beam at the incident angle. As the antenna directivity in the direction of reflected signals decreases, only shallow valley appears at the signal fading curve.

Another example is the monopulse combination method by which sum signals and elevation error signals of the monopulse antenna are combined in an optimal

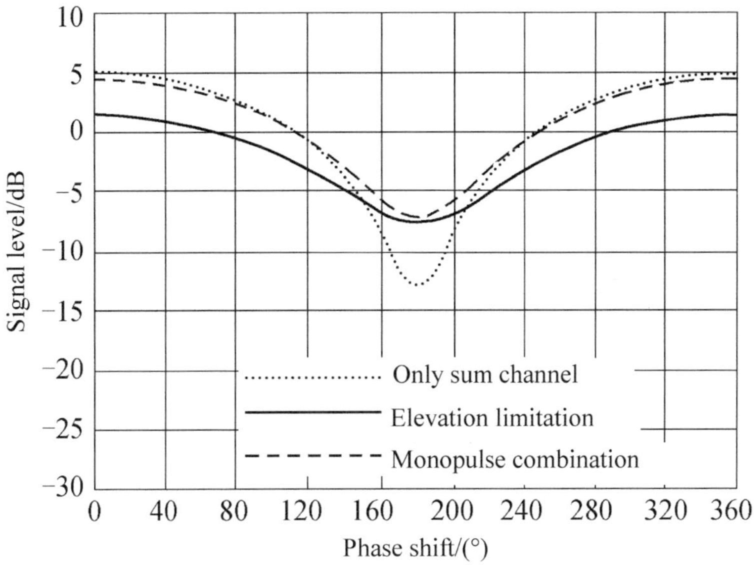

Fig. 5.28 Relation between angle diversity signal level and phase shift between direct/reflected signals

manner. It is equivalent to combine two wide-beam antennas with little difference of vertical interval. This forms a 2-array-element phased array antenna with adaptively adjustable pattern. No matter how accurate the antenna mechanical pointing is (i.e., self-tracking), the strongest signals can always be obtained. This combines the antenna with functions of the optimal receiver/combiner, making its electrical axis point to the direction of incident waves and thereby realizing real-time tracking of incident waves.

The relation between signal level and phase shift between direct/reflected signals under conditions of standard sum channel receiving, the elevation limitation method, and the monopulse combination method is shown in Fig. 5.28. It can be seen that relevantly deep valley appears when the sum channel receiving method is used, while the other two methods only give two shallow valleys. Moreover, valley depth of the latter two methods is close.

Combination of angle diversity and height diversity surpasses any abovementioned diversity technique applied alone in terms of effects of improving multipath fading. Total received energy under conditions of angle diversity, height diversity, and the combination of the two is shown in Fig. 5.29.

Of the five diversity techniques, polarization diversity is the most realizable; hence, it is popularized in C&T system application. In addition to the usage of resisting multipath interference, the polarization diversity technique boasts other strengths, which is the reason it is used in circumstances where there is no multipath interference.

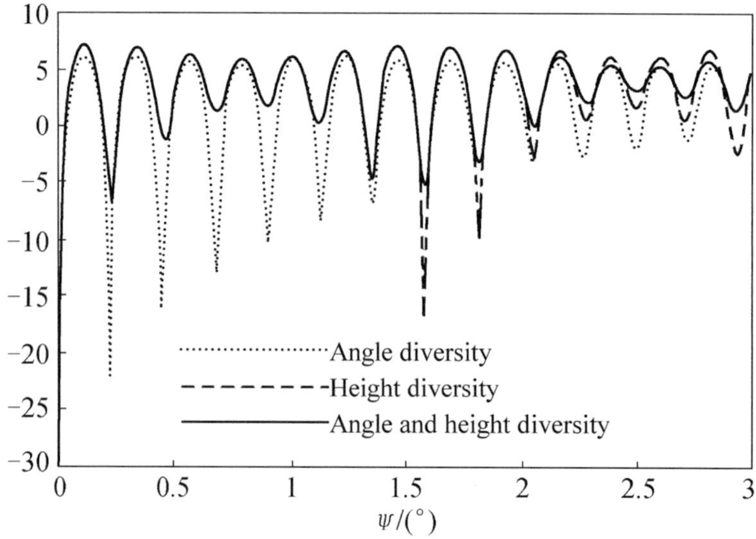

Fig. 5.29 Relation between total received energy and grazing angle for angle diversity and height diversity

5.4 New Methods for Simulation and Calibration

5.4.1 Dynamic Simulation Method Based on Motion Equation

Dynamic simulation for TT&C signals includes the simulations of echo signal's time delay (i.e., distance), Doppler frequency shift (i.e., velocity), and amplitude (i.e., various path losses). Currently, the main technical challenges are how to achieve the accurate correlation between distance and velocity and the simulation of extreme long distance. For example, in Mars exploration, the time delay will reach to approx. 3,000 s as the distance is 401.3×10^6 km, so it is impractical to simulate time delay from extreme long distance through range storage.

Since the relationship between moving object's distance and its velocity is strictly defined by its motion equation, the dynamic simulation based on the motion equation is able to present the accurate correlation of distance and velocity. Besides, the ranging of CW TT&C signal is achieved by phase measuring based on a fact that the distance change of a moving object can be represented by the phase change of its arrived echo, and the corresponding echo Doppler frequency shift is ω_d, trans-reception signal phase difference is $\phi = \omega_d t$ which will change continuously with time (i.e., object's distance is change continuously). In fact, the dynamic simulation is used to generate an analog signal with a Doppler frequency shift ω_d, which brings a phase difference $\omega_d t$ from the local ranging signal with the same phase change as that caused by the range change, and so the actual receiving signal is simulated by the analog signal. If the simulated time t is long enough, it can simulate an extreme distant range and so provides a potential solution for deep-space distant simulation.

In kinematics, the motion equation for the radius vector of point target is

$$R = R_0 + vt + \frac{1}{2}at^2 + \cdots \tag{5.83}$$

The "radius vector" in kinematics refers to "radial direction" in radio telemetering. As for the expression above, R_0 is the initial distance at the radial direction, ν is the radial velocity of the target, and α is the radial acceleration of the target.

The meaning of Expression (5.83) in mathematics is that a power series is used to represent a motion track varying with the time. In general, it can be approximated by taking only the first three terms, where the first term is the component of initial distance, the second term is the component of velocity, and the third term is the acceleration. Learned from the expression above, the range and the velocity is strictly correlated in physics, that is, the change of radial range R is caused by ν and α. The dynamic simulation is so used to implement the motion equation aforesaid. For radio telemetering:

$$R_0 = \frac{1}{2}C\tau_0 \tag{5.84}$$

$$v = -\frac{C}{2}\frac{f_{0d}}{f_0} = -\frac{C}{2}\frac{f_{Rd}}{f_R} \tag{5.85}$$

$$a = -\frac{C}{2}\frac{f^{\circ}{}_{0d}}{f_0} = -\frac{C}{2}\frac{f^{\circ}{}_{Rd}}{f_R} \tag{5.86}$$

$$\tau_0 = \frac{\varphi_0}{2\pi f_R}$$

where τ_0 is the initial time delay of echo; ϕ_0 is the corresponded phase (with ambiguity); f_0 is the carrier frequency; f_R is the frequency of each ranging signal; f_{0d} and f_{Rd} are the Doppler frequency of the carrier and the ranging signal, respectively; and $f^{\circ}{}_{0d}$ and $f^{\circ}{}_{Rd}$ are the change rate of their Doppler frequency shifts.

By substituting Expression (5.84), (5.85), and (5.86) into Expression (5.83), we can obtain

$$R = \frac{C}{2}\left(\tau_0 - \frac{f_{Rd}}{f_R}t - \frac{1}{2}\frac{f^{\circ}{}_{Rd}}{f_R}t^2\right) \tag{5.87}$$

The expression above is the range expression with considerations of velocity and acceleration. The dynamic simulation is used to generate an analog signal with frequency of f_R as well as Doppler frequency f_{Rd}, change rate of Doppler frequency $f^{\cdot}{}_{Rd}$, and phase ϕ_0. Since the expression above is highly correlated, if the right side of the equation is simulated, so is the other side correlatively. For example, once the f_{Rd} and $f^{\circ}{}_{Rd}$ are simulated, the range will be strictly simulated too. As long as the velocity exists, then R will be changed all the time. In theoretically speaking, the infinite distant range could be reached. Therefore, it can be used to simulate the extreme distant range in deep-space TT&C. This method takes the Doppler frequency as an independent variable. It is simpler than time delay simulation with shift register and can simulate farther distance with higher accuracy. Its accuracy is determined by the simulation accuracy of the Doppler frequency, while the accuracy of frequency control is higher than that of time delay. Understanding from the physical sense, it means that the frequency difference $f_d(t)$ between the local ranging signal and the returned ranging signal corresponds a phase difference of $\varphi(t) = \int f_d(t)\mathrm{d}t$, which establishes a correlation relationship between the changes of frequency (i.e., the relationship between the range and the velocity). Therefore, if the simulation of $f_d(t)$ is implemented, so is the change of range with time.

In simulating ranging accuracy, f_R is the fine tone or code clock; if the ambiguity resolving is required, f_R is the tone at each order or the ranging code.

The specific scheme is as shown in Fig. 5.30.

In Fig. 5.30, f_S is the reference frequency (e.g., 5 MHz or 10 MHz); f_{sd} and $f^{\circ}{}_{sd}$ are to-be-simulated Doppler frequency shift and its change rate, respectively, converted to the reference frequency, which can be calculated through Expressions (5.85) and (5.86) with target radial velocity and acceleration specified in the proposal; f_0 is the frequency of receiving carrier, $M = f_0/f_S$; f_{0d} and $f^{\circ}{}_{0d}$,

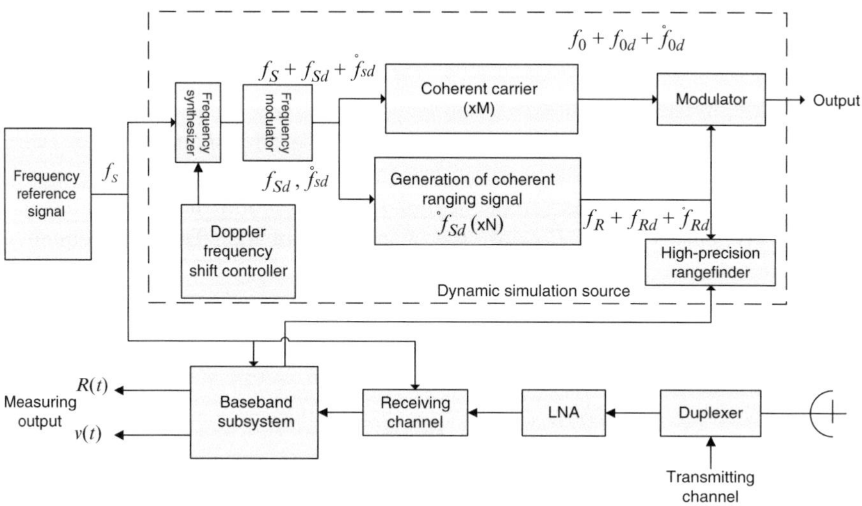

Fig. 5.30 BIT simulation source

respectively, are the Doppler frequency shift and its change rate at the frequency of receiving carrier, $f_{0d} = Mf_{sd}$, $f^{\circ}{}_{0d} = Mf^{\circ}{}_{sd}$; f_R is the frequency of ranging signal, $N = f_R/f_S$; and f_{Rd} and $f^{\circ}{}_{Rd}$, respectively, are the Doppler frequency shift and its change rate of the ranging signal, $f_{Rd} = Nf_{sd}$, $f^{\circ}{}_{Rd} = Nf^{\circ}{}_{sd}$.

It can be seen from Fig. 5.30 that all of the carrier and the ranging signal output by the modulator carry their Doppler frequencies, respectively. They are coherently generated by the same reference frequency f_S, so the velocity (measured by carrier signal) and the range (measured by ranging signal) are strictly coherent. The signal can be applied to the receiving system via the coupler before LNA to check the dynamic performances of the system. The "frequency modulator" shown in the Fig. 5.26 also can be used to simulate the component of frequency change caused by acceleration, which can be realized by integral phase modulation and reduce the difficulty of "frequency synthesizer." The video "high-precision range finder" shown in this figure is used to measure the range value and the great number n simulated by the ranging analog signal. Its ranging accuracy must be higher than the accuracy of the equipment to be measured, so that it can be served as a reference for correcting analog signal resource and ensure the corrected accuracy will not bring big effect on the test results of the measured equipment. Following measures shall be taken to improve the ranging accuracy of the video "high-precision range finder": select wide band instead of full-digitalized tone ring, and such range finder is required to operate under high SNR, constant amplitude, and longer capture time. According to this, its accuracy can be higher than the equipment to be measured. Since the signal in the transmitting channel has no Doppler frequency, the transmission subsystem may be excluded to the dynamic simulation BIT. The well-developed static BIT methods may be used if the static performance of the transmitting subsystem is required to inspect.

The initial range R_0 in Expression (5.83) is also to be simulated. The simulation methods are described as follows:

(1) Simulation by early/late generated ranging signal: that is, the early signal generated by the transmitted ranging signal (e.g., early gate triggered by transmitted code clock) and the late signal generated by the received ranging signal (e.g., late gate triggered by code clock) are used to simulate the initial time delay under static status or make a fine adjustment as t_0 to improve the accuracy.
(2) Simulation by velocity: insert any f_d to change the range, and then output one trigger signal when the range is R_0 to trigger t_0 or to remove f_d, keeping R_0 range, as shown in Fig. 5.31.

    For the error caused by this method, the changeable delay line for short delay is used for fine adjustment (digital or analog). After it exceeds one non-ambiguity range or further, the difference may be calculated by "accumulation of great values" with great range values, in which the great value n is measured by the high-precision range finder.

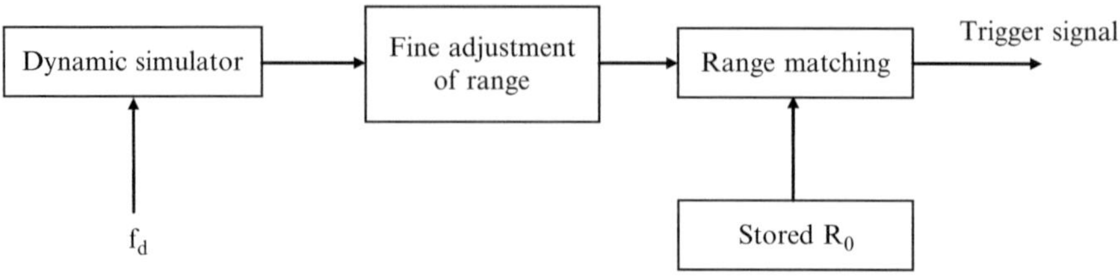

Fig. 5.31 Generation of initial range R_0

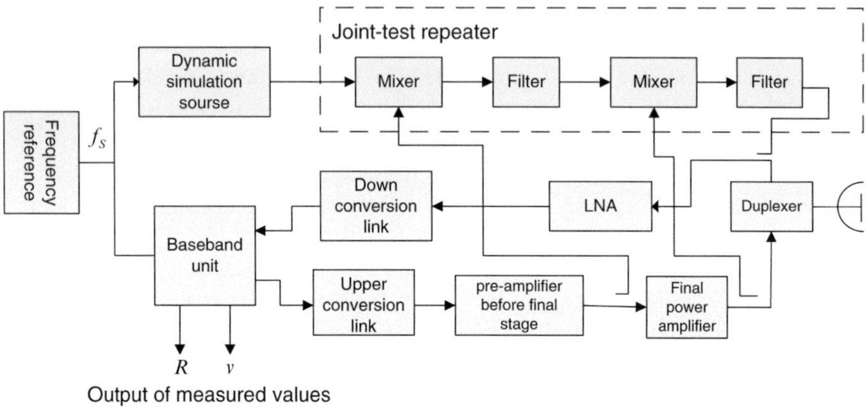

Fig. 5.32 Simulation scheme with uplink channel

(3) Simulation by time delay memory: for instance, simulation may be implemented by the shift register. However, with increasing of R_0, its accuracy is reduced, and the quantization noise will affect the operation of narrowband PLL; if the range is too long, the memory capacity is required to be very large and the simulation will not be able to implement.

The scheme as shown in Fig. 5.32 may be used for simulation with the uplink channel.

As shown in Fig. 5.32, the simulation of dynamic performance is much easier to implement in the event that there is no ranging signal being put into the uplink channel. If the ranging signal is required to put into the uplink channel, the uplink ranging signal shall be demodulated in the joint-test repeater and coherently converted to a downlink ranging signal with downlink Doppler frequency, and then the carrier is modulated.

The method above takes the Doppler frequency as an independent variable without the need of time delay simulation, so it is more simple and feasible. Since the phase difference and the frequency difference between T/R signals is accurately related, so the corresponding range and velocity shall be accurately related based on $\varphi(t)=\int f_d(t)\,dt$ with higher simulation accuracy. It provides a new method for dynamic simulation of TT&C signals, which has been applied to the engineering field.

5.4.2 *Phase Calibration Using Radio Star Noise [12, 13]*

5.4.2.1 Basic Principle

The high-precision angle-tracking system often employs the amplitude monopulse technique, which requires "phase calibration" to make the phase difference between the sum signal and the differential signal become zero. The traditional method for phase calibration is "tower phase calibration," that is, the signal beacon is placed on a boresight tower. However, there is no boresight tower under many practical circumstances, so phase calibration without boresight tower is required. In fact, "phase calibration with the radio star (RS)" is a better way to solve such problem. It uses the radio star in the space to replace the beacon resource (when the selected radio star is approximated as a point source) on the tower for phase calibration.

But, the following issues will be introduced when the single-frequency sine signal source is replaced by the noise of the radio star:

(1) The RS noise is wideband white noise with fluctuant phase and amplitude, but the amplitude of the sine wave signal is constant and its phase variation follows ωt rule.
(2) The time delay τ and phase φ of the sine wave signal can be converted by $\varphi = \omega\tau$; therefore, the time delay τ may be adjusted by adjusting φ. But only the single spectral line in the wideband noise satisfies the above relationship, rather than the whole wideband noise; that is, for a certain time delay τ, the phase shift of each spectral component is different, so φ can not be used to adjust the time delay.
(3) For sine wave signal, the demodulation of angle error signal can be implemented by a phase discriminator; but for RS noise signal, the demodulation of angle error signal can only be realized by the cross-correlation technique of the sum signal and the differential signal.

Therefore, "time delay calibration" is needed and phase calibration will cause following consequences:

(1) For getting an undistorted wideband signal, two conditions shall be satisfied: linear phase characteristics and constant amplitude characteristics, that is, the phase shift value is varied linearly with the frequency. However, the shift value of a phase shifter is usually not varied with the frequency, which will cause distortion of the signal in the channel (sum channel generally) being adjusted by such shifter, further resulting reduction of relevant peaks of the sum channel and the differential channel, and the reduction of output value Δ as well.
(2) The shift value of a phase shifter normally is less than or equal to 2π; if the phase difference corresponding to the time delay difference of the sum channel and the differential channel is greater than 2π, adjusting to $\tau = 0$ is impossible and, as a result, the relevant peak can not be obtained and the Δ is decreased.

Therefore, the time delay consistency shall be controlled in angle tracking of the wideband signals. In addition, another factor causing inconsistent phase shift is the inconsistency of each LO signal in the sum channel and the differential channel. As LO signal is a single frequency signal, this kind of inconsistency can be calibrated by phase shifter.

As mentioned above, a higher tracking accuracy for wideband signal will be achieved with proper design, and in case of constant signal spectral density, the wider the band, the greater the cross-correlation value, that is, the higher the *S/N*, the less the angle error of the thermal noise.

5.4.2.2 "Phase Calibration" With RS Wideband

For wideband signal, phase calibration is actually done through time delay calibration. Here to continue to use the conventional term, phase calibration with double quotation marks is used to represent this kind of calibration.

In "phase calibration" with the radio star, the electrical axis of antenna points to the radio star and its input signal is wideband white noise, where the correlation function of the sum signal and the differential signal may be expressed by the same expression deduced for wideband interferometer of the radio star in deep-space TT&C [14]:

$$C_{xy}(\tau) = 2B\sqrt{T_{ax}T_{ay}}\cos\left[(2\pi f_0 + \pi B)\left(\tau + \tau_g\right)\right]\frac{\sin\pi\left(\tau + \tau_g\right)}{\pi B\left(\tau + \tau_g\right)} \tag{5.88}$$

where $C_{xy}(\tau)$ is the cross-correlation function of the x station and the y station of the interferometer; T_{ax} and T_{ay} are the equivalent noise temperature of the radio star measured at the X station and Y station, respectively; τ_g is the delay difference from the radio star to the two stations; f_0 is the center frequency of the receiving bandwidth; and B is the receiving bandwidth. It can be seen from the expression that the wider the B value is, the greater the correlation value will be.

The "phase calibration" has some differences with the interferometer case:

(1) The X station and Y station of the interferometer are substituted by the sum channel Σ and the difference channel Δ of angle tracking.
(2) B is the bandwidth of the tracking receiver to be calibrated.
(3) τ_g is the delay difference between the sum channel Σ and the difference channel Δ.
(4) $2B\sqrt{T_{ax}T_{ay}}$ can be substituted by $U_\Sigma U_\Delta K$, where U_Σ and U_Δ, respectively, are the RMS values of the sum signal and the difference signal at the correlator input end; K is the transmission coefficient of the correlator; and $U_\Sigma U_\Delta K$ is related to the noise intensity of a radio star source and the bandwidth B it is taken from.

By considering these factors above, the expression above can be converted to the cross-correlation function expression suitable for "phase calibration" of the radio star:

$$C_{\Sigma\Delta}(\tau) = U_{\Sigma}U_{\Delta}K\cos\left[(2\pi f_0 + \pi B)(\tau + \tau_g)\right]\frac{\sin \pi B(\tau + \tau_g)}{\pi B(\tau + \tau_g)} \tag{5.89}$$

To make $C_{\Sigma\Delta}(\tau)$ achieve its peak value, an additional delay difference $(-\tau_g)$ shall be added between sum channel and difference channel to make $(\tau + \tau_g) = 0$; at this time, its peak value is

$$C_{\Sigma\Delta}(-\tau_g) = U_{\Sigma}U_{\Delta}K \tag{5.90}$$

This correlation value can be normalized to $KE^2\frac{U_{\Delta}}{U_{\Sigma}}$ (E is the AGC comparison level of sum channel), which is in direct proportion to the angle deviation but independent of the range, that is, the angle error signal up to its maximum value. The normalization of the sum signal and the difference signal can be achieved either by conventional AGC or by sampling U_{Δ} and U_{Σ} then calculating U_{Δ}/U_{Σ}.

According to above analysis, the block diagram of wideband "phase calibration" by the radio star can be shown in Fig. 5.33.

Since there is no carrier signal in radio noise, the conventional carrier phase-lock loop method can not be used. Instead, an angle error extraction method is used, in which the sum signal and difference signal are in direct correlation.

In Fig. 5.33, the delay adjuster is used to calibrate inconsistency of sum channel and difference channel, while phase shifter is used to calibrate inconsistency of phase shift in each LO, since wideband filtering requires consistency of each channel's amplitude/phase frequency characteristics. The calibration criterion is to make the output voltage of correlator reach the maximum value.

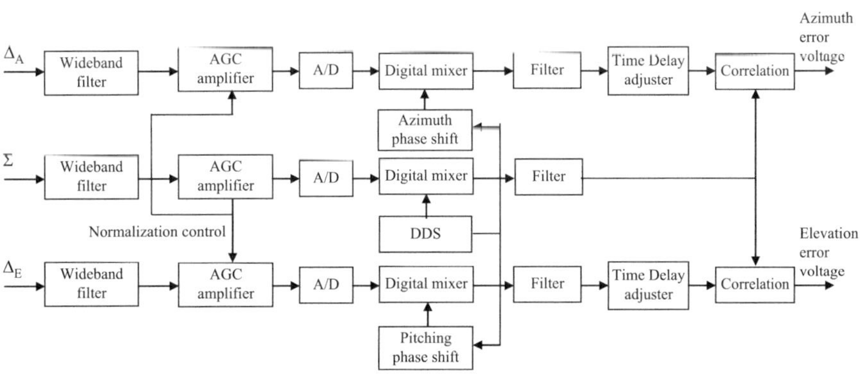

Fig. 5.33 Block diagram of tri-channel monopulse wideband "phase calibration" by the radio star

5.4.2.3 Radio Star Narrowband Phase Calibration

When the wideband filter in Fig. 5.33 changes into narrowband, it called phase calibration with narrowband, which is common in TT&C. In this case, the respective radio noises $n(t)$ output by the sum channel and difference channel with the same narrowband filter are both narrowband Gaussian noise, which can be expressed as follows:

$$\begin{aligned} n_\Sigma(t) &= R_{n\Sigma}(t)[\omega_0 t + \Phi_n(t)] \\ n_\Delta(t) &= R_{n\Delta}(t)[\omega_0 t + \Phi_n(t)] \end{aligned} \tag{5.91}$$

where $R_{n\Sigma}(t)$ and $R_{n\Delta}(t)$ are random varied envelopes and $\Phi_n(t)$ is their random varied phase. According to the same reason as above discussed, the $R_n(t)$ and $\Phi_n(t)$ in sum channel and difference channel are correlated, that is, the correlation of $\Phi_n(t)$ is in phase, while that of $R_{n\Sigma}(t)$ and $R_{n\Delta}(t)$ shows the same change rule with different amplitude. Learned from the expression above, the phase calibration has now changed to the calibration of single-frequency ω_0 signal.

If there are different time delay τ_Σ and τ_Δ in sum channel and difference channel, the sum signal is

$$\begin{aligned} n_\Sigma(t) &= R_{n\Sigma}(t)\cos\left[\omega_0 t + \Phi_n(t) + \omega_0\tau_\Sigma\right] \\ &= R_{n\Sigma}(t)\cos\left[\omega_0 t + \Phi_n(t) + N\cdot 2\pi + \Delta\Phi_\Sigma\right] \\ &= R_{n\Sigma}(t)\cos\left[\omega_0 t + \Phi_n(t) + \Delta\Phi_\Sigma\right] \end{aligned} \tag{5.92}$$

where N is $\omega_0\tau_\Sigma/2\pi$ (round off to integer), $\Delta\Phi_\Sigma = \omega_0\tau_\Sigma - N\cdot 2\pi$.

Similarly, the difference signal is

$$n_\Delta(t) = R_{n\Delta}(t)\cos\left[\omega_0 t + \Phi_n(t) + \Delta\Phi_\Delta\right] \tag{5.93}$$

Learned from Expression (5.91) and (5.92), there is phase difference between the sum channel and the difference channel, $\Delta\Phi = \Delta\Phi_\Sigma - \Delta\Phi_\Delta$. The phase calibration is done by introducing an additional phase shift ($\Delta\Phi$) in the sum channel (or $-\Delta\Phi$ in difference channel) so that the final phase difference between the sum signal and the difference signal is equal to zero.

Narrowband phase calibration can be done without using "delay calibration" (but can be done by appropriate delay compensation). The reason is that since the sine wave is periodic (so is the corresponding correlation peak), after $N\cdot 2\pi$ phase shift by time delay, the remaining phase fraction $\Delta\Phi$ can be calibrated by the phase shifter to make the sum signal and the difference signal in phase, and the phase discriminator can output the maximum error voltage. However, it should be noted that the envelope of the narrowband Gaussian noise changes very slowly, as shown in Fig. 5.34.

But for single-frequency sine signal, the envelope is constant, which is the difference between them. Therefore, the amplitude normalization in-phase

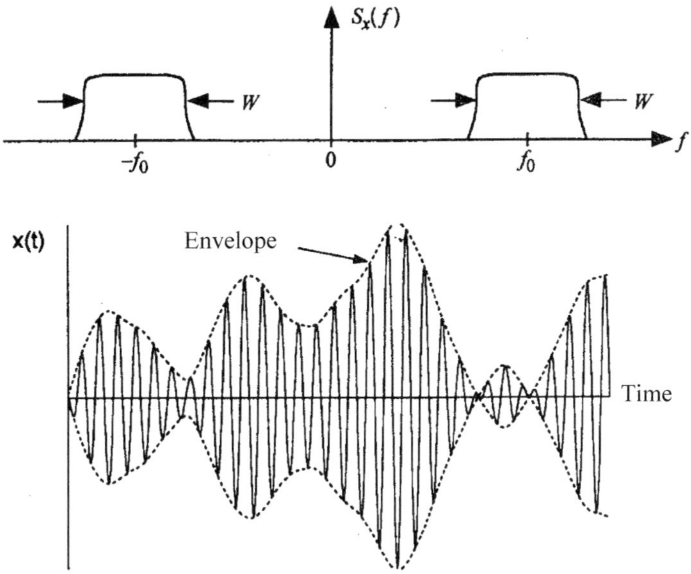

Fig. 5.34 Spectrum of narrowband Gaussian noise and change of its envelope

calibration by the radio star is more important, otherwise the error will be introduced in amplitude/phase calibration. Such error can be reduced by normalizing with wideband AGC, normalizing with direct samples in amplitudes of sum/difference signals, or increasing of smoothing time. If the delay difference between the sum channel and the difference channel is not great enough, its error can be reduced to the minimum extent. The change of envelop can be dealt with AGC normalization.

According to analysis above, the block diagram for narrowband "phase calibration" by the radio star is as shown in Fig. 5.35.

In Fig. 5.35, DDS is used to adjust the phase shift, and the phase discriminator is used to obtain the voltage of angle error, in order to achieve the maximum output after calibration.

This method has been used in TT&C system, especially applicable to such situation that no calibration tower is allowed to build, for example, if the antenna with large aperture is used, and the corresponding far field is very far away, then a higher calibration tower (e.g., deep-space station) is required to be built; or such situation that no calibration tower can be built on the TT&C station borne by some special carriers like survey vessel. In addition, the radio star also is used for astral calibration, measurement of G/T value or G value, etc. But for this method, it is very important to select an appropriate radio star according to the aperture of antenna, in order to maintain strong and stable flux intensity and prevent from "extended source."

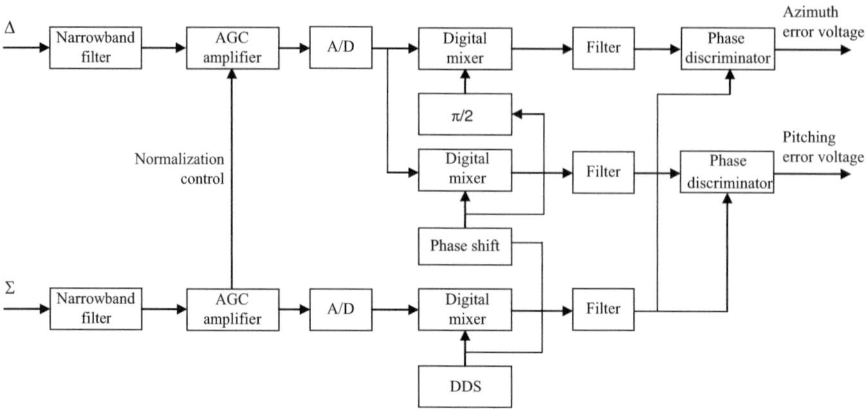

Fig. 5.35 Block diagram of dual-channel monopulse narrowband phase calibration by the radio star

5.4.3 On-Orbit Phase Calibration by Measuring "Cross-Coupling" Value

Learned from the Expression (2.243) hereof, the "crossing coupling" value K_{CE} in dual-channel monopulse is

$$K_{\mathrm{CE}} = \mathrm{tg}\,\phi$$

therefore,

$$\phi = \mathrm{arctg}\ K_{\mathrm{CE}} \tag{5.94}$$

In Expression (5.94), Φ is the phase inconsistency of the sum signal and the difference signal, which is able to be calculated only if K_{CE} is known, thus to carry out phase calibration. Reference [15] specifically describes this scheme, and its Principle is as follows:

It assumes that the azimuth signal and the elevation signal output by the antenna are completely orthogonal with same difference pattern, and when the antenna directs to any point P_1, P_1 is deviated X_1 degrees from the electrical boresight in azimuth and Y_1 degrees in elevation, that is, P1 (X_1, Y_1). At that time, the sum signal/difference signal of "dual-channel monopulse" for phase orthogonal multiplexing is

$$\begin{aligned} u_{\Sigma}(t) &= k\cos\omega t \\ u_{\Delta 1}(t) &= E_1\sin(\omega t + \varphi) + A_1\cos(\omega t + \varphi) \end{aligned} \tag{5.95}$$

where Φ is the phase inconsistency of sum signal and difference signal after the antenna feeder is introduced, that is, the phase to be compensated during "phase calibration"; K is the amplitude of sum signal; A_1 is the amplitude of azimuth difference signal; and E_1 is the amplitude of elevation difference signal. The phases in the Expression are based on the phase of sum channel.

Theoretically, the monopulse demodulation is in multiplying of the signal $k\cos\omega t$ and signal of difference channel $u_{\Delta 1}(t)$:

$$\begin{aligned}&\left(E_1\sin(\omega t+\phi)+A_1\cos\left(\omega t+\phi\right)\right)\times\left(k_1\cos\omega t\right)=\\&\frac{k}{2}(E_1\sin(2\omega t+\phi)+E_1\sin\phi)+\frac{k}{2}(A_1\cos(2\omega t+\phi)+A_1\cos\phi)\end{aligned}\tag{5.96}$$

Set the gain of low-pass filter and the multiplicator as 1, then the azimuth error voltage $VAz1$ at P1 is

$$V_{AZ1}=\frac{k}{2}(E_1\sin\phi+A_1\cos\phi)\tag{5.97}$$

Similarly, the sum signal $k\sin\omega t$ with phase shift π/2 is multiplied by the difference signal $u_{\Delta 1}(t)$ at first, then the elevation error voltage is obtained after the resulting value aforesaid passed through the low-pass filter:

$$V_{EL1}=\frac{k}{2}(E_1\cos\phi+A_1\sin\phi)\tag{5.98}$$

When the antenna azimuth is directly deviated from $P_2(X_2,Y_2)$ in λ degrees, since the azimuth signal output by the antenna is completely orthogonal to the elevation signal, then the amplitude of azimuth signal at P_2 is $A_2=A+A_I$ (A is the added amplitude of azimuth signal from deviation angle λ); and the amplitude of elevation signal remains the same, i.e., E_1, so the difference signal at P_2 shall be

$$u_{\Delta 2}(t)=A_2\cos(\omega t+\phi)+E\sin(\omega t+\phi)$$

According to Expression (5.96) and Expression (5.97), the azimuth error voltage $VAz2$ and the elevation error voltage $VEl2$ are obtained as follows:

$$V_{AZ2}=\frac{k}{2}(E_1\sin\phi+A_2\cos\phi)\tag{5.99}$$

$$V_{El2}=\frac{k}{2}(E_1\cos\phi+A_2\sin\phi)\tag{5.100}$$

Subtract the error voltage at P_1 from that at P_2:

$$\Delta V_{AZ}=V_{AZ2}-V_{AZ1}=\frac{k}{2}A\cos\phi=\frac{k}{2}A\sin\left(\frac{\pi}{2}+\phi\right)\tag{5.101}$$

$$\Delta V_{EL}=V_{EL2}-V_{EL1}=\frac{k}{2}A\sin\phi=\frac{k}{2}A\cos\left(\frac{\pi}{2}+\phi\right)\tag{5.102}$$

$$\frac{\Delta V_{AZ}}{\Delta V_{EL}} = \frac{\frac{k}{2}A\sin\left(\frac{\pi}{2} + \phi - \theta\right)}{\frac{k}{2}A\cos\left(\frac{\pi}{2} + \phi - \theta\right)} = \mathrm{tg}\left(\frac{\pi}{2} + \phi - \theta\right)$$

$$\mathrm{tg}\,\phi = \frac{\Delta V_{\mathrm{EL}}}{\Delta V_{\mathrm{AE}}} = K_{\mathrm{CE}} \tag{5.103}$$

$$\phi = \mathrm{arctg}\, K_{CE} \tag{5.104}$$

As mentioned above, K_{CE} is the cross coupling, and Φ is the phase inconsistency of sum channel and difference channel. The phase calibration is performed according to the phase modulation of sum channel by Φ.

The azimuth error sensitivity at the output point of the multiplicator is

$$K_A = \frac{\Delta V_{AZ0}}{\lambda} \tag{5.105}$$

where λ is the azimuth deviation angle and ΔV_{AZ0} is the increment of azimuth error voltage after phase calibration. It is also used to calibrate the angle error sensitivity. In addition, the K_{A} value may be calculated by the angle error voltage before calibration because of Φ, then it is obtained according to the vector relationship of $\Delta V_{AE0} = \Delta V_{AE}/\cos\phi$:

$$K_A = \frac{\Delta V_{AZ}}{\lambda\cos\phi} = \frac{\sqrt{\Delta V_{AE}{}^2 + \Delta V_{EL}{}^2}}{\lambda} \tag{5.106}$$

The advantages of this method are:

(1) On-orbit phase calibration: This method is performed without source calibration which is used to perform on-orbit phase calibration during "tracking while calibrating" in the course of on-orbit flight of the flight vehicle, using the signal transmitted by the flight vehicle itself, or other satellites at the same band. The detailed calibration method describes as follows: in case of digital guidance from ground station or tracking flight vehicle, the electrical boresight is made to deviate twice (i.e., Point P_1 and Point P_2); then you may use the error voltage values measured from these two points to calculate the result and finish the calibration. Refer to "Deviation for tracking" as described in Sect. 5.3.6.1 hereof for the method of electrical boresight deviation during tracking.
(2) Rapid calibration: if previous mentioned "calibration with radio star" or "calibration with calibration tower" is used, then additional calibration signals are required and the "phase calibration" shall be carried out according to following steps: ① the electrical boresight of the antenna adjusts to the direction of calibrated signal source to find the zero point of difference lobe; ② set the deviated elevation angle of the antenna as 0 and the azimuth deviates from a specified angle, then find the phase θ_{Az} and the calibrated value of gain k_{Az} after

rough calibration and fine calibration of the azimuth phase for receiver; ③ set the deviated azimuth angle of the antenna as 0 and the elevation angle deviates from a specified angle, then find the phase θ_{El} and the calibrated value of gain k_{El} after rough calibration and fine calibration of the elevation phase for receiver; ④ check cross coupling.

For these steps, the first step for adjusting the electrical boresight to the calibrated source is the most important and difficult. If the zero point of the difference lobe is not adjusted correctly, the phase shift value and the error gain will be affected. Normally, the closed-loop tracking is used for looking into the zero point. For calibration with tower, there are few steps consisting of rough calibration of receiver → auto-tracking for adjustment → fine calibration of receiver, sometimes the zero point of difference lobe only can be traced after multiple times of auto-tracking. The whole process is very complicated and takes a lot of time and requires more manual judgments, which are not beneficial for auto-calibration. Nevertheless, this method could suppress these shortcomings above and only takes one time for calibration successfully.

5.4.4 Geometric Optics Application for Range Calibration

5.4.4.1 Geometric Optics and Physical Optics

The physical optics method also refers to as surface current integral method, that is, the Kirchhoff diffraction integral formula is used to analyze and calculate the antenna on the basis of Maxwell equation for electromagnetic field, which is used to calculate the scattered field from the metal reflection surface. The current of scattered field for simulation is conducted by an incident wave from the feeder that imposes on the conductive surface.

Geometric optics method refers to as ray method, which means the ray is used to describe the transmission, refraction, reflection, etc. of electromagnetic wave, and to analyze and calculate the antenna based on such results. It is the zero wave-length approximation of physical optics.

Both methods are approximation methods, where the prerequisite for physical optics approximation is $\rho^2 \geq 100\lambda^2$ (ρ is the curvature radius of reflection point) and that for geometric optics is $\sqrt{\rho} \geq 100\sqrt{\lambda}$; as a result, the physical optics is more accurate.

In application of range calibration, the impacts on ranging accuracy is the main issue concerned for comparison of these two methods, in another hand, the distance for CW ranging is relevant to the phase Φ of signal; therefore, the two methods have following differences with respect to the impacts on RF signal phase:

(1) The physical optics takes the impact of diffraction into account while the geometric optics does not. The diffraction will result in change of equiphase surface of electromagnetic field, and the change of RF phase will cause the change of ranging distance.

(2) In geometric optics, the electric wave is considered as an ideal plane wave and the beam is a straight line perpendicular to that plane (which is the ray). In fact, the results practically analyzed and calculated by the physical optics show that the antenna has a beam with certain width and the equiphase surface is not an ideal plane. As a result, both analytical methods have the phase error.
(3) The geometric optics method assumes that the reflection surface is an ideal conductor and produces specular reflection without phase delay, on another hand, the physical optics method shows that the incident wave has induced the current on the parabolic plane and then a scattered field has been simulated by such current. Therefore, there is phase delay and it will be affected by distortion of parabolic plane and materials.
(4) In assumption of geometric optics, the radio wave is assumed to transmit in the vacuum and no shelter or other mediums will affect the transmission as it passes through. Actually, such factors affecting radio wave transmission exist.

As mentioned above, the physical optics method is more accurate than the geometric optics. Such accuracy reflects the RF phase Φ, for the range error, $\Delta R = \Delta\Phi/\lambda$, where λ is the RF wavelength. Since λ at RF is so small that the introduced ΔR is small as well, and the ranging is implemented by LF tone ranging, consequently, the geometric optics method introduces small errors for range calibration and is simple and easy to practice.

5.4.4.2 Geometric Relationship of Parabolic Antenna

For an ideal parabolic antenna, assuming that the surface of reflector is an ideal conductor and the phase center of point-source feed is located on the focal point of the parabolic plane. According to the geometric optics principle, the spheric wave from the focal point to the parabolic plane will convert to the plane wave transmitting in axial direction along the parabolic plane after reflection, with the same phase field, so as to form a sharp beam for the antenna at far field.

For common Cassegrain antenna, its geometric optics relationships are as shown in Fig. 5.36.

In Fig. 5.36, the two focal points on the hyperboloid, respectively, are the virtual focus O_1 that overlaps with the focal point of the parabolic plane and the actual focus O_2 that overlaps with the phase center of the feed source. According to the characteristic of hyperboloid, the range from any point K on the hyperboloid to the two focuses is equal to a constant m_1, that is,

$$O_2K - KO_1 = m_1$$

Another characteristic shows that when any electromagnetic wave O_2K transmitted from the feed source of the actual focus O_2 has been reflected through the hyperboloid (the surface of hyperboloid locates the far field of the feed source), the

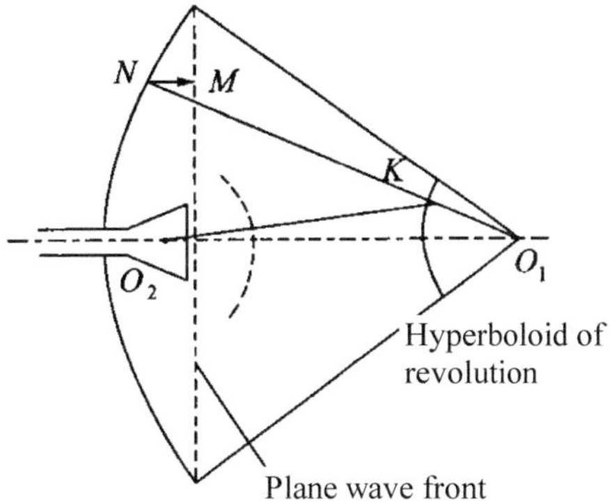

Fig. 5.36 Geometric relationship of Cassegrain antenna

direction of the reflected wave is on the extension line KN of the beam O_1K transmitted from the virtual focus O_1 on the hyperboloid.

Based on the characteristic of the parabolic plane, when each electromagnetic wave (O_1KN) transmitted from the focal point O_1 of the parabolic plane has been reflected by the parabolic plane, the reflected wave NM is parallel to the symmetry axis O_2O_1 of the parabolic plane and is a constant.

$$O_1K + KN + NM = O_1N + NM = m_2 \tag{5.107}$$

then,

$$\begin{aligned} m &= m_1 + m_2 = O_2K - KO_1 + KO_1 + KN + NM \\ &= O_2K + KN + NM \end{aligned} \tag{5.108a}$$

Therefore, the radio wave from the phase center O_2 of the feed source has the same wave path to the aperture of parabolic plane as it is reflected twice by hyperboloid and parabolic plane, and the time delay of radio wave introduced by such wave path is the ranging time delay introduced by the antenna (i.e., the time delay from the phase center to the aperture of antenna), which is the theoretical basis of applied geometric optics for range calibration. Such length m of the one-way wave path of the geometric ray may be calculated by Expression (5.108) during design or measured on the actual antenna for its geometric dimension, and then two-way time delay T_A corresponding 2 m is so obtained.

5.4.4.3 Offset-Feed Zero-Range Calibration Based on Geometric Optics

In TT&C system, the range values measured by the ground station include the time delay caused by the ground station and the transponder of the spacecraft, which must be deducted to determine the range from the reference position of the ground antenna to that of satellite antenna. The actual range in Fig. 5.37 means the range

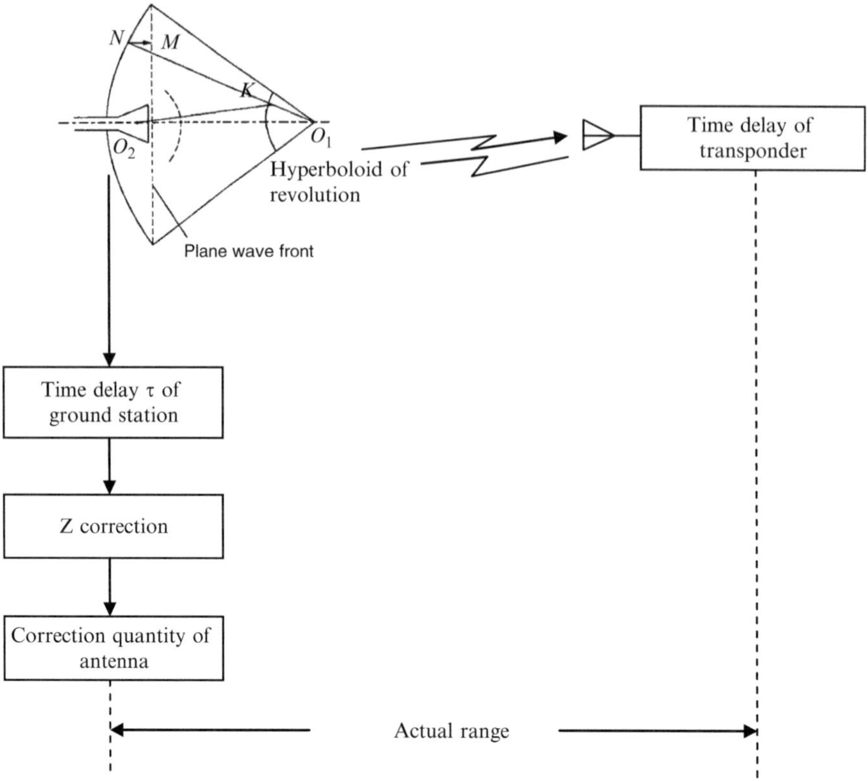

Fig. 5.37 Correction of time delay

between the reference position of the ground station and that of the spacecraft, where the reference position of the ground station is considered as the phase center of antenna, and from Fig. 5.37, we can obtain that:

(Time delay in actual range) = (Time delay measured by TT&C system) – (Time delay of transponder) – (Measured time delay of ground station) + (Z correction).

Transponder delay is provided by transponder product. An additional corrected value shall be added for some antenna.

In Fig. 5.37:

(1) Time delay of ground station τ_s

Time delay of ground station defines the sum of the delay between the uplink ranging equipment (including channel and baseband) and the range-calibrated coupler and the delay between the range-calibrated coupler and the downlink ranging equipment (including channel and baseband). Apparently, the defined time delay of ground station does not include other delay generated in the ground station, for example, there is time delay between the range-calibrated coupler and the reference position of ground station.

Time delay of ground station is relevant to different ground stations and the configuration thereof, which shall be measured each time while ranging. Such measurement includes calibration before tracking and calibration after tracking. The former is performed automatically at the beginning of ranging, and the latter only will be performed when there are some changes of the equipment during tracking or the calibration is not conducted before the tracking.

A test repeater is used to measure the time delay of ground station and approximate to the actual ranging configuration. A range-calibrated coupler is employed between the T/R "duplexer" and the feed source, which is used to transmit the uplink signal to the test repeater that converts the signal to the downlink frequency, and then the signal returns back to the receiving channel via range-calibrated coupler. Such signal is transmitted to the ranging terminal via LNA and downlink channel for range calibration, and the time delay of ground station measured at this moment includes the delay of test repeater.

If the uplink and the downlink operated at different bands, the repeater with different band shall be used so as to enable the downlink signals transmitted from such repeater to couple to the receiving channel before LNA at the same band and remain as close as possible to the feed.

(2) Z correction

It is not enough only to subtract transponder delay and ground station delay measured above, because there are two deviations:

1) Measured ground station delay includes the delay of the test transponder and the delay of cables connecting test transponder and range-calibration coupler. This extra delay is not included in the channel in actual spacecraft ranging.
2) Ground station delay does not include two-way delay between range-calibration coupler and reference position of ground station.

Z correction includes two deviations, that is, Z correction equals the result by subtracting two-way delay between range-calibration coupler and reference position of ground station from the delay of the test transponder and its cables. Z correction is symbolized, which is a minus when two-way delay between range-calibration coupler and reference position of ground station is big.

The principles of range calibration are described as follows:

- Measure the delay of ground station (before or after tracking): including the delay of the test transponder and the delay of cables connecting test transponder and ranging coupler.
- Connect "ZDD" to "ranging coupler" and "edge of aperture plane" (i.e., simulate a transponder on the aperture plane), respectively, in order to measure the time delay of radio wave from the ranging coupler to the aperture plane.
- Actual range delay between the aperture plane and reference position can be measured geometrically.
- Determine the delay of the test transponder (and its cables): Connect a ZDD to the transponder. Although it is known that ZDD is zero delay, it has a delay

actually, which has been measured in the laboratory. The delay of the test transponder and its cables can be obtained by subtracting the ground station delay measured by ZDD from the delay of the test transponder and then adding known ZDD delay.

- Through a simple geometric transformation, an expression about Z correction can be obtained:

Z = (test transponder delay) – (total two-way delay between ranging coupler and aperture plane) – (two-way delay T_A between aperture plane and reference position as calculated in geometric optics principles)

It is required to determine Z correction regularly or redetermine Z if signal path is affected by any change of equipment. It is also required to provide a list of Z correction for the purpose of ranging.

In order to make this method more approximate to geometric optics, zero-range antennas of ZDD should be placed on the edge of the antenna aperture and kept far away from the sub-reflector support pole, thereby achieving the following advantages: ① closer to space electric field outside the aperture; ② farther away from the feeder; ③ more accurate T_A; ④ a reduction of shielding, multipath, and diffraction effects of the support pole; and ⑤ a little impact on antenna performance. If zero-range antenna beam becomes narrower and closer to a ray beam, it will be closer to geometric optics and can reduce multipath interference, and thus zero-ranging accuracy will be higher.

The advantage of this method is that with no need to calibrate range zero value by the calibration tower, it is only required to calculate additional time delay of the antenna in "geometric optics method." And actual range can be obtained by deducting this time delay and other error from measured range value. As for the giant antenna deep-space TT&C station which could not establish a calibration tower and some mobile TT&C stations, the method is a simple approach to obtain actual range.

5.4.5 *Effect of Radio Wave Propagation Characteristic on Orbit Determination Accuracy*

It is generally acknowledged that in spacecraft orbit determination, the orbital error caused by radio wave refraction correction residuals may account for about one third of total orbit determination error. It is also generally accepted that the atmosphere below 60 km is troposphere, and above 60 km is ionosphere. Radio waves used to measure the orbit of the spacecraft mainly pass through the troposphere and ionosphere. The atmospheric refractive index in the troposphere changes with altitude (it is generally assumed that an atmospheric layered model is used), and a refraction effect will take place when radio waves pass through the troposphere. Any ranging, angle-measuring or velocity-measuring error caused by the refraction effect is directly correlated to the refractive index. In the atmosphere below 3 km, a sharper change in atmospheric refractive index and a bigger radio wave refraction correction residual will bring bigger error in measurement elements and affect the orbit accuracy. Any change of electron concentration profile in the ionosphere would curve radio waves passing through the ionosphere, thus producing measurement element error. If the orbit determination accuracy demands less, the effects of ionosphere would not be considered generally. However, with increasing requirements for measuring accuracy, apart from tropospheric refraction error, ionospheric refraction error should be corrected. Tropospheric refraction corrections can be made after a refractive index is calculated by measuring the temperature, humidity, and pressure in the troposphere. Ionospheric refraction corrections can be made by measuring electron concentration in the ionosphere or total electron contents in the radio wave path. Due to complex changes of the atmosphere, unsatisfactory detection means and inconsistency between correction model and actual conditions in atmosphere, the residuals of radio wave correction are still higher.

The space surrounding the Earth's surface and containing gas molecules, electrons, and ions is called an atmosphere. According to the height above ground, physical and electrical characteristics, the Earth's atmosphere can be, in general, divided into four layers: (1) troposphere between the Earth's surface and about 12 km above the ground, (2) stratosphere between the tropopause and about 60 km above the ground, (3) ionosphere between about 60–2,000 km above the ground, and (4) magnetosphere between 2,000 km and a 10,000 or 100,000 km above the ground.

There is no obvious boundary between layers. Also, the number and height of layers are different.

The troposphere is the lowest region of the atmosphere. The tropopause at poles of the Earth is about 9 km high and increases gradually as the latitude decreases. In equatorial zone, it can reach about 17 km. The troposphere is a mixture of various gases (including nitrogen, oxygen, hydrogen, and carbon dioxide) and water vapor. Atmospheric density in the troposphere decreases with increasing altitude. The heat radiated to the ground by the Sun warms the troposphere by air vertical convection. Generally, the humidity, pressure, and vapor pressure in the troposphere decreases with the increasing altitude, but in some areas, the humidity would occasionally increase with the increasing altitude, which is, the occurrence of a thermal inversion layer. The dielectric property in the troposphere changes with time and space. Therefore, radio wave propagation in the troposphere is different from that in free space. The propagation path curves, and its velocity is different from the velocity of light in vacuum. As a result, radio waves have a refraction effect in the atmosphere.

According to many experiments of flight vehicle orbit determination, when the target distance is about 100 km and the elevation angle is 5–10°, the range error from the refraction may reach tens of meters, the velocity error 0.1 m/s and elevation error 1 arc-minute. Therefore, radio wave propagation must be corrected. Figure 5.38 shows a curve of range error (ΔR) before correction.

As shown in Fig. 5.38, t_0 is a crossing point at which a flight vehicle flies over the ground TT&C station from front to rear. Because at this point, the distance from the flight vehicle is shortest and the elevation angle is highest, the range error ΔR_0 is

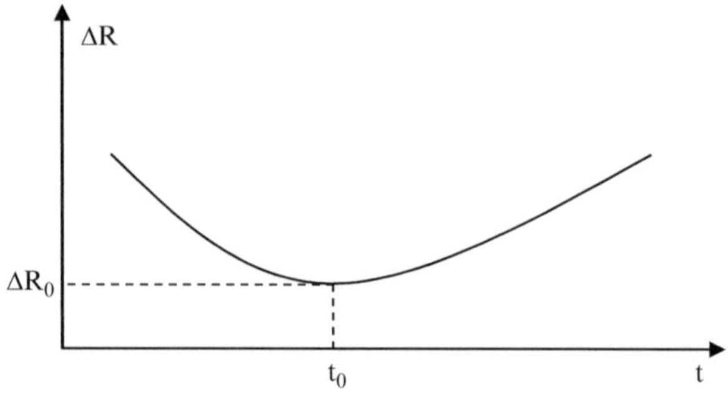

Fig. 5.38 Range error curve caused by atmospheric refraction

smallest. As the distance between the flight vehicle and the TT&C station increases and the elevation angle decreases, the path in which radio waves pass through the atmosphere grows longer and range error from the refraction also grows bigger. In the calibration flight test, a curve as shown above may be caused by atmospheric propagation.

In case of Doppler velocity measuring, a relation between velocity error ΔV and other element error is

$$\frac{\Delta V}{V} = -\frac{\Delta C}{C} - \frac{\Delta f_d}{f_d} + \frac{\Delta f_0}{f_0} \tag{5.108b}$$

where $\Delta C/C$ is the error resulted by inaccurate light velocity measurement and ΔC is the error of radio wave propagation velocity. Due to effect of uneven air density in the troposphere and electron density in the ionosphere on radio wave propagation in the atmosphere, the propagation velocity is not a constant but has a margin of error.

$\Delta f_d/f_d$ is Doppler frequency measurement error, including Doppler frequency error from radio wave propagation, Doppler frequency shift error caused by noise, and Doppler frequency measurement error from measuring equipment.

$\Delta f_0/f_0$ is the error from transmitting frequency instability.

According to the above analysis, both the first two terms in velocity-measuring error are relevant to the effect of radio wave refraction. In order to guarantee velocity-measuring accuracy of the system, any error from radio wave refraction must be corrected.

Atmospheric refractivity in the troposphere, also called refractive modulus N (dimension is expressed in N unit), is

$$N = (n-1)\times 10^8 = \frac{77.6}{T}\left(P + \frac{4801P_e}{T}\right) \tag{5.109}$$

where n is the refractive index, $n = c/v$, where c is the velocity of light in vacuum and v is the velocity of light in real medium; P is total pressure (mb); P_e is partial pressure of water vapor (mb); and T is temperature (°K).

According to the expression (5.109), tropospheric refractivity is related to such factors as temperature, atmospheric pressure, and water vapor pressure which can be measured by a sonde. Thus, it can be seen that refraction effect on velocity measuring is relevant to flight height, time, climate, etc.

Atmospheric refractivity *N* and atmospheric refractive index *n* can be converted by using the above expression. *N* and n are of the same physical essence. It is more convenient to use them in different scenarios. We must know *n* or *N* if we want to calculate the effect of radio wave propagation on measuring accuracy. Therefore, we have researched many atmospheric refractive index models.

The research on climate effect on radio wave propagation focuses on the changes of atmospheric refractive index n with time and place and its characteristics. According to many data statistics, atmospheric refractive index n has a very small annual change rate which is negligible. The research focuses on seasonal change, diurnal change, and changes with height and horizon. These climatic characteristics are called radio climatology.

(1) Horizontal change of atmospheric refractivity *N*

The refractivity is horizontally uniform but is obviously different in a wide range. Figure 5.39 shows annual average isolines of ground surface refractivity N_s.

There are large difference between seal level refractivity N_{sea} and ground surface refractivity N_s. Such differences are caused by complex terrain. Ground surface refractivity N_s is usually used in atmospheric refraction research.

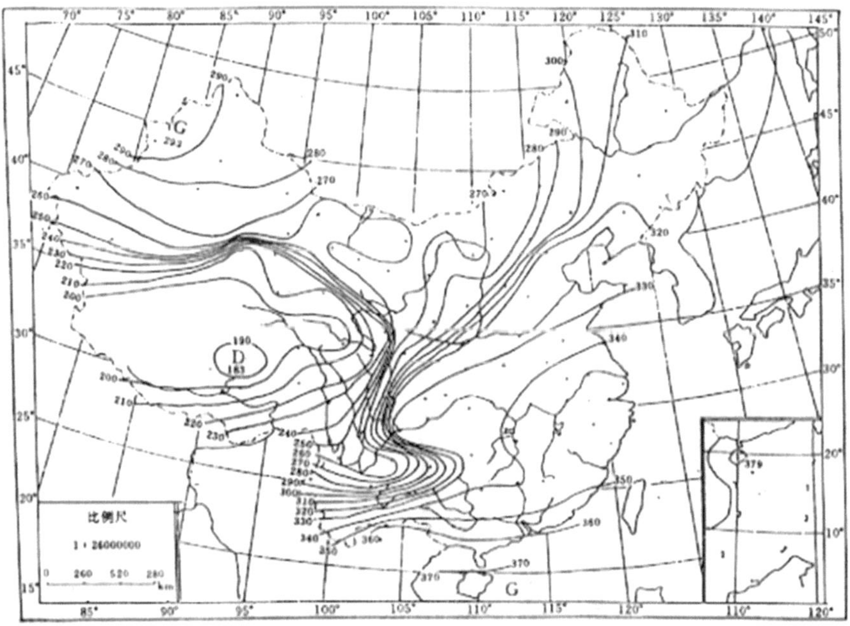

Fig. 5.39 Annual average isolines of ground surface refractivity N_s in China

(2) Change of atmospheric refractivity with altitude

In general, the refractivity N decreases with increasing altitude. According to many years of statistical data, the average refractivity N changed with altitude as a negative exponential function:

$$N(h) = N_s \exp[-c_\alpha (h - h_0)] \quad (5.110)$$

where $N(h)$ is the refractivity (N unit) at an altitude of h, N_s is ground surface refractivity (N unit), h is the height above sea (km), h_0 is ground altitude (km), and c_α is exponential decay coefficient (km^{-1}).

The refractivity N in lower atmosphere layer in different areas may be significantly different, but the difference decreases with increasing altitude. At 9 km, 105 N units may be used in most parts of the world. This book takes China as an example.

It is lower atmosphere that produces refraction effect on radio waves. Therefore, the research on refractivity change with altitude focuses on the change of N values within 1 km above ground, that is, $\Delta N_1 = N_s - N(1\text{ km})$. Table 5.10 gives annual mean of ΔN_1 at 7:00 a.m. in January and July in ten areas including Beijing.

(3) Seasonal change of atmospheric refractivity N

N obviously changes with seasons in one year, although average refractivity N has a small annual change rate (ground surface refractivity N_0 has 2–3 N units in annual change). Table 5.11 gives average ground surface refractivity N_s values of ten areas including Beijing in different seasons (January, April, July, and October).

It can be seen from the above table that the refractivity in southeast coasts and middle reaches of Yangtze River changes greatly, because these areas features a big humidity change throughout the year. In Ürümqi, temperature changes much, but annual refractivity change is still very small due to less rainfall, arid climate, and small humidity change in the whole year.

Table 5.10 Annual mean of ΔN_1 at 07:00 a.m. in areas including Beijing (N unit)

Area	January	July
Beijing	37	54
Shanghai	41	70
Urumchi	26	57
Guangzhou	38	58
Lanzhou	28	37
Kunming	34	44
Wuhan	46	62
Chengdu	32	39
Lhasa	21	31
Hohhot	32	37

Table 5.11 N_s values in areas including Beijing in different seasons (*N* unit)

Area	January	April	July	October	Annual mean
Beijing	301	301	361	311	320
Shanghai	316	333	383	342	343
Urumchi	292	281	290	285	285
Guangzhou	323	362	380	353	353
Lanzhou	253	246	272	265	278
Kunming	259	270	299	280	279
Wuhan	314	334	383	337	342
Chengdu	303	320	351	320	320
Lhasa	192	194	220	202	202
Hohhot	277	260	292	272	272

Table 5.12 Diurnal change of N_s in Beijing (*N* unit)

Month	01:00	07:00	13:00	19:00
January	308	313	300	307
February	308	312	291	300
March	369	367	357	362
April	325	322	310	321

(4) Diurnal change of atmospheric refractivity *N*

According to experimental data statistics, atmospheric refractivity *N* has a very obvious diurnal change, with obvious periodicity. Refractivity *N* is highest before dawn and smallest around 14:00 p.m. The following table gives statistics of ground surface refractivity N_s in Beijing at 01:00, 07:00, 13:00, and 19:00 every day in January, April, July, and October (Table 5.12).

According to the table above, diurnal changes of *N* values in most areas are usually within 10–20 N units or may be higher if the weather is anomalous.

60–2,000 km above the surface is the ionosphere where a great number of charged ions are contained and unevenly distributed. If propagating in this inhomogeneous medium, radio waves are not only refracted but have changes in phase velocity *V* (propagation velocity in equiphase surface).

Calculation formula of ionospheric refractive index is

$$n = 1 - \frac{40.3\ N_e}{f^2} \tag{5.111}$$

where f is transmitting signal frequency (Hz) and N_e is electron concentration (electron/m^3). According to the expression, if the frequency increases, the refraction index will decrease.

Distribution change of the ionosphere is shown in the Table 5.13.

It can be seen from the above table that altitude distribution of electron concentration in the ionosphere changes with the area, day and night, season, latitude, and

Table 5.13 Distribution change of ionosphere

Layer	Altitude (km)	Max. electron concentration (Nr/cm^3)	Change
D	70–90	10^3–10^4	This layer disappears at night
E	100–120	2×10^5	Electron concentration of this layer is high in the daytime and low at night
F_1	160–180	2×10^5	This layer mostly exists in the day in summer
F_2	300–450	1×10^6	In summer, electron concentration of this layer is high in the daytime and low at night
	250–350	2×10^6	In winter, electron concentration is high in winter and low in summer

solar activity. The effect of the ionosphere on velocity-measuring accuracy is not constant, but changes with factors such as time, place, and operating frequency.

From the above, it can be seen that the effect of radio wave refraction on orbit determination accuracy changes with factors such as time, place, and operating frequency. Therefore, it is necessary to correct ionospheric effects in real time.

5.4.6 *Tropospheric Radio Wave Refraction Error Correction*

5.4.6.1 Range Error Correction

The range formula for spheric layering of the troposphere is

$$R_e = \int_{r_0}^{r_T} \frac{n^2 r dr}{p^{1/2}} \tag{5.112}$$

$$p = \left(n^2 r^2 - n_0^2 r_0^2 \cos\theta_0\right) \tag{5.113}$$

where n is refractive index (measured by a height indicator), n changes with r, and n_0 is ground refractive index of the TT&C station.

θ_0 is the apparent elevation.

$r_{T,}$ r_0 refer to the geocentric distance of the target and TT&C station, respectively.

R_e is the apparent range (namely, the distance measured by the TT&C station).

R_0 is the real range.

The geocentric φ between the TT&C station and the target is

$$\Phi = A_0 \int_{r_0}^{r_T} \frac{dr}{\sqrt{n^2 r^2 - A_0^2}} \tag{5.114}$$

The real elevation α_0 of the target is

$$\alpha_0 = \arctan \frac{r_T \cos \Phi - r_0}{r_T \sin \Phi} \tag{5.115}$$

Therefore, elevation correction ε_y and range correction ΔR are

$$\varepsilon_y = \theta_0 - \alpha_0 \tag{5.116}$$

$$\Delta R = R_e - R_0 \tag{5.117}$$

The velocity correction can be obtained by differentiating the range refraction error ΔR, namely,

$$\Delta R = \frac{d(\Delta R)}{dt} \approx \frac{\Delta(\Delta R)}{\Delta t} \tag{5.118}$$

An approximate calculation expression is

$$\Delta R = 0.07\,\mathrm{N_s} \csc\theta_0 \tag{5.119}$$

where N_s is ground surface atmospheric refractivity of the TT&C station.

After correction, there is residual error which is mainly caused by n error.

The expression for residual error after correction is

$$\sigma_{\Delta R} = \frac{\int \Delta n dh}{\sin \theta_0} \tag{5.120}$$

where Δn is measuring error of the refractive index.

5.4.6.2 Correction of Angle Measuring Accuracy Error from Tropospheric Refraction

Tropospheric refraction may cause elevation angle-measuring error, but the refractive index is irrelevant to frequency.

Atmospheric refractive index in the troposphere changes with altitude. The result of refraction from radio wave propagation is that the apparent elevation θ_0 of the target measured on the ground is bigger than its real elevation α_0, and elevation angle-measuring error occurs $\Delta E = \theta_0 - \alpha_0$.

The following reduction formula is often used in engineering for an approximate evaluation of error from tropospheric refraction and for error correction.

(1) Target in the troposphere

When the apparent elevation of the target θ_0 is higher than 5°, there is

$$\Delta E = \frac{h}{h+k} N_s \cot \theta_0 10^{-3} \quad \text{(mrad)} \tag{5.121}$$

When $\theta_0 \leq 5°$, there is

$$\Delta E = (bN_s + a)\frac{h}{h+k} \cot \theta_0 10^{-3} \quad \text{(mrad)} \tag{5.122}$$

$$b = (1 - 0.005/\theta_0)10^{-3} \cot \theta_0 \quad (\text{mrad}/N) \tag{5.123}$$

(N is the representation unit of atmospheric refractive index, N unit)

$$a = (0.15/\theta_0)(0.02/\theta_0 - 1)10^{-3} \cot \theta_0 \quad \text{(mrad)} \tag{5.124}$$

where

N_S is ground surface atmospheric refractive index of the TT&C station. In standard scenario, $N_S \approx 313$.
θ_0 is the apparent elevation of antennas.
h is target altitude (km).
k is a fixed constant, 10.675 km.

(2) Target outside the troposphere

When $\theta_0 > 5°$, there is

$$\Delta E = N_s \cot \theta_0 10^{-3} \quad \text{(mrad)} \tag{5.125}$$

When $\theta_0 = 2°–5°$, there is

$$\Delta E = (bN_s + a) \cot \theta_0 10^{-3} \quad \text{(mrad)} \tag{5.126}$$

Corrections are made to measured elevation value by using the above ΔE. After such corrections, there is residual error $\sigma_{\Delta E}$, which is mainly caused by ΔN_s, N_s error. If $\theta_0 > 5°$, it approximates:

$$\sigma_{\Delta E} \approx \Delta N_s \times \cot \theta_0 \times 10^{-3} \text{ (mrad)} \tag{5.127}$$

5.4.6.3 Tropospheric Refraction Velocity Error Correction

The approximate calculation formula of the apparent velocity (measured by the TT&C station) is

$$\dot{R}_e = n_r\left(\cos \Delta \dot{E} \dot{R}_0 - \sin \Delta \dot{E} \dot{a}_0\right) \tag{5.128}$$

where ΔE is elevation error and $\Delta E = \theta_0 - \alpha_0$.

α_0 is real elevation (at the target).

Correction $\Delta\dot{R} = \dot{R}_e - \dot{R}_0$.

If measurement correction level is

$$\int \Delta n dh = 0.05(m) \tag{5.129}$$

The ranging accuracy by tropospheric correction can be estimated.

For example, if $\theta_0 = 5°$ (with respect to real elevation, given minimum precision-keeping elevation angle is 5°), there is

$$\sigma_{\Delta R}(5°) = 0.57(m) \tag{5.130}$$

If the elevation is 30°, there is

$$\sigma_{\Delta R}(30°) = 0.1(m)$$

$$\sigma_{\Delta\dot{R}} = \frac{\int \Delta n dh}{\operatorname{tg}\theta_0 \sin\theta_0} \frac{d\theta_0}{dt} \tag{5.131}$$

where Δn is measuring error of the refractive index.

According to the above expression, velocity-measuring correction accuracy has a close relationship with geometric parameters of the target relative to the TT&C station and is also in direct proportion to the target velocity. This error may be bigger.

5.4.7 Ionospheric Refraction Correction Methods

There are two methods for real-time correction to ionospheric refraction error. One is to calculate correction by using measured or predicted ionospheric characteristic parameters (electron concentration profile or total electron content in the ionosphere); the other is that the system itself has means to eliminate ionospheric refraction error, such as dual frequency ranging, dual frequency Doppler velocity measuring, or measurement of both group delay and phase delay.

Ionospheric conditions present significantly random in time domain, frequency domain, and space domain, and many countries have made immerse amounts of researches, measurements, and statistics. However, it is difficult to meet a high-accuracy requirement due to its poor accuracy in long-term forecasting. The second method can be used to detect ionospheric refraction effect in the propagation path in real time. Thus, it has a much higher accuracy than the first method.

(1) Calculation correction method

Orbit determination requires that the straight-line range R_0 of radio wave propagation and its rate of change $\dot{R}_0$ should be given, but what the velocity-measuring system measures is the rate of change $\dot{R}_e$ of apparent propagation range R_e. After an approximate treatment $R_e \approx R_0$, the following can be obtained:

$$R_e = \int_{R_e} ndl \approx \int_{R_0} (1 - 40.3N_e/f_T^2)dl = R_0 - \frac{40.3}{f_T^2}\int_{R_0} N_e dl \tag{5.132}$$

Error correction term $\Delta R = R_e - R_0 = -\frac{40.3}{f_T^2}\int_{R_0} N_e dl.$

Doppler frequency measured by the velocity-measuring system is

$$f_d = -(f_T/c)\dot{R}_e \approx -(f_T/c)\dot{R}_e + \dot{a}(t)/f_T \tag{5.133}$$

where the first term $-(f_T/c)\dot{R}_0$ is an error-free Doppler frequency value and the second term $\dot{a}(t)/f_T$ is velocity-measuring error term caused by ionospheric refraction, that is,

$$\dot{a}(t) = \frac{40.3}{c}\frac{d}{dt}\int_{R_0} N_e dl \tag{5.134}$$

(2) Dual-frequency Doppler method

The measurement accuracy of ionospheric model is limited (to 60–85 %), so the effect of refraction correction would not be good while using computation correction for correcting the error of radio wave refraction of ionosphere. Therefore, the dual-frequency Doppler ionospheric refraction error in real-time correction shall be adopted to eliminate the primary effect of ionosphere, with descriptions as the following:

Expression of group path:

$$R_e = \int_{r_0}^{r_{\text{ionosphere}}} \frac{n^2 r dr}{p^{1/2}} + \int_{\text{ionosphere}}^{r_T} \frac{r dr}{p^{1/2}} \tag{5.135}$$

where $r_{\text{ionosphere}}$ is the height in the bottom of ionosphere; the first term in the expression is troposphere and the second term is ionosphere.

The additional group delay caused by the ionosphere is related to the frequency, which can be expressed in power series of frequency:

$$R_e = R_a + \frac{A_1}{f^2} + \frac{A_2}{f^3} + \frac{A_3}{f^4} + \cdots \tag{5.136}$$

where R_a is the additional delay for the distance from the target to the measured station against the troposphere, without effect of additional delay by ionosphere.

f is frequency. A_1, A_2, A_3... are the structural coefficients of ionosphere, the second term herein is the additional group path of ionosphere, and A_2 is the effect of Earth's magnetic field.

Only the first two terms shall be taken from the expression above if the second effect is ignored:

$$R_e = R_a + \frac{A_1}{f^2} \tag{5.137}$$

The Doppler frequency resulted by target motion is affected by troposphere and ionosphere during the radio wave passing through atmosphere, where the effect of ionosphere is related to frequency, so that the expression of Doppler frequency can be expressed with the similar method above:

$$f_d = f_{da} + \frac{D_1}{f} \tag{5.138}$$

where f_{da} is the target motion Doppler including troposphere effect, D_1 is the effect factor of ionosphere, and the second term D_1/f is the additional Doppler frequency shift caused by the ionosphere. This expression only takes the primary effect into account and ignores the effect of ionosphere by secondary level or above.

The principle of refraction error self-correction is described as follows:

When the f_1 signal that is transmitted from the ground station arrives at the target, of which the received signal group delay path:

$$R_g = R_a + \frac{A_1}{f^2} \tag{5.139}$$

Received frequency:

$$f_{RT} = f_1 + f_{1da} + \frac{D_1}{f_1} \tag{5.140}$$

Additional Doppler frequency shift:

$$f_{1D} = f_{1da} + \frac{D_1}{f_1} \tag{5.141}$$

The retransmission downlink frequency of transponder f_2 and f_3, respectively, are

$$f_2 = \alpha_2 f_{RT} = \alpha_2 f_1 + \alpha_2 f_{1da} + \frac{\alpha_2 D_1}{f_1} \tag{5.142}$$

and

$$f_3 = \alpha_3 f_{RT} = \alpha_3 f_1 + \alpha_3 f_{1da} + \frac{\alpha_3 D_1}{f_1} \tag{5.143}$$

The signals at the above two frequencies received by the ground station have the following features.

Group path of f_2 modulation signal:

$$R_{2g} = 2R_a + \frac{A_1}{f_1^2} + \frac{A_1}{f_2^2} = 2R_a + \frac{A_1}{f_1^2} + \frac{A_1}{\alpha_2^2\left(f_1 + f_{1da} + \frac{D_1}{f_1}\right)^2} \tag{5.144}$$

Carrier frequency for receiving f_2 signal:

$$f_{2R} = f_2 + f_{2da} + \frac{D1}{f_2} \tag{5.145}$$

That is,

$$f_{2R} = \alpha_2 f_1 + \alpha_2 f_{1d} + f_{2da} \frac{D1}{f_2} \tag{5.146}$$

Then the Doppler frequency of f_2 signal is

$$\begin{aligned} f_{2D} &= \alpha_2 f_{1d} + f_{2d} + \frac{D_I}{f_2} = \alpha_2 f_{Ida} \frac{\alpha_2 D_I}{f_I} + f_{2da} + \frac{D_I}{f_2} \\ &= (\alpha_2 f_{Ida}) + \frac{\alpha_2 D_I}{f_I} + \frac{D_I}{\alpha_2 (f_I + f_{Id})} = (\alpha_2 f_{Ida} + f_{2da}) + \frac{\alpha_2 D_I}{f_I} + \frac{D_I}{\alpha_2 f_I \left(I + \frac{f_{Id}}{f_I}\right)} \end{aligned} \tag{5.147}$$

By substituting into expression (5.62):

$$f_{2d} = (\alpha_2 f_{1da} + f_{2da}) + \frac{a_2 D_1}{f_1} + \frac{D_1}{\alpha_2 f_1 \left(\frac{f_{1da}}{f_1} + \frac{D_1}{f_1^2}\right)} \tag{5.148}$$

where $\alpha_2 f_{1da} + f_{2da}$ is the target motion Doppler, including convection effect, and the second and third terms are the additional Doppler frequency caused by ionosphere.

Make

$$f_{2do} = \alpha_2 f_{1da} + f_{2da} \tag{5.149}$$

then expression (5.148) is converted to

$$f_{2D} = f_{2do} + \frac{\alpha_2 D_1}{f_1} + \frac{D_1}{\alpha_2 f_1 \left(1 + \frac{f_{1da}}{f_1} + \frac{D_1}{f_1^2}\right)} \tag{5.150}$$

Similarly, the Doppler value of downlink f_3 signal can be measured:

$$f_{3D} = f_{3do} + \frac{\alpha_3 D_1}{f_1} + \frac{D_1}{\alpha_3 f_1 \left(1 + \frac{f_{1da}}{f_1} + \frac{D_1}{f_1^2}\right)} \tag{5.151}$$

where

$$f_{3do} = \alpha_3 f_{1da} + f_{3da} \tag{5.152}$$

by considering

$$\frac{f_{2da}}{f_{3da}} = \frac{\alpha_2}{\alpha_3} \tag{5.153}$$

then

$$f_{2do} = \frac{\alpha_2}{\alpha_3} f_{3do} \tag{5.154}$$

As mentioned above, f_{2do}, f_{3do}, and D_1 are calculated by Expression (5.150), (5.152), and (5.154). The resulting f_{2do} and f_{3do} are the values of the Doppler target motion with convection effect that eliminate the ionosphere effect. By correcting the refraction error of troposphere, then the Doppler velocity for the target motion can be solved. The solved D_1 value is used to estimate the additional Doppler frequency shift caused by ionosphere.

In addition,

$$D_1 = \frac{1}{C} \frac{dA_1}{dt} \tag{5.155}$$

Therefore, A_1 value is solved from the expression above and the additional group path of ionosphere by distance measurement can be corrected.

What is mentioned above is the correction process by real-time self-correction method for ionosphere refraction error via two-direction coherent dual-frequency Doppler values, with the following possible error sources:

1) Ignore the secondary effect of ionosphere refraction, including group path delay:

$$\sigma_R = \frac{A_2}{f^3} + \frac{A_3}{f^4} + 0 \tag{5.156}$$

Add Doppler frequency shift in secondary level or above:

$$\sigma_{\Delta fd} = \frac{D_2}{f^2} + \frac{D_3}{f^3} + 0 \tag{5.157}$$

Normally, in case of high frequency, for example, 2, 4, and 6 GHz, the ionosphere secondary effect is less two orders of magnitude than the primary effect.

2) System measurement errors, which are the errors Δf_{2D} and Δf_{3D} corresponding to f_{2d} and f_{3d} that are measured by the measurement system. The effects of them can be directly estimated by the linear equation set of (5.150) and (5.151).

5.4.8 Factors Affecting Correction Accuracy

The following are some factors affecting the correction accuracy of radio wave refraction:

(1) Assumption error of expression: the existing refraction correction expression assumes that the atmospheric pressure is uniform, that is, it can be layered spherically, but in fact the atmospheric structure is uneven.
(2) Random fluctuation of atmosphere: Some noise effects caused by high frequency can be smoothed by the radio measurement system itself and data processing, but the LF fluctuation of atmosphere can not be filtered, which is the main reason for failure of accuracy requirements by velocity measurement system and has great influence on accuracy of velocity measurement.
(3) Errors of atmosphere measuring instrument and conversion error of refraction index expression: the refraction index of atmosphere used in correction expression is calculated by the empirical expression below, in which the measured temperature of radiosonde is T, pressure is P, and humidity is e_w:

$$N = \frac{77.6}{T}\left(P + 4,810\frac{e_w}{T}\right) \tag{5.158}$$

The accuracy of this expression is 0.5 % at 0–30 GH_Z. Radiosonde has measurement errors on temperature, pressure, and humidity, respectively 0.2 °C, 1 hP_a and 3 %. Under 15 °C, the error of refraction index caused by these measurement errors is 2–3 N.
(4) Errors by non real-time measurement for atmosphere structure: the atmospheric profile for refraction correction is not the one that measured in real time under operating state of radio measurement system but operating before 30–60 min or after 30–60 min. The sounding detection takes 20–30 min once, since the atmosphere structure varies with time, so that the detected data shall not represent the actual state instantaneously and will also cause the error. The atmosphere structure changes sharply at morning and sunset but has least

changes around afternoon. Thus, the operating time for radio measurement system shall be selected properly to improve the correction accuracy.

(5) Original parameters error: there have been errors on the original parameters used for radio wave correction, which will affect refraction correction, such as range, elevation, range rate of change, local Earth radius, station address, etc.

For system with high measurement accuracy, the distribution of accuracy requirements shall focus on correction of radio refraction error (with troposphere and ionosphere correction) in overall design, and ratio of it and other part in the system shall not be less than 1:1. In spite of this, the radio correction has great difficulties, especially for troposphere refraction correction. To improve the accuracy of troposphere correction, the following two newly methods shall be considered: one is error correction of troposphere refraction by self-adjustment of GPS base station; the other one adopts range correction by microwave radiator, namely, Macor technique.

References

1. Liu Jiaxing (2011) Spacecraft TT&C and information transmission technology [M]. National Defence Industry Press, Beijing
2. Liu Jiaxing (2008) Ideas on development of Ka-band TT&C system [J]. J Astronaut 29(6):1685–1688
3. Kang Jian, Wang Yufei (2006) Rain attenuation distribution characteristic of satellite communication link in China. J Commun 27(9):78–81
4. Shengmei Q (1996) Distribution of minute rainfall in China. J Commun 17(3):79–83
5. Xin Peiquan (2004) Estimation of rain attenuation in spaceborne SAR system design. Mod Radar 26(9):15–19
6. Chen Fangyun (1992) Satellite TT&C manual. Science Press, Beijing
7. Wu Weiling (2000) Key techniques in mobile communication [M]. Beijing University of Posts and Telecommunications Press, Beijing
8. Yang Dacheng (2005) Advanced techniques in modern mobile communication. China Machine Press, Beijing
9. Radar handbook (3rd edn) (2010) chief editors: Merrill, skolnik, (trans: China Nanjing Electronic Technology Research Institute). Electronic Industry Press, Beijing
10. Ippolitolouis LJ Jr (2012) (trans: Baosun S) Satellite Communication System Engineering, National Defence Industry Press, Beijing
11. Zlemer RE (2005) Fundamentals of digital communication (trans: Yin Changchuan, Hao Jianjun). China Machine Press, Beijing
12. Liu Jiaxing (2010) Towerless phase calibration by using radio star noise. Telecommun Eng 50(6):1–4
13. Qiu san-shan, Wang Yuan-ling, Yang Hong-jun (2010) Feasibility analysis of phase calibration for tracking receiver using radio star in deep space TT&C system. Telecommun Eng 50(8):22–25
14. Yu Zhijian (2009) Deep space C&T system. National Defense Industry Press, Beijing
15. Li Ke (2007) A rapid phase calibration method of dual-channel tracking. Telecommun Eng 47(6):99–101

MIX
Papier aus verantwortungsvollen Quellen
Paper from responsible sources
FSC® C105338

If you have any concerns about our products,
you can contact us on
ProductSafety@springernature.com

In case Publisher is established outside the EU,
the EU authorized representative is:
Springer Nature Customer Service Center GmbH
Europaplatz 3, 69115 Heidelberg, Germany

Printed by Libri Plureos GmbH
in Hamburg, Germany